Reproduction in Aquatic Animals

Manabu Yoshida • Juan F. Asturiano
Editors

Reproduction in Aquatic Animals

From Basic Biology to Aquaculture Technology

Springer

Editors
Manabu Yoshida
Misaki Marine Biological Station
The University of Tokyo
Miura, Kanagawa, Japan

Juan F. Asturiano
Institute for Animal Science and Technology
Universitat Politècnica de València
Valencia, Spain

ISBN 978-981-15-2289-5 ISBN 978-981-15-2290-1 (eBook)
https://doi.org/10.1007/978-981-15-2290-1

This Springer imprint is published by the registered company Springer Nature Singapore Pte Ltd.
The registered company address is: 152 Beach Road, #21-01/04 Gateway East, Singapore 189721, Singapore

Preface

Even though there are many animals in terrestrial habitats, there is great biodiversity to be found in aquatic ecosystems. Furthermore, aquatic animals show various reproductive systems: many animals perform external fertilization, others show internal fertilization without mating, some are viviparous, etc. This means that the reproduction systems of aquatic animals are highly diverse, and they are excellent models for studying adaptive evolution and species-specificity of fertilization. In addition, many aquatic animals such as fish, crustaceans, and mollusks are important as fishery and aquaculture resources. Nevertheless, their reproductive systems are also diverse, resulting in difficulties in cultivation, especially in the production of juveniles. Thus, comprehensive knowledge of the reproductive systems of various aquatic animals will help us understand the systems of each animal, resulting in breakthroughs in the research areas and aquaculture technologies. However, only a few books overviewed the reproductive systems of aquatic animals from invertebrates to fishes since many researchers focused their animals of interest. Therefore, our aim with this book was to overview the various reproductive systems of aquatic animals.

The idea for this book was initially conceived in the International Symposium on "AQUAGAMETE: Reproduction of Aquatic Animals" held in the Joint Meeting of the 22nd International Congress of Zoology and the 87th meeting of the Zoological Society of Japan, which was held from 14th to 19th November 2016 in Okinawa, Japan. Three years have passed since the initial planning, and we have developed the book ideation. In order to introduce up-to-date knowledge on the reproduction systems of various aquatic animals from basic biology to aquaculture technology, we invited up-and-coming researchers to contribute. This book consists of 17 chapters and a foreword that details the history of spermatology to be read before the main chapters. Finally, the book covers the reproductive systems of both sperm and egg in cnidarians, annelids, arthropods, mollusks, echinoderms, ascidians, elasmobranchs, teleosts, and amphibians. Four chapters focus on the technological and aquaculture aspects, in particular relating to fishes.

This book is designed for people who are neither experts/well-read/knowledgeable in the field of reproductive biology nor aquaculture. The assumed readers are graduate students and postgraduates in biology and agricultural sciences and also non-academics who are interested in the field.

We hope that this book will be useful to many readers, particularly scientists and technicians in the field of reproductive biology and fishery science area.

Finally, we would like to thank all the authors and contributors who made this book a reality.

Miura, Japan
Valencia, Spain
September 2019

Manabu Yoshida
Juan F. Asturiano

Contents

Foreword: A Brief History of Spermatology

Jacky Cosson

This foreword mostly aims to introduce, from a historical stand point, how the notions of gametes emerged, and to describe the tortuous approach by the pioneer scientists who first discovered and explored the functions and structure of aquatic animal gametes and their interactions. Sperm cells most likely became of interest to scientists due to the fact that they hold the key to life and have an incredible ability to move, in spite of their small dimensions.

It is commonly acknowledged that *spermatology* is a scientific domain dealing with the structure and function of spermatozoa. For this reason, it can be supposed that the history of spermatology began in 1677 with Leeuwenhoek's description of the spermatozoon, the male entity, responsible for animal procreation and rendered visible for the first time through his microscope. Therefore, it is considered that spermatology starts at this date as biologists enjoy to attribute a structure to a function. For obvious reasons, this foreword mostly covers the last three-and-a-half centuries, if we accept Leeuwenhoek to be the "inventor" of spermatozoa.

It is out of the present topic to discuss the history of human reproduction, in its medical aspects. Instead, in this book, we will concentrate more specifically on the gametes of aquatic animals with our main aim being to trace how the study of aquatic animals can be so important in the understanding of the mechanisms of gamete interaction.

J. Cosson (✉)
Faculty of Fisheries and Protection of Waters, Research Institute of Fish Culture and Hydrobiology, University of South Bohemia in České Budějovice, Vodnany, Czech Republic

M. Yoshida, J. F. Asturiano (eds.), *Reproduction in Aquatic Animals*,
https://doi.org/10.1007/978-981-15-2290-1_1

Let Us Go Back to/Return to the Seventeenth Century

Long before the identification of the individual elements later called "spermatozoa," Leonardo da Vinci, by reference to Hippocrates, wrote: "The origin of our semen is located in the brain and in the lungs and it is in the testis of ancestors that the final mixing occurred." This premonitory statement contains quite a lot of veracity that modern science would finally establish as true: the **brain** definitely contributes "psychologically" to the delivery of sperm to the female for reproduction; the **lungs** constitute an organ making a large use of cilia which are homologue to flagella in their constitution and the **testis** contributes to the efficient mixing of the ancestor's genomes during the meiotic process of spermatogenesis.

Already in 1623, L. Gardinius (L. du Gardin) assumed that there were fertilizing particles in male semen, but it wasn't until 1677 that they were in fact observed by the human eye and described. However, the priority comes to Antonie van Leeuwenhoek, who in that year found spermatozoa in the semen of fish, frogs, and mammals, thanks to one of his rudimentary microscopes made of a single spheroidal lens. His publication to the Royal Society *De Natis E Semine Genitali Animalculis* (1678) remains famous and frequently cited.

In his letter (1677), he described that there are "living animalcules" in human semen: "The size of animalcules is ten thousand times less than a water louse. They move like a snake or like an eel swimming in water, have globule at the end, and are very flexible." He supported his letter with a drawn picture of what had been seen under the microscope (see Fig. 1). He continued to observe many other animals in his later works and showed that the animalcules were produced by the testes. Leeuwenhoek knew his discovery was important to the understanding of reproduction and insisted that "a sperm cell was the only thing that made an embryo, and that the egg and uterus merely nourished it as it grew." At that time, the prevailing view on reproduction was that the embryo grew from the egg alone, after the semen added a "volatile spirit" to spark its development.

In the context of his epoch, he called them animalcula and interpreted them in Aristotelean terms that could be considered nowadays as quite male chauvinist. "Life comes from the male whereas the female produces nutrition for it in the egg." Two sentences from Leeuwenhoek's letter read as follows: "What I investigate is

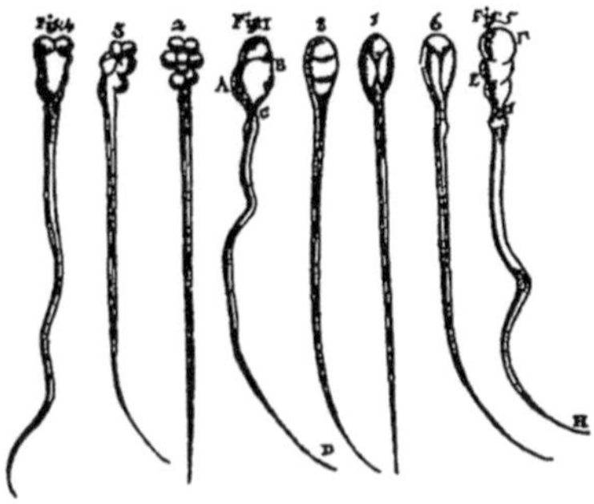

Fig. 1 Drawn picture of spermatozoa by Leeuwenhoek. (From Leeuwenhoek, *Phil. Transact. Roy. Soc.*, 12: 1040–1046). This figure is Public Domain

only what, without sinfully defiling myself, remains as a residue after conjugal coitus. And, if your Lordship should consider that these observations may disgust or scandalize the learned, I earnestly beg your Lordship to regard them as private and to publish or destroy them, as your Lordship thinks best." Evidently, the Royal Society did not regard the topic to be indecent as they published the letter. One hundred years later, scientists were perhaps more prudish, as exemplified by this statement by Herman Schützercranz, physician of the Swedish king: "You cannot and ought not know whatever happens at fertilization," which denotes a lack of openness for a medical doctor. Whereas Leeuwenhoek's famous letter undoubtedly is the first description of spermatozoa, the events around his discovery are more complex.

According to Cole (1930), Leeuwenhoek's letter to the Royal Society in November 1677 wasn't published until 1679 and was preceded by a communication to the Académie Française (French Academy) by Christiaan Huygens, dated July 30th 1678. Huygens describes in this communication small animals similar to tadpoles in the semen of a dog. His comments, after translation into English read as this premonitory sentence: "This discovery seems very important and should give employment to those interested in the generation of animals," predicting the advent of *artificial propagation of animals*. Nevertheless, in a letter dated March 26th 1678, Huygens admits to having seen and read Leeuwenhoek's letter of 1677. In August 1678, Nicolas Hartsoeker published a letter in the Journal des Savants (drafted by Huygens because of Hartsoeker's inability to write in French), in which he describes animals similar to little eels in the semen of the cock; the latter differed thus in shape from the tadpole-like animalcula of the dog. It is amazing to note that this is the first example of comparative spermatology! In conclusion, two investigators published data on spermatozoa in the year 1678; both did so during the time span needed for Leeuwenhoek's letter to be translated from Dutch to Latin (in three different versions) and printed by the Royal Society. Such huge delay in the transmission of information seems incredible in today's internet era!

Furthermore, Leeuwenhoek himself attributed the discovery of the animalcula to a certain "Dominus Ham," that is Mr. Ham, a person who never published anything on semen nor its content. This man is commonly believed to be Ludwig van Hammen of Danzig, but according to Cole (1930) it is more likely to be Johan Ham, a Dutchman from Arnhem, born in 1650 or 1651, a student at the time of his discovery, and who later became a Doctor in Arnhem. Apparently, Johan Ham was the first person to see mammalian spermatozoa and Leeuwenhoek was informed by him; thus, Huygens became the first to publish data on mammalian spermatozoa and Hartsoeker the first to publish data on avian spermatozoa.

Leeuwenhoek later studied and described spermatozoa from other classes of animals.

All observations on animalcula were met with great interest. Robert Hooke (the first man to use the word "cell") thus had to demonstrate the existence of spermatozoa to King Charles II, who expressed his delight to see the animalcula. Yet, the significance of the animalcula remained obscure. To some philosophers, the huge number of animalcula made no sense for any idea of conception. According to

Leeuwenhoek: "Eventually, thousands of those animacules were agitating in a tiny space of a sand grain size." And after all, Leeuwenhoek had found a multitude of small creatures when he examined scrapings from the teeth (probably bacteria). To others, the existence of small swarming creatures validated the idea that offspring comes from the male. The man-like homunculi depicted by Hartsoeker and Dalenpatius are famous and classically used as illustrations in Fig. 2.

Some investigators went so far as to claim that they could see horse animalcula in horse semen and donkey animalcula in donkey semen and that the donkey animalcula had longer ears. Still others claimed that they could see male and female animalcula and even mating and childbirth among these!

The humunculus also seemed to confirm the preformation concept, that is, the belief that everything is present in the seed although in a miniaturized form and that development merely consists of an increase in size. The preformationists could either be animalculist or ovist; the latter believed that the animal is already formed in the egg. An ovist would thus claim that he could see the chick in the unfertilized egg. The preformationist theory has the merit that it explains original small men were contained already in the organs of Adam and Eve. It also has the consequence that the human race will become extinct when the stock of seed is exhausted.

The ovist school can be said to have begun with the publication by Harvey in 1651 of his influential book *De Generatione* with its prophetic quote on the frontispiece "Ex ovo omnia," "all (animals) from eggs." Harvey thus believed that the male (semen) played no part in the formation of the fetus.

During an experimental dissection of a mated roe deer, he could find no spermatozoa in the uterus. The debate between believers in epigenesis (i.e., the embryo and its parts undergo differentiation of initially undifferentiated entities) and believers in preformation went on for several centuries. If a vote had been taken in the seventeenth century, the preformationists would have won by a wide margin although some thinkers, such as Descartes, were supporters of epigenesis.

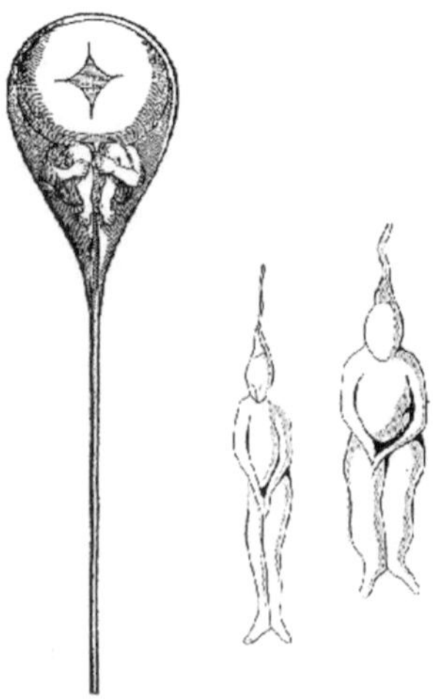

Fig. 2 Drawing of homunculi in sperm by Hartsoeker. (Left: from Essai de dioptrique 1694) and by Dalenpatius (right: from *Nouvelles de la République des Lettres*, 1699). This figure is Public Domain

The structure and meaning of the animalcula was also debated at the end of that century. P. Dionis (1698) asked for further inquiries as he believed that they are formed by minute fibers in semen exposed to air. M. Lister (1698) also inquiring about the origin of the seminal animalcula concluded (free translation from Latin): "Homunculi in great numbers: when I reflect upon it, I leave it to be cared for by others, to me it is a fairy tale."

Leeuwenhoek's importance as a microscopist is widely recognized. Less known is the fact that he also tried to investigate inheritance by an experiment. He mated a gray rabbit buck with a white rabbit doe and noted that all the young were gray—another "proof" of the validity of the "seed-dominant" concept. Evidently, he did not perform—or at least did not report—a control, that is a cross between a white male and a gray female.

Continued in the Eighteenth Century

The uncertainty continued into the eighteenth century. E. F. Geoffroy and C. du Cerf (1704) observed numerous, but non "fully mature" animalcula in boys, well developed and active ones in adults, few and feeble ones in old men, and no animalcula in sterile individuals. They would conclude that animalcula are needed for reproduction and can hence be considered the founders of *andrological spermatology*. Other opinions also prevailed; M. Schurig (1720) in his *Spermatologia Historico Medica* considered the animalcula to be only an "active portion of the semen agitated in a viscid mass." A. Vallisnieri (1721) and P. L. M. de Maupertuis (1744) admit that animalcula exist but claim that they have no direct relation to reproduction: they are *entozoa* (tapeworms) and keep the semen fluid. The philosopher J. M. Gestner (1737) accepts seminal animalcula as a fact but claims that their discovery is to be credited to Hippocrates who, according to Gesner, was able to see them by his "enormous force of reason" rather than by using a microscope!

During this eighteenth century, the great naturalists were against the idea of animalcula playing a role in reproduction. Carl Linnaeus (1746) believed them to be inert "corpuscules," P. Lyonet (1751) to be "entozoic" parasites, G. L. L. Buffon (1752) and J. T. Needham (1749) to be aggregates of living organic molecules derived from the mucilaginous part of the semen. The entozoa hypothesis seems to have been very popular, and several attempts were made to include them in the classification of animals: according to Hill (1752), the animalcula are *infusoria* (protists) and deserve a genus name, *Macrocercus*, related to Vorticella and Euglena. Spallanzani (1776) ranks them among the animals and Blumenbach (1779) again among the "infusorial" animals, with the genus and species name *Chaos spermaticum*. Cuvier (1817) classifies them in the genus *Cercaria*. Bory de Saint-Vincent (1827) similarly regards them as belonging to the family *Cercariae* and invents a new genus name, *Zoospermos*. Carl Ernst von Baer (1827) modified that name to *Spermatozoon*, a word that caught on and is still in use today.

It was inevitable that artificial insemination would sooner or later be performed and that the outcome of such studies would influence the thinking on the role of spermatozoa. The first such experiment in Europe was performed by M. Jacobi in 1763, when he discovered how to fertilize fish eggs with milt. It must be noted that the artificial propagation of fish had been developed in China many years before by Fan Li (born in 517 B.C.), with no understanding of the fact that sperm cells were the fertilizing elements present in milt. Not long after Jacobi, Lazzaro Spallanzani succeeded in performing artificial fertilization not only in fish and frogs but also in a bitch (Gabriel and Vogel 1955; Castellani 1973; Sandler 1973). He also filtered the semen in 1784 or 1785 and noted that frog eggs were fertilized by the seminal fraction that could be squeezed out of the filter paper, but that no fertilization occurred when the filtrate was added to the eggs.

Experiments of this kind would eventually become decisive to our understanding of the role of spermatozoa. Yet, Spallanzani himself did not draw the correct conclusion. He had previously performed some experiments where he had added frog semen, which he supposed was devoid of spermatozoa, to eggs and these had then developed. He concluded from his various experiments that it is the "seminal aura" outside the animalcula that is capable of fertilization. It was only much later that this type of experiment was repeated and that the correct conclusion was drawn. The priority thus goes to Povost and Dumas, who in 1824 published their data and interpretations. The technique of artificial insemination may have a much older history however. There is a rumor that an Arabic person, named Hegira, in 1332 had a stallion ejaculate on a cloth that he then inseminated in his mare's vagina and that a foal was born after the expected period (Adlam 1980). The funny part of the story is that the stallion was not his own and the semen was a theft from a competitor and was performed in the darkness of the night by exposing the stallion to the smell of a mare's vaginal secretion. There may even be records of artificial insemination in the Talmudic books. These records may, however, refer to legends rather than to actual experiments.

Then in the Nineteenth Century

Not long after Spallanzani's experiments, artificial insemination had even been practiced in humans (reviewed in Adlam 1980). The first records are from the end of the eighteenth century. More important from a scientific point of view were the observations performed by Koelliker in 1844. He examined semen from many species of marine animals and also performed some insemination experiments. He could, among others, draw three fundamental conclusions: (1) semen of all animals contain spermatozoa, (2) these are formed from cells in the testes, and (3) spermatozoa have to come into contact with eggs for a fertile union to occur.

Fifteen years later, Koelliker could also conclude from more insemination experiments that it is the sperm head that is essential for fertilization to occur and after yet another 20 years, Hertwig (1892) made a statement in his doctoral thesis that made him famous: "Fertilization is the union of sexually differentiated nuclei." (Die Befruchtung beruht auf die Verschmelzung van geschlechtlich differenzierten Zellkernen.) By this definition of "fertilization," the important events in reproduction are those which involve the nuclei and their contents, a concept which has been fully verified by cytogenetics, the branch of biology that developed at the beginning of the last century as a result of the fusion of genetics and cytology.

It is interesting however to note that Hertwig advanced his thesis on the fertilization events without having been able to see fertilization occur. The first person to watch sperm entry into an egg (that of a sea urchin) actually was Hermann Fol during experiments conducted in the *Zoological Station in Villefranche-sur-Mer* (France). Two of the publications of Fol (1878, 1879) contain the very first description of the ability of a spermatozoon to fertilize and penetrate the egg of an echinoderm. The first person to see a mammalian egg was C. E. von Baer (1827) and the priority of transferring fertilized mammalian eggs and embryos from the biological mother to a surrogate mother (a rabbit doe) belongs to an Englishman, Walter Heap.

Finally, it was only in 1870 that the observations of Schweiger-Scidel and La Valette allowed the spermatozoon to acquire the status of fertilizing cell and these notions are confirmed etymologically, as the appellation of "spermatozoon" literally means "semen looking like an animal."

And Now Reaching the Twentieth Century

Improved microscopes and improved microscopical techniques were of decisive importance for further investigation of the spermatozoa. In an effort to approach "comparative spermatology," initiated by Leeuwenhoek and further developed by Koelliker, there were prominent investigators, such as La Valette St. George and Emil Ballowitz, who published some of their observations in the last decades of the nineteenth century. Somewhat later, Gustaf Retzius became a leading spermatologist.

He described the detailed structure of several hundred animal species, including many rare animals from all six continents. This is a unique investigation that could never be repeated. He noted (as others had done before him) that related species tend to have spermatozoa of similar structure and that it is possible to draw phylogenetic conclusions from sperm data. The fact that pangolin, echidna, and platypus have spermatozoa of the reptilian (sauropsid) type, whereas marsupials and other mammals have not, is thus an indication that the eutherian mammals branched off the mammalian stem before the appearance of the marsupials, and that the pangolins are the most primitive extant eutherians. He also noted that coelenterates, polychete worms, and mussels have small spermatozoa of a characteristic shape, which he referred to as "primitive" spermatozoa. Half a century later, Franzn showed that "primitive spermatozoa characterize animals that broadcast their spermatozoa into the ambient water," usually for external fertilization. The shape of the spermatozoa is thus dependent both on the reproductive biology and on the phylogenetic position.

The contribution of B. Baccetti and coll. in Sienna (Italy) and D. W. Fawcet (1970) illustrates the continuity of the investigation field of "evolutive spermatology."

By the turn of the nineteenth century, Frank Lillie in the famous *Woods Hole Lab* observed that during fertilization sperm are controlled by a substance released from the sea-urchin egg and was thus establishing the basis of chemotaxis as a sperm guidance mechanism. Using sea water with a high potassium content, he also observed parthenogenesis.

The only organelle that is sperm specific and quite widely distributed in the animal kingdom is the acrosome. It was first described by Valentine, given its present name "acrosome" by Lenhossck and shown to contain the lytic enzymes acrosin, initially termed lysin by Hibbard and Tyler, and hyaluronidase by Leuchtenberger and Schrader.

The rather dramatic transformation of the acrosome upon contact with the egg or with egg-water was first described by Jean Dan (1952) and termed the acrosome reaction. Localized under the acrosome is the sub-acrosomal material, termed the *perforatorium* by its discoverer, Waldeyer. He also undergoes experiments on the acrosome reaction as was described by Arthur L. Colwin and Laura H. Colwin (1955).

Other sperm organelles are those that are also found in somatic cells. This is true, for instance, of mitochondria, flagellar axonemes, microtubular arrays, etc. In some cases, our knowledge of these organelles has come from the study of the somatic cells, in other cases spermatozoa have been shown to be the ideal study object. It would take too much space to explore extensively here all the various sperm organelles with morphological and biochemical tools. The book that is dealing with several aspects of aquatic animals spermatology provides large information on these topics.

Obviously, because of the global topic of the book devoted to *Reproduction in Aquatic Animals: From Basic Biology to Aquaculture Technology* published by Springer Nature and edited by Manabu Yoshida and Juan F. Asturiano, readers of this book will also be able to access important information that deals with the sperm/egg interaction that involves the egg partner of various aquatic species.

References

Adlam JP (1980) J Sex Med 56:1–14

Baer C-E, Lin I (1827) De ovi mammalium et hominis genesi. Lipsiae

Blumenbach JF (1779) Handbuch der Naturgeschichte. Th.I, Goettingen

Bory de Saint-Vincent JBGM (1827) Zoosperme, zoospermos. In: Encyclopédic méthodique. T. II, Paris, pp 795–816

Buffon (1752) Découverte de la liqueur séminale dans les femelles vivipares, vol 1748. Paris, pp 211–228.

Castellani C (1973) J Hist Biol 6:37–68

Cole FJ (1930) Early theory of sexual generation. Clarendon Press, Oxford

Colwin AL, Colwin LH (1955) J Biophys Biochem Cytol 10:211–230

Cuvier G (1817) Le Règne animal. L. Hauman Ed., Bruxelles

Dan J (1952) Studies on the acrosome. 1. Reaction to egg water and other stimuli. Biol Bull (Woods Hole) 103:54–66
de Maupertuis PLM (1744) Dissertation physique à l'occasion du nègre blanc. Venus Physique Part I, Leyde. no editor
Dionis P (1698) Dissertation sur la génération de l'homme, avec des réflexion nouvelles sur plusieurs faits singuliers. L d'Houry Ed., Paris
Fawcet DW (1970) A comparative view of sperm ultrastructure. Biol Reprod Supp 2:90–127
Fol H (1878) Mém Soc Phys Hist Nat Genève 31:89. (quoted in: Buscaglia M, & Duboule, 2002)
Fol H (1879) Recherches sur la Fécondation et le commencement de L'Hénogénie chez divers animaux. Mém Soc Phys Hist Nat Genève 26:134
Gabriel ML, Vogel S (eds) (1955) Great experiments in biology. Prentice Hall, Englewood Cliffs
Geoffroy EF, Du CC (1704) Qaestio medica: An hominis primordia, vermis? In: Geoffroy E. F. Tractatus de materia medica. 8o. Parisiis, 1741
Gestner JM (1737) De diaeta disputatae, Liber1, Goettingen
Giardinius L (1623) De animatione foetus, quaestio, in qua ostenditur, quod anima rationalisante organizationem non infundatur. Duaci
Hartsoeker N (1678) Jour Savans Paris, le 29 Août
Harvey W (1651) Exercitationes de generatione, quibus accedunt quaedam de partu, de membranis ac humoribus, uteri et de conceptione. O. Pulleyn, London, 302 p. (The Generation of Animals, G. Whitteridge, trans., Blackwell, Oxford, 1981, 502 p)
Hertwig WAO (1892) Altere und neuere Entwickelungstheorieen. Berlin
Hill J (1752) An history of animals. Thomas Osborne, London
Huygens C (1678) Jour Savants, Paris, le 15 Aôut
Linnaeus C (1746) Sponsalia plantarum. Edited by Wahlhom, Stockholmiae
Lyonet P (1751) Schreiben die samenthierchen betreffend. In: Physikal Belustigung., Bd., Berlin
Needham JT (1749) Observations on the generation, composition, and decomposition, of animal and vegetable substances. Philos Trans R Soc Lond 45(1748):480–485.
Sandler I (1973) J Hist Biol 6:193–223
Schurig M (1720) Spermatologia historico-medica, h.e. seminis humani consideratio. Francofurti ad Moenum
Spallanzani L (1776) Opuscoli di fisica animale e vegetabile T I & II Modena
Vallisnieri A (1721) Historia della generazione dell'uomo e degli animali, se sia da' vermicelli spermatici, o dalle uova. Venezia
van Leeuwenhoek A (1677) Observationes de natis e semine genitali animalculis. Phil Transact Roy Soc 12:1040–1046
von Koelliker RA (1844) Beiträge zur Kenntniss der Geschlechtsverhtnisse und der Samen-Flssigkeit wirlbelloser Thiere und die Bedeutung de sogetlatltietl sogenannten Samenthiere. In: "Die Selbständigkeit", Berlin

Part I
Overview

Chapter 1
Overview: Reproductive Systems in Aquatic Animals

Manabu Yoshida

Abstract Many animals live in aquatic habitats. Regarding reproduction, all terrestrial animals perform internal fertilization, whereas aquatic animals show various reproductive systems: internal fertilization without mating, external fertilization, viviparous, oviparous, and parthenogenesis. In this chapter, I would like to provide an overview of the reproductive systems of aquatic animals and introduce each chapter in this book.

Keywords Fertilization · Reproductive systems · Internal fertilization · External fertilization · Viviparous · Oviparous · Hermaphrodite · Oocyte maturation · Sperm function · Assisted reproductive technology

1.1 Introduction

In the animal kingdom, there are about 30 phyla. Major animal groups (phyla and classes in Vertebrata) are shown in Fig. 1.1. Many animals belonging to the two highly evolved phyla—Vertebrata and Arthropoda—and some invertebrates, e.g., earthworms (Annelida) and snails (Mollusca), live in terrestrial habitats, and some parasitic animals, e.g., Ascaris (Nematoda) and tapeworms (Platyhelminthes) live in the body of other animals. Almost all other animals live in the aquatic habitat. Although they belong to Vertebrata or Arthropoda, many fishes and crustaceans live in water. In fact, only one group of animals does not live in the aquatic environment—phylum Onychophora, a small group related to Arthropoda. Corresponding to the diversity in aquatic animals, their reproductive system is also highly diverse. All terrestrial animals perform internal fertilization, whereas aquatic animals show

M. Yoshida (✉)
Misaki Marine Biological Station, School of Science, The University of Tokyo, Miura, Kanagawa, Japan
e-mail: yoshida@mmbs.s.u-tokyo.ac.jp

M. Yoshida, J. F. Asturiano (eds.), *Reproduction in Aquatic Animals*,
https://doi.org/10.1007/978-981-15-2290-1_2

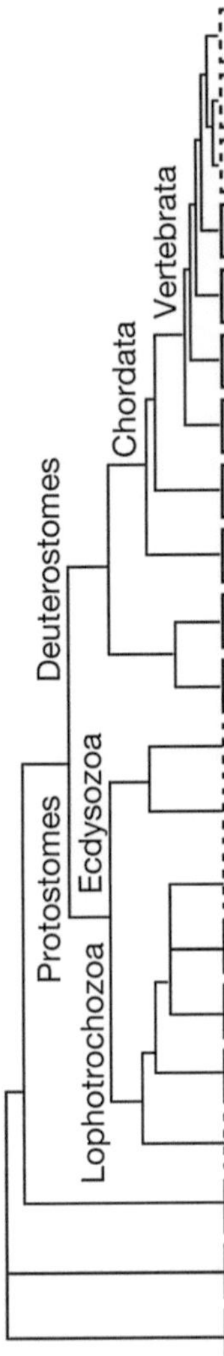

	aquatic life	style of fertilization	viviparity
Mammalia	some species	internal	mainly viviparous
Bird	some species. Reproduction was performed at terrestrial	internal	oviparous
Reptilia	some species. Reproduction was performed at terrestrial	internal	mainly oviparous
Amphibia	many species especially in juvinile	internal/extarnal	mainly oviparous
Tereostei	most species	internal/extarnal	mainly oviparous
Chondrichthyes	all species	internal	oviparous/ viviparous
Agnatha	all species	internal	oviparous
Urochordata	all species	external/internal w/o mating	oviparous/ viviparous
Cephalochordata	all species	external	oviparous
Echinodermata	all species	mainly external	mainly oviparous
Hemichordata	all species	external	oviparous
Arthropoda	many species especially in pancrastaceans	mainly internal	oviparous/ viviparous
Nematoda	some free-living species	internal	oviparous/ viviparous
Annelida	many species	internal/external	mainly oviparous
Mollusca	many species	internal/external	oviparous/ viviparous
Nemertea	most species	internal/external	oviparous/ viviparous
Ectoprocta	all species	internal/external	oviparous/ viviparous
Platyhelminthes	many species	mainly internal	oviparous/ viviparous
Cnidaria	all species	external/internal w/o mating	oviparous/ viviparous
Ctenophora	all species	external/internal w/o mating	oviparous/ viviparous
Porifera	all species	external/internal w/o mating	oviparous/ viviparous

Fig. 1.1 Major animal groups (phyla and classes in Vertebrata), habitat, and style of fertilization

various reproductive systems; some are internal fertilizers with or without mating and many aquatic animals perform external fertilization (see Fig. 1.1).

In this chapter, I would like to provide an overview of the reproductive systems of aquatic animals and to help the readers understand the focus on specific animals in each chapter.

1.2 Taxa and Living Habitats

The salinity of water is the most important factor affecting life of all animals and plants. There are three habitats in the aquatic environment: seawater (saline water), freshwater, and brackish water.

1.2.1 Seawater

Seawater, or saline water, is water whose salinity is approximately 3.5% (35 g/L) and mainly located in the oceans which cover about 70% of the surface of the earth. As the major diversifications of modern Metazoa, e.g., the Cambrian explosion, occurred in the ocean, most animal phyla are still seen in the seawater habitat. Despite several difficulties in evaluation, the total number of seawater animal species is estimated at approximately 200,000–250,000 species (Bouchet 2006). The major animals are the mollusks (approximately 52,000–75,000 species), arthropods (approximately 40,000–50,000 species), fish (approximately 15,000 species), flatworms (approximately 15,000 species), annelids (approximately 12,000 species), and nematodes (approximately 12,000 species) (Bouchet 2006). Interestingly, even though insects are the major group on the earth, and in terrestrial and freshwater habitats, there are only a few insect species in the seawater habitat.

1.2.2 Freshwater

Freshwater is water containing <0.5 g salts per 1 kg water. Freshwater is located on land, and it covers only 0.8% of the surface of the earth (Dudgeon et al. 2006). Because low osmolality induces an influx of water and delivers a fatal blow to an animal cell, which has no cell wall, the freshwater habitat is a more severe environment than the seawater habitat. The total number of described freshwater animal species is approximately 126,000 species. Many of the animals are hexapods (insects and collembolans) (approximately 76,000 species) and other arthropods (approximately 18,000 species), and another major group is the vertebrates (approximately 18,000 species) (Balian et al. 2008), which have well-developed systems for osmotic regulation and water-resistant skins.

In contrast, 43% of all fish species (approximately 14,000 species) predominantly live in the freshwater habitat (Nelson et al. 2016); one reason being that teleost fish seem to be derived from a freshwater ancestor (Carrete Vega and Wiens 2012).

Amphibians also live mainly in the freshwater habitat. Amphibians comprise approximately 6000 species, and about 80% of these are aquatic living or water dependent (Vences and Kohler 2008). Among them, only 144 species are salt-tolerant (Hopkins and Brodie 2015). See Chap. 11 for details.

1.2.3 Brackish Water

Brackish water is water whose salinity is between that of fresh water and seawater. It exists in areas where fresh water and seawater are mixed, such as estuaries. As salinity of the area changes irregularly, animals living in brackish water should have a tolerance for changing osmolarity. In addition to euryhaline species, locally adapted stenohaline species, which are typically found in seawater habitats, are also found in these habitats (Cognetti and Maltagliati 2000). Some teleost fishes such as mullet, seabass, flatfish, and eel are highly adapted to brackish water, and some bivalves, annelids, and crustaceans also live in these areas.

1.3 Reproduction Systems

As described in Sect. 1.1, the reproduction systems of aquatic animals are diverse, and they are excellent models for studying the adaptive evolution and species specificity of fertilization. In this section, I would like to introduce the types of the reproduction systems in aquatic animals (also see Fig. 1.1).

1.3.1 Asexual Reproduction

Although I do not want to focus on asexual reproduction because the main theme of the book is "sexual reproduction," asexual reproduction is a popular system in animals, and I should discuss this system first. Asexual reproduction is a type of reproduction producing offspring from a single parent without meiosis; that is, the offspring does not arise from gametes but from a part of the parent's body. Budding and fragmentation are types of asexual reproduction in metazoans. Many aquatic invertebrates including starfishes (Echinodermata), ascidians (Urochordata), planarians (Platyhelminthes), and medusae, and corals (Cnidaria) reproduce in this manner. For example, many hydromedusae have two life stages: one is the asexual stage of the polyp that produces polyps and medusae by budding and fragmentation, and the other is the sexual stage of the medusa that produce their gametes by meiosis. Some planarians switch their reproduction systems between the sexual and asexual stages (Kobayashi et al. 2012). As costs of reproduction are lower and producing offspring is faster than that by sexual reproduction, many asexually reproducing animals build a colony of clone individuals, for example, corals, bryozoans, colonial ascidians, and sponges.

1.3.2 *Parthenogenesis*

Parthenogenesis is also a type of reproduction producing offspring from a single parent, and it is usually considered as one type of asexual reproduction. However, in this system, the offspring arises from a gamete (usually the egg) without fertilization. Therefore, I consider that parthenogenesis is not an "asexual" (non-sexual) but a "unisexual" reproduction system. Parthenogenesis is also observed in several aquatic animals including vertebrates such as some amphibians, sharks, and teleosts (Neaves and Baumann 2011).

1.3.3 *Internal or External Fertilization*

As cells in all animals, including gametes, must live in some aquatic conditions, all terrestrial animals perform fertilization internally. Quite a few aquatic animals also perform internal fertilization, such as many crustaceans, snails, elasmobranches, and amphibians. In contrast, many aquatic animals perform external fertilization (see Fig. 1.1). Interestingly, some animals, such as jelly fishes and bryozoans, perform internal fertilization without mating; they spawn their sperm into the surrounding aquatic environments, their sperm swim and go into the female body, and finally they find the egg, and fertilization occurs.

1.3.4 *Viviparity or Oviparity*

All animals showing external fertilization lay their eggs in the external environment, and their embryos develop outside the body of the parent. Some animals showing internal fertilization also release their fertilized eggs and the embryos develop in the external circumstances. This developmental style is called "oviparity." In contrast, some animals showing internal fertilization keep their fertilized egg in the body, and the embryos develop inside the mother's body until larvae or juveniles. This developmental style is called "viviparity." Embryos of viviparous animals are usually supplied with additional nutrition from the mother (matrotrophy), but that of some viviparous animals use only the yolk of the eggs and are not supplied with additional nutrition. The latter style is sometimes distinguished from viviparity and called "ovoviviparity." The most famous viviparous animals are mammalians, but interestingly, there are viviparous animals in almost all taxa. Viviparous animals are considered to have evolved from oviparous animals (Blackburn 1999; Lode 2012), and the evolutionary transition from oviparity to viviparity occurred many times; for example, viviparity has evolved between 98 and

129 times considering only the squamate reptiles (Bystroff 2018). Among aquatic animals, many elasmobranches are viviparous and ovoviviparous, and many researchers have studied the variation in developmental styles. In Chap. 10, we focus on the developmental styles observed in elasmobranches.

1.3.5 Hermaphrodite or Dioecious

One of the important features is hermaphroditism; many fishes can change their sexes and many invertebrates are hermaphrodites. Although some hermaphrodite animals, like the nematode *Caenorhabditis elegans*, perform self-fertilization, that is, fertilize an egg and a spermatozoon from the same animal (one type of parthenogenesis); many hermaphrodite animals prevent self-fertilization as seen in many plants (Sawada et al. 2014). In the internal fertilizers like the sea hare, spermatozoa, and eggs are separated and only non-self-sperm received by mating can access the egg. However, in the case of external fertilizers, gametes themselves should recognize self and non-self-partners. The ascidians (Urochordata) are well-known hermaphrodite animals, and the system of self/non-self-recognition between gametes has been investigated (Sawada et al. 2014). In Chap. 9, the system of ascidians is reviewed thoroughly.

1.4 Sexual Behaviors

Sexual behavior is one of the interesting subjects in reproductive biology and depends on the reproductive system.

Fishes are one of the major, developed animal group in aquatic animals. Irrespective of internal fertilization or external fertilization, many fishes show mating behavior, and fertilization occurs just after spawning. Usually, the fish sperm, especially freshwater fish sperm, have very short lives; the lifetime of these sperm is around 30 s to a few minutes (Cosson et al. 2008). Thus, mating behavior is adapted and optimized to perform fertilization. Furthermore, motility of the sperm is regulated precisely by environmental elements: osmolarity, ions, and egg-derived factors among others (Cosson et al. 2008). Regulation of motility in the fish sperm is reviewed in Chap. 4.

Another interesting mating behavior has been observed in squids (Mollusca), which is one among alternative reproductive strategies. Usually, the female squid receives spermatophores from the male partner (called a “consort”) by mating and stores them in the storage organ in her body (Iwata et al. 2011). The spermatozoa await ovulation in the storage organ, and after ovulation/spawning of the egg, the spermatozoa reach the egg on/near the oviduct in the mantle cavity. However, the male “sneaker” comes and releases his sperm near the mouth of the female, and some sperm succeed in fertilizing the egg. Interestingly, the sneaker spermatozoa behave differently from the consort spermatozoa; the sneaker spermatozoa form a cluster even though they can swim freely (Hirohashi et al. 2013). The alternative reproduction tactics of the squids and sperm behaviors are reviewed in Chap. 13.

The benthic invertebrates showing external fertilization, e.g., sea urchins, ascidians, and corals, usually spawn their gametes directly into the external water. In these cases, it is a hard mission for the sperm to find its partner egg. Thus, the egg often releases sperm activation substances and attractants to ensure fertilization. This topic is reviewed in Chaps. 3 (general introduction and ascidians), 8 (corals), and 12 (sea urchins).

1.5 Behavior of Gametes

1.5.1 Behavior of the Egg

In general, mature oocytes are arrested at the prophase I of meiosis with the large nucleus (germinal vesicle) in the ovary of the female body. After stimulation for ovulation, the oocytes re-enter meiosis, induce germinal vesicle breakdown, and become "fertilization-eligible" eggs. Progression of meiosis after stimulation of ovulation is varied and species-dependent; e.g., the unfertilized eggs of many vertebrates, including amphibians and fish, are arrested at metaphase II, whereas sea urchin eggs finish meiosis completely before fertilization. Maturation of oocytes is described in detail in Chaps. 6 (fish) and 7 (invertebrates).

The ovulated "fertilization-eligible" eggs usually have vitelline coats (often called "chorion"), and they often impede the entry of the sperm into the egg. Especially in the fish, the chorion is too thick and too hard for the sperm to penetrate it. Instead, there is a tiny passage on the chorion, named "micropyle," to enable the sperm to access the egg (Yanagimachi et al. 2013). Moreover, some guides and/or guidance molecules are located on the chorion around the micropyle (Yanagimachi et al. 2013; Yanagimachi et al. 2017). Chapter 5 reviews the mechanism of sperm guidance toward the micropyle in the fish egg.

1.5.2 Behavior of the Sperm

Considering the sperm, the spermatozoa are usually immotile while stored in the male body and initiate their motility when they are ejaculated or spawned from the body (Yoshida et al. 2008). As described in Sect. 1.4, the sperm of many external fertilizers show chemotactic behavior toward the egg to find the conspecific egg. The initiation of sperm motility and sperm chemotaxis are reviewed in Chaps. 2 (overview), 3 (invertebrates), and 4 (fish). When the spermatozoa approach the egg, spermatozoa of many animals show acrosome reaction. Interestingly, the teleost spermatozoa have no acrosome, and they can enter the egg without acrosome reaction. In this context, the fish sperm cannot penetrate the chorion of the egg, so it looks for the micropyle to access the egg (see Sect. 1.5.1).

1.5.3 Polyspermy Block

The egg should receive only one spermatozoon for fertilization that is the fusion between female and male pronuclei, to maintain the genome in the embryo. Thus, almost all eggs prevent multiple sperm entry. The system is called "polyspermy block," which is one of the interesting aspects of research on fertilization (Dale 2014). It is reviewed in Chaps. 6 (fish), 7 (invertebrates), and 11 (amphibians).

Interestingly, the egg of some amphibian species like newts, receive multiple sperms during fertilization, but for fusion of the pronuclei, only one male pronucleus is selected from the spermatozoa that enter (see Chap. 11 for details).

1.6 Issues of Reproductive Biology for Aquaculture

In aquatic animals, fish, mollusks, and crustaceans are useful as food and are a target for farming (aquaculture). In this book, we also focus on the topics of reproductive biology in aquaculture. Especially, the establishment of the complete culture technology of some high-value fish, like tuna, eel, and sturgeon, is demanded as these fish have become endangered because of overfishing. However, as described above, reproduction systems of aquatic animals are varied and species-dependent. Reproducing juveniles is one of the difficulties in establishing aquaculture methods. In Chap. 14, the practical case of the European eel is reviewed.

Furthermore, an assisted reproductive technology is also demanded in aquaculture for the conservation and propagation of the animals. One of the technologies is cryopreservation of gametes; the technique provides flexibility in the production of embryos/juveniles and stocking of elite broodstock and/or endangered animals. Chapter 15 reviews the cryopreservation of sperm in aquatic animals. Another important technique in assisted reproductive technology is germ cell transplantation, which enables surrogate production. In Chaps. 16 and 17, there are reviews of studies on the technology used in the sturgeon and teleosts, respectively.

1.7 Conclusion

As overviewed in this chapter, reproduction systems in aquatic animals are highly diverse, and it is difficult to grasp the whole picture of reproduction. We hope this book helps readers understand features of the reproductive systems in each aquatic animal.

References

Balian EV, Segers H, Leveque C, Martens K (2008) The freshwater animal diversity assessment: an overview of the results. Hydrobiologia 595:627–637. https://doi.org/10.1007/s10750-007-9246-3

Blackburn DG (1999) Viviparity and oviparity: evolution and reproductive strategies. In: Knobil E, Neill JD (eds) Encyclopedia of reproduction, vol 4. Academic, London, pp 994–1003

Bouchet P (2006) The magnitude of marine biodiversity. In: Duarte CM (ed) The exploration of marine biodiversity: scientific and technological challenges. Fundación BBVA, Bilbao, pp 31–62

Bystroff C (2018) Intramembranal disulfide cross-linking elucidates the super-quaternary structure of mammalian CatSpers. Reprod Biol 18:76–82. https://doi.org/10.1016/j.repbio.2018.01.005

Carrete Vega G, Wiens JJ (2012) Why are there so few fish in the sea? Proc R Soc B Biol Sci 279:2323–2329. https://doi.org/10.1098/rspb.2012.0075

Cognetti G, Maltagliati F (2000) Biodiversity and adaptive mechanisms in brackish water fauna. Mar Pollut Bull 40:7–14. https://doi.org/10.1016/S0025-326x(99)00173-3

Cosson J, Groison AL, Suquet M, Fauvel C, Dreanno C, Billard R (2008) Marine fish spermatozoa: racing ephemeral swimmers. Reproduction 136:277–294. https://doi.org/10.1530/REP-07-0522

Dale B (2014) Is the idea of a fast block to polyspermy based on artifact? Biochem Biophys Res Commun 450:1159–1165. https://doi.org/10.1016/j.bbrc.2014.03.157

Dudgeon D et al (2006) Freshwater biodiversity: importance, threats, status and conservation challenges. Biol Rev 81:163–182. https://doi.org/10.1017/S1464793105006950

Hirohashi N et al (2013) Sperm from sneaker male squids exhibit chemotactic swarming to CO_2. Curr Biol 23:775–781. https://doi.org/10.1016/j.cub.2013.03.040

Hopkins GR, Brodie ED (2015) Occurrence of amphibians in saline habitats: a review and evolutionary perspective. Herpetol Monogr 29:1–27

Iwata Y, Shaw P, Fujiwara E, Shiba K, Kakiuchi Y, Hirohashi N (2011) Why small males have big sperm: dimorphic squid sperm linked to alternative mating behaviours. BMC Evol Biol 11:236. https://doi.org/10.1186/1471-2148-11-236

Kobayashi K, Maezawa T, Nakagawa H, Hoshi M (2012) Existence of two sexual races in the planarian species switching between asexual and sexual reproduction. Zool Sci 29:265–272. https://doi.org/10.2108/zsj.29.265

Lode T (2012) Oviparity or viviparity? That is the question …. Reprod Biol 12:259–264. https://doi.org/10.1016/j.repbio.2012.09.001

Neaves WB, Baumann P (2011) Unisexual reproduction among vertebrates. Trends Genet 27:81–88. https://doi.org/10.1016/j.tig.2010.12.002

Nelson JS, Grande TC, Wilson MVH (2016) Fishes of the world, 5th edn. Wiley, Hoboken

Sawada H, Morita M, Iwano M (2014) Self/non-self recognition mechanisms in sexual reproduction: new insight into the self-incompatibility system shared by flowering plants and hermaphroditic animals. Biochem Biophys Res Commun 450:1142–1148. https://doi.org/10.1016/j.bbrc.2014.05.099

Vences M, Kohler J (2008) Global diversity of amphibians (Amphibia) in freshwater. Hydrobiologia 595:569–580. https://doi.org/10.1007/s10750-007-9032-2

Yanagimachi R et al (2013) Sperm attractant in the micropyle region of fish and insect eggs. Biol Reprod 88(47):01–11. https://doi.org/10.1095/biolreprod.112.105072

Yanagimachi R et al (2017) Chemical and physical guidance of fish spermatozoa into the egg through the micropyle 1. Biol Reprod 96(4):780–799. https://doi.org/10.1093/biolre/iox015

Yoshida M, Kawano N, Yoshida K (2008) Control of sperm motility and fertility: diverse factors and common mechanisms. Cell Mol Life Sci 65:3446–3457. https://doi.org/10.1007/s00018-008-8230-z

Part II
Basic Knowledge of Male Gametes in Aquatic Animals

Chapter 2
Introduction to Sperm Motility of Aquatic Animals

Jacky Cosson

Abstract The purpose of this chapter is to present a historical point of view on sperm motility, showing how pioneer investigations using sea urchins as a model species gradually radiated to other aquatic species because of their ability to approach more specific questions.

Keywords Flagellum · Axoneme · Fish · Shellfish · Echinoderms · Jellyfish

2.1 Introduction

During evolution, aquatic animals have preceded terrestrial ones, mostly because life appeared initially in water and, the latter remains absolutely necessary for the maintenance of living creatures. If one tries to trace the evolution of scientific knowledge in the field of sperm, eggs, and their association with fertilization, it also appears that basic notions have been acquired through observations of aquatic species.

In this regard, this book flies over a great diversity of species that will be briefly explored in the present paragraphs: it is conceived as an overview so as to avoid overlap with the detailed and more specific chapters coming afterward.

2.2 Interest to the Reproduction of Aquatic Animals

The first human interest to aquatic animals has probably been associated with the possibility of rising some species, using aquaculture as a substitute to fishing, a more hazardous activity. An overview of the main aquatic species considered as

J. Cosson (✉)
Faculty of Fisheries and Protection of Waters, Research Institute of Fish Culture and Hydrobiology, University of South Bohemia in České Budějovice, Vodnany, Czech Republic

M. Yoshida, J. F. Asturiano (eds.), *Reproduction in Aquatic Animals*,
https://doi.org/10.1007/978-981-15-2290-1_3

important in aquaculture partly overlaps the panel of species that have been used to establish important steps in reproductive biology.

A non-exhaustive list includes: (1) fish species, either marine species (e.g., sea bream, sea bass) as alternative to fishing or traditional freshwater species such as those of central Europe (e.g., carp, trout); (2) shellfish such as oysters, commercially important in Europe or Japan; (3) other species of interest for sport-fishing or fishkeeping, both popular hobbies, for high-value products such as caviar from sturgeons or jewels from black pearl oysters; and (4) research and conservation of species as detailed afterward in the present book.

2.3 Importance of Marine Stations

These marine stations are places of predilection, mostly because they offer a source of great diversity of biological material from the sea and are enjoyable spots of scientific melting pot where to organize meetings or summer schools. As examples, let us mention *Woods Hole* in Cape Cod (USA), the *Zoological Station* in Napoli (Italy), the *Misaki Marine Biological Station* in Misaki (Japan), etc.

2.4 Various Aquatic Species Used Historically in Reproductive Biology

2.4.1 Sea Urchin

First of all is *sea urchin*, a historically very important species: sea urchins allow basic observations on sperm motility. Mostly the advantages of sea urchins are that they provide a huge amount of milt per individual, a crucial property for early biochemical studies, and they are gravid most of the year with mature sperm cells able to swim for very long periods. As examples early studies on sea urchins were developed by F. Lillie in the *Marine Biological Laboratory* (MBL) in *Woods Hole* (USA) then later by Barbara and Ian Gibbons (1981) in the *Kewalo Marine Lab* (Hawaii) and by Charles Brokaw (Brokaw and Gibbons 1975) in the *Caltech's Kerckhoff Marine Lab* (USA) while very long ago, fertilization of sea urchin eggs was observed for the first time by H. Fol (1879) in the Marine Station of *Villefranche sur mer* (France).

The studies of Sir Gray (1955) on the description of sea urchin sperm movement remain a fundamental pillar that underpins the resistance force theory (RFT) and furnishes the physical basis of flagellar movement. Mostly, flagella develop waves that initiate close to the head/tail junction and propagate backwardly towards the distal tip. The viscosity of the swimming solution offers a resistance to the wave propagation which, by reaction, provokes the forward translation of the whole sperm cell. The demonstration that molecular motors (dyneins) actuate the sea urchin sperm flagellum has been elegantly demonstrated and are summarized in the historical review of Ian Gibbons (2017). The mechanism of sliding/bending of flagellar microtubules (Brokaw 1990; Brokaw and Gibbons 1975; Satir 1974) that is fundamental to flagellar waves initiation and propagation was mainly established on sea urchin sperm flagella observations, while the bridge for applications from sea urchins to mammal sperm studies was developed since long (Gagnon 1995, 1998); more details on these aspects will be found in the update of Chap. 12 in this book.

2.4.2 Ascidians

Studies on ascidians species (e.g., *Ciona*) were developed by Masaaki Morisawa's group (Yoshida et al. 1993) in *Misaki Marine Biological Station* (Japan) mostly dealing with sperm/egg interaction, so-called chemotaxis. Among the main results, the chemoattractants and the mechanisms underlying sperm attraction were clearly established in the studies carried on ascidia (Yoshida et al. 1993). The chemical structure of the attracting molecules as well as the tracking of sperm cells swimming under control of the substance diffusing from the egg were brilliantly described in a series of papers of Morisawa and colleagues published in the 1990s. More details on the update are found in Chaps. 3 and 9 of the present book.

2.4.3 Jellyfish

In parallel, the description of fundamental features of chemical sperm guidance to the egg were run on **cnidaria** (siphonophores) by Christian Sardet's group (Cosson et al. 1986) in the *Villefranche Marine Station* and other **jellyfish** by Richard Miller (1985) on the coral reefs of Australia (see also Chap. 8 of the present book). The importance of calcium ions in the mechanism that governs sperm chemoattraction was emphasized by the results obtained on cnidaria sperm (Cosson 2015; Miller 1985). The book of Jacky Cosson (2015) provides an overview of the present knowledge on sperm guidance in various animal species.

2.4.4 Fish

Studies on spermatozoa of fish (teleost and chondrostean) were developed by Morisawa group (Morisawa 1985) in *Misaki Marine Biological Station* and by Cosson's group (Cosson et al. 2008) in the *Villefranche Marine Station*, both dealing with the understanding of the motility activation step and sperm guidance in teleost, mostly in herrings, the latter by Gary Cherr and colleagues (Yanagimachi et al. 1992) in the *Bodega Marine Station* (USA).

The main role of osmolality and ions in the mechanism of fish sperm activation were clearly established first on trout and carp spermatozoa and also on that of marine fish species (Morisawa 1985; Cosson 2004; Cosson et al. 2008). The successive steps of signaling cascade going from the signal perception to the activation of the flagellar machinery are nowadays well described in several fish species (Dzyuba and Cosson 2014) and a more detailed description can be found in the Chap. 4 of the present book.

Studies on energetic aspects of many fish sperm motility demonstrated that ATP stores drastically limit the duration of the motility period (Dzyuba et al. 2017). Among fish species, it is presently only in herring that a mechanism of sperm chemoattraction has been clearly established (Yanagimachi et al. 1992): an update can be found in the Chap. 5 of the present book.

2.4.5 Other Animals

Worth to mention as important in the history of gamete interaction and also detailed in this book are studies on **eel** sperm, their regulation and reproduction control (Chap. 14), on **frogs and amphibian** fertilization (Chap. 11), and on **squids** (Chap. 13).

2.5 More Practical Aspects

The accumulation of basic knowledge on the *gametes* of various aquatic animal species has been, in a more recent past, paralleled by more practical studies dealing with sperm cryopreservation aspects some being presented in Chap. 15 as well as recent germ cell transplantation experiments on sturgeon and marine fish species as examples (see Chaps. 16 and 17) that lead to very important applications for the conservation of aquatic species.

2.6 Future Developments

The variety of topics covered by the present book confirms how the knowledge in the field of gametes and their interaction owe to the studies on aquatic animal species and constitutes a positive augury that will prolong, in the future, the brilliant way traced by the pioneering "classical" biologists thanks to the input of other disciplines such as molecular biology, biochemistry, mathematics, or physics of fluids and that of modern technologies such that more and more sophisticated visualization is attainable by optical and electron microscopy.

References

Brokaw CJ (1990) The sea urchin spermatozoon. BioEssays 12(9):449–452. https://doi.org/10.1002/bies.950120910. [PMID: 2175174]

Brokaw CJ, Gibbons IR (1975) Mechanisms of movement in flagella and cilia. In: Wu TY-T, Brokaw CJ, Brennan C (eds) Swimming and flying in nature. Plenum Publishing Corporation, New York, pp 89–126

Cosson J (2004) The ionic and osmotic factors controlling motility of fish spermatozoa. Aquac Int 12:69–85. https://doi.org/10.1023/B:AQUI.0000017189.44263.bc

Cosson J (2015) Flagellar mechanics and sperm guidance. Bentham Books Publisher, 424 pages. isbn:978-1-68-108-128-1, 978-1-68-108-129-8. (e-Book 2015)

Cosson J, Carré D, Cosson M-P (1986) Sperm chemotaxis in siphonophores: identification and biochemical properties of the attractant. Cell Motil Cytoskeleton 6:225–228. https://doi.org/10.1002/cm.970060222

Cosson J, Dreanno C, Fauvel C, Groison A-L, Suquet M, Billard R (2008) Marine fish spermatozoa: racing ephemeral swimmers. Reproduction 136:277–294

Dzyuba V, Cosson J (2014) Motility of fish spermatozoa: from external signaling to flagella response. Reprod Biol 14(3):165–175. https://doi.org/10.1016/j.repbio.2013.12.005.vivi

Dzyuba B, Bondarenko O, Gazo I, Prokopchuk G, Fedorov P, Cosson J (2017) Energetics of fish spermatozoa: the proven and the possible. Aquaculture 472:60–72

Fol H (1879) Recherches sur la Fécondation et le commencement de L'Hénogénie chez divers animaux. Mém Soc Phys Hist Nat Genève 26:134

Gagnon C (1995) Regulation of sperm motility at the axonemal level. Reprod Fertil Dev 7:189–198

Gagnon C (1998) The male gamete: from basic knowledge to clinical applications. Cache River Press, Vienna

Gibbons IR (1981) Cilia and flagella of eukaryotes. J Cell Biol 91(3 Pt 2):107s–124s. https://doi.org/10.1083/jcb.91.3.107s. [PMID: 6459326]

Gibbons IR (2017) Discovery of dynein and its properties: a personal account. In: Dyneins. Academic, New York, pp 1–87. https://doi.org/10.1016/B978-0-12-809471-6.00001-2

Gray J (1955) The movement of sea-urchin spermatozoa. J Exp Biol 32:775–801

Miller RL (1985) Sperm chemo-orientation in the metazoa. In: Metz CB, Monroy A (eds) Biology of fertilization. Biology of the sperm. Academic, New York, pp 275–337. https://doi.org/10.1016/B978-0-12-492602-8.50015-2

Morisawa M (1985) Initiation mechanism of sperm motility at spawning in teleosts. Zool Sci 2:605–615

Satir P (1974) The present status of the sliding microtubule model of ciliary motion. In: Sleigh MA (ed) Cilia and flagella. Academic, New York, pp 131–141

Yanagimachi R, Cherr GN, Pillai MC, Baldwin JD (1992) Factors controlling sperm entry into the micropyles of salmonid and herring eggs. Develop Growth Differ 34:447–461. https://doi.org/10.1111/j.1440-169X.1992.00447.x

Yoshida M, Inaba K, Morisawa M (1993) Sperm chemotaxis during the process of fertilization in the ascidians *Ciona savignyi and Ciona intestinalis*. Dev Biol 157(2):497–506. https://doi.org/10.1006/dbio.1993.1152. [PMID: 8500656]

Chapter 3
Sperm Activation and Chemotaxis in Invertebrates

Jumpei Ikenaga and Manabu Yoshida

Abstract The gametes of aqua-living animals are equipped with several systems to ensure fertilization. Sperm activation and chemotaxis toward the conspecific eggs—some factors from the eggs or female organs activate and attract conspecific sperm—are the first two steps of authentication between sperm and eggs. Phenomena and molecular mechanisms underlying sperm activation and chemotaxis are varied, even in the same taxa. Considering species specificity of sperm chemotaxis, the system may prevent crossbreeding. It is an interesting point why the system of sperm activation and chemotaxis has been evolved with such a high diversity. In this chapter, we reviewed the sperm activation and chemotaxis in aquatic invertebrates.

Keywords Sperm chemotaxis · Fertilization · Species specificity

3.1 Introduction

In cases of aqua-living animals, many of them perform external fertilization. Even in animals undergoing internal fertilization, many of their males spawn spermatozoa into surrounding water without mating (see Chap. 1 for details). Thus, it is challenging for spermatozoa to find conspecific eggs, and they are equipped with several processes to overcome difficulties in the aquatic circumstances such as seawater. First, the spermatozoa start swimming when spawned into water around the male body (initiation). When the spermatozoa get close to the egg, the motility of the sperm is activated by chemicals from the egg or the female reproductive organs (activation). In many cases, the spermatozoa change their swimming pattern due to chemicals from the egg and are guided toward the egg (chemotaxis). Even in mammals, the process of changing sperm motility (initiation, activation, and chemotaxis)

J. Ikenaga · M. Yoshida (✉)
Misaki Marine Biological Station, School of Science, The University of Tokyo, Miura, Kanagawa, Japan
e-mail: yoshida@mmbs.s.u-tokyo.ac.jp

M. Yoshida, J. F. Asturiano (eds.), *Reproduction in Aquatic Animals*,
https://doi.org/10.1007/978-981-15-2290-1_4

has been observed. In this chapter, we reviewed the sperm activation and chemotaxis in aquatic invertebrates.

3.2 Sperm Activation

Spermatozoa are usually immotile while stored in the male body and become motile when they are ejaculated or spawned from the male. Since most invertebrates living in aquatic circumstances are external fertilizers or internal fertilizers without mating, sperm motility is activated in aquatic conditions. Initiation of sperm motility in aquatic vertebrates showing external fertilization, such as fishes and amphibians, is mediated by changes of osmolarity (cf. Chapters 2, 4, and 11). On the other hand, in marine invertebrates, there is no difference in osmolarity between the seminal plasma and the aquatic circumstances around the male body, and inducer of sperm motility is almost unknown. The spermatozoa of several other animals are almost quiescent when they are spawned, and some factors released from eggs or female organs activate their motility. These female-derived "sperm-activating factors" are proposed in many marine invertebrates, but identification of these factors have been performed in only a few animals. Thus, the molecular mechanisms of sperm activation are almost unknown.

3.3 Sperm Chemotaxis

In many animals and plants, it is widely observed that spermatozoa sense substances released from the eggs or female organs and are led toward the direction to the egg. This phenomenon is called "sperm chemotaxis" and guarantees that conspecific gametes successfully come together. Sperm chemotaxis was already observed in the nineteenth century in ferns and mosses (Pfeffer 1884), and in animals, when J.C. Dan first observed it in the hydrozoan *Spirocodon saltatrix* (Dan 1950). Similar to the studies on sperm activation, sperm chemotaxis has been studied in marine invertebrates, taking advantage of external fertilization. In fact, Miller studied sperm chemotaxis extensively in many metazoan animals such as Cnidaria, Mollusca, and Echinodermata (see review; Miller 1985b). For now, the mechanisms of sperm chemotaxis are particularly researched in sea urchins and ascidians (see Sect. 3.4).

Although sperm chemotaxis in mammals does not seem to be species-specific, sperm chemotaxis of many other animals, especially marine invertebrates, is species-specific (Table 3.1). Thus, sperm chemotaxis may prevent crossbreeding, even though little is known about the molecular mechanisms of species specificity in sperm chemotaxis.

Sperm movement and its flagellar beating during chemotactic response have been observed in several marine invertebrate species. On the surface of a glass slide, the spermatozoon usually swims in a circular track with low asymmetric flagellar

Table 3.1 Observed sperm activation and sperm chemotaxis

Phylum	Class	Style of fertilization	Sperm activation	Sperm chemotaxis and species-specificity[a]
Cnidaria	Anthozoa	External	+ (Morita et al. 2006)	+/+ (Coll et al. 1994; Coll et al. 1995; Morita et al. 2006)
	Hydrozoa	External	n.d.	+/+ (Carré and Sardet 1981; Dan 1950; Miller 1966; Miller 1973; Miller 1979a; Noda and Kanai 1981)
	Staurozoa	External	n.d.	+? (Miller 1985b)
Mollusca	Bivalvia	External	+ (Alavi et al. 2014)	n.d.
	Polyplacophora	External	n.d.	+/− (Miller 1977)
	Gastropoda	External	n.d.	+ (Riffell et al. 2002)
	Cephalopoda	Internal (mating)	+ (Tosti et al. 2001)	+ (Hirohashi et al. 2013; Zatylny et al. 2002)
Annelida		External	+ (Lillie 1913b)	+? (Miller 1985b)
Bryozoa		Internal	n.d.	+? (Miller 1985b)
Arthropoda		Internal (mating)	n.d.	− (Miller 1985b)
Echinodermata	Echinoidea	External	+ (Hansbrough and Garbers 1981; Lillie 1913a; Ohtake 1976a; Suzuki 1990)	+ (Guerrero et al. 2010; Ward et al. 1985)
	Asteroidea	External	+ (Nishigaki et al. 1996)	+/+ (Miller 1985a)
	Holothuroidea	External	+ (Morita et al. 2009)	+/+ (Miller 1985a; Miller 1997; Morita et al. 2009)
	Ophiuroidea	External	n.d.	+/+ (Miller 1985a; Miller 1997)
Chordata	Ascidiacea	External	+ (Minganti 1951; Yoshida et al. 1994)	+/+ (Matsumori et al. 2013; Miller 1975; Miller 1982; Minganti 1951; Yoshida et al. 2013; Yoshida et al. 1993; Yoshida et al. 2002)
	Larvacea	External	n.d.	+ (Miller and King 1983)

[a]Initial plus or minus means existence of sperm chemotaxis and second plus or minus means that of species specificity. Even in class in which species specificity is "+", not all species show species specificity. "+?" means that the group has species specificity but less evidence

beating in the absence of eggs nor sperm attractants. On the other hand, when the spermatozoon shows chemotactic behavior, it draws a distinctive track: periodical quick turn and straight swimming (see review; Yoshida and Yoshida 2011). In the pathway, the spermatozoon dynamically regulates its flagellar beating pattern. This regulates sperm direction, resulting in the spermatozoon's approach toward the egg.

In the regulation of flagellar beating, transient Ca^{2+} increase is a cue signal (see Sect. 3.5.1 for details).

3.4 Knowledge of Sperm Activation and Chemotaxis in Aquatic Invertebrates

As described in Sects. 3.2 and 3.3, phenomena of sperm activation and chemotaxis have been observed in many animals. We show an overview of sperm activation and chemotaxis in Table 3.1, and detailed descriptions are shown in Sects. 3.4.1–3.4.6.

3.4.1 Cnidarians

Studies on sperm activation and chemotaxis in animals were initially performed on Cnidarian species. The first observation of sperm chemotaxis was on the sperm of hydrozoan, *Spirocodon saltatrix* (Dan 1950), and Miller had studied extensively on hydrozoan species (Freeman and Miller 1982; Miller 1966, 1970, 1973, 1979a, 1979b). Sperm chemotaxis was observed in many hydrozoan species, even though they are sessile or planktonic, or internal or external fertilizers (Miller 1985b). The sperm chemotaxis in hydrozoan is highly species-specific: Miller examined chemotaxis in 32 species of marine hydromedusae, and only 13 heterospecific cross-reactions were found (Miller 1979a). The attractant seems to be released from the egg itself (Miller 1985b) or from an extracellular structure localized around the animal pole of the egg called the cupule (Cosson et al. 1984).

The sperm attractants in the hydrozoan species seem to be proteins or peptides: the sperm attractant of *Hippopodius hippopus* seemed to be a protein with a molecular mass of 25 kDa and pI 3.5, even though it was still not identified (Cosson et al. 1986). On the other hand, the sperm attractant in the scleractinian coral *Montipora digitata* was found to be the unsaturated fatty alcohol dodeca-2,4-diynol (Coll et al. 1994). In the species, three fatty alcohols were isolated as sperm activating substances, but only dodeca-2,4,-diynol had sperm-attracting activity (Coll et al. 1994). Coll and his colleagues also found that the native sperm attractant of the soft coral *Lobophytum crissum* is the macrocyclic diterpene alcohol (-)-epi-thunbergol (Coll et al. 1995). Other coral species in the genus *Acropora* also showed sperm chemotaxis toward the egg (Morita et al. 2006), even though the sperm attractants were still not elucidated.

In addition to that in the hydorozoan and anthozoan species, it is thought that sperm chemotaxis is present in the Schyphozoan species.

Like in other animals, the chemotactic movement of the Cnidarian spermatozoa is mediated by Ca^{2+} (Cosson et al. 1983, 1984; Morita et al. 2006).

3.4.2 Mollusks

In protostome animals, fertilization of mollusks, especially in bivalves, has been studied well due to their economic importance. Alavi and his co-workers studied sperm activation of some bivalves (Bivalvia): Manila clam (*Ruditapes philippinarum*), Pacific oyster (*Crassostrea gigas*), and Japanese scallop (*Patinopecten yessoensis*) (Alavi et al. 2014). They showed that the sperm of the bivalves initiate its motility when released into sea water, and in addition, the movement of sperm is activated by 5-hydroxytryptamine (5-HT), one of the physiologically active substances inducing spawning and oocyte maturation in bivalves (effects of 5-HT on reproduction of mollusks are described in Chap. 7, Sect. 7.3 for details). During the initiation of sperm motility, 5-HT induces a Ca^{2+} influx via voltage-dependent ion channels associated with K^{+} efflux, resulting in the activation of CaM-dependent flagellar beating (Alavi et al. 2014) (see Fig. 3.1). On the other hand, there is no report showing the chemotactic response of the bivalve sperm.

Sperm chemotaxis has been observed in the primitive mollusk chitons (Polyplacophora): Miller showed that the ethanol extracts of the eggs of several chitons attract the spermatozoa (Miller 1977). Interestingly, sperm chemotaxis in chitons is not species-specific (Miller 1977, 1985b). Sperm activation and chemotaxis is also observed in cephalopods. Spermatozoa of the octopus *Octopus vulgaris* are stored in a female genital tract prior to fertilization and seem to be pre-activated by progesterone. Progesterone-treated sperm show a breakdown of outer membrane around the acrosomal region, which is like a process of acrosome reaction in mammals (Tosti et al. 2001). Sperm chemotaxis in cephalopods is observed in *Octopus vulgaris* and the cuttlefish *Sepia officinalis*, and their sperm attractants are identified as the peptides named Octo-SAP (De Lisa et al. 2013) and SepSAP (Zatylny et al. 2002), respectively.

Concerning snails (Gastropoda), one of the major groups of mollusks, little is known about their sperm activation and chemotaxis since most of them are internal fertilizers. However, abalone, a primitive snail, is an external fertilizer, and its sperm shows chemotactic behavior (Riffell et al. 2002, 2004). The sperm attractant of the red abalone (*Haliotis rufescens*) is identified as the amino acid L-tryptophan (Riffell et al. 2004).

Sperm chemotaxis in mollusks plays roles not only in the success of conspecific fertilization but also in reproductive strategy. Males of some squids (Cephalopoda) contend on females for mating, and the winner acts as a consort male, the loser a sneaker male. In other words, the consort male accompanies the female (see the Chap. 13). The consort male deposits his spermatophore around the opening of the oviduct on the female mantle cavity, and the sperm will get first access to the egg capsules. On the other hand, the sneaker male gets close to the female and scatters

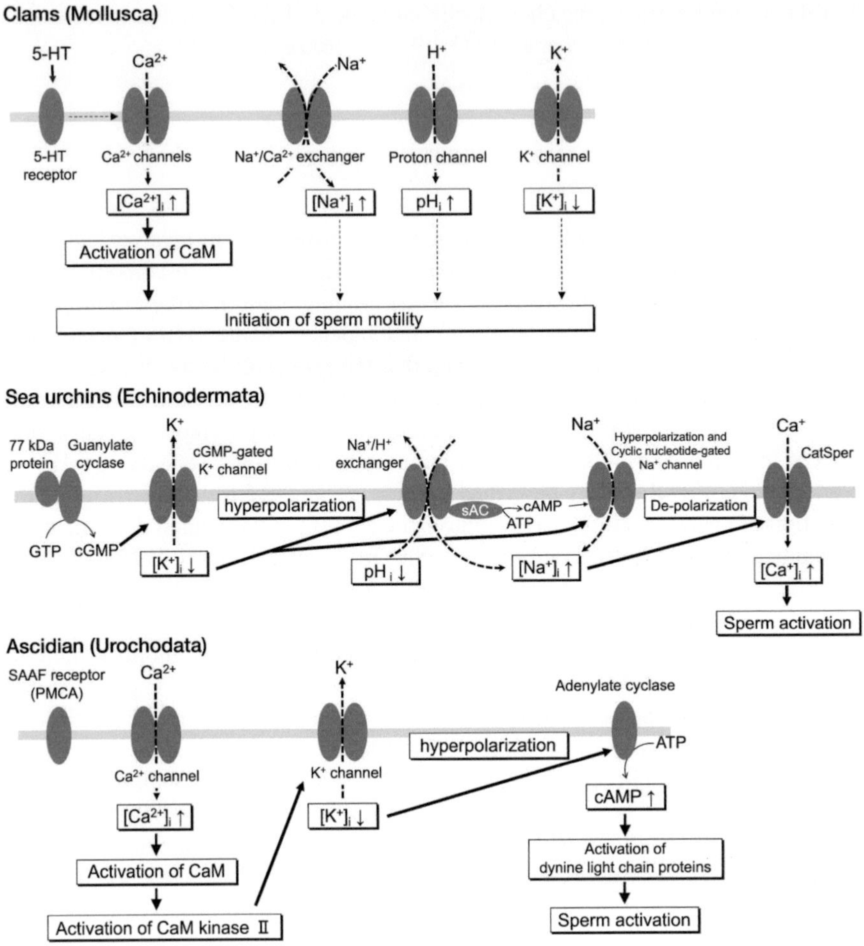

Fig. 3.1 Signaling cascades of sperm activation in animals: clams, sea urchins, and ascidians

its sperm when the female holds the egg capsules on her arm to lay it on the substrate, i.e., the male is "sneaking" to fertilize the eggs. Interestingly, in the squid *Heterololigo bleekeri*, spermatozoa from sneaker males make a cluster after spawning, whereas spermatozoa from consort males do not (Hirohashi et al. 2013). The clustering of sperm is caused by sperm chemotaxis toward CO_2 (Hirohashi et al. 2013). Higher sensitivity of the sneaker male sperm to pH seems to cause a chemotactic response only in the sneaker sperm (Hirohashi et al. 2013; Iida et al. 2017). For details, see Chap. 13, Sect. 13.3.4.

3.4.3 Arthropods and Other Protostomes

Knowledge of sperm activation and chemotaxis in protostomes other than mollusks is minimal, even in Arthropoda. One of the few studies on arthropod sperm was done in the horseshoe crab (Xiphosura) *Limulus polyphemus*. The *Limulus* spermatozoa are almost immotile when spawned in seawater and are activated if they encounter an egg-derived factor (Clapper and Brown 1980a). The egg-derived factor activating sperm release (also called a sperm motility initiating factor (SMI)) is still not identified, but it appears to be a hydrophobic peptide with the molecular mass of 500–2000 (Clapper and Brown 1980b; Clapper and Epel 1982). Fertilization of arthropods are reviewed in Chap. 7, Sect. 7.2.3.

In other protostomes, existence of sperm activation and chemotaxis is suggested in Annelida and Bryozoa (Miller 1985b), but these should be confirmed.

3.4.4 Sea Urchins (Echinodermata, Echinoidea)

Sea urchins, a group of echinoderms, have been studied well for a century, and their sperm activation and chemotaxis are also well known. Activation of sea urchin spermatozoa by some factors associated with the eggs was observed a century ago (Lillie 1913b). The sperm of the sea urchin is a traditional model for research on sperm activation, although sea urchin spermatozoa are usually highly activated after spawning in sea water. First, Ohtake showed that the jelly layer of eggs of *Hemicentrotus pulcherrimus* contains sperm-activating substances, which activate sperm motility and respiratory in acidic sea water (Ohtake 1976b). After that, the decapeptide named speract and another 14-amino-acid peptide named resact were identified as the sperm-activating substances from *Strongylocentrotus purpuratus* (Hansbrough and Garbers 1981) and *Arbacia punctulata* (Suzuki et al. 1984), respectively. Moreover, resact was found to be not only the sperm-activating substance but also the sperm attractant (Ward et al. 1985). Suzuki and his co-workers expanded the work in the various sea urchin species and finally found 74 sperm-activating peptides from 17 species distributed over five orders (Suzuki 1995).

Receptors and signaling mechanisms of the sperm-activating peptides, especially speract and resact, have been investigated for many years. Speract and resact bind to their receptors on the sperm membrane and activate guanylyl cyclase, which is the enzyme for the production of cGMP (Ramarao and Garbers 1985). The receptor of resact is a membrane-type guanylate cyclase (Shimomura et al. 1986), and as such, resact directly activates the guanylyl cyclase and induces cGMP production. On the other hand, the speract receptor is not a guanylyl cyclase: Speract binds to the 77 kDa membrane protein that seems to be associated with a guanylate cyclase (Bentley et al. 1988; Dangott and Garbers 1984; Dangott et al. 1989). In any case, increase of cGMP induces the activation of cyclic nucleotide-gated K^+-selective channel and K^+ efflux, resulting in the hyperpolarization of membrane potential

(Babcock et al. 1992; Galindo et al. 2000). Then, Ca^{2+} efflux and Na^+ influx via a Na^+/Ca^{2+} exchanger and alkalization via a Na^+/H^+ exchanger occur (Lee and Garbers 1986; Nishigaki et al. 2004). Finally, increase of Ca^{2+} in the sperm head and flagella is observed (Böhmer et al. 2005; Guerrero et al. 2010; Kaupp et al. 2003; Wood et al. 2005) (see Fig. 3.1). An additional pathway, cAMP can induce Ca^{2+} increase (Cook and Babcock 1993). In the sperm of *S. purpuratus*, the Na^+/H^+ exchanger seems to associate with a soluble adenylate cyclase, which is the enzyme for the production of cAMP (Nomura and Vacquier 2006). Thus, this may be activated by the signal pathway initiated from speract.

Interestingly, sperm chemotaxis in sea urchins other than *Arbacia* is unclear. Chemotaxis of *A. punctulata* sperm has been observed in 1985 (Ward et al. 1985) and that of *A. lixula* has also been detected (Yoshida, unpublished data). On the other hand, spermatozoa of other sea urchins, including *S. purpuratus* and *H. pulcherrimus*, have been considered not to show chemotactic behavior toward the sperm-activating peptides for a long time (Cosson 1990; Darszon et al. 2008; Miller 1985b). However, Guerrero and his collaborators have shown that the *Lytechinus pictus* spermatozoa displayed Ca^{2+} responses similar to a sperm showing chemotactic behavior, and finally the sperm has shown chemotactic behavior toward speract even though the phenomenon is not obvious (Guerrero et al. 2010) (see Chap. 12 for details).

Regardless, chemotactic response of the sea urchin sperm other than *Arbacia* is not obvious. Why do only a small number of sea urchin species conserve sperm chemotaxis? This is a difficult and unsolved question. A possible hint to solve the question is that resact is the only 14-amino-acid peptide, while other sperm-activating peptides, including speract, are mostly 10-amino-acid peptides (Suzuki 1995). In addition, the receptor of resact is different from that of other sperm-activating peptides.

3.4.5 Echinoderms Other than Sea Urchin (Starfish, Sea Cucumber, and Brittle Star)

Like in the sea urchin, the spermatozoa of many other echinoderms have their motility in sea water. Thus, study on sperm activation is little. In exceptional cases, the sperm of the sea cucumber *Holothuria atra* is quiescent in sea water and activated by substances from the egg (Morita et al. 2009).

Sperm chemotaxis in echinoderms other than Echinoidea are observed in starfishes (Asteroidea), sea cucumbers (Holothuroidea), and brittle stars (Ophiuroidea) (Miller 1985a). In the experiments, species specificity of sperm chemotaxis has been observed between the used species in the brittle stars, but some hetero-specific cross reactivities were observed in starfishes (Miller 1985a). Moreover, in sea cucumbers, hetero-specific cross-reactivities were observed in broad genus and specificity of sperm chemotaxis was observed in limited genus (Miller 1985a). In the sea cucumber, sperm attractants seem to be released from the egg cell and retained at the vitelline membrane (Morita et al. 2009). In the starfish, sperm attractants seem to exist in the jelly layer: the sperm-activating peptides named Asterosaps that was purified from the jelly layer of the starfish *Asterias amurensis* (Nishigaki et al. 1996) has sperm-attracting activity (Böhmer et al. 2005). Asterosaps are 3.8-kDa glutamine-rich polypeptides and have an intramolecular disulfide linkage between 8C and 32C (Nishigaki et al. 1996). In another starfish, *Pycnopodia helianthoides*, the sperm attractant was identified as the 12-kDa protein named Startrak, and the sequence of N-terminal 34 amino acids has been decoded (Miller and Vogt 1996; Punnett et al. 1992). Interestingly, the synthetic 32-amino-acid peptide, which is part of the decoded region of Startrak, has a stronger sperm-attracting activity than the purified Startrak (Miller and Vogt 1996). Furthermore, the N-terminus sequence of Startrak (xxAELGLCIARVRQQNQGQDDVSIYQAIMSQCQS) has a high degree of homology with the sequences of Asterosaps (e.g., sequence of Asterosap P15: GGTQFGVCIARVRQQHQGQDEASIFQAILSQCQS) (Böhmer et al. 2005). Therefore, the region is important for the chemotaxis of starfish sperm.

Like resact in the sea urchin *A. punctulata*, Asterosaps bind to a membrane-type guanylyl cyclase on the sperm membrane and increase $[Ca^{2+}]_i$ via the increase in cGMP (Matsumoto et al. 2003; Nishigaki et al. 2000).

3.4.6 Urochordate

Urochordate (tunicate) is one of the primitive groups (subphylum) of chordates, and it consists of ascidians (Ascidiacea), salps (Thaliacea), and larvaceans (Larvacea). In tunicates, ascidian species have been used as materials for developmental biology for a century, and fertilization of ascidians has been well investigated. The first observation of sperm behavior around the egg was observed in the ascidian *Styela partita* a century ago; Conklin observed that the spermatozoa accumulated at the vegetal pole of the egg (Conklin 1905). Several decades later, Miller studied sperm activation and chemotaxis in the ascidians (Miller 1975, 1982; Minganti 1951). In ascidians, the sperm activators and attractants are released from egg cells and pass through the vitelline membranes (Yoshida et al. 1993). In ascidians, species specificity is observed in not only sperm chemotaxis but also sperm activation (Yoshida 2014; Yoshida et al. 2013), probably because the sperm attractant has both sperm-activating and sperm-attracting activities (Yoshida et al. 1994, 2002). Thus, the attractant of ascidians was called sperm-activating and -attracting factor (SAAF) (Yoshida et al. 1994). Species specificity of sperm chemotaxis is not so strict: spermatozoa of some species show chemotactic behavior toward egg extracts from congeneric species (Miller 1982, 1985b; Yoshida et al. 2013). Actually, the SAAF in the ascidians *Ciona intestinalis* (type A; also called *C. robusta*) and *Ciona savignyi* is the same molecule: the polyhydroxysterol sulfoconjugate (25S)-3α,4β,7α,26-tetrahydroxy-5α-cholestane-3,26-disulfate (*Ciona*-SAAF) (Oishi et al. 2004;

Yoshida et al. 2002). Furthermore, we identified the SAAF of another ascidian, *Ascidia sydneiensis,* as 3α,7α,8β,26-tetrahydroxy-5α-cholest-22-ene-3,26-disulfate (*Assydn*-SAAF) (Matsumori et al. 2013; Watanabe et al. 2018). Differences of SAAF between the two genus (*Ciona* and *Ascidia*) comprise only of the position of the OH group and the double bond. Such small differences in the SAAFs may be sufficient for species-specific responses.

In the activation of the *Ciona* sperm, SAAF binds to the receptor, and Ca^{2+} channel is activated followed by Ca^{2+} influx (Yoshida et al. 1994). Then, Ca^{2+} binds to CaM and activation of CaM/CaM kinase pathway occurs (Nomura et al. 2004). Through the regulation of K^+ channel by CaM kinase, hyperpolarization occurs, and adenylated cyclase is activated followed by an increase in the concentration of cAMP (Izumi et al. 1999). Finally, dynein light chain and axonemal protein are phosphorylated, triggering the sperm activation (Nomura et al. 2000) (see Fig. 3.1).

On the other hand, in spermatozoa showing chemotactic behavior, transient increase in the intracellular Ca^{2+} concentration ($[Ca^{2+}]_i$) is observed periodically (Ca^{2+} burst), and the Ca^{2+} bursts play a key role in the regulation of flagellar beating (Shiba et al. 2008). Calaxin, a Ca^{2+}-sensor protein associating axonemal dynein (Mizuno et al. 2009), mediates the pattern of the sperm flagellar beating during chemotactic behavior. Calaxin and Ca^{2+} may regulate dynein-mediated microtubule sliding in the axonemes, resulting in the control of the propagation of asymmetric flagellar bending (Mizuno et al. 2012).

Recently, the SAAF receptor in the sperm of *C. intestinalis* was identified as a plasma membrane Ca^{2+}-ATPase (PMCA) (Yoshida et al. 2018). PMCA may probably be continuously activated by SAAF to keep $[Ca^{2+}]_i$ at low levels, and when the spermatozoon detects a decrease in SAAF, SAAF may detach from PMCA and become inactivated, resulting in the Ca^{2+} bursts (Yoshida et al. 2018).

Sperm chemotaxis of tunicates other than ascidians has also been observed in the larvacean *Oikopleula dioica* (Miller and King 1983).

3.5 Molecular Mechanisms of Sperm Activation and Chemotaxis

As described in Sect. 3.4, sperm activation and chemotaxis are highly diverse systems in animals. Molecular mechanisms controlling sperm motility are well investigated in mammalian species, especially in mouse and human, but those in aquatic animals have been studied only in a few species. Since there are many review papers describing molecular mechanisms in sperm activation and chemotaxis (Darszon et al. 2008; Kaupp et al. 2008; Kaupp and Strunker 2017; Nishigaki et al. 2014; Yoshida et al. 2008; Yoshida and Yoshida 2011, 2018), in this section we describe only the essence of common mechanisms. Summary of the signaling cascades of sperm activation in the three animal groups—clams, sea urchins, and ascidians—was shown in Fig. 3.1.

3.5.1 Intracellular Ca^{2+} Concentration Is the Most Important Factor in the Sperm Activation and Chemotaxis

The most important player for regulating sperm activation and chemotaxis is Ca^{2+}. In a spermatozoon showing chemotactic responses, requirement of extracellular Ca^{2+} is seen in many animals (Yoshida and Yoshida 2011), and transient Ca^{2+} increases have been observed in the sea urchins and the ascidian (Böhmer et al. 2005; Guerrero et al. 2010; Shiba et al. 2008; Wood et al. 2003). Thus, Ca^{2+} mediates sperm flagellar movement during chemotactic behavior. However, the precise role of Ca^{2+} in flagellar movement during sperm chemotaxis remains poorly understood. In the ascidian, Calaxin, which is a Ca^{2+}-sensor protein associating axonemal dynein, appears to mediate Ca^{2+}-induced asymmetrical beating of the sperm flagellum (Mizuno et al. 2009, 2012), and the protein may be a global player in spermatozoa of other animals.

How are the Ca^{2+} transients controlled by the sperm attractants? In mammals, it is known that a sperm-specific Ca^{2+} channel, CatSper, plays a crucial role in regulating sperm function (Lishko and Mannowetz 2018). Sperm chemotaxis of the sea urchin sperm also seems to be mediated by CatSper (Seifert et al. 2015). On the other hand, some taxon in deuterostomes, including bony fishes and amphibians, and all protostomes seems to lack CatSper (Cai and Clapham 2008). Furthermore, in the ascidian sperm chemotaxis, the sperm attractant seems to mediate intracellular Ca^{2+} via the Ca^{2+} pump, even though the role of CatSper in the ascidian sperm is not elucidated (Yoshida et al. 2018). Interestingly, the Ca^{2+} transients in the sperm are observed when the spermatozoon swims away from the egg (Böhmer et al. 2005; Guerrero et al. 2010; Shiba et al. 2008). The model in the ascidian easily explains the phenomena. On the other hand, in the sea urchin sperm, the Ca^{2+} transients seem to be induced by increases in the attractant after an appropriate delay (Böhmer et al. 2005; Kashikar et al. 2012). It seems that the sensing system of the sperm attractant may be diverse in the animals.

3.5.2 Other Factors

Changes in membrane potential and pH are also important events in signaling in sperm activation and chemotaxis even though these are not molecules; these two events modulate the Ca^{2+} signals.

Change of intracellular pH—in many cases alkalization was observed in spermatozoa—appears to involve a sperm-specific Na^+/H^+ exchanger (Wang et al. 2007), and alkalization drives the CatSper channel (Kirichok et al. 2006). In contrast, molecules regulating membrane potential in spermatozoa are varied, although hyperpolarization of the sperm membrane is induced by an efflux of K^+. In the activation of sea urchin sperm, hyperpolarization of the sperm was induced by cyclic nucleotide-gated K^+-selective channel (Babcock et al. 1992; Galindo et al. 2000). Although other players mediating K^+ efflux have not been identified in aquatic invertebrates, it is known in mammalian sperm that some K^+ channels, such as Slo3, are involved in hyperpolarization and Ca^{2+} regulation (Chavez et al. 2014; Schreiber et al. 1998). For details, see the reviews (Nishigaki et al. 2014; Ritagliati et al. 2018; Yoshida and Yoshida 2018).

References

Alavi SM, Matsumura N, Shiba K, Itoh N, Takahashi KG, Inaba K, Osada M (2014) Roles of extracellular ions and pH in 5-HT-induced sperm motility in marine bivalve. Reproduction 147:331–345. https://doi.org/10.1530/REP-13-0418

Babcock DF, Bosma MM, Battaglia DE, Darszon A (1992) Early persistent activation of sperm K^+ channels by the egg peptide speract. Proc Natl Acad Sci U S A 89:6001–6005

Bentley JK, Khatra AS, Garbers DL (1988) Receptor-mediated activation of detergent-solubilized guanylate cyclase. Biol Reprod 39:639–647

Böhmer M et al (2005) Ca^{2+} spikes in the flagellum control chemotactic behavior of sperm. EMBO J 24:2741–2752

Cai X, Clapham DE (2008) Evolutionary genomics reveals lineage-specific gene loss and rapid evolution of a sperm-specific ion channel complex: CatSpers and CatSperbeta. PLoS One 3:e3569. https://doi.org/10.1371/journal.pone.0003569

Carré D, Sardet C (1981) Sperm chemotaxis in siphonophores. Biol Cell 40:119–128

Chavez JC et al (2014) SLO3 K^+ channels control calcium entry through CATSPER channels in sperm. J Biol Chem 289:32266–32275. https://doi.org/10.1074/jbc.M114.607556

Clapper DL, Brown GG (1980a) Sperm motility in the horseshoe crab, *Limulus polyphemus* L. I. Sperm behavior near eggs and motility initiation by egg extracts. Dev Biol 76:341–349

Clapper DL, Brown GG (1980b) Sperm motility in the horseshoe crab, *Limulus polyphemus* L. II. Pertial characterization of a motility initiating factor from eggs and the effects of inorganic cations on motility initiation. Dev Biol 76:350–357

Clapper DL, Epel D (1982) Sperm motility in the horseshoe crab. III. Isolation and characterization of a sperm motility initiating peptide. Gamete Res 6:315–326

Coll JC et al (1994) Chemical aspects of mass spawning in corals. I. Sperm-attractant molecules in the eggs of the scleractinian coral *Montipora digitata*. Mar Biol 118:177–182

Coll JC et al (1995) Chemical aspects of mass spawning in corals. II. (-)-*Epi*-thunbergol, the sperm attractant in the eggs of the soft coral *Lobophytum crassum* (Cnidaria, Octocorallia). Mar Biol 123:137–143. https://doi.org/10.1007/Bf00350332

Conklin EG (1905) The organization and cell-lineage of the ascidian egg. J Acad Natl Sci Phil 13:1–126

Cook SP, Babcock DF (1993) Selective modulation by cGMP of the K^+ channel activated by speract. J Biol Chem 268:22402–22407

Cosson MP (1990) Sperm chemotaxis. In: Gagnon C (ed) Controls of sperm motility: biological and clinical aspects. CRC Press, Boca Raton, pp 104–135

Cosson MP, Carré D, Cosson J, Sardet C (1983) Calcium mediates sperm chemotaxis in siphonophores. J Submicrosc Cytol 15:89–93

Cosson MP, Carré D, Cosson J (1984) Sperm chemotaxis in siphonophores. II. Calcium-dependent asymmetrical movement of spermatozoa induced by attractant. J Cell Sci 68:163–181

Cosson J, Carré D, Cosson MP (1986) Sperm chemotaxis in siphonophores: identification and biochemical properties of the attractant. Cell Motil Cytoskeleton 6:225–228

Dan JC (1950) Fertilization in the medusan, *Spirocodon saltatrix*. Biol Bull 99:412–415

Dangott LJ, Garbers DL (1984) Identification and partial characterization of the receptor for speract. J Biol Chem 259:13712–13716

Dangott LJ, Jordan JE, Bellet RA, Garbers DL (1989) Cloning of the mRNA for the protein that crosslinks to the egg peptide speract. Proc Natl Acad Sci U S A 86:2128–2132
Darszon A, Guerrero A, Galindo BE, Nishigaki T, Wood CD (2008) Sperm-activating peptides in the regulation of ion fluxes, signal transduction and motility. Int J Dev Biol 52:595–606. https://doi.org/10.1387/ijdb.072550ad
De Lisa E, Salzano AM, Moccia F, Scaloni A, Di Cosmo A (2013) Sperm-attractant peptide influences the spermatozoa swimming behavior in internal fertilization in *Octopus vulgaris*. J Exp Biol 216:2229–2237. https://doi.org/10.1242/jeb.081885
Freeman G, Miller RL (1982) Hydrozoan eggs can only be fertilized at the site of polar body formation. Dev Biol 94:142–152
Galindo BE, Beltran C, Cragoe EJ Jr, Darszon A (2000) Participation of a K^+ channel modulated directly by cGMP in the speract-induced signaling cascade of *Strongylocentrotus purpuratus* sea urchin sperm. Dev Biol 221:285–294
Guerrero A, Nishigaki T, Carneiro J, Tatsu Y, Wood CD, Darszon A (2010) Tuning sperm chemotaxis by calcium burst timing. Dev Biol 344:52–65. https://doi.org/10.1016/j.ydbio.2010.04.013
Hansbrough JR, Garbers DL (1981) Speract. Purification and characterization of a peptide associated with eggs that activates spermatozoa. J Biol Chem 256:1447–1452
Hirohashi N et al (2013) Sperm from sneaker male squids exhibit chemotactic swarming to CO_2. Curr Biol 23:775–781. https://doi.org/10.1016/j.cub.2013.03.040
Iida T, Iwata Y, Mohri T, Baba SA, Hirohashi N (2017) A coordinated sequence of distinct flagellar waveforms enables a sharp flagellar turn mediated by squid sperm pH-taxis. Sci Rep 7:12938. https://doi.org/10.1038/s41598-017-13406-z
Izumi H, Márian T, Inaba K, Oka Y, Morisawa M (1999) Membrane hyperpolarization by sperm-activating and -attracting factor increases cAMP level and activates sperm motility in the ascidian *Ciona intestinalis*. Dev Biol 213:246–256
Kashikar ND et al (2012) Temporal sampling, resetting, and adaptation orchestrate gradient sensing in sperm. J Cell Biol 198:1075–1091. https://doi.org/10.1083/jcb.201204024
Kaupp UB, Strunker T (2017) Signaling in sperm: more different than similar. Trends Cell Biol 27:101–109. https://doi.org/10.1016/j.tcb.2016.10.002
Kaupp UB et al (2003) The signal flow and motor response controlling chemotaxis of sea urchin sperm. Nat Cell Biol 5:109–117
Kaupp UB, Kashikar ND, Weyand I (2008) Mechanisms of sperm chemotaxis. Annu Rev Physiol 70:93–117. https://doi.org/10.1146/annurev.physiol.70.113006.100654
Kirichok Y, Navarro B, Clapham DE (2006) Whole-cell patch-clamp measurements of spermatozoa reveal an alkaline-activated Ca^{2+} channel. Nature 439:737–740
Lee HC, Garbers DL (1986) Modulation of the voltage-sensitive Na^+/H^+ exchange in sea urchin spermatozoa through membrane potential changes induced by the egg peptide speract. J Biol Chem 261:16026–16032
Lillie FR (1913a) The mechanism of fertilization. Science 38:524–528. https://doi.org/10.1126/science.38.980.524
Lillie FR (1913b) Studies of fertilization. V. The behavior of the spermatozoa of Nereis and Arbacia with special reference to egg-extractives. J Exp Zool 14:515–574. https://doi.org/10.1002/jez.1400140403
Lishko PV, Mannowetz N (2018) CatSper: a unique calcium channel of the sperm flagellum. Curr Opin Physiol 2:109–113. https://doi.org/10.1016/j.cophys.2018.02.004
Matsumori N et al (2013) A novel sperm-activating and attracting factor from the ascidian *Ascidia sydneiensis*. Org Lett 15:294–297. https://doi.org/10.1021/ol303172n
Matsumoto M et al (2003) A sperm-activating peptide controls a cGMP-signaling pathway in starfish sperm. Dev Biol 260:314–324
Miller RL (1966) Chemotaxis during fertilization in the hydroid *Campanularia*. J Exp Zool 162:23–44. https://doi.org/10.1002/jez.1401620104
Miller RL (1970) Sperm migration prior to fertilization in the hydroid *Gonothyrea loveni*. J Exp Zool 175:493–504

Miller RL (1973) The role of the gonomedusa and gonangium in the sexual reproduction (fertilization) of the hydrozoa. Publ Seto Mar Biol Lab 20:367–400. https://doi.org/10.5134/175769
Miller RL (1975) Chemotaxis of the spermatozoa of *Ciona intestinalis*. Nature 254:244–245
Miller RL (1977) Chemotactic behavior of chitons (Mollusca: Polyplacophora). J Exp Zool 202:203–212
Miller RL (1979a) Sperm chemotaxis in the hydromedusae. I. Species-specificity and sperm behavior. Mar Biol 53:99–114
Miller RL (1979b) Sperm chemotaxis in the hydromedusae. II. Some chemical properties of the sperm attractants. Mar Biol 53:115–124
Miller RL (1982) Sperm chemotaxis in ascidians. Amer Zool 22:827–840
Miller RL (1985a) Demonstration of sperm chemotaxis in echinodermata: Asteroidea, Holothuroidea, Ophiuroidea. J Exp Zool 234:383–414. https://doi.org/10.1002/jez.1402340308
Miller RL (1985b) Sperm chemo-orientation in metazoa. In: Metz CB, Monroy A (eds) Biology of fertilization, vol 2. Academic, New York, pp 275–337
Miller RL (1997) Specificity of sperm chemotaxis among great barrier reef shallow-water holothurians and ophiuroids. J Exp Biol 279:189–200. https://doi.org/10.1002/(Sici)1097-010x(19971001)279:2<189::Aid-Jez10>3.0.Co;2-B
Miller RL, King KR (1983) Sperm chemotaxis in *Oikopleula dioica* FOL, 1872 (UROCHODATA: Larvacea). Biol Bull 165:419–428
Miller RL, Vogt R (1996) An N-terminal partial sequence of the 13 kDa Pycnopodia helianthoides sperm chemoattractant 'startrak' possesses sperm-attracting activity. J Exp Biol 199:311–318
Minganti A (1951) Esperienze sulle fertilizine nelle ascidie. Publl Staz Zool Napoli 23:58–65
Mizuno K, Padma P, Konno A, Satouh Y, Ogawa K, Inaba K (2009) A novel neuronal calcium sensor family protein, calaxin, is a potential Ca2+-dependent regulator for the outer arm dynein of metazoan cilia and flagella. Biol Cell 101:91–103. https://doi.org/10.1042/BC20080032
Mizuno K et al (2012) Calaxin drives sperm chemotaxis by Ca^{2+}-mediated direct modulation of a dynein motor. Proc Natl Acad Sci U S A 109:20497–20502. https://doi.org/10.1073/pnas.1217018109
Morita M, Nishikawa A, Nakajima A, Iguchi A, Sakai K, Takemura A, Okuno M (2006) Eggs regulate sperm flagellar motility initiation, chemotaxis and inhibition in the coral *Acropora digitifera, A. gemmifera* and *A. tenuis*. J Exp Biol 209:4574–4579. https://doi.org/10.1242/jeb.02500
Morita M, Kitamura M, Nakajima A, Sri Susilo E, Takemura A, Okuno M (2009) Regulation of sperm flagellar motility activation and chemotaxis caused by egg-derived substance(s) in sea cucumber. Cell Motil Cytoskeleton 66:202–214. https://doi.org/10.1002/cm.20343
Nishigaki T, Chiba K, Miki W, Hoshi M (1996) Structure and function of asterosaps, sperm-activating peptides from the jelly coat of starfish eggs. Zygote 4:237–245
Nishigaki T, Chiba K, Hoshi M (2000) A 130-kDa membrane protein of sperm flagella is the receptor for asterosaps, sperm-activating peptides of starfish *Asterias amurensis*. Dev Biol 219:154–162
Nishigaki T et al (2004) A sea urchin egg jelly peptide induces a cGMP-mediated decrease in sperm intracellular Ca^{2+} before its increase. Dev Biol 272:376–388
Nishigaki T, Jose O, Gonzalez-Cota AL, Romero F, Trevino CL, Darszon A (2014) Intracellular pH in sperm physiology. Biochem Biophys Res Commun 450(3):1149–1158. https://doi.org/10.1016/j.bbrc.2014.05.100
Noda K, Kanai C (1981) Light and electron-microscopic studies on fertilization of *Pelmatohydra robusta*. 1. Sperm entry to a specialized region of the egg. Dev Growth Differ 23:401–413
Nomura M, Vacquier VD (2006) Proteins associated with soluble adenylyl cyclase in sea urchin sperm flagella. Cell Motil Cytoskeleton 63:582–590. https://doi.org/10.1002/cm.20147
Nomura M, Yoshida M, Inaba K, Morisawa M (2000) Purification and identification of 26 kDa axonemal phosphoprotein regulating SAAF-induced sperm activation in ascidian, Ciona intestinalis. Zool Sci (Tokyo) 17:34

Nomura M, Yoshida M, Morisawa M (2004) Calmodulin/Calmodulin-dependent protein kinase II mediates SAAF-induced motility activation of ascidian sperm. Cell Motil Cytoskeleton 59:28–37. https://doi.org/10.1002/cm.20020

Ohtake H (1976a) Respiratory behaviour of sea-urchin spermatozoa. I. Effect of pH and egg water on the respiratory rate. J Exp Zool 198:303–311

Ohtake H (1976b) Respiratory behaviour of sea-urchin spermatozoa. II. Sperm-activating substance obtained from jelly coat of sea-urchin eggs. J Exp Zool 198:313–322

Oishi T, Tsuchikawa H, Murata M, Yoshida M, Morisawa M (2004) Synthesis and identification of an endogenous sperm activating and attracting factor isolated from eggs of the ascidian *Ciona intestinalis*; an example of nanomolar-level structure elucidation of novel natural compound. Tetrahedron 60:6971–6980. https://doi.org/10.1016/j.tet.2004.02.075

Pfeffer W (1884) Locomotorische Richtungsbewegungen durch chemische Reize. Untersuchungen aus dem botanischen Institut in Tübingen 1:363–482

Punnett T, Miller RL, Yoo B-H (1992) Partial purification and some chemical properties of the sperm chemoattractant from the forcipulate starfish *Pycnopodia helianthoides* (Brandt, 1835). J Exp Zool 262:87–96

Ramarao CS, Garbers DL (1985) Receptor-mediated regulation of guanylate cyclase activity in spermatozoa. J Biol Chem 260:8390–8396

Riffell JA, Krug PJ, Zimmer RK (2002) Fertilization in the sea: the chemical identity of an abalone sperm attractant. J Exp Biol 205:1439–1450

Riffell JA, Krug PJ, Zimmer RK (2004) The ecological and evolutionary consequences of sperm chemoattraction. Proc Natl Acad Sci U S A 101:4501–4506

Ritagliati C, Baro Graf C, Stival C, Krapf D (2018) Regulation mechanisms and implications of sperm membrane hyperpolarization. Mech Dev 154:33–43. https://doi.org/10.1016/j.mod.2018.04.004

Schreiber M, Wei A, Yuan A, Gaut J, Saito M, Salkoff L (1998) Slo3, a novel pH-sensitive K^+ channel from mammalian spermatocytes. J Biol Chem 273:3509–3516

Seifert R et al (2015) The CatSper channel controls chemosensation in sea urchin sperm. EMBO J 34:379–392. https://doi.org/10.15252/embj.201489376

Shiba K, Baba SA, Inoue T, Yoshida M (2008) Ca^{2+} bursts occur around a local minimal concentration of attractant and trigger sperm chemotactic response. Proc Natl Acad Sci U S A 105:19312–19317. https://doi.org/10.1073/pnas.0808580105

Shimomura H, Dangott LJ, Garbers DL (1986) Covalent coupling of a resact analogue to guanylate cyclase. J Biol Chem 261:15778–15782

Suzuki N (1990) Structure and function of sea urchin egg jelly molecules. Zool Sci 7:355–370

Suzuki N (1995) Structure, function and biosynthesis of sperm-activating peptides and fucose sulfate glycoconjugate in the extracellular coat of sea urchin eggs. Zool Sci 12:13–27. https://doi.org/10.2108/zsj.12.13

Suzuki N, Shimomura H, Radany EW, Ramarao CS, Ward GE, Bentley JK, Garbers DL (1984) A peptide associated with eggs causes a mobility shift in a major plasma membrane protein of spermatozoa. J Biol Chem 259:14874–14879

Tosti E, Di Cosmo A, Cuomo A, Di Cristo C, Gragnaniello G (2001) Progesterone induces activation in *Octopus vulgaris* spermatozoa. Mol Reprod Dev 59:97–105. https://doi.org/10.1002/mrd.1011

Wang D, Hu J, Bobulescu IA, Quill TA, McLeroy P, Moe OW, Garbers DL (2007) A sperm-specific Na^+/H^+ exchanger (sNHE) is critical for expression and in vivo bicarbonate regulation of the soluble adenylyl cyclase (sAC). Proc Natl Acad Sci U S A 104:9325–9330. https://doi.org/10.1073/pnas.0611296104

Ward GE, Brokaw CJ, Garbers DL, Vacquier VD (1985) Chemotaxis of *Arbacia punctulata* spermatozoa to resact, a peptide from the egg jelly layer. J Cell Biol 101:2324–2329

Watanabe T et al (2018) Synthesis and complete structure determination of a sperm-activating and -attracting factor isolated from the ascidian Ascidia sydneiensis. J Nat Prod 81:985–997. https://doi.org/10.1021/acs.jnatprod.7b01052

Wood CD, Darszon A, Whitaker M (2003) Speract induces calcium oscillations in the sperm tail. J Cell Biol 161:89–101

Wood CD, Nishigaki T, Furuta T, Baba SA, Darszon A (2005) Real-time analysis of the role of Ca^{2+} in flagellar movement and motility in single sea urchin sperm. J Cell Biol 169:725–731

Yoshida M (2014) Sperm chemotaxis: the first authentication events between conspecific gametes before fertilization. In: Sawada H, Inoue N, Iwano M (eds) Sexual reproduction in animals and plants. Springer, Tokyo, pp 3–11. https://doi.org/10.1007/978-4-431-54589-7_1

Yoshida M, Yoshida K (2011) Sperm chemotaxis and regulation of flagellar movement by Ca^{2+}. Mol Hum Reprod 17:457–465. https://doi.org/10.1093/molehr/gar041

Yoshida M, Yoshida K (2018) Modulation of sperm motility and function prior to fertilization. In: Kobayashi K, Kitano T, Iwao Y, Kondo M (eds) Reproductive and developmental strategies. Diversity and commonality in animals. Springer, Tokyo, pp 437–462. https://doi.org/10.1007/978-4-431-56609-0_21

Yoshida M, Inaba K, Morisawa M (1993) Sperm chemotaxis during the process of fertilization in the ascidians *Ciona savignyi* and *Ciona intestinalis*. Dev Biol 157:497–506. https://doi.org/10.1006/dbio.1993.1152

Yoshida M, Inaba K, Ishida K, Morisawa M (1994) Calcium and cyclic AMP mediate sperm activation, but Ca^{2+} alone contributes sperm chemotaxis in the ascidian, *Ciona savignyi*. Dev Growth Differ 36:589–595. https://doi.org/10.1111/j.1440-169X.1994.00589.x

Yoshida M, Murata M, Inaba K, Morisawa M (2002) A chemoattractant for ascidian spermatozoa is a sulfated steroid. Proc Natl Acad Sci U S A 99:14831–14836. https://doi.org/10.1073/pnas.242470599

Yoshida M, Kawano N, Yoshida K (2008) Control of sperm motility and fertility: Diverse factors and common mechanisms. Cell Mol Life Sci 65:3446–3457. https://doi.org/10.1007/s00018-008-8230-z

Yoshida M, Hiradate Y, Sensui N, Cosson J, Morisawa M (2013) Species-specificity of sperm motility activation and chemotaxis: a study on ascidian species. Biol Bull 224:156–165

Yoshida K, Shiba K, Sakamoto A, Ikenaga J, Matsunaga S, Inaba K, Yoshida M (2018) Ca^{2+} efflux via plasma membrane Ca^{2+}-ATPase mediates chemotaxis in ascidian sperm. Sci Rep 8:16622. https://doi.org/10.1038/s41598-018-35013-2

Zatylny C, Marvin L, Gagnon J, Henry J (2002) Fertilization in Sepia officinalis: the first mollusk sperm-attracting peptide. Biochem Biophys Res Commun 296:1186–1193

Chapter 4
Fish Sperm Maturation, Capacitation, and Motility Activation

Luz M. Pérez

Abstract Motility activation mechanism in fish sperm has been studied only in a few species, and there are many hypotheses about the mechanism for spermatozoa activation that need further confirmation. This review summarizes the current knowledge on sperm acquisition of capability to move, happening in the sperm duct (sperm capacitation), and the ionic fluxes related to motility activation, as well as their relation with the sperm membrane potential changes.

Keywords Teleost · Ion fluxes · Membrane potential

4.1 Introduction

The word "Fish" is not a taxonomical term, as it includes very different animals like sharks (class Chondrichthyes), lungfish (clade Sarcopterigians, subclass Dipnoi), sturgeons (subclass Chondrostei), gars (infraclass Holostei), and the teleost fish (infraclass Teleostei). Teleost fish, with more than 25,000 species, are the largest and most diverse group of vertebrates, and that is reflexed in their reproductive strategies. Although they are generally external fertilizers, some species show internal fertilization. Spawning can be a massive event or a small release of gametes; it can happen in the water column or on different substrates. Sperm can then be released to freshwater, to seawater, or to the female oviduct. In general, sperm is immotile inside the testis and sperm duct and is activated by freshwater or seawater, but in some species, like some Cottidae, sperm can be already motile inside the sperm duct (Koya et al. 1993). Also, some species show sperm which is immotile both inside the sperm duct and in seawater and is activated by factors released by the eggs, like the sperm of the pacific herring *Clupea pallasi* (Vines et al. 2002).

L. M. Pérez (✉)
Aquaculture and Biodiversity Research Group, Institute for Animal Science and Technology, Universitat Politècnica de València, Valencia, Spain
e-mail: mlpereig@dca.upv.es

M. Yoshida, J. F. Asturiano (eds.), *Reproduction in Aquatic Animals*,
https://doi.org/10.1007/978-981-15-2290-1_5

When the spermatozoa are released to the surrounding water, it experiences an osmotic shock, which could be either hypoosmotic (in freshwater spawners) or hyperosmotic (in seawater spawners). Internal fish body fluids have an osmolality around 300 mOsm/kg, higher than freshwater (Allan 1995; Wetzel 2001) and lower than seawater (around 1000 mOsm/kg).

This situation is completely different from other marine animals, like sea urchins, ascidians, or even sharks, whose internal salinity is similar to the surrounding sea, and is called osmoconformers (Eckert and Randall 1989). In such cases, sperm motility is not activated by an osmotic shock experienced at spawning.

The water chemistry where the fish sperm is released can also be very different: the major ions are Na^+ and Cl^- in seawater, or Ca^{2+} and HCO^{3-} in most freshwater bodies (Table 4.1).

Sperm cells could have been evolved to face all these different environments and reproductive strategies.

Fish sperm motility can be divided into several processes:

1. Sperm maturation and capacitation.
2. Spermatozoa activation: happens in milliseconds, the sperm cells change from immotile to motile state. The external signal is transduced through the sperm membrane.
3. Sperm navigation: sperm cells swim towards the egg, use energy, and enter into the micropyle.
4. Motility stop: cessation of movement.

In this review, we focus only on the first two topics.

4.2 Sperm Maturation and Capacitation

Sperm maturation has been defined as the stage in which the non-functional gametes become mature spermatozoa, with the ability to move and fertilize. It includes physiological but not morphological changes (Schulz et al. 2010) and occurs in the sperm duct (Morisawa and Morisawa 1988; Miura et al. 1992). In this chapter, we use the term *sperm maturation* for the whole process that enables spermatozoa to move and fertilize and the term *sperm capacitation* for the ultimate process of matu-

Table 4.1 Major ions in natural waters (Allan 1995; Wetzel 2001)

	Seawater	Freshwater
Ca^{2+} (mg/L)	3.4	63–68
Mg^{2+} (mg/L)	17.6	17–25
Na^+ (mg/L)	77.2	4.5–15.7
K^+ (mg/L)	1.6	1.9–3.4
Cl^- (mg/L)	90.2	9.9–10.1
SO_4^- (mg/L)	9.3	10.0–16.6
HCO_3^- (mg/L)	0.4	73.9–85.4

ration, mediated by ionic changes in the sperm duct, which allows the sperm to be motile in the presence of appropriate stimulus.

4.2.1 Hormonal Regulation of Sperm Maturation

It is known that fish sperm maturation and spermiation are regulated by progestins, which could be DHP (17α,20β-dihydroxy-4-pregnen-3-one), 20βS (17,20β,21-trihydroxy-4-pregnen-3-one), or both, depending on the fish species (reviewed by Scott et al. 2010). In Japanese eel (*Anguilla japonica*), it has been reported that DHP regulates sperm maturation through an increase in seminal plasma pH, which in turn increases the sperm content in cAMP, which allows it to acquire motility (Miura et al. 1991, 1992, 1995). DHP also increases seminal plasma pH in masu salmon (*Oncorhynchus masou*; Miura et al. 1992) and rabbitfish (*Siganus argenteus*; Rahman et al. 2003), with an increase in sperm motility. In contrast, rainbow trout (*Oncorhynchus mykiss*) treated with DHP implants did not show altered sperm pH, sperm motility, or altered concentrations of Na^+ or K^+ in semen (Milla et al. 2008). The mechanism involved in the increase of the seminal plasma pH by DHP in eel is still unclear, although the presence of carbonic anhydrases (eSRS22/CA) in the spermatids and spermatozoa suggest its involvement in such function (Miura and Miura 2003, 2011).

As commented, other progestin, 20βS, is involved in the sperm maturation in other fish species: three Scienidae (Atlantic croacker, *Micropogonias undulatus*; red drum, *Sciaenops ocellatus*; and spotted seatrout *Cynoscion nebulosus*) and a flatfish species (southern flounder, *Paralichthys lethostigma*), where it also increases sperm motility or hypermotility (Thomas et al. 2009). It was proposed that 20βS acts on the sperm membrane through the membrane progestin receptor alpha (mPRα). Its action was associated with a rapid increase in the sperm intracellular Ca^{2+} (Thomas 2003) and with a transient increase in cAMP levels in the sperm cells (Thomas et al. 2004). In flounder sperm, it was suggested that progestins stimulate sperm motility by activating an mPRα/stimulatory G protein/membrane adenylyl cyclase pathway (Tan et al. 2014).

Different from DHP, a relation between 20βS and seminal plasma pH has not been reported as far as we now.

Progestins mediate the sperm maturation inside the seminal plasma fluid. Seminal plasma is a heterogeneous and complex protein-rich fluid in which the sperm cells are diluted. It is principally comprised by mineral compounds (Na^+, K^+, Mg^{2+}, and Ca^{2+}) and characterized by low concentrations of proteins as well as other organic substances, such as hormones and pheromones, cholesterol, glycerol, vitamins, free amino acids, sugars, citric acid, and lipids (Ciereszko et al. 2000; Cosson 2004). The composition of seminal plasma in freshwater fish species was extensively reviewed by Alavi and Cosson (2006) and Ciereszko (2008).

Some relations between DHP and the ionic composition of seminal plasma, apart from pH, have been reported in several fish species. In rainbow trout DHP injections in spermiating males significantly raised the K^+/Na^+ ratio of the seminal plasma (Scott and Baynes 1982). When OHP (17α-hydroxyprogesterone), a DHP-related compound, was injected to Japanese eel, seminal plasma K^+ decreased concomitantly with a decrease in sperm motility (Miura et al. 2013). In brook trout (*Salmo fontinalis*), it was observed that sperm duct absorbed Na^+ and secreted K^+ (Marshall et al. 1989a, b), creating a seminal plasma with high ratio K^+/Na^+, but in this work DHP did not affect the ion levels.

Summarizing, fish sperm maturation is mediated by changes in the sperm duct pH, K^+, and Na^+ in some species, which can be related with the steroid hormone DHP. In contrast, the progestin 20βS mediates sperm maturation in other fish species by modulating the intracellular Ca^{2+} and cAMP levels in the sperm cells.

4.2.2 *Sperm Capacitation*

Fish sperm capacitation, the ultimate step of sperm maturation for acquiring motility, is mediated by changes in seminal plasma pH, HCO_3^-, and K^+, and in some species through changes in seminal plasma Na^+ or Ca^{2+}.

Testicular sperm, which is sperm that has not yet passed by the sperm duct, is unable to be activated by freshwater or seawater at least in rainbow trout and Japanese eel (Morisawa and Morisawa 1986; Miura et al. 1995; Ohta et al. 1997).

In rainbow trout, it was demonstrated that the spermatozoa collected from the testis (not from the sperm duct or ejaculated) were immotile, whereas they gradually became able to be activated (they *capacitate*) when they were diluted in artificial seminal plasma (Morisawa and Morisawa 1986). Then, rainbow trout sperm capacitation (or potential to move) was acquired during incubation of intratesticular spermatozoa in an artificial medium, mimicking the seminal plasma and containing, besides other ionic components, 40 mM K^+ (which also maintained the spermatozoa immotile) and 20 mM HCO_3^- at pH 8.2 (Morisawa and Morisawa 1988). According to these authors, HCO_3^- and pH were the most important factors promoting the trout sperm capacitation.

In eel species sperm capacitation depends on seminal plasma K^+, pH, HCO_3^-, and Na^+. Like in rainbow trout, in Japanese eel, sperm extracted from the testis without passing by the sperm duct is not able to acquire motility (is not *capacitated*), unless it is incubated in solutions containing HCO_3^- and/or high pH (Miura et al. 1995; Ohta et al. 1997).

In natural conditions, Japanese eel testicular sperm is surrounded by a low pH (pH 7.5) and high K^+ in the seminal plasma, but when it moves to the sperm duct, it experiences a pH increase (up to 8.0), as well as a decrease in seminal plasma K^+ concentration (Ohta et al. 1997). When Japanese eel sperm was incubated in artificial seminal plasma (ASP) containing HCO_3^-, sperm motility increased proportionally to the HCO_3^- concentration (up to 30 mM HCO_3^-; Ohta et al. 1997).

Seminal plasma pH was also important for Japanese eel sperm capacitation, as intratesticular sperm incubated in HCO_3^--free ASP (*incapacitating* condition) was capacitated when pH levels were increased up to 8.4–8.7, higher than seminal plasma physiological pH levels (pH 8.1). The incapacitating effect of HCO_3^--free ASP was also observed in ejaculated eel sperm, and it could be induced and reversed several times (Ohta et al. 2001). This indicates that in Japanese eel both HCO_3^- and extracellular or intracellular pH (pH_e or pH_i) are involved in eel sperm capacitation.

Sperm intracellular pH in a related species, the European eel (*Anguilla anguilla*), is 7.2 in immotile stage (Vílchez et al. 2017) while seminal plasma pH in this species is 8.5 (Asturiano et al. 2004). This reveals an important pH gradient in European eel sperm cells in quiescent stage. In this species, the acidification of the pH_i (by sodium acetate) did not affect further sperm motility, whereas alkalinization of the pH_i by NH_4Cl inhibited further motility (Vílchez et al. 2017). Then, European eel sperm capacitation was possible at low but not at high pH_i.

In both Japanese and European eels, it was observed that the K^+ removal from the artificial seminal plasma (ASP) induced a rapid decrease in motility, which was reversible (Ohta et al. 2001; Vílchez et al. 2017). Therefore, K^+ ions in seminal plasma are also involved in sperm capacitation in eel species.

Seminal plasma sodium is also important for European eel sperm capacitation (Vílchez et al. 2016). In the sperm from this species, in quiescent stage, there is an Na^+ equilibrium outside/inside the spermatozoa (97 vs 109 mM Na^+; Vílchez et al. 2016; Asturiano et al. 2004). When Na^+ was removed from seminal plasma, eel sperm motility (after seawater activation) was suppressed (Vílchez et al. 2016).

As far as we know, there are no studies in other fish species regarding the role of pH or other ions in sperm capacitation. Despite this, variations in sperm motility related to the ionic composition of seminal plasma have been reported in other fish species.

In common bleak (*Alburnus alburnus*), sperm motility was positively correlated with seminal fluid pH, K^+, and Na^+ (Lahnsteiner et al. 1996).

Seasonal variations in the ionic composition related to sperm capacitation have been observed in several species.

Variations in Na^+ and specially Mg^{2+} were observed in ocean pout (*Macrozoarces americanus*) seminal plasma during reproductive season (Wang and Crim 1997). In that study, higher Mg^{2+} and lower Na^+ levels were found in mid-spermiating season, when sperm showed the highest motility, in comparison with early season.

In European eel, low Ca^{2+} and Mg^{2+} concentrations were observed in seminal plasma of high motility samples, while K^+ levels increased with sperm motility (Asturiano et al. 2004). Despite this, *capacitation* of Japanese eel sperm was not affected by Ca^{2+} or Mg^{2+} (Ohta et al. 1997).

In summary, fish sperm capacitation depends on the ionic composition of seminal plasma. The variations in the ionic composition in seminal plasma observed in different species probably reflects changes in the resting membrane potential of the sperm cells; this topic is discussed in the next section.

4.2.3 *Resting Membrane Potential in Fish Sperm Cells*

The membrane potential (V_m) is the relative difference of ionic (electric) charges across a membrane. It is created by differential ion concentrations and maintained by ion channels and ion transporters. In many cells, K^+, Na^+, and Cl^- are the main contributors to the membrane potential (www.physiologyweb.com). The numerical value of the membrane potential is generally negative, meaning that the inside of the cell is negative with respect to the outside solution, which is taken as the reference or zero value.

In mammalian sperm, V_m depends on ions K^+, Na^+, and Cl^-, but also from intracellular pH_i (or H^+) (Navarro et al. 2007); and intracellular alkalinization induced a rapid sperm membrane hyperpolarization, mediated by a pH-sensitive K^+ current originated from the sperm flagellum.

Resting membrane potential (V_m) should then be essential for further excitation, and it depends of the difference of ions (inorganic and organic) between the sperm cell and the seminal plasma.

Resting V_m (that is, V_m in quiescent, immotile sperm) has been measured only in the sperm from a few fish species. Krasznai et al. (2003a) showed that carp sperm cells were depolarized in the seminal plasma ($V_m = -2.6$ mV), and they hyperpolarize upon hypoosmosis-induced activation of motility to $V_m = -29$ mV.

In salmonids species, the absolute V_m has not been quantified, but it has been observed, like in carp, that sperm is more depolarized in quiescent stage than in activated stage, and it *hyperpolarizes* at activation (Boitano and Omoto 1991; Kho et al. 2001). In trout sperm, the relation between seminal plasma K^+ and V_m was demonstrated by Gatti et al. (1990) and Boitano and Omoto (1991). Resting V_m was shown to be dependent from seminal plasma K^+, H^+, and Na^+ (the higher concentrations inducing a more depolarized state).

Summarizing, fish sperm resting membrane potential depends on seminal plasma ions; at least K^+, pH, and Na^+ are involved in V_m in several fish species. Then, it is possible to think that sperm capacitation depends on the resting membrane potential.

4.3 Sperm Activation

It is considered that "normal" fish sperm activation in the external environment can occur in two different ways: activation by external ions, called *ionic mode* of activation, or activation by osmotic shock, or *osmotic mode* of activation (Bondarenko et al. 2013).

Ionic mode is present in salmonids and sturgeons; sperm is activated by the efflux of intracellular K^+ when it is released to freshwater, due to the reduced amount of K^+ in freshwater. *Osmotic mode* is present in other freshwater species and in marine fish, in which sperm activation is due to hypo- or hyperosmotic shock experienced by the sperm when released to freshwater or seawater.

It is clear that the osmotic change is responsible for the *osmotic mode* of activation, as even nonionic solutions with appropriate osmotic level can activate the sperm motility in a large number of species (Morisawa 2008; Alavi and Cosson 2006). In this sense, hypotonic non-electrolite solutions (as sucrose, mannitol, etc.) can activate the sperm motility of many fish species from cyprinids to Tetraodontidae (Table 4.2).

The optimal osmolality for marine sperm motility activation can vary between 480 and 1100 mOsm/kg, except for turbot, which sperm can be activated between 300 and 1100 mOsm/kg (reviewed by Alavi and Cosson 2006). According to this review, the optimal osmolality for sperm motility in freshwater fish is 150–200 mOsm/

Table 4.2 Fish species with sperm able to be activated in nonionic solutions

Common name	Scientific name	Family	Authors
Freshwater fish			
Rainbow trout	*Onchorhynchus mykiss*	Salmonidae	Morisawa et al. (1983a)
Masu salmon	*Onchorhynchus masou*	Salmonidae	Morisawa et al. (1983a)
Goldfish	*Carassius auratus*	Cyprinidae	Morisawa et al. (1983b)
Carp	*Cyprinus carpio*	Cyprinidae	Morisawa et al. (1983b)
Crucian carp	*Carassius carassius*	Cyprinidae	Morisawa et al. (1983b)
Dace	*Tribolodon spp.*	Cyprinidae	Morisawa et al. (1983b)
Vimba	*Vimba vimba*	Cyprinidae	Alavi et al. (2010)
Zebrafish	*Danio rerio*	Cyprinidae	Wilson-Leedy et al. (2009)
Streaked prochilod	*Prochilodus lineatus*	Prochilodontidae	Gonçalves et al. (2013)
Pirapitinga	*Brycon orbingyanus*	Characidae	Gonçalves et al. (2013)
Northern pike	*Esox lucius*	Esocidae	Alavi et al. (2009)
Ayu	*Plecoglossus altivelis*	Plecoglossidae	Morisawa et al. (1983a)
Marine fish			
Bogue	*Boops boops*	Sparidae	Lahnsteiner and Patzner (1998)
White seabream	*Diplodus sargus*	Sparidae	Lahnsteiner and Patzner (1998)
Gilthead seabream	*Sparus aurata*	Sparidae	Zilli et al. (2008)
Stripped seabream	*Lithognatus mormyrus*	Sparidae	Zilli et al. (2008)
Red mullet	*Mullus barbatus*	Mullidae	Lahnsteiner and Patzner (1998)
Horse mackerel	*Trachurus mediterraneus*	Carangidae	Lahnsteiner and Patzner (1998)
European seabass	*Dicentrarchus labrax*	Dicentrarchidae	Dreanno et al. (1999)
Halibut	*Hippoglossus hipoglossus*	Pleuronectidae	Billard et al. (1993)
Senegalese sole	*Solea senegalensis*	Pleuronectidae	Martínez-Pastor et al. (2008)
Flathead mullet	*Mugil cephalus*	Mugilidae	Lee et al. (1992)
Pufferfish	*Takifugu niphobles*	Tetraodontidae	Morisawa and Suzuki (1980)
European eel	*Anguilla anguilla*	Anguillidae	Pérez et al. pers. observation

kg (carp), up to 300 mOsm/kg (salmonidae), but is lower for sturgeons and paddlefish sperm, between 0 and 120 mOsm/kg, which also showed a low osmolality in their seminal plasma. In the euryhaline freshwater acclimated tilapia (*Sarotherodon melanotheron*), the optimal osmolality varies between 150 and 300 mOsm/kg, but it varies from 300 to 800 when fish were acclimated to seawater (Legendre et al. 2016).

As nonionic hyperosmotic solutions can activate marine sperm motility (Table 4.2), it can be deduced that the influx of ions from the external media should not be very important for sperm motility activation, nor the ionic composition of the external media. That hypothesis was checked in European eel sperm, as sperm activation in Ca-free, Na-free, or K-free media was similar to activation in seawater (Pérez et al. 2016; Vílchez et al. 2016, 2017), although in some species, like pufferfish, some parameters can be reduced in non-ionic solutions, like sperm velocities in pufferfish sperm activated in sucrose (Gallego et al. 2013).

Thus, if any ion exchange is involved in marine sperm activation, an *ion efflux*, but not an *ion influx*, could be related to sperm activation. Ion fluxes during sperm activation are reviewed in the next section.

4.3.1 Changes in Sperm Membrane Potential and Potassium Efflux at Activation

Early studies by Morisawa's and Tanaka's research groups already suggested that a sperm membrane hyperpolarization, caused by a K^+ efflux from the sperm cell, induced the sperm motility activation in salmonids. Tanimoto and Morisawa (1988) showed that K^+ channel blockers inhibited trout sperm motility when sperm was activated in a K-free activation solution, and inferred that trout sperm activation was dependent from a K^+ efflux, which in turn, would change the sperm membrane potential (V_m). Tanimoto et al. (1994) demonstrated the K^+ efflux from the sperm from rainbow trout and masu salmon, by measuring the K^+ concentrations in the seminal plasma before and after sperm activation. Their results confirmed that a K^+ efflux occurred when sperm motility was initiated by a decrease in external K^+.

Inhibitors of K^+ channel blockers have shown to inhibit sperm motility in several fish species. The voltage-gated K^+ channel inhibitor 4-AP inhibited sperm motility in both Japanese and European eels (Tanaka et al. 2004; Vílchez et al. 2017), Atlantic croaker (*Micropogonias undulatus*; Detweiler and Thomas 1998), and common carp (Krasznai et al. 1995). Barium and TEA (tetra-ethyl ammonium), two nonspecific K^+ channel inhibitors, inhibited 50% sperm motility in rainbow trout (Tanimoto and Morisawa 1988), and 25% in the European eel sperm (Vílchez et al. 2017), but did not inhibited gilthead seabream motility (*Sparus aurata*; Zilli et al. 2008). TEA also inhibited 50% of motility in rainbow trout sperm, but in contrast, it did not affect European eel sperm motility (Vílchez et al. 2017).

The first direct measurement of fish spermatozoa membrane potential (V_m) was done by Gatti et al. (1990) in rainbow trout sperm. Sperm V_m depended on the concentration of K, H^+, and Na^+ in the seminal plasma.

Boitano and Omoto (1991) also measured directly the spermatozoa membrane potential in rainbow trout and demonstrated that K^+ contributed to rapid changes in the Vm. The dependence of rainbow trout spermatozoa V_m on K^+ ions was also confirmed by Kho et al. (2001, 2003), which demonstrated a gradual decrease in sperm V_m (a hyperpolarization) concomitantly with a gradual decrease in K^+ concentration in the activation solution. Thus, at spawning in freshwater, a decrease in environmental K^+ may cause a spontaneous efflux of K^+ through a K^+ channel, thereby leading to membrane hyperpolarization.

The first quantitative study of fish sperm V_m was performed by Krasznai et al. (2003a). They showed that carp sperm cells were in a depolarized state in the seminal plasma ($V_m = -2.6$ mV), and they hyperpolarize upon hypoosmosis-induced activation of motility ($V_m = -29$ mV). Same authors demonstrated that a voltage-gated potassium channel blocker (4-aminopyridine, 4-AP) eliminated the hyperpolarization of the sperm cells after hypoosmotic shock, and inhibited sperm motility. This finding suggested that an increase in potassium permeability (and an efflux of K^+_i) was responsible for the hyperpolarization observed in carp spermatozoa at sperm activation.

In recent years, potassium flux has been measured in the spermatozoa from other cyprinid species such as in zebrafish (*Danio rerio*). Fechner et al. (2015) characterized the potassium selective DrCNGK channel as responsible for K^+ fluxes in zebrafish sperm. This potassium channel was modulated (gated) by the intracellular pH, which is consistent with previous findings from Gatti et al. (1990). The authors observed that zebrafish sperm intracellular alkalinization (which happens at sperm activation) strongly activated the CNGK channel. The K^+ current was blocked by external TEA. The presence of genes codifying CNGK channels was explored in several fish genomes. Apart from zebrafish, the CNGK gene was also present in rainbow trout, spotted gar, and coelacanth genomes (Fechner et al. 2015).

Regarding marine fish, the changes in the spermatozoa membrane potential at activation have been studied in only one marine fish species, the pacific herring (Vines et al. 2002). At activation by the egg SMIF (sperm motility initiation factor), sperm showed depolarization, contrary to that observed in trout and carp sperm at activation. Besides depolarization, activation also involved an elevation in intracellular pH and $[Ca^2]_i$ and a decrease in $[Na^+]_i$. The sperm membrane depolarization (a) was induced by the Na^+/Ca^{2+} exchanger localized in the plasma membrane (Vines et al. 2002) or (b) was due to a K^+ influx into the sperm cell, through a hypothetical K channel or K^+/H^+ exchanger (Yanagimachi et al. 2017).

Vines et al. (2002) also demonstrated that herring spermatozoa depolarization activated *voltage-sensitive Ca^{2+} channels* and that extracellular Ca^{2+} was also required for sperm activation, which is different from the sperm from other marine fish species. When the egg activator SMIF was not present, herring sperm could be fully activated in diluted seawater without sodium but not in seawater containing Na^+. Then, the authors stated that sperm depolarization, sodium efflux, and calcium influx were unequivocally related to motility initiation in herring sperm (Vines et al. 2002).

Apart from the CNKG potassium channel described for zebrafish, other K^+ channels could potentially be involved in fish sperm motility. The presence of Ksper/Slo3, a sperm-specific K^+ channel essential for male fertility in mammals (Santi et al. 2010; Zeng et al. 2011), has been demonstrated in fish genomes by Vicens et al. (2017). Ksper/Slo3 gene was present in spotted gar (*Lepisosteus oculatus*), herring (*Clupea harengus*), and Atlantic salmon (*Salmo salar*) genomes. However, it was absent in coelacanth, zebrafish, and *Tetraodon* species.

Apart from Pacific herring, the membrane potential has not been measured in the spermatozoa of any other marine fish species. It is not known if the depolarization observed in the herring sperm at activation is a common fact for marine fish sperm, or if the hyperpolarization observed in salmonid and cyprinid species also happens in the sperm cells from other freshwater fish families. Then, measurements on spermatozoa membrane potential in quiescent and activated stage are needed for other representative species from seawater and freshwater environments.

4.3.2 Cell Volume Changes, Water Flow, Stretch-Activated Channels

When sperm cells are released in media with a low osmolality, they swell due to water influx, and when they are released in seawater, they shrink due to hyperosmotic shock and water efflux. This osmotic behavior has been observed in carp spermatozoa (Krasznai et al. 2003a) and other fish species and is common to most cell types. This change in cell volume is supposedly due to the influx or efflux of water, mediated through transmembrane proteins called *aquaporins*. These water pores facilitate the transepithelial flow of water through cellular membranes. For example, they are also present in gills, intestine, or kidney. The role of aquaporins in fish sperm activation has been recently reviewed by Cerdá et al. (2017) and is not studied in this review.

It has been proposed that the osmotic signal in fish sperm is transduced in the sperm membrane through *stretch-activated channels (SACs)* (Krasznai et al. 2003b). Synonyms for SACs are *mechanosensitive channels*, *mechanosensitive ion channels*, or *stretch-gated ion channels,* that is, mechanosensitive channels present in the spermatozoa membrane. They will be activated when the sperm cell membrane swells or shrinks in response of hypo- or hyperosmotic shock (Krasznai et al. 2003b).

The presence of this type of channels in fish sperm cells has been proved by using a previously considered specific SAC inhibitor, the gadolinium ion (Gd^{3+}). In carp and pufferfish, gadolinium decreases sperm motility in a dose-dependent manner (Krasznai et al. 2003b), but not in human or *Ciona* (invertebrate) sperm, thus indicating a role for SACs on sperm activation in carp and pufferfish.

However, using gadolinium to identify the presence of SACs has been questioned by Caldwell et al. (1998) and others, as Gd^{3+} can also block several types of Cl^- channels, it can bind to other anions present in physiological salt solutions, and it is known that certain SACs are insensitive to Gd^{3+} (reviewed by Caldwell et al. 1998).

Also, mechanosensitivity is considered as a phenotypic character of ion channels, and many different ion channels can be sensitive to mechanical stimulus (reviewed by Sachs 2010). Channels previously labeled as "voltage-gated" or "ligand gated" are also mechanically sensitive (Sachs 2010). The only requirement for mechanical sensitivity is that the channels change its shape between closed and open states and that the membrane stress can reach the channels. There are two basic types of mechanosensitive channels (MSCs or SACs): those gated by stress and those gated by tension in the lipid bilayer (Sachs 2015).

The presence of this SAC or MSC channel in fish sperm deserves further confirmation.

4.3.2.1 Cell Volume Changes at Sperm Activation

It has been observed that sperm cells from many freshwater fish increase their volume (became swollen) at activation by hypoosmotic media (carp; Krasznai et al. 2003a; Bondarenko et al. 2013), and sperm cells from marine fish reduced its volume by activation in hyperosmotic seawater (Cosson et al. 2008b). Until recently, this reduction in size has not been measured in a marine species, the European eel (Vílchez et al. 2016, 2017). In this species, changes in cell volume (sperm head area) were also observed in quiescent stage when sperm was diluted in Na-free or in K-free media; in those conditions, further motility in seawater was inhibited (Vílchez et al. 2016, 2017). Thus, seminal plasma Na^+ and K^+ are both preserving sperm cell size and motility in quiescent sperm from European eel.

Sperm volume changes in relation to the osmotic environment have been studied in sperm from a few fish species. Hypoosmotic shock induced sperm head swelling in common carp or rainbow trout (Perchec et al. 1996; Takei et al. 2015; Bondarenko et al. 2013); however, sterlet (*Acipenser ruthenus*) and brook trout (*Salvelinus fontinalis*) sperm did not changed their cell volume in response to hypoosmotic motility activation (Bondarenko et al. 2013). In European seabass, a marine fish, sperm head became swollen after hyperosmotic activation (Dreanno et al. 1999).

Thus, sperm volume changes as a response to environmental osmolality seem to be species-specific and need further research.

4.3.2.2 Transient Receptor Potential Channels

Some of the SACs hypothetically present in fish sperm cells could be part of the transient receptor potential (TRP) channels. TRP channels are involved in different responses to the external environment, including mechanosensation, temperature, and pressure, among others, and appear in sensory neurons and many other cell types (reviewed by Plant 2014).

One of the four ion channels already characterized in human spermatozoa is a TRP, specifically TRPV4 (Mundt et al. 2018). This channel is involved in the initial step of the sperm hyperactivation. The other three ion channels in human sperm are (1) CatSper, the specific Ca^{2+} channel; (2) Ksper, spermatozoa specific K^+ cannel controlling membrane potential; and (3) the proton channel H_v1, mediating intracellular pH alkalinization (reviewed by Mundt et al. 2018).

Some TRP channels have been found in fish sperm or in fish genomes. Majhi et al. (2013) demonstrated the presence of transient receptor potential Vanilloid family member subtype 1 (TRPV1) in the sperm cells from freshwater teleost fish, *Labeo rohita*. Activation of this channel increased the quality and duration of fish sperm movement. TRPV1 gene was conserved in various fish lineages, but they showed different number of copies of the gene (Majhi et al. 2013). Functional studies on this channel in fish sperm would clarify if it has a role on sperm motility, like TRPV4 in mammals.

4.3.3 Calcium Sperm Channels and Calcium Fluxes

A Ca^{2+}-specific channel of sperm, CatSper, is essential for the initiation of the hyperactive motility of mammalian spermatozoa prior to fertilization (Kirichok et al. 2006; Darszon et al. 2004), mediating $Ca^{2+}{}_i$ influx and then motility hyperactivation. While the gene codifying Catsper is absent in teleost fish (Cai and Clapham 2008), the presence of a CatSper-like protein (44 kD) was detected by Yanagimachi et al. (2017) in the mid piece of the spermatozoa of flounders, medaka, herring and trout. However, spermatozoa from loach, zebrafish, and goldfish did not react to the antibody (Yanagimachi et al. 2017).

Some fish sperm need extracellular calcium to activate sperm motility. That has been observed in herring, tilapia, and some salmonid species (Yanagimachi 1957a, b; Yanagimachi and Kanoh 1953; Baynes et al. 1981; Cosson et al. 1989, Legendre et al. 2016).

Euryhaline fish, like tilapia species, are a special case regarding Ca^{2+}, and they can reproduce in both freshwater and marine water. Tilapia sperm behavior regarding Ca^{2+} is different in freshwater-acclimated fish than in seawater-acclimated fish. In the first ones, sperm (activated in hypoosmotic water) does not need external Ca^{2+} for activation (Linhart et al. 1999; Morita et al. 2003; Legendre et al. 2016), but in seawater-acclimated tilapias sperm activation in hyperosmotic media needs extracellular Ca^{2+}.

In several fish species, high concentration of external Ca^{2+} in the activation media inhibits sperm motility. This has been observed in striped bass (*Morone saxatilis*; He et al. 2004), Java carp (*Puntius javanicus*; Morita et al. 2006), and muskellunge *Esox masquinongy* (Lin et al. 1996), but this effect was not observed in other species like catfish (*Clarias batrachus*) or goby (*Oxyeleotris marmorata*) (Morita et al. 2006).

4.3.3.1 Role of Intracellular Calcium

Studies about fish sperm motility including measurements of intracellular Ca^{2+} levels in spermatozoa have been performed only in a few fish species: rainbow trout (Cosson et al. 1989; Boitano and Omoto 1991; Tanimoto et al. 1994; Takei et al. 2012), carp (Krasznai et al. 2000; Krasznai et al. 2003a), tilapia (*Oreochromis mossambicus*, Morita et al. 2003, 2004), redtail splitfin (*Xenotoca eiseni*; Liu et al. 2018), and some marine fish species: pufferfish (*T. niphobles*; Oda and Morisawa 1993; Gallego et al. 2013), Pacific herring (Cherr et al. 2008), and European eel (Gallego et al. 2013; Pérez et al. 2016). In general, in such studies, it was observed that there was an increase in sperm intracellular Ca^{2+} at activation, but in some cases, it was found that in Ca-free activation media the sperm was activated but intracellular calcium did not increase, as in pufferfish (Gallego et al. 2013) and European eel (Pérez et al. 2016). Also, some authors found a decrease of $[Ca^{2+}]_i$ at activation in some specific conditions in trout sperm (pretreated with glycerol, Takei et al. 2012).

In pufferfish (Oda and Morisawa 1993), it was observed that $[Ca^{2+}]_i$ increased at the osmolality-dependent initiation of sperm motility; such increase happened even in Ca^{2+}-free conditions. Sperm motility was activated even in isosmotic conditions when a Ca^{2+} was introduced in the sperm cells by a Ca^{2+} ionophore. Gallego et al. (2013) in the same species, and the same laboratory, also found that $[Ca^{2+}]_i$ increased when sperm was activated in hyperosmotic media, but when sperm was activated in Ca-free media, $[Ca^{2+}]_i$ did not increase, but motility was activated. Gallego et al. (2013) pointed that Oda and Morisawa (1993) did not use a Ca^{2+} chelator in the activation media, thus trace amounts of this ion could be masking their results.

Like pufferfish, European eel sperm experienced an increase in $[Ca^{2+}]_i$ during the normal sperm activation in seawater (Gallego et al. 2014). However, if seminal plasma was depleted of Ca^{2+} and sperm is activated without Ca^{2+}, the increase in $[Ca^{2+}]_i$ was not produced, but the sperm motility was fully activated (Pérez et al. 2016). For that reason, it is considered that an increase in $[Ca^{2+}]_i$ is not necessary for the sperm activation in the European eel.

An indirect evidence of the importance of Ca^{2+} fluxes on fish sperm motility comes from studies with calcium channel inhibitors. In marine species, inhibitors of voltage-gated calcium channels reduced suppressed sperm motility; this was observed in Atlantic croaker (Detweiler and Thomas 1998) and Pacific herring (Vines et al. 2002). Also, inhibitors of voltage-gated calcium channels inhibited sperm motility in different freshwater species, as the bluegill (*Lepomis macrochirus*, Zuccarelli and Ingermann 2007) and sterlet (*Acipenser ruthenus*; Alavi et al. 2011).

In Pacific herring with sperm activated by the egg molecule SMIF, the $[Ca^{2+}]_i$ increase at sperm motility activation is due to two mechanisms: (1) reverse Na^+/Ca^{2+}

exchange (Ca^{2+} influx, Na^+ efflux) and (2) activation of voltage-gated calcium channels by the depolarization of sperm membrane induced by Na^+/Ca^{2+} exchanger (Vines et al. 2002).

Apart from promoting or inducing sperm motility, Ca^{2+} can inhibit sperm motility in some conditions. When trout sperm (which usually needs extracellular Ca^{2+}) was demembranated, relatively low concentrations of Ca^{2+} inhibited sperm motility (Okuno and Morisawa 1989). To explain this apparent paradox, authors hypothesized that the influx of Ca^{2+} could happen transiently, to synthesize cAMP for the flagellar motility. While trout spermatozoa was demembranated, in the intact spermatozoa from another freshwater fish species, bluegill, elevated Ca^{2+} levels inhibit sperm motility but low levels permit or promote sperm motility (Zuccarelli and Ingermann 2007).

The increase in $[Ca^{2+}]_i$ post-activation has been attributed to an influx from the extracellular media or to the release from internal stores. In rainbow trout and carp sperm (Cosson et al. 1989; Krasznai et al. 2000), the increase in $[Ca^{2+}]_i$ required an influx from the external medium, as sperm cells were immotile in Ca-free activator. In other studies, the increase in $[Ca^{2+}]_i$ post-activation was observed even in the absence of external Ca^{2+}, indicating that the increase in $[Ca^{2+}]_i$ is due to the release from intracellular stores (rainbow trout, Boitano and Omoto 1991; pufferfish, Oda and Morisawa 1993; tilapia, Morita et al. 2003).

Besides the results in pufferfish and European eel (Gallego et al. 2013; Pérez et al. 2016), Ca^{2+} is considered an important second messenger for fish sperm activation. Zilli et al. (2012) summarized the mechanisms proposed for physiological roles of Ca^{2+} as second messenger in fish sperm motility initiation:

(a) Ca^{2+} acts directly on the axonemal structures (European sea bass and tuna, Cosson et al. 2008a, b).
(b) Ca^{2+} regulates Ca^{2+}/calmodulin-dependent protein phosphorylation which in turn activates the axoneme (pufferfish or seawater-acclimated euryhaline tilapia; Krasznai et al. 2003a; Morita et al. 2004).
(c) Ca^{2+} leads to a cAMP-dependent protein phosphorylation that activates axoneme in gilthead sea bream and stripped sea bream (Zilli et al. 2008).

Two signal transduction pathways are now hypothesized to be involved in sperm motility initiation in fish: cAMP/PKA and Ca^{2+}/CaM/CaMK (reviewed by Zilli et al. 2017).

4.3.4 *Motility in Seminal Plasma*

Three Cottidae species (*Alcichthys alcicornis, Blepsias cirrhosus, Hexagrammos octogrammos*) have *internal gametical association,* where spermatozoa arriving at ovarian cavity could enter into the micropyle of ovulated eggs, but not into the egg cytoplasm for subsequent fertilization until the eggs are released to seawater (Munehara et al. 1989).

In those species, sperm is already motile when it is taken from the sperm duct (Koya et al. 1993). In those species, motility in seminal plasma was not inhibited nor by K^+ (seminal plasma 16–20 mM K^+) or osmolality (values 290–330 mOsm/kg). Thus, quiescence was not observed in seminal plasma. However, sperm motility depended on ions; when diluted in mannitol instead of ionic media, sperm was immotile. When diluted in hyperosmotic seawater (1000 mOsm/kg), sperm from *B. cirrhosus* and *H. octogrammus* was immotile, while in *A. alcycornis* motility was observed.

Motility in seminal plasma was also observed in *Cottus hangionensis* (Ohta and Shinriki 1998, cited in Koya et al. 2002), and in *Aptocylus ventricosus*. All of them are Scorpaeniformes. In those species, ion Na^+ was important for isosmotic sperm activation, but pH had little effect, with a motility decrease only at $pH < 6.0$ (Koya et al. 1993).

More studies on those species are needed to understand the whole activation process in fish sperm.

References

Alavi SMH, Cosson J (2006) Sperm motility in fishes. (II). Effects of ions and osmolality: a review. Cell Biol Int 30:1–14. https://doi.org/10.1016/j.cellbi.2005.06.004

Alavi SMH, Rodina M, Viveiros ATM, Cosson J, Gela D, Boryshpolets S, Linhart O (2009) Effects of osmolality on sperm morphology, motility and flagellar wave parameters in northern pike (*Esox lucius* L.). Theriogenology 72:32–43. https://doi.org/10.1016/j.theriogenology.2009.01.015

Alavi SMH, Kozak P, Hatef A, Hamackova J, Linhart O (2010) Relationships between reproductive characteristics in male *Vimba vimba* L. and the effects of osmolality on sperm motility. Theriogenology 74:317–325. https://doi.org/10.1016/j.theriogenology.2010.02.017

Alavi SMH, Gela D, Rodina M, Linhart O (2011) Roles of osmolality, calcium- potassium antagonist and calcium in activation and flagellar beating pattern of sturgeon sperm. Comp Biochem Physiol A Mol Integr Physiol 160:166–174. https://doi.org/10.1016/j.cbpa.2011.05.026

Allan JD (1995) Stream ecology. In: Structure and function of running waters. 2nd edn. Springer, The Netherlands. xiv, 436 pp

Asturiano JF, Pérez L, Garzón DL, Marco-Jiménez F, Peñaranda DS, Vicente JS, Jover M (2004) Physio-chemical characteristics of seminal plasma and development of media and methods for the cryopreservation of European eel sperm. Fish Physiol Biochem 30:283–293. https://doi.org/10.1007/s10695-005-1553-x

Baynes SM, Scott AP, Dawson AP (1981) Rainbow trout, *Salmo gairdnerii* Richardson, spermatozoa: effects of cations and pH on motility. J Fish Biol 19(3):259–267. https://doi.org/10.1111/j.1095-8649.1981.tb05830.x

Billard R, Cosson J, Crim L (1993) Motility of fresh and aged halibut sperm. Aquat Living Resour 6:67–75. https://doi.org/10.1051/alr:1993008

Boitano S, Omoto CK (1991) Membrane hyperpolarization activates trout sperm without an increase in intracellular pH. J Cell Sci 98:343–349

Bondarenko O, Dzyuba B, Cosson J, Gunes Y, Prokopchuk G, Psenicka M, Linhart O (2013) Volume changes during the motility period of fish spermatozoa: interspecies differences. Theriogenology 79:872–881. https://doi.org/10.1016/j.theriogenology.2013.01.005

Cai X, Clapham DE (2008) Evolutionary genomics reveals lineage-specific gene loss and rapid evolution of a sperm-specific ion channel complex: CatSpers and CatSperb. PLoS One 3(10):e3569. https://doi.org/10.1371/journal.pone.0003569

Caldwell RA, Clemo HF, Baumgarten CM (1998) Using gadolinium to identify stretch-activated channels: technical considerations. Am J Physiol 275(2 Pt 1):C619–C621. https://doi.org/10.1152/ajpcell.1998.275.2.C619

Cerdá J, Chauvigné F, Finn RN (2017) The physiological role and regulation of aquaporins in teleost germ cells. Adv Exp Med Biol 969:149–171. https://doi.org/10.1007/978-94-024-1057-010

Cherr GN, Morisawa M, Vines CA, Yoshida K, Smith EH, Matsubara T, Pillai MC, Griffin FJ, Yanagimachi R (2008) Two egg-derived molecules in sperm motility initiation and fertilization in the Pacific herring (*Clupea pallasi*). Int J Dev Biol 52(5–6):743–752. https://doi.org/10.1387/ijdb.072566gc

Ciereszko A (2008) Chemical composition of seminal plasma and its physiological relationship with sperm motility, fertilizing capacity and cryopreservation in fish. In: Alavi SMH, Cosson JJ, Coward K, Rafiee G (eds) Fish spermatology. Alpha Science International, Oxford, pp 215–240

Ciereszko A, Glogowski J, Dabrowski K (2000) Biochemical characteristics of seminal plasma and spermatozoa of freshwater fishes. In: Tiersch TR, Mazik PM (eds) Cryopreservation in aquatic species. World Aquaculture Society, Louisiana, pp 20–48

Cosson J (2004) The ionic and osmotic factors controlling motility of fish spermatozoa. Aquac Int 12(1):69–85. https://doi.org/10.1023/B:AQUI.0000017189.44263.bc

Cosson MP, Billard R, Letellier L (1989) Rise of internal Ca^{2+} accompanies the initiation of trout sperm motility. Cell Motil Cytoskeleton 14:424–434. https://doi.org/10.1002/cm.970140312

Cosson J, Groison A, Suquet M, Fauvel C (2008a) Motility characteristics of spermatozoa in cod (*Gadus morhua*) and hake (*Merluccius merluccius*). Cybium 32:176–177

Cosson J, Groison A, Suquet M, Fauvel C, Dreanno C, Billard R (2008b) Studying sperm motility in marine fish: an overview on the state of the art. J Appl Ichthyol 24:460–486. https://doi.org/10.1111/j.1439-0426.2008.01151.

Darszon A, Guerrero A, Galindo BE, Nishigaki T, Wood CD (2004) Sperm-activating peptides in the regulation of ion fluxes, signal transduction and motility. Int J Dev Biol 52:595–606. https://doi.org/10.1387/ijdb.072550ad

Detweiler C, Thomas P (1998) Role of ions and ion channels in the regulation of Atlantic croaker sperm motility. J Exp Zool 281:139–148. https://doi.org/10.1002/(SICI)1097-010X(19980601)281:2<139::AID-JEZ8>3.0.CO;2-P

Dreanno C, Cosson J, Suquet M, Cibert C, Fauvel C, Dorange G, Billard R (1999) Effects of osmolality, morphology perturbations and intracellular nucleotide content during the movement of sea bass (*Dicentrarchus labrax*) spermatozoa. J Reprod Fertil 116:113–125. https://doi.org/10.1530/jrf.0.1160113

Eckert R, Randall D (1989) Animal physiology, mechanisms and adaptations, 3rd edn. Interamericana-McGraw-Hill Ed, Madrid. 683 pp

Fechner S, Alvarez L, Bönigk W, Müller A, Berger T, Pascal R, Trötschel C, Poetsch A, Stolting G, Siegfried KR, Kremer E, Seifert R, Kaupp UB (2015) A K^{+}-selective CNG channel orchestrates Ca^{2+} signalling in zebrafish sperm. eLife 4:e07624. https://doi.org/10.7554/eLife.07624

Gallego V, Pérez L, Asturiano JF, Yoshida M (2013) Study of pufferfish (*Takifugu niphobles*) sperm: development of methods for short-term storage, effects of different activation media and role of intracellular changes in Ca^{2+} and K^{+} in the initiation of sperm motility. Aquaculture 414–415:82–91. https://doi.org/10.1016/j.aquaculture.2013.07.046

Gallego V, Martínez-Pastor F, Mazzeo I, Peñaranda DS, Herráez MP, Asturiano JF, Pérez L (2014) Intracellular changes in Ca^{2+} , K^{+} and pH after sperm motility activation in the European eel (*Anguilla anguilla*): preliminary results. Aquaculture 418–419:155–158. https://doi.org/10.1016/j.aquaculture.2013.10.022

Gatti JL, Billard R, Christen R (1990) Ionic regulation of the plasma membrane potential of rainbow trout (*Salmo gairdneri*) spermatozoa: role in the initiation of sperm motility. J Cell Physiol 143:546–554. https://doi.org/10.1002/jcp.1041430320

Gonçalves ACS, Nascimento AF, Costa AC, Leal MC, Viveiros ATM (2013) Initiation and suppression of sperm motility isosmolality-dependent in two south American fish species: streaked prochilod (*Prochilodus lineatus*) and Piracanjuba (*Brycon orbignyanus*). Anim Reprod 10:62–70

He S, Jenkins-Keeran K, Woods LC (2004) Activation of sperm motility in striped bass via a cAMP-independent pathway. Theriogenology 61(7–8):1487–1498. https://doi.org/10.1016/j.theriogenology.2003.08.015

Kho KH, Tanimoto S, Inaba K, Oka Y, Morisawa M (2001) Transmembrane cell signaling for the initiation of trout sperm motility: roles of ion channels and membrane hyperpolarization for cyclic AMP synthesis. Zool Sci 18:919–928

Kho KH, Morisawa M, Choi KS (2003) Membrane hyperpolarization increases cAMP to induce the initiation of sperm motility in salmonid fishes, rainbow trout and masu salmon. J Microbiol Biotechnol 13:833–840

Kirichok Y, Navarro B, Clapham DE (2006) Whole-cell patch-clamp measurements of spermatozoa reveal an alkaline-activated Ca^{2+} channel. Nature 439:737–740. https://doi.org/10.1038/nature04417

Koya Y, Munehara H, Takano K, Takahashi H (1993) Effects of extracellular environments on the motility of spermatozoa in several marine sculpins with internal gametic association. Comp Biochem Physiol 106:25–29. https://doi.org/10.1016/0300-9629(93)90033-Z

Koya Y, Munehara H, Takano K (2002) Sperm storage and motility in the ovary of the marine sculpin *Alcichtys alcicornis* (Teleostei: Scorpaeniformes). J Exp Zool 292:45–155. https://doi.org/10.1002/jez.1150

Krasznai Z, Márián T, Balkay L, Gáspár R, Trón L (1995) Potassium channel regulate hypo-osmotic shock induced motility of common carp (*Cyprinus carpio*) sperm. Aquaculture 125:123–128. https://doi.org/10.1016/0044-8486(94)00234

Krasznai Z, Márián T, Izumi H, Damjanovich S, Balkay L, Trón L, Morisawa M (2000) Membrane hyperpolarization removes inactivation of Ca^{2+} channels, leading to Ca^{2+} influx and subsequent initiation of sperm motility in the common carp. Proc Natl Acad Sci U S A 97:2052–2057. https://doi.org/10.1073/pnas.040558097

Krasznai Z, Morisawa M, Morisawa S, Krasznai Z, Trón L, Gáspár R (2003a) Role of ion channels and membrane potential in the initiation of carp sperm motility. Aquat Living Resour 16:445–449. https://doi.org/10.1016/S0990-7440(03)00054-8

Krasznai Z, Morisawa M, Krasznai ZT, Morisawa S, Inaba K, Bazsáné ZK, Rubovszky B, Bodnár B, Borsos A, Márián T (2003b) Gadolinium, a mechano-sensitive channel blocker, inhibits osmosis-initiated motility of sea- and freshwater fish sperm, but does not affect human or ascidian sperm motility. Cell Motil Cytoskeleton 55:232–243. https://doi.org/10.1002/cm.10125

Lahnsteiner F, Patzner RA (1998) Sperm motility of the marine teleosts *Boops boops, Diplodus sargus, Mullus barbatus* and *Trachurus mediterraneus*. J Fish Biol 52:726–742. https://doi.org/10.1111/j.1095-8649.1998.tb00816.x

Lahnsteiner F, Berger B, Weismann T, Patzner A (1996) Motility of spermatozoa of *Alburnus alburnus* (Cyprinidae) and its relationship to seminal plasma composition and sperm metabolism. Fish Physiol Biochem 15:167–179. https://doi.org/10.1007/BF01875596

Lee CS, Tamaru CS, Kelley CD, Moriwake A, Miyamoto GT (1992) The effect of salinity on the induction of spawning and fertilization in the stripped mullet, *Mugil cephalus*. Aquaculture 102:289–296. https://doi.org/10.1016/0044-8486(92)90155-E

Legendre M, Alavi SMH, Dzyuba B, Linhart O, Prokopchuk G, Cochet C, Dugué R, Cosson J (2016) Adaptations of semen characteristics and sperm motility to harsh salinity: Extreme situations encountered by the euryhaline tilapia *Sarotherodon melanotheron heudelotii* (Dumeril, 1859). Theriogenology 86(5):1251–1267. https://doi.org/10.1016/j.theriogenology.2016.04.066

Lin F, Liu L, Dabrowski K (1996) Characteristics of muskellunge spermatozoa I: ultrastructure of spermatozoa and biochemical composition of semen. Trans Am Fish Soc 125(2):187–195. https://doi.org/10.1577/1548-8659(1996)125<0187:COMSIU>2.3.CO;2

Linhart O, Walford J, Sivaloganathan B, Lam TJ (1999) Effects of osmolality and ions on the motility of stripped and testicular sperm of freshwater- and seawater-acclimated tilapia, *Oreochromis mossambicus*. J Fish Biol 55:1344–1358. https://doi.org/10.1111/j.1095-8649.1999.tb02080.x

Liu Y, Cheng H, Tiersch TR (2018) The role of alkalinization-induced Ca^{2+} influx in sperm motility activation of a viviparous fish Redtail Splitfin (*Xenotoca eiseni*). Biol Reprod 99(6):1159–1170. https://doi.org/10.1093/biolre/ioy15

Majhi RK, Kumar A, Yadav M, Swain N, Kumari S, Saha A, Pradhan A, Goswami L, Saha S, Samanta L, Maity A, Nayak TK, Chattopadhyay S, Rajakuberan C, Kumar A, Goswami C (2013) Thermosensitive ion channel TRPV1 is endogenously expressed in the sperm of a fresh water teleost fish (*Labeo rohita*) and regulates sperm motility. Channels (Austin) 7(6):483–492. https://doi.org/10.4161/chan.25793

Martinez-Pastor F, Cabrita E, Soares F, Anel L, Dinis MT (2008) Multivariate cluster analysis to study motility activation of *Solea senegalensis* spermatozoa: a model for marine teleosts. Reproduction 135(4):449–459. https://doi.org/10.1530/REP-07-0376

Marshall WS, Bryson SE, Idler DR (1989a) Gonadotropin stimulation of K^+ secretion and Na^+ absorption by brook trout (*Salvelinus fontinalis*) sperm duct epithelium. Gen Comp Endocrinol 75(1):118–128. https://doi.org/10.1016/0016-6480(89)90016

Marshall WS, Bryson SE, Idler DR (1989b) Control of ion transport by the sperm duct epithelium of brook trout (*Salvelinus fontinalis*). Fish Physiol Biochem 7(1–6):331–336. https://doi.org/10.1007/BF00004725

Milla S, Terrien X, Sturm A, Ibrahim F, Giton F, Fiet J, Prunet P, Le Gac F (2008) Plasma 11-deoxycorticosterone (DOC) and mineralocorticoid receptor testicular expression during rainbow trout *Oncorhynchus mykiss* spermiation: implication with 17alpha,20-beta-dihydroxy-progesterone on milt fluidity. Reprod Biol Endocrinol 6:1–13. https://doi.org/10.1186/1477-7827-6-19

Miura C, Miura T (2003) Molecular control mechanisms of fish spermatogenesis. Fish Physiol Biochem 28(1–4):181–186. https://doi.org/10.1023/B:FISH.0000030522.71779.47

Miura C, Miura T (2011) Analysis of spermatogenesis using an eel model. Aqua BioSci Monogr 4(4):105–129. https://doi.org/10.5047/absm.2011.00404.0105

Miura T, Yamauchi K, Takahashi H, Nagahama Y (1991) Involvement of steroid hormones in gonadotropin-induced testicular maturation in male eels (*Anguilla japonica*). Biomed Res 12:241–248. https://doi.org/10.2220/biomedres.12.241

Miura T, Yamauchi K, Takahashi H, Nagahama Y (1992) The role of hormones in the acquisition of sperm motility in salmonid fish. J Exp Zool 261(3):359–363. https://doi.org/10.1002/jez.1402610316

Miura T, Kasugai T, Nagahama Y, Yamauchi K (1995) Acquisition of potential for sperm motility in vitro in Japanese eel (*Anguilla japonica*). Fish Sci 61:533–534. https://doi.org/10.2331/fishsci.61.533

Miura A, Nomura K, Imaizumi H, Jinbo T, Masuda Y, Tanaka H, Ohta H (2013) Administration of 17α-hydroxyprogesterone into mature male Japanese eel reduces sperm motility by decreasing potassium ion concentrations in the seminal plasma. Aquaculture 414–415:217–223. https://doi.org/10.1016/j.aquaculture.2013.08.011

Morisawa M (2008) Adaptation and strategy for fertilization in the sperm of teleost fish. J Appl Ichthyol 24:362–370. https://doi.org/10.1111/j.1439-0426.2008.01126.x

Morisawa S, Morisawa M (1986) Acquisition of potential for sperm motility in rainbow trout and chum salmon. J Exp Biol 126:89–96

Morisawa S, Morisawa M (1988) Induction of potential for sperm motility by bicarbonate and pH in rainbow trout and chum salmon. J Exp Biol 136:13–22

Morisawa M, Suzuki K (1980) Osmolality and potassium ion: their roles in initiation of sperm motility in teleosts. Science 210:1145–1147. https://doi.org/10.1126/science.7444445

Morisawa M, Suzuki K, Morisawa S (1983a) Effects of potassium and osmolality on spermatozoan motility of salmonid fishes. J Exp Biol 107:105–113

Morisawa M, Suzuki K, Shimizu H, Morisawa S, Yasuda K (1983b) Effects of osmolality and potassium on motility of spermatozoa from freshwater cyprinid fishes. J Exp Biol 107:95–103

Morita M, Takemura A, Nakajima A, Okuno M (2006) Microtubule sliding movement in tilapia sperm flagella axoneme is regulated by Ca^{2+}/calmodulin-dependent protein phosphorylation. Cell Motil Cytoskeleton 63(8):459–470. https://doi.org/10.1002/cm.20137

Morita M, Takemura A, Okuno M (2003) Requirement of Ca^{2+} on activation of sperm motility in euryhaline tilapia *Oreochromis mossambicus*. J Exp Biol 206(5):913–921. https://doi.org/10.1242/jeb.00153

Morita M, Takemura A, Okuno M (2004) Acclimation of sperm motility apparatus in seawater-acclimated euryhaline tilapia *Oreochromis mossambicus*. J Exp Biol 207(2):337–345. https://doi.org/10.1242/jeb.00748

Mundt N, Spehr M, Lishko PV (2018) TRPV4 is the temperature-sensitive ion channel of human sperm. eLife 7:e35853. https://doi.org/10.7554/eLife.35853

Munehara H, Takano K, Koya Y (1989) Internal gametic association and external fertilization in the Elkhorn sculpin, *Alcichthys alcicornis*. Copeia 1989(3):673–678. https://doi.org/10.2307/1445484

Navarro B, Kirichok Y, Clapham DE (2007) KSper, a pH-sensitive K current that controls sperm membrane potential. PNAS 104(18):7688–7692. https://doi.org/10.1073/pnas.0702018104

Oda S, Morisawa M (1993) Rises of intracellular Ca^{2+} and pH mediate the initiation of sperm motility by hyperosmolality in marine teleosts. Cell Motil Cytoskeleton 25:171–178. https://doi.org/10.1002/cm.970250206

Ohta H, Shinriki Y (1998) Changes in osmotic pressure that trigger the initiation of sperm motility in the river sculpin *Cottus hangiongensis*. Fish Physiol Biochem 18:29–35. https://doi.org/10.1023/A:1007773429318

Ohta H, Ikeda K, Izawa T (1997) Increases in concentrations of potassium and bicarbonate ions promote acquisition of motility in vitro by Japanese eel spermatozoa. J Exp Zool 277:171–180. https://doi.org/10.1002/(SICI)1097-010X(19970201)277:2%3C171::AID-JEZ9%3E3.0.CO;2-M

Ohta H, Kagawa H, Tanaka H, Unuma T (2001) Control by the environmental concentration of ions of the potential for motility in Japanese eel spermatozoa. Aquaculture 198:339–351. https://doi.org/10.1016/S0044-8486(00)00597-4

Okuno M, Morisawa M (1989) Effects of calcium on motility of rainbow trout sperm flagella demembranated with triton X-100. Cell Motil Cytoskeleton 14(2):194–200. https://doi.org/10.1002/cm.970140206

Perchec G, Cosson MP, Cosson J, Jeulin C, Billard R (1996) Morphological and kinetic changes of carp (*Cyprinus carpio*) spermatozoa after initiation of motility in distilled water. Cell Motil Cytoskeleton 353:113–120. https://doi.org/10.1002/(SICI)1097-0169(1996)35:2<113::AID-CM4>3.0.CO;2-B

Pérez L, Vílchez MC, Gallego V, Morini M, Peñaranda DS, Asturiano JF (2016) Role of calcium on the initiation of sperm motility in the European eel. Comp Biochem Physiol A Mol Integr Physiol 191:98–106. https://doi.org/10.1016/j.cbpa.2015.10.009

Plant TD (2014) TRPs in mechanosensing and volume regulation. Handb Exp Pharmacol 223:743–766. https://doi.org/10.1007/978-3-319-05161-1_2

Rahman MS, Morita M, Takemura A, Takano K (2003) Hormonal changes in relation to lunar periodicity in the testis of the forktail rabbitfish, *Siganus argenteus*. Gen Comp Endocrinol 131:302–309. https://doi.org/10.1016/S0016-6480(03)00025-X

Sachs F (2010) Stretch-activated ion channels: what are they? Physiology (Bethesda) 25(1):50–56. https://doi.org/10.1152/physiol.00042.2009

Sachs F (2015) Mechanical transduction by ion channels: a cautionary tale. World J Neurol 5(3):74–87. https://doi.org/10.5316/wjn.v5.i3.74

Santi CM, Martínez-López P, de la Vega-Beltrán JL, Butler A, Alisio A, Darszon A, Salkoff L (2010) The SLO3 sperm-specific potassium channel plays a vital role in male fertility. FEBS Lett 584(5):1041–1046. https://doi.org/10.1016/j.febslet.2010.02.005

Schulz RW, de França LR, Lareyre JJ, Le Gac F, Chiarini-García H, Nóbrega RH, Miura T (2010) Spermatogenesis in fish. Gen Comp Endocrinol 165(3):390–411. https://doi.org/10.1016/j.ygcen.2009.02.013

Scott AP, Baynes SM (1982) Plasma levels of sex steroids in relation to ovulation and spermiation in rainbow trout (*Salmo gairdneri*). In: Richter CJJ, Goos HJT (eds) Reproductive physiology of fish. Pudoc, Wageningen, pp 103–106

Scott AP, Sumpter JP, Stacey N (2010) The role of the maturation-inducing steroid, 17,20β-dihydroxy-4-pregnen-3-one, in male fishes: a review. J Fish Biol 76:183–224. https://doi.org/10.1111/j.1095-8649.2009.02483.x

Takei GL, Mukai C, Okuno M (2012) Transient Ca^{2+} mobilization caused by osmotic shock initiates salmonid fish sperm motility. J Exp Biol 215(4):630–641. https://doi.org/10.1242/jeb.063628

Takei GL, Mukai C, Okuno M (2015) Regulation of salmonid fish sperm motility by osmotic shock-induced water influx across the plasma membrane. Comp Biochem Physiol A 182:84–92. https://doi.org/10.1016/j.cbpa.2014.12.013

Tan W, Aizen J, Thomas P (2014) Membrane progestin receptor alpha mediates progestin-induced sperm hypermotility and increased fertilization success in southern flounder (*Paralichthys lethostigma*). Gen Comp Endocrinol 200:18–26. https://doi.org/10.1016/j.ygcen.2014.02.003

Tanaka S, Utoh T, Yamada Y, Horie N, Okamura A, Akazawa A, Mikawa N, Poka H, Kurokura H (2004) Role of sodium bicarbonate on the initiation of sperm motility in the Japanese eel. Fish Sci 70:780–787. https://doi.org/10.1111/j.1444-2906.2004.00871.x

Tanimoto S, Morisawa M (1988) Roles for potassium and calcium channels in the initiation of sperm motility in rainbow trout. Develop Growth Differ 30(2):117–124

Tanimoto S, Kudo Y, Nakazawa T, Morisawa M (1994) Implication that potassium flux and increase in intracellular calcium are necessary for the initiation of sperm motility in salmonid fishes. Mol Reprod Dev 39:409–414. https://doi.org/10.1002/mrd.1080390409

Thomas P (2003) Rapid, nongenomic steroid actions initiated at the cell surface: lessons from studies with fish. Fish Physiol Biochem 28:3–12. https://doi.org/10.1023/B:FISH.000

Thomas P, Pang Y, Zhu Y, Detweiler C, Doughty K (2004) Multiple rapid progestin actions and progestin membrane receptor subtypes in fish. Steroids 69:567–573. https://doi.org/10.1016/j.steroids.2004.05.004

Thomas P, Tubbs C, Garry VF (2009) Progestin functions in vertebrate gametes mediated by membrane progestin receptors (mPRs): identification of mPRα on human sperm and its association with sperm motility. Steroids 74:614–621. https://doi.org/10.1016/j.steroids.2008.10.020

Vicens A, Andrade-López K, Cortez D, Gutiérrez RM, Treviño CL (2017) Premammalian origin of the sperm-specific Slo3 channel. FEBS Open Bio 7:382–390. https://doi.org/10.1002/2211-5463.12186

Vílchez MC, Morini M, Peñaranda DS, Gallego V, Asturiano JF, Pérez L (2016) Sodium affects the sperm motility in the European eel. Comp Biochem Physiol A Mol Integr Physiol 198:51–58. https://doi.org/10.1016/j.cbpa.2016.04.008

Vílchez MC, Morini M, Peñaranda DS, Gallego V, Asturiano JF, Pérez L (2017) Role of potassium and pH on the initiation of sperm motility in the European eel. Comp Biochem Physiol A Mol Integr Physiol 203:210–219. https://doi.org/10.1016/j.cbpa.2016.09.024

Vines CA, Yoshida K, Griffin FJ, Pillai MC, Morisawa M, Yanagimachi R, Cherr GN (2002) Motility initiation in herring sperm is regulated by reverse sodium-calcium exchange. Proc Natl Acad Sci U S A 99(4):2026–2031. https://doi.org/10.1073/pnas.042700899

Wang Z, Crim LW (1997) Seasonal changes in the biochemistry of seminal plasma and sperm motility in the ocean pout, *Macrozoarces americanus*. Fish Physiol Biochem 16:77–83. https://doi.org/10.1007/BF00004542

Wetzel RG (2001) Limnology; lake and river ecosystems, 3rd edn. Elsevier Academic, London. 1006 pp

Wilson-Leedy JG, Kanuga MK, Ingermann RL (2009) Influence of osmolality and ions on the activation and characteristics of zebrafish sperm motility. Theriogenology 71:1054–1062. https://doi.org/10.1016/j.theriogenology.2008.11.006

Yanagimachi R (1957a) Some properties of the sperm-activating factor in the micropyle area of the herring egg. Annot Zool Japon 30:114–119

Yanagimachi R (1957b) Studies of fertilization in *Clupea pallasi*. VI. Fertilization of the egg deprived of the membrane. Jpn J Ichthyol 6:41–47

Yanagimachi R, Kanoh Y (1953) Manner of sperm entry in herring egg, with special reference to the role of calcium ions in fertilization. J Fac Sci Hokkaido Univ Ser VI Zool 11(3):487–494

Yanagimachi R, Harumi T, Matsubara H, Yan W, Yuan S, Hirohashi N, Iida T, Yamaha E, Arai K, Vines C, Cherr GN (2017) Chemical and physical guidance of fish spermatozoa into the egg through the micropyle. Biol Reprod 96(4):780–799. https://doi.org/10.1093/biolre/iox015

Zeng X-H, Yang C, Kim ST, Lingle CJ, Xia X-M (2011) Deletion of the Slo3 gene abolishes alkalization-activated K+ current in mouse spermatozoa. Proc Natl Acad Sci U S A 108:5879–5884. https://doi.org/10.1073/pnas.1100240108

Zilli L, Schiavone R, Storelli C, Vilella S (2008) Molecular mechanisms determining sperm motility initiation in two sparids (*Sparus aurata* and *Lithognathus mormyrus*). Biol Reprod 79:356–366. https://doi.org/10.1095/biolreprod.108.068296

Zilli L, Schiavone R, Vilella S (2012) Molecular mechanism regulating axoneme activation in marine fish: a review. Int Aquat Res 4:2. https://doi.org/10.1186/2008-6970-4-2

Zilli L, Schiavone R, Vilella S (2017) Role of protein phosphorylation/dephosphorylation in fish sperm motility activation: state of the art and perspectives. Aquaculture 472:73–80. https://doi.org/10.1016/j.aquaculture.2016.03.043

Zuccarelli MD, Ingermann RL (2007) Calcium-induced quiescence of sperm motility in the bluegill (*Lepomis macrochirus*). J Exp Zool A Ecol Genet Physiol 307(10):590–599. https://doi.org/10.1002/jez.414

Chapter 5
Sperm Guidance into Teleost Fish Egg

Ryuzo Yanagimachi

Abstract Micropyle is a thin canal in the envelope (chorion) of teleost egg. This is the only place in the chorion that permits sperm entry. Spermatozoa swimming freely in water have one- or three-dimensional trajectory. Once in contact with egg's chorion, spermatozoa initiate thigmotactic (sliding) motion along the chorion surface. Although this "two-dimensional" sperm movement facilitates sperm's search for the micropyle, there is a specific glycoprotein around the outer opening of the micropyle which directs spermatozoa into the canal. It is called the "micropylar sperm attractant" or MISA. In herring, there is another known factor called the sperm "motility-initiation factor" or SMIF which renders intrinsically motionless spermatozoa motile. For herring, both SMIF and MISA are necessary for sperm entry into the micropyle. The herring micropyle is a thin, manhole-like canal with a slight depression of chorion around the micropyle. This is called **Type I** micropyle. Other fish with Type I micropyle, with or without a chorionic depression around the micropyle, include flounder, pollack, and munmichog. Some other fish (e.g., salmon, cod, and pufferfish) have a funnel-like micropyle with a wide, conical mouth (**Type II**). In fish with Type I and II micropyles, except for herring, sperm entry into the micropyle is possible without MISA; however, entry becomes inefficient as most spermatozoa swim over the micropyle. In another group of fish (e.g., goldfish, loach, and zebrafish), the chorion around the micropyle is deeply indented like a large sinkhole or has radially or spirally arranged grooves (**Type III** micropyle). MISA is absent from the chorion of Type III fish eggs. For fish with Types I and II micropyles, chemical interactions between spermatozoa and chorion around the micropyle assist sperm entry into the micropyle, whereas in those with Type III micropyles, sperm entry is purely physical. In this case, physical configurations of the chorion around the micropyle "directs" spermatozoa toward the micropyle.

Keywords Egg · Fertilization · Fish · Micropyle · Sperm

R. Yanagimachi (✉)
University of Hawaii, Honolulu, HI, USA
e-mail: yana@hawaii.edu

M. Yoshida, J. F. Asturiano (eds.), *Reproduction in Aquatic Animals*,
https://doi.org/10.1007/978-981-15-2290-1_6

5.1 Introduction

Fertilization is the union of male and female gametes. Since it is a critical moment of the life cycle of animal, Mother Nature provides all means of warranting its success. In some fish, fertilization occurs outside of female's body, while in others it takes place inside of the body (oviparity vs viviparity). The mechanism by which the sperm–egg union is warranted may be different in these two groups, but there must be something in common to all species of fish. Fish is unique in that the fertilizing spermatozoon can enter the egg only through the micropyle, which is a thin canal in the chorion. How do fish spermatozoa "find" such a tiny micropyle? This is the subject of this review.

5.2 Sperm Guidance into Egg Through the Micropyle

5.2.1 Does Ovarian Fluid Guide Spermatozoa to the Egg?

In both oviparous and viviparous fish, fully mature eggs that are ready to be spawned are in the ovarian (or coelomic) cavity filled with a viscous fluid. This fluid, commonly called the ovarian fluid (OVF), contains a protease inhibitor which prevents eggs from spontaneous (auto-) activation (Minin and Ozerova 2008). During spawning of oviparous fish, OVF is released into water along with the eggs. While OVF can enhance sperm motility (e.g., Butts et al. 2012; Lehnert et al. 2017; Devigili et al. 2017) and/or extends sperm's motile life (e.g., Yeates et al. 2013; Makiguchi et al. 2016), it is unlikely that OVF, which is rapidly dispersing into water, "guides" spermatozoa to the egg's micropyle. It seems that the prime function of OVF is the prolongation of fertile life of mature eggs and easing the release of eggs from female's cloaca. To understand fish fertilization, it is important to recall the behavior of male and female fish during spawning. In fish that pawns in rapidly running water (e.g., salmon and trout), a female first releases eggs and is followed by an immediate discharge of spermatozoa by a male. In other fish (e.g., goldfish, zebrafish, and pollack), female and male release eggs and spermatozoa simultaneously while swimming rapidly side by side in close approximation. In a third example (e.g., rice fish or medaka and clown fish), the female lays eggs first (several eggs at one time) and a few or several seconds later, the male pours spermatozoa onto the egg while agitating water vigorously using its anal fin. There is no evidence that OVF stays around the egg and "directs" spermatozoa into the micropyle. Since fish eggs washed thoroughly with a physiologically balanced salt solution (e.g. Ringer's) remain fully fertile, there must be mechanisms independent of OVF that warrant sperm entry into the micropyle. The mechanism by which an egg "guides" a fertilizing spermatozoon into the micropyle may be different in different species of fish (Yanagimachi et al. 2017). Here, I discusses a few examples of sperm entry into the fish egg.

5.2.2 Sperm Guidance in Herring

Herring (*Clupea pallasii*) is unique in that spermatozoa are intrinsically motionless in seawater. Thus, it naturally raises the question as to how they fertilize eggs. Yanagimachi and Kanoh (1953) observed that spermatozoa became vigorously motile on contact with the egg's chorion around the micropyle before entering the canal. Here, it is important to emphasize that herring has an unusual spawning behavior. A school of males approach the spawning ground first to release milt in the seawater, producing the so-called white water. This attracts and induces spawning of females as well as other males (Hay 1985; Carolsfeld et al. 1997). Apparently, eggs are released into sperm-suspending seawater before attaching to solid objects such as seaweeds and rocks. Readers are referred to photographs of "white water" and of kelp with multiple layers of eggs attached to both sides of the blade (cf. Google: keywords: herring spawning white water; herring roe on kelp). I myself examined several blades of kelp and found that all or almost all of the eggs on the blade were fertilized and developed normally (Yanagimachi, unpublished observation). If eggs first attach to solid objectives (e.g., seaweeds and other eggs), many of them would be left unfertilized as the outer opening of micropyle becomes blocked.

We previously examined and reported on sperm entry in herring egg using the following approach: (1) Hundreds of mature unfertilized eggs were put in a dish, with care being taken not to overlap eggs. Eggs stacked firmly to the bottom of the dish when either fish Ringer's solution or seawater was added to the dish. Alternatively, eggs were spread on a dish of Ringer's solution or seawater using a spatula. Ringer's solution (Hirano et al. 1971) and 1/2 diluted seawater (a mixture of 1 part of seawater and 1 part of distilled water) were chosen because they maintain the eggs and spermatozoa fertile for a much longer time than natural seawater. Eggs were fertilized and developed normally in these media. (2) Using a low-power ordinary objective lens (×10–20), we looked for eggs with clearly visible micropyles with top or semi-profile views being preferable. (3) After the micropyle was brought into the focus of a water-immersion objective lens (×40–50), freshly prepared sperm suspension (in Ringer's solution or 1/2 seawater) was added from the side of the objective lens. The use of a high-density sperm suspension was avoided because it made observation of fertilizing spermatozoon into the micropyle difficult.

Using this method, we tracked sperm behavior before and after insemination (Yanagimachi and Kanoh 1953; Yanagimachi 1957a; Yanagimachi et al. 1992). We observed that in either Ringer's solution or seawater, the vast majority of herring spermatozoa were motionless. A few did a rapid frisky (jumping around) motion for few seconds or less. In some cases, roughly 2–3% or more of spermatozoa exhibited such motions, but again the majority were motionless at any instance. Agitating the medium by pipetting or shaking the dish temporarily increased the number of spermatozoa that jumped around. Most spermatozoa remained motionless even after insemination, but when the water current brought them into contact with the chorion around the micropyle, they suddenly became very active. They swim along the chorion surface and entered the micropyle one by one, resulting in a quick filling of the micropylar canal with many spermatozoa (Fig. 5.1a, b). For a movie of herring spermatozoa gathering around the outer opening of micropyle, see Fig. 7 of Yanagimachi et al. (2013). The spermatozoon that entered the micropyle first entered the egg cytoplasm (Fig. 5.1b). Within 30–40 min after insemination, spermatozoa within the micropylar canal were suddenly pushed out of the canal by a colloidal material of cortical granule origin (Fig. 5.1c) (Yanagimachi and Kanoh 1953; Yanagimachi 1957b).

Figure 5.2a shows tracks of the sperm heads that entered the micropyle. Note that all spermatozoa move along the surface of chorion counter-clockwise. Figure 5.2b shows the degree of sperm motion on and near egg surface during the first 1 min after insemination. In water away from eggs, spermatozoa were motionless (−). A few displayed a "jumping around " motion sporadically (±). In contrast, very active continuous sperm movement was seen in the micropyle region (++). When the same egg was re-examined 10–20 min later, many more spermatozoa were still swarming around the outer opening of the micropyle (+++), but none were able to enter the micropyle canal because micropylar canal was already filled with spermatozoa. Spermatozoa were "trying" to repeatedly enter the micropyle before going away (Fig. 5.2c). At this time, a fairly large numbers of spermatozoa were seen actively swimming along chorion surface as well as in water away from the eggs (Fig. 5.2d). They were spermatozoa that had been activated in the micropylar region. Close examinations (Fig. 5.2e and e′) revealed that many of activated spermatozoa swam along the chorion surface, reaching the opposite to the micropyle, while others swam along chorion surface, "lifted up," and swam for a while before stopping. Another subset returned to the micropylar region to resume an active, thigmotactic movement. The presence of a "flock" of spermatozoa swimming actively in the micropylar region several or more minutes after insemination can give us a false

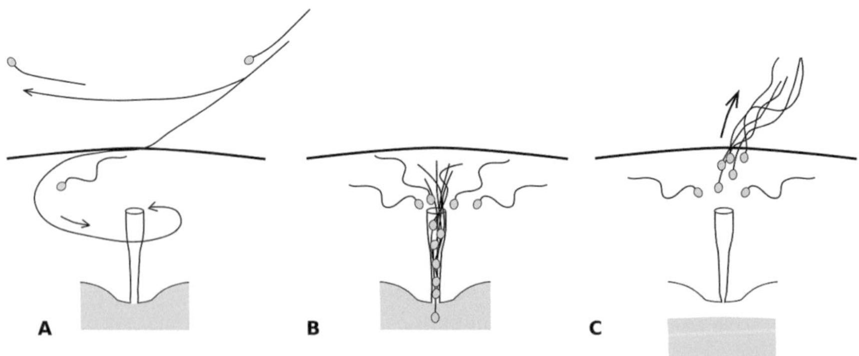

Fig. 5.1 Herring spermatozoa and egg's micropyle. (**a**) Herring spermatozoa are intrinsically motionless in seawater. However, they become very active upon contact with the chorion around the micropyle. They slide along chorion surface to enter the micropyle. (**b**) The micropylar canal is soon filled with many spermatozoa. It is the first spermatozoon that enters the egg cytoplasm. (**c**) Sooner or later, all supernumerary spermatozoa are pushed out of the canal by a flow of colloidal material of cortical granule origin as a result of egg activation

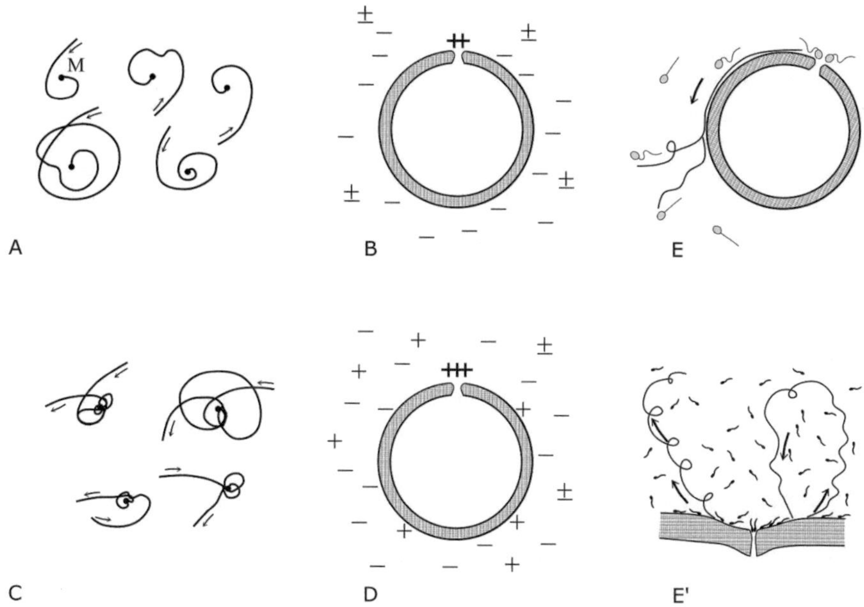

Fig. 5.2 Behavior of herring spermatozoa on chorion surface. (**a**) Track of the first (fertilizing) spermatozoon before entering the micropylar canal. Note that the direction of thigmotactic movement is counter-clockwise. (**b**) Within the first few minutes after insemination, many active spermatozoa are seen around the micropyle (++), but not in other areas of the egg (−). Few spermatozoa in seawater show sporadic movement (±). (**c**) Track of supernumerary spermatozoa that are unable to enter the micropyle which is already filled with spermatozoa. Supernumerary spermatozoa often "try to" enter the micropyle repeatedly before moving away. (**d**) >10 min after insemination, very many vigorously moving spermatozoa are seen in the micropylar region (+++). Some are moving actively (+) along the surface of chorion away from the micropyle. (**e** and **e′**) Spermatozoa activated in the micropylar region swim along chorion surface, lift up, and stop (**a**). Some others (**b**) return to the micropylar region to continue active movement. (**a**, **c**, and **e′** are reproduced from Yanagimachi et al. (1992))

impression that the micropyle attracts spermatozoa from a distance. However, we can say that herring spermatozoa brought by chance to the chorion surface near the micropyle are activated and enter it in a chemotactic fashion. Since herring spermatozoa can live in seawater for days, it is conceivable that spermatozoa that failed to fertilize first drift in water for days before fertilizing eggs of other females. In fact, one experiment I did supported this view. Spermatozoa moving actively around the micropyle of an egg already fertilized were sucked in a capillary tubing. They kept swimming for a few minutes before stopping. When they were added to fresh eggs 24 h later, they resumed active movement around the micropyle to fertilize (Yanagimachi, unpublished observation).

How then do herring eggs activate spermatozoa? According to Oda et al. (1995), herring spermatozoa are not activated by ovarian fluid, but they are activated by seawater in which mature eggs are "washed" for 40–50 min. This water contains an 8 kDa protein called the herring sperm-activating protein or HSAP. It is homologous to the kazal-ype trypsin inhibitor (Oda et al. 1998). HSAP is synthetized by ovarian follicular cells and is "absorbed" in the outer (adhesive) layer of the chorion except in the area where the micropyle is present. Although Oda et al. (1995) maintain that HSAP is released from eggs into seawater to activate herring spermatozoa, this protein is apparently not essential for fertilization because herring eggs thoroughly rinsed in Ringer's solution or seawater remain fully fertile. However, it is possible that HSAP increases the incidence of "jumping around" spermatozoa to enhance sperm–egg collisions (Cherr et al. 2008). What is essential for herring sperm activation and fertilization is a glycoprotein around the outer opening of the micropyle (Yanagimachi 1957a, b; Yanagimachi et al. 1992, 2017). We named it the "sperm motility initiation factor" (SMIF) (Pillai et al. 1993). Observation of the chorions isolated from eggs indicates that SMIF is bound tightly to the outer surface of the chorion around the micropyle. Although it cannot be removed from the chorion surface even by vigorous washings, it is removed readily by treatment with trypsin or weak acid (pH 3.5) (Yanagimachi 1957a). We identified SMIF as a 105 kDa glycosylated polypeptide (Pillai et al. 1993; Griffin et al. 1996). It is localized on the chorion around the micropyle (Fig. 5.3a). Sperm's SMIF receptor is in the midpiece region of the head (Griffin et al. 1996). This observation contrasted the claim by Yoshida et al. (1999) that the HSAP receptor is in sperm tail.

We later reported another glycoprotein that directs activated herring spermatozoa into the micropyle. This glycoprotein is on the surface of chorion immediately around the outer opening of the micropyle and can be stained dark by Coomassie blue (a nonspecific protein stain) (Fig. 5.3b). We called it micropylar sperm attractant or "MISA" (Yanagimachi et al. 2017). Similar materials were found around the outer opening of micropyles of several other fish species (e.g., flounders, trout, and pufferfish) (Yanagimachi et al. 2017). Treatment with trypsin or acidic Ringer solution (pH 3.5) removes both SMIF and MISA, rendering eggs infertile. At present, we are unable to remove or inactivate SMIF and MISA separately from herring eggs.

The chemical nature of herring MISA as well as its relationship of SMIF are yet to be investigated. These two could be separate molecules or two domains of a

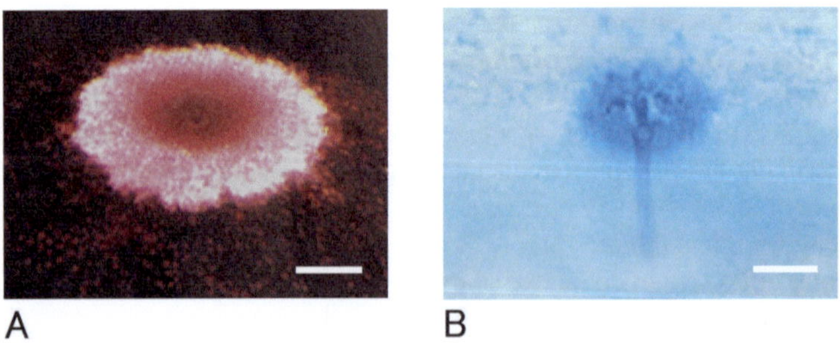

Fig. 5.3 Herring SMIF and MISA. (**a**) SMIF as demonstrated by anti-SMIF antibody. Note that the chorion in the immediate vicinity of micropyle is less strongly reacted to the antibody. Bar = 200 μm. (Reproduced from Griffin et al. 1996). (**b**) MISA stained by Coomassie blue. Bar = 10 μm

single molecule. To understand the nature of SMIF and MISA, the origin of these molecules is of particular interest. It is most likely that both are products of a giant micropylar cell and nearby follicular cells. The ultra-structures of micropylar cells in the herring and other fish have been described in detail (Ohta and Takano 1982; Ohta and Teranishi 1982; Kobayashi and Yamamoto 1985; Nakanishi and Iwamatsu 1989, 1994), and the essential role of Hippo-signaling pathway in micropyle formation in the zebrafish (*Danio rerio*) was reported by Yi et al. (2018).

5.2.3 Sperm Guidance in Flounder, Salmon, and Trout

Unlike herring spermatozoa, spermatozoa of most other fish begin to move actively upon their release from the male into the surrounding water (Morisawa 1994; Alavi and Cosson 2006). For example, spermatozoa from flounder *Pleuronectus obscurus* begin to swim very fast in seawater upon contact with seawater. The active motility last about 30 s, then the motility declines rapidly. By 1 min, only a few (5–10%) are barely motile. Salmonid spermatozoa also have a short motile life in freshwater.

MISA with staining affinity to Coomassie blue and lectins can be seen around the outer opening of micropyle of flounder egg (Fig. 5.4a). Spermatozoa coming close to the micropyle enter it (Fig. 5.4b) (for a movie of flounder sperm entering the micropyle, see Fig. 2 of Yanagimachi et al. 2013). Although not all the spermatozoa coming close to the micropyle enter it, the micropyle is quickly filled by many spermatozoa (Fig. 5.4c). It is the first spermatozoon that enters the egg cytoplasm to fertilize. The remaining spermatozoa are pushed out of the canal sooner or later, as it happens in the herring. In Ca^{2+}-free seawater, spermatozoa appear to not recognize the micropyle and swim over the micropyle (Fig. 5.4d). Even though few may enter the micropyle "accidentally," none are able to advance into the egg cytoplasm (Yanagimachi et al. 2017).

A brief treatment of unfertilized flounder eggs with a very low concentration of trypsin is enough to remove MISA. In the absence of MISA, much fewer eggs are fertilized than untreated control eggs (Yanagimachi et al. 2017). Apparently, flounder spermatozoa can enter the micropyle without MISA, but this process is much less efficient than when it is present. Like herring spermatozoa, flounder spermatozoa have Ca^{2+}-channel, CatSper (Yanagimachi et al. 2017). In Ca^{2+}-free seawater, flounder spermatozoa swim actively, but most swim over the micropyle with only a few entering it "accidentally" (Yanagimachi et al. 2017).

When barfin flounders (*Verasper moseri*) were kept in captivity (tank) for years, they often failed to spawn (release eggs into water) despite ovulating at regular intervals. When eggs were squeezed out of such females soon after the expected time of ovulation, two kinds of eggs were observed. Some looked very clear while others were opaque or translucent. The latter showed various signs of degeneration. It is likely that the former were ovulated recently, and the latter ones were from the previous ovulation cycle. Of course, degenerated and degenerating eggs were not fertilized, but even normal-looking ones were fertilized very poorly or not fertilized

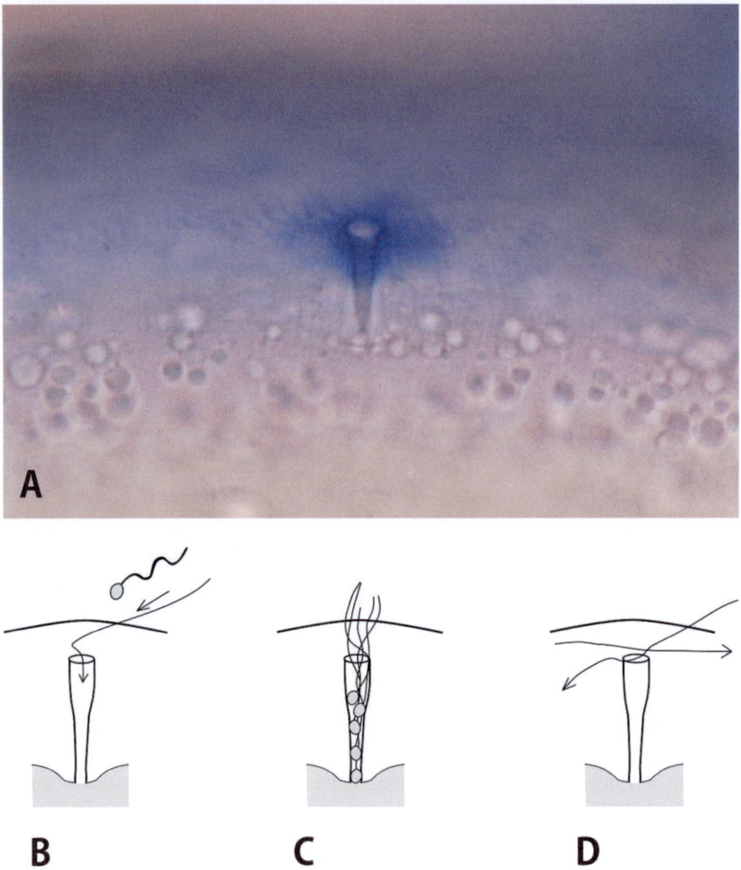

Fig. 5.4 Micropyle and sperm entry into flounder egg. (**a**) Micropyle stained with Coomassie blue. The diameter of the outer opening of micropyle is about 6 μm. (**b** and **c**) Sperm entry in normal seawater. (**d**) In Ca^{2+}-free seawater, spermatozoa swim over the micropyle even though few may enter it "accidentally"

at all. We then found that spermatozoa "ignored" micropyles of eggs with and without obvious signs of degradation. One likely explanation was that catepsin released from dead or dying eggs removed MISA from the recently ovulated eggs, thus rendering them infertile (Yanagimachi et al. 2017).

To examine sperm entry in salmon/trout eggs, we preferred to use isolated chorions because we could follow the behavior of spermatozoa on the chorion surface much more clearly and easily in isolated chorion than with intact eggs (Yanagimachi et al. 1992). We found that spermatozoa of rainbow trout (*Oncorhynchus mykiss*) and silver salmon (*Oncorhynchus kisutch*) enter the micropyle quickly, filling the canal within roughly 15 s. In Ca^{2+}-free media, none or sometimes one or a few entered the micropyle, whereas in normal media the micropyle was filled with 20 or more spermatozoa within 15 s. When chorions or rainbow trout eggs were treated with acidic Ringer's solution (pH 3.5) for 2 min, Coomassie blue-affinity material (SMIF) disappeared from the vestibule of the micropyle and sperm entry into the micropylar canal became much less efficient than in controls (Yanagimachi, unpublished data).

5.2.4 Sperm Guidance in Loach, Zebrafish, and Goldfish

In loach (*Misgurnus anguillicaudatus*) and zebrafish, the chorion around the micropyle is deeply indented like a large sink hole (Fig. 5.5a, b), whereas that of goldfish (*Carassius auratus auratus*) has many radially arranged grooves around the micropyle (Yanagimachi et al. 2017). MISA is absent on the chorion of all of these fish eggs. We observed many goldfish spermatozoa swimming along chorion's grooves before entering the micropyle. Goldfish eggs treated with trypsin remained perfectly fertile (Yanagimachi et al. 2017). We also observed loach and zebrafish spermatozoa entering the micropyle after swimming along the surface of chorion with a deep depression around the micropyle. This happened even in Ca^{2+}-free water, even though the spermatozoon reaching the bottom of micropylar canal was unable to fuse with the egg properly.

According to Pereira-Santos et al. (2017), characid spermatozoa have a long (>75 s) motile life in freshwater. Their eggs are small (~0.7 mm in diameter), and egg's chorion has grooves as well as a large micropyle (7.5 μm in diameter). These features seem to increase the chance of fertilization in this freshwater fish. According to Creech et al. (1998), the chorion of eggs of fathead minnow (*Pimephales promelas*) have radially arranged grooves around the micropyle (Type III micropyle). This region of the chorion carries nitric oxides synthase which generates nitric oxide to stimulate and guide spermatozoa into the micropyle. However, no one has confirmed it in this and other fish.

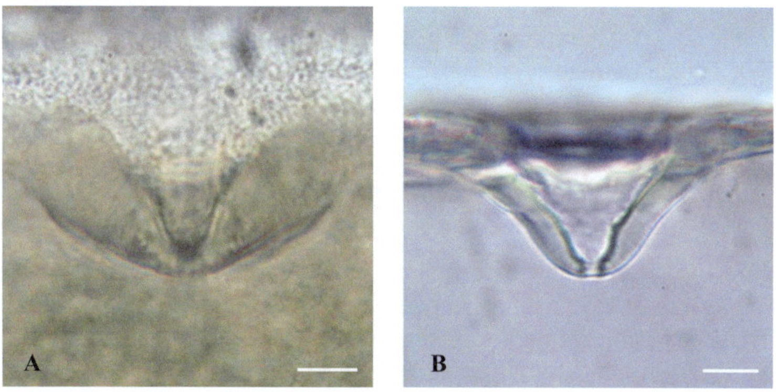

Fig. 5.5 Micropyles of zebrafish (**a**) and loach (**b**). Bar =10 μm. (Reproduced from Yanagimachi et al. (2017))

5.2.5 Sperm Guidance in Viviparous Fish

Not much is known about fertilization in viviparous fish primarily due to difficulty in catching the right moment of fertilization. Fertilization in the sea sculpin, *Alcichthys alcicornis* (Koya et al. 2002) is of particular interest. This fish copulates during the beginning of breeding season. Spermatozoa migrate into female's ovarian capsule where they are then stored during entire the breeding season. When the female ovulates at intervals, spermatozoa enter the egg's micropyle. However, they are unable to fuse with the egg properly, perhaps due to a low (0.41 mM) Ca^{2+} concentration in the "ovarian" fluid. Sperm entry and egg development begin only after eggs are shed in Ca^{2+}-rich seawater (Munehara et al. 1989). Since the smooth lumpsucker (*Aptocyclus ventricosus*) which is close to sculpins has Type II micropyle (Yanagimachi, unpublished data), it is likely that sculpins also have this type of micropyle. Whether MISA is involved in the sperm entry in the micropyle of this fish remains to be investigated.

While this article was in preparation, I tried to observe micropyles of guppy (*Poecilla reticulata*) and platyfish (*Xiphophorus maculatus*). "Mature" ovarian eggs were examined, and neither a distinct depression nor radial grooves in the chorion were found. The guppy has a Type I micropyle with a very wide (~20 μm) canal and sperm head width is around ~1–2 μm (Yanagimachi, unpublished data). It is unknown whether guppy egg has (or needs) MISA around its micropyle.

The female guppy has sperm-storage "micropockets" in the ovarian tissue right above each of growing eggs (Kobayashi and Iwamatsu 2002). It is likely that these micropockets "pour" spermatozoa onto the micropylar region of egg during ovulation. Since guppy spermatozoa move actively in isotonic Ringer's solution (Yanagimachi, unpublished observations), there must be mechanisms preventing spermatozoa from full activation until they are released from the sperm-storage pockets. In the platyfish, spermatozoa are stored in female's ovarian ducts (Potter and Kramer 2000). Many spermatozoa are seen between cells of ovarian pockets and pits, apparently without free space for tail beatings. I speculate that (a) ovarian pocket or pit cells are loosely packed before mating, (b) spermatozoa enter the space between the cells, (c) the space is then "closed" until ovulation begins, and (d) freed spermatozoa then fertilize eggs (Yanagimachi, personal view).

5.2.6 Addendum: Examination of Sperm Entry in Micropyle

The following describes how we examined fish spermatozoa moving into the micropyle. First, we must have media that permit eggs and spermatozoa to maintain their viability for a significant period of time such that we can treat/manipulate them before mixing (insemination). Second, we must have media allowing normal fertilization and embryo development. Eggs and spermatozoa of many fish quickly lose their fertility in water where they spawn. For example, salmon eggs become infertile within few minutes after being shed in freshwater due to "activation" by hypotonic water. Salmon sperm in water move actively for only half minute or so. Thus, Ringer's solution is better than freshwater for examining sperm entry into eggs. In this medium, salmon eggs remain fertile for hours, and spermatozoa can enter eggs normally. However, inseminated eggs must be transferred to freshwater to allow them to activate and develop. In the rice fish (medaka, *Oryzias latipes*), freshwater is the natural media in which fertilization and embryo development take place. However, Ringer's solution is a better medium for insemination and embryo development because eggs remain fertile much longer than in freshwater. In the herring, fertilization and embryo development occur normally in Ringer's solution. Ringer's solution or half strength seawater (1:1 dilution of seawater with distilled water) is the medium of choice. Herring eggs remain fertile much longer than in natural seawater and fertilization, and embryo development can occur normally. Whenever we use a new species of fish for fertilization studies, we must first select (or prepare novel) solutions that allow eggs and spermatozoa to remain fertile for a significant period of time. We then find media supporting normal fertilization and embryo development. In short, seawater or freshwater in which fish normally spawn are not necessarily the best to examine sperm entry into eggs. Media suitable for egg and sperm handling, insemination, and embryo development must be specifically developed for each species. Table 5.1 lists media we have been using for our studies of fish fertilization.

To examine sperm entry in the egg, we prepare the following with slight variations depending on species-specific characteristics of eggs: plastic or glass dishes (~50–90 mm in diameter) (Fig. 5.6a), microscope slides with four spots of vaseline–paraffin–beeswax (9:1:0.5) mixture (Fig. 5.6b) (Yanagimachi 2014), and microscope slides with two small pieces of glass glued to the slide (Fig. 5.6c). The thicknesses of the wax spot and the glass pieces varies dependent on the diameter of the egg to be examined. Eggs with "sticky" chorion are allowed to stick to the bottom of the dish. Rinse eggs with Ringer's solution to remove debris and ovarian fluid. Non-sticky eggs can be stuck to the dish by coating slide or coverslip with poly-l-lysin prior to use (Andoh et al. 2008). While eggs are in Ringer's solution, examine the eggs using a low-power objective lens (×10–20). Look for the micropyle with a proper viewing angle (top or semi-profile view) (Fig. 5.6d). Once it is found, change the objective lens to a water-immersion objective lens (×40–50) for clearer viewing of the micropyle. Replace Ringer's solution with seawater or freshwater (depending on species of fish) and pour freshly prepared sperm suspension from the side of the objective lens. The use of a dense sperm suspension must be avoided in order to see the movement of individual spermatozoa clearly. The micropyle can also been seen after compressing eggs lightly between a slide and coverslip (Fig. 5.6e). Buoyant (floating) eggs may stick to the coverslip coated with poly-l-lysin (Sigma) or Cell-Tak (Discovery Lab, Corning). The use of isolated chorions may be preferable for examining sperm entry in very large eggs (such as trout eggs) (Fig. 5.6f, Yanagimachi et al. 1992).

Sperm entry into the micropyle can also be seen after mounting eggs between a slide and coverslip after lightly compressing eggs under the coverslip. The eggs of

Table 5.1 Media recommended for gamete pretreatment, insemination, and examination of sperm entry in micropyle

Species	Treatment before insemination		Medium that initiates sperm movement	Medium for insemination and examination of sperm entry in micropyle
	Eggs	Sperm		
Herring	Ringer's solution	Ringer's solution or ASW[a]		ASW[a] or Ringer's solution
Flounder	Ringer's solution	Ringer's solution	ASW[a]	ASW[a]
Trout/salmon	Ringer's solution	K-rich Ringer's solution	Ringer's solution or 1/100 Ringer's solution[b]	Ringer's solution or 1/100 Ringer's solution[b]
Pufferfish	Ringer's solution	Ringer's solution	ASW[a]	ASW[a]
Medaka	Ringer's solution	–[c]	Ringer's solution or 1/100 Ringer's solution	Ringer's solution or 1/100 Ringer's solution
Zebrafish/goldfish	Ringer's solution	Ringer's solution[d]	1/100 Ringer's solution	1/100 Ringer's solution

[a]Natural seawater is, of course, a good medium, but artificial seawater (ASW) with known compositions is preferable for analytical studies of fertilization processes
[b]River water or tap water can be used, but for analytical studies of fertilization, 1/100 Ringer's solution is preferable. 1/100 Ringer's solution denotes the mixture of 1 part of Ringer's solution and 99 parts of distilled water
[c]At present, we do not know artificial medium which keeps medaka sperm alive without movement
[d]Zebrafish sperm remain motionless in Ringer's solution without losing their fertility. Goldfish sperm, in contrast, begin to move fairly actively in Ringer's solution, but in 1–2 min they become motionless. They display very active movement when transferred to 1/100 Ringer's solution

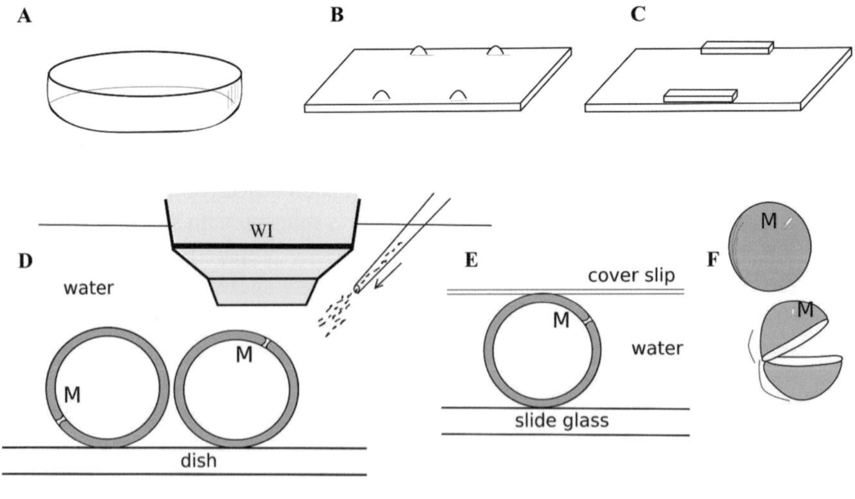

Fig. 5.6 Dish and slides necessary for the examination of egg's micropyle and insemination. For explanations, see the text

some species (like those of the eel) do not stick to glass/plastic even after PLL coating. This method should work. Replacement of the medium (e.g., from Ringer's solution to seawater) as well as insemination can be done by running a medium (or sperm suspension) under the coverslip. The speed of the fluidic flow can be controlled through the proper use of a piece of filter paper applied to an edge of the coverslip.

5.3 Variations in the Structure and Sperm Attractant of Micropyle

Although different types of micropyle in fish eggs have been classified in several different ways (e.g., Riehl and Schulte 1977; Kunz 2004), we classified fish egg micropyles into three (Fig. 5.7) (Yanagimachi et al. 2017).

Type I: manhole-like. The chorion around the outer opening of micropylar canal is "flat" with or without shallow indentation around the outer opening of micropyle. Micropyles of the herring, flounder, pollack, seabream, and mummichog (*Fundulus heteroclitus*) are examples of this type. Here, MISA is present on the surface of chorion immediately around the outer opening of the micropyle.

Type II: funnel-like. The micropyle is like a funnel with a wide, conical mouth and a narrow long stem, with or without an indentation in the chorion around the outer opening of micropyle, which is commonly called the "vestibule." MISA is in the conical mouth of this type of micropyle. The micropyles of the salmon, trout, cod, pufferfish, and medaka are examples of this type.

Type III: *sinkhole like*. The chorion around the micropyle is deeply indented or have radially or spirally arranged grooves. Here, MISA is absent. Micropyles of goldfish, loach, zebrafish, and rosy barb are examples.

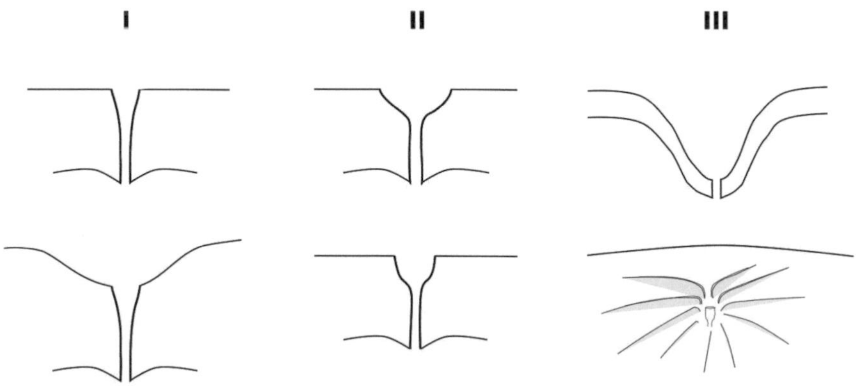

Fig. 5.7 Diagrams of three types of micropyle in fish egg. Type I, manhole-like, with or without shallow depression of chorion around the micropyle. Type II, funnel-like. Type III, sink-hall-like or radial/spiral groove bearing

Table 5.2 Characteristics of egg and spermatozoa of various fish

Eggs Micropyle type	Sperm Presence (+) or absence (−) of CatSper channel	Examples of fish
I (manhole-like)	+	Herring, cod, eel, sea-bream, flying fish, mummichog, puffer
II (funnel-like)	+	Salmon, trout, anchovy, medaka, rockfish, lump sucker, sandfish
III (sinkhole-like or grooves in chorion around micropyle)	−	Zebrafish, carp, goldfish, loach, dace, smelt, stickleback

When these three types of micropyles are mapped on a fish phylogenetic tree, Type I is found in species of relatively ancient clads. Type II micropyle appeared in recently diversified species (>130 million years ago). Type II and III micropyles seem to have emerged independently from the phylogenetic lineage and perhaps as the result of quick adaptation of each species to a particular environment (Yanagimachi et al. 2017). It would be of particular interest to know what types of micropyle the gar (*Xenentodon cancila*) and the bowfin (*Amia calva*) have. Ancestors of these fish diverted from those of all other teleosts 390 million years ago.

Physical (structural) and chemical characteristics of the micropyle and of the nearby region of the chorion must be determined by specific genes which are presently unknown. According to mitogenomic study by Ramsden et al. (2003), Esociformes (pika, pickerels, and mudminnows) is the sister group of Salmonidae (salmon, trout), not of Osmeroidei (smelt, galaxiids, and icefishes). As shown in Table 5.2, salmon has Type II micropyle, whereas smelt has Type III micropyle. Whether there is any relationship between phylogeny of fish and micropyle type of eggs remains to be investigated. Variations in the structure of micropyle in different species may appear vast, but the fundamental structure is the same in all fish (Fig. 5.8). One or a few groups of genes may be responsible in determining the structure of micropyle and its adjacent area.

5.4 Importance of Thigmotactic Movement of Spermatozoa in Fish Fertilization

Spermatozoa swim along interphase of water-air and water-solid. All kinds of spermatozoa exhibit this contact reaction (Lillie 1919; Cosson et al. 2003). Dewitz (1885) seemed to be the first who stated this clearly: "Cockroach spermatozoa maintain contact with slide bodies and free surfaces of liquid; thigmotactic rotations on the surface of the egg finally result in the penetrations of this micropyle of this hard-shelled egg" (cited from Lillie). In fish, this two-dimensional sperm movement (instead of one or three dimensional) highly increases the chances of spermatozoa reaching the vicinity of the micropyle (Cosson et al. 2008; Ishimoto et al. 2018). A

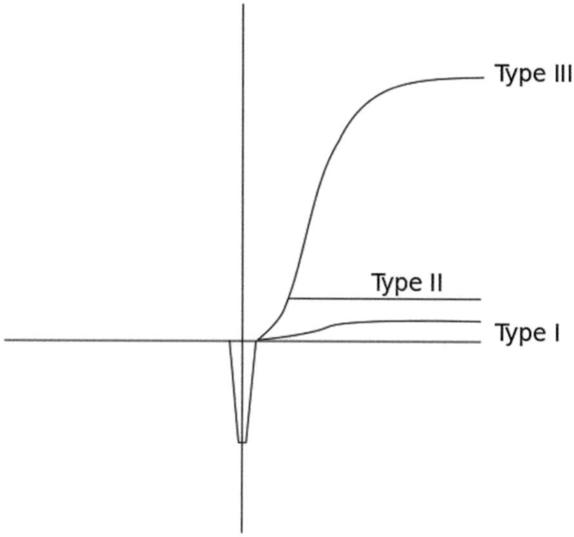

Fig. 5.8 Geometrical comparison of three types of micropyle

deep-depression of chorion around the micropyle or the presence of radially or spirally arranged grooves in the chorion around the micropyle, as seen in goldfish and zebrafish, further enhances thigmotactic sperm approach to the micropyle. In many other fish (e.g., the herring, flounder, salmon, pufferfish), the chorion does not have a large depression or grooves. Instead, they have MISA around the micropyle which acts to "guide" sliding spermatozoa into the canal. Sperm entry into the micropyle (fertilization) is possible without MISA, but it is far less efficient than when it presents.

5.5 Specificity of Sperm Entry in Micropyle

The occurrence of live hybrids between closely related fish is well known, suggesting that sperm–egg interaction in fish is not strictly species-specific. As previously stated, herring spermatozoa are intrinsically immotile in seawater and become motile (activated) by SMIF around the egg's micropyle. Thus far, eggs of all other marine fish I have tested were unable to activate herring spermatozoa.

When eggs of three species of flounders (black flounder (*Pleuronectus obscurus*), starry flounder (*Platichthys stellatus*) and barfin flounder (*Verasper moseri*)) were cross-inseminated in dish, sperm entry into the micropyle of homologous species was always far more efficient than that between heterologous species (Yanagimachi, unpublished observation). Almost all spermatozoa of heterologous species run over the micropyle. However, some entered the micropyle "accidentally." In the case of barfin flounder spermatozoa added to black flounder egg, the head of the barfin flounder spermatozoon was too large to pass through the lower half of the micropylar canal of the black flounder egg. The widths of sperm head and micropylar canal as well as interaction between spermatozoa and egg's MISA seem to be two major factors limiting hybridization in fish. The time and site of spawning as well as behavioral differences between species seem to also be the major factors preventing/reducing hybridization.

I purposely inseminated isolated chorions of fish eggs with sea urchin spermatozoa. Chorions had been kept frozen or stored in 10% formalin. When they were thoroughly rinsed and inseminated with sea urchin spermatozoa in seawater, I observed spermatozoa entering the micropylar canal. The chorions tested include those of the black flounder, rainbow trout, zebrafish, and medaka (Yanagimachi, unpublished data). Here, I want to emphasize again that fish spermatozoa can enter the micropyle by a "random" movement, but their entry is maximized by MISA or sperm-guiding structure of the chorion around the micropyle.

5.6 Sperm Guidance into Micropyle and Possible Role of Sperm Trypsin in Fish Fertilization

Even though fish spermatozoa can enter the micropyle "accidentally," both physical and chemical factors around the outer opening of micropyle seem to maximize their success. In fish with Types I and II micropyles, MISA (micropylar sperm attractant) tightly bound to the chorion surface around the micropyle maximizes the chance of successful sperm entry into the egg. At present, we know little about the molecular basis of chemical interactions between spermatozoa and MISA. Herring which I studied for many years is an excellent material for studying sperm–egg interaction, but this fish may not be readily accessible to many investigators. Trout and salmon, on the other hand, are much more readily accessible in many countries and may be better suited for this kind of study. Fortunately, MISA of salmon eggs has affinity to FITC-conjugated lectins like wheat germ agglutinin (WGA) (see Figs. 2 and 4 of Yanagimachi et al. 2017). After eggs are treated with such lectins, MISA can be separated from other components of the chorion for further purification and chemical identification. Sperm's receptor for MISA can then be isolated and identified.

It is puzzling that fish spermatozoa have trypsinogen or trypsin-like proteinase (Miura et al. 2006; Ohta et al. 2008; Rajapakse et al. 2014). It is membrane-bound (Ohta et al. 2008). Since trypsin (proteinase) is an acrosomal enzyme in other vertebrates, it is natural to raises a question as to why acrosome-less fish spermatozoa have an enzyme like trypsin? Is this merely developmental or evolutional relics of spermatogenesis? However, it is possible that sperm trypsin serves for both sperm entry into the egg and egg activation. At least in the herring, soybean trypsin inhibitor (0.2%), benzamidine (1 mM), and TLCK (1 mM) in insemination medium completely blocks herring fertilization by preventing sperm activation in the micropylar region. Fertilization block is reversible with soybean trypsin inhibitor and benzamidine, but not with TLCK (Yanagimachi et al. 2017). The incidence of fertilization in the flounder is also largely reduced by soybean trypsin inhibitor, TLCK and benzamidine (Yanagimachi, unpublished data). In fact, Miura et al. (2006) and Ohta et al. (2008) previously reported that the rate of fertilization in the eel and bass was reduced by anti-eel-trypsin antibody and serine protease inhibitors. It is very puzzling that FITC-conjugated soybean trypsin inhibitor binds very strongly to the heads of herring, and flounder spermatozoa only after sperm death or spermatozoa were killed by membrane-disrupting agents (Yanagimachi, unpublished data). The site and role of trypsin-like proteinase in fish fertilization remain to be investigated.

5.7 Chemical Basis of Sperm-MISA Interaction in Herring

As stated already, herring spermatozoa are unique in that they are intrinsically immotile in seawater. They may move sporadically for a few seconds, but remain motionless until they come in contact with the chorion surface around the micropyle. It is water current, not the egg, that brings the eggs and spermatozoa in contact. Sperm motility-initiation factor (SMIF) is bound tightly on chorion surface and is not diffusible. It is tightly bound to chorion surface around the micropyle and is not readily removable. It is very specific. Herring spermatozoa are not activated by the chorion around the micropyle of any other fish species thus far tested (flounders, salmon, pufferfish, and sea urchin; Yanagimachi, unpublished). Except for Na^+-free medium (Vines et al. 2002), the only agent that is known to activate herring spermatozoa is Ca^{2+} ionophore which permits Ca^{2+} entry into the cell or allows the release of intra Ca^{2+} release from internal store. However, active movement continues for only a few minutes. In contrast, herring spermatozoa around the micropyle kept moving for hours.

In herring, intra- and extracellular Na^+, K^+, and Ca^{2+} all play essential roles in sperm activation in the micropyle region of egg (Yanagimachi et al. 2017). When we preloaded herring spermatozoa with Fluo-4 AM (fluorescent indicator intracellular Ca^{2+}), we observed some spermatozoa "sparkling" for a second or so. On the chorion around the micropyle, hundreds of "green" spermatozoa were soon observed, many of them passing one by one through micropylar canal of mechanically isolated chorion (see a movie Fig. 7b of Yanagimachi et al. 2013). Rising in intracellular Ca^{2+} is obviously an important component of sperm activation in the micropylar region. Our current working hypothesis of the sequence of events orchestrating activation of herring spermatozoa around the micropyle involves: intra- and extracellular Na^+, K^+, H^+, and Ca^{2+}, adenylyl cyclase, and CatSper Ca^{2+} channels (Fig. 5.9).

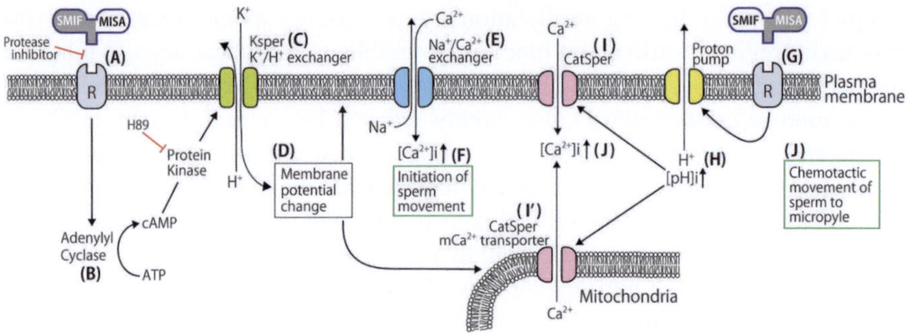

Fig. 5.9 Working model of events leading to sperm entry into herring egg (reproduced from Yanagimachi et al. 2017). Here SMIF and MISA are considered a single molecule with two active domains. Binding of SMIF-MISA to sperm receptor (**a**) activates adenylyl cyclase (**b**) that causes an increase in intracellular cAMP of spermatozoa. This in turn activates a presumptive KSper channel and K^+/H^+ exchanger (**c**), allowing an influx of K^+ from seawater and an elevation of intracellular pH. This causes a marked change in the membrane potential of the plasma membrane and mitochondrial membrane (**d**), resulting in the activation of a Na^+/Ca^{2+} exchanger (**e**) in the plasma membrane. An increase in intercellular Ca^{2+} brings immotile spermatozoa into an active state (**f**). When motile spermatozoa come in contact with SMIF-MISA (**g**) around the micropyle, protein pumps in the sperm plasma membrane are activated. Further increase of intracellular pH (**h**) activates CatSper (**i**) in the plasma membrane or CatSper/Ca^{2+} channel in the outer mitochondrial membrane (**i′**), resulting in further increase of intracellular Ca^{2+} concentration (**j**) by Cainflux from seawater and/or Ca^{2+} efflux from mitochondria. This secondary increase of intracellular Ca^{2+} concentration alters sperm's tail beating pattern, which facilitates sperm entry into micropylar canal

5.8 Herring and Shad

Both herring (*Clupea sp.*) and shad (*Alosa sp.*) are fish of the family Clupeidae. Herring lives and spawns in ocean or brackish water. Shad, like salmon, spends most of their life in sea but spawns in freshwater exclusively. Characteristics of egg and spermatozoa of these two must be quite different. I know that the chorion of herring egg swells rapidly in freshwater, micropyle is closed, and eggs disintegrate (Yanagimachi, unpublished data). The fertilization and embryo development of the Pacific herring (*Clupea pallasii*) occurs best in water with salinity of 16–20 ppt (ocean water has 35–37 ppt) (Griffin et al. 1998). It would be interesting to know if shad eggs can survive, be fertilized, and develop in seawater with a salinity of 16–35 ppt. Perhaps, not. It is interesting to consider why the eggs and spermatozoa of herring and shad are so different in their salinity tolerance? How does shad micropyle look like? Does shad need SMIF and MISA for fertilization, like herring? What sort of genes and gene activities are responsible for such differences in these two fish? These are just a few of interesting questions to be answered.

Acknowledgments I would like to thank Dr. Hiroyuki Munehara (Hokkaido University) for valuable information on internal fertilization in sculpin and other fish. I am grateful to Dr. Michael Ortega (University of Hawaii) for his active interest and giving me valuable advice on fish spawning behavior. His revising English in the original manuscript as well as preparation of many figures are greatly appreciated. I also thank Dr. Carol Vines (University of California), Dr. Tatsuo Harumi (Asahikawa Medical University), and Dr. Atsushi Sugawara (University of Hawaii) for generous assistance in the preparation of figures. Special appreciation is expressed to the University of Hawaii Foundation for providing me with Post-retirement Research Supporting Fund. Acknowledgements are due to Oxford University Press for letting me use figures published in *Biology of Reproduction* 96:780–799 (2017) and to Blackwell Publishing for figures in *Development, Growth and Differentiation* 34:447–461 (1992) and 38:193–202 (1996).

References

Alavi SM, Cosson J (2006) Sperm motility in fishes. (II) Effects of ions and osmolality. A review. Cell Biol Int 30:1–4

Andoh T, Matsubara T, Harumi T, Yanagimachi R (2008) The use of poly-l-lysin to facilitate examination of sperm entry into pelagic, non-adhesive fish eggs. Int J Dev Biol 52:753–757

Butts IA, Johnson K, Wilson CC, Pitcher TE (2012) Ovarian fluid enhances sperm velocity based on relatedness in lake trout, *Salvelinus namaycush*. Theriogenology 78:2015–2019.e1

Carolsfeld J, Tester M, Kreiberg H, Sherwood NM (1997) Pheromone-induced spawning of Pacific herring. I. Behavioral characterization. Hormones Behav 31:256–268

Cherr GN, Morisawa M, Vines CA, Yoshida K, Smith EH, Matsubara T, Pillai MC, Griffin FJ, Yanagimachi R (2008) Two egg-derived molecules in sperm motility and fertilization in the Pacific herrin. Int J Dev Biol 52:743–752

Cosson J, Huitorei P, Gagnon C (2003) How spermatozoa come to be confined to surfaces. Cell Motil Cytoskeleton 54:56–63

Cosson J, Groison AL, Suquet M, Fauvel C, Dreanno C, Billard R (2008) Studying sperm motility in marine fish: an overview of the state of the art. J Appl Ichthyol 24:460–486

Creech M, Arnold EV, Boyles B, Muzinich MC, Mortville C, Bohle DS, Atherton RW (1998) Sperm motility enhancement by nitric oxide produced by t he oocyte of fathead minnow, *Pimephelas promelas*. J Androl 19:667–674

Devigili A, Fitzpatrick JL, Gasparini JC, Ramnarine IW, Pilastro A, Evans JP (2017) Possible glimpses into early speciation: the effect of ovarian fluid on sperm velocity with post-copulatory isolation between two guppy populations. J Evol Biol 30:66–74

Dewitz J (1885) Ueber die vereinigung der spermatozoen mit dem ei. Arch Ges Physiol 37:219–223. (Cited from Lillie, 1910)

Griffin FJ, Vines CA, Pillai MC, Yanagimachi R, Cherr GN (1996) Sperm motility initiation factor is a minor component of the Pacific herring egg chorion. Dev Growth Differ 38:193–202

Griffin FJ, Pillai MC, Vines CA, Kaaria J, Hibbard-Robbins T, Yanagimachi R, Cherr GN (1998) Effects of salinity on sperm motility, fertilization, and development in the Pacific herring, *Clupea pallasi*. Biol Bull 194:25–35

Hay D (1985) Reproductive biology of Pacific herring (*Clupea harengus pallasi*). Can J Fish Aquat Sci 42(Suppl 1):111–126

Hirano T, Johnson DW, Bern HA (1971) Control of water movement in flounder urinary bladder by prolactin. Nature 230:469–471

Ishimoto K, Cosson J, Gaffiney EA (2018) A simulation study of sperm motility hydrodynamics near fish egg and spheres. J Theor Biol 389:187–197

Kobayashi H, Iwamatsu T (2002) Fine structure of the storage micropocket of spermatozoa in the ovary of the guppy *Poecilla reticulata*. Zool Sci 19:545–555

Kobayashi W, Yamamoto TS (1985) Fine structure of the micropylar cell and its change during oocyte maturation in the chum salmon, *Oncorhynchus keta*. J Morphol 184:263–276

Koya Y, Munehara H, Tanano K (2002) Sperm storage and motility in the ovary of the marine sculpin *Alcichthys alcicornis* (Teloeti: Scorpaeniformes), with internal gametic association. J Exp Zool 292:145–155

Kunz YW (2004) Developmental biology of teleost fishes. Springer, Berlin. isbn:978-1-4020-2997-4

Lehnert SJ, Butts IAE, Flannery EW, Peters KM, Heath DD, Pitcher TE (2017) Effects of ovarian fluid and genetic differences on sperm performance and fertilization success of alternative reproductive tactics in Chinook salmon. J Evol Biol 30:1236–1245

Lillie FR (1919) Problems of fertilization, vol 38. University of Chicago Press, Chicago, p 246

Makiguchi Y, Torap M, Kojima T, Pitcher TE (2016) Reproductive investment patterns and comparison of sperm quality in the presence and absence of ovarian fluid in alternative reproduction tactics of masu salmon *Oncorhynchus maso*. Theriogenology 86:2189–2193

Minin AA, Ozerova SG (2008) Spontaneous activation of fish eggs is abolished by protease inhibitors. Ontogenez 39:362–366. (in Russian)

Miura C, Ohta T, Ozaki Y, Miura T (2006) Trypsin is a multiple factor in spermatogenesis. Proc Natl Acad Sci U S A 106:20972–20977

Morisawa M (1994) Cell signaling mechanisms for sperm motility. Zool Sci 11:647–662

Munehara H, Takano K, Koya Y (1989) Internal gametic association and external fertilization in the Elkhorn scuplin, *Alcichthys alcicornis*. Copeia 1989:673–678

Nakanishi S, Iwamatsu T (1989) Ultrastructural changes in micropylar cells and formation of the micropyle during oogenesis in the medaka, *Oryzias latipes*. J Morphol 202:339–349

Nakanishi S, Iwamatsu T (1994) Ultrastructural changes in micropylar and granulosa cells during in vitro oocyte maturation in the medaka, Oryzias latipes. J Exp Zool 270:547–555

Oda S, Igarashi Y, Ohtake H, Sakai M, Shimizu N, Morisawa M (1995) Sperm-activating proteins from unfertilized eggs of the Pacific herring, *Clupea pallasii*. Dev Growth Differ 37:257–261

Oda S, Igarashi Y, Manaka K, Koibuchi N, Sakai-Sawada M, Sakai K, Morisawa M, Ohtake H, Shimizu N (1998) Sperm-activating protein obtained from the herring eggs are homologous to trypsin inhibitors and synthetized in follicle cells. Dev Biol 204:55–63

Ohta H, Takano K (1982) Ultrastructure of micropyle cells in the pre-ovulatory follicles of Pacific herring, *Clupea pallasi* velennciennes. Bull Fac Fish Hokkaido Univ 33:57–64

Ohta H, Teranishi T (1982) Ultrastructure and histochemistry of granulose and micropylar cells in the ovary of the loach, *Misgurnus anguillicaudatus*. Bull Fac Fish Hokkaido Univ 33:1–8

Ohta T, Miura C, Shimizu Y, Mizuno K, Matsubara H, Tanaka H, Miura T (2008) The role of trypsin-like protease in sperm on fertilization in fish. Cybium 32(2 Suppl):166

Pereira-Santos M, Shimoda E, de Andrade AFC, Silva LA, Fujimoto T, Senhorini JA, Yasui GS, Nakaghi SO (2017) Grooves around the micropyle decreases the insemination dose in fish. Zygote 25:731–739

Pillai MC, Shields T, Yanagimachi R, Cherr GN (1993) Isolation and partial characterization of the sperm motility initiation factor from eggs of the Pacific herring, *Clupea pallasii*. J Exp Zool 265:336–342

Potter H, Kramer CR (2000) Ultrastructural observations on sperm storage in the ovary of platyfish, *Xiphophorus maculatus* (Teleosti: Poeciliidae): the role of the duct epithelium. J Morphol 245:110–129

Rajapakse S, Ogiwara K, Takahashi T (2014) Characterization and expression of trypsinogen and trypsin in medaka testis. Zool Sci 82:840–848

Ramsden SD, Brinkmann H, Hawryshyn CW, Taylor JS (2003) Mitogenetics and the sister of salmonidae. Trends Ecol Evol 18:607–610

Riehl R, Schulte E (1977) Licht-und elektronenmikroskopoische untersuchungen an den micropulen ausgewehlter susswasser -teleostee. Arch Fiserwiss 28:95–107

Vines CA, Yoshida K, Griffin FJ, Pillai MC, Morisawa P, Yanagimachi R (2002) Motility initiation in herring sperm is regulated by reverse sodium-calcium exchange. Proc Natl Acad Sci U S A 99:2026–2031

Yanagimachi R (1957a) Some properties of sperm-activating factor in the micropyle area of the herring egg. Annot Zool Japon 30:114–119

Yanagimachi R (1957b) Studies of fertilization in *Clupea pallasii*. III. Manner of sperm entrance into the egg. Zool Mag (Tokyo) 66:226–233

Yanagimachi R (2014) Germ cells and fertilization: why I studied these topics and what I learned along the path of my study. Andrology 2:787–793

Yanagimachi R, Kanoh Y (1953) Manner of sperm entry in the herring egg, with special reference to the role of calcium ions in fertilization. J Fac Sci Hokkaido Univ Series VI Zool 11:487–498

Yanagimachi R, Cherr GN, Pillai MC, Baldwin JD (1992) Factors controlling sperm entry into the micropyles of salmonid and herring eggs. Dev Growth Differ 34:447–461

Yanagimachi R, Cherr GN, Matsubara T, Andoh T, Harumi T, Vines C, Pillai M, Griffin F, Matsubara H, Weatherby T, Kaneshiro K (2013) Sperm attractant in the micropyle region of fish and insect eggs. Biol Reprod 88(2):47, 1–11

Yanagimachi R, Harumi T, Matsubara H, Yan W, Yuan S, Hirohashi N, Iida T, Yamaha E, Arai K, Matsubara T, Andoh A, Vines C, Cherr GN (2017) Chemical and physical guidance of fish spermatozoa into the egg through the micropyle. Biol Reprod 96(4):780–799

Yeates SE, Diamond SE, Emerson BC, Holt WW, Gage MJ (2013) Cryptic choice of conspecific sperm controlled by the impact of ovarian fluid on sperm swimming behavior. Evolution 67:3523–3536

Yi X, Yu J, Ma C, Dong G, Shi W, Li L, Luo L, Sampath K, Ruan H, Huang H (2018) The effector of Hippo signaling, Taz, is required for formation of the micropyle and fertilization in zebrafish. BioRxiv. https://doi.org/10.1101/319475

Yoshida K, Inaba K, Ohtake H, Morisawa M (1999) Purification and characterization of prolyl enodopeptidase from Pacific herring, *Clupea pallasi*, and its role in the activation of sperm motility. Dev Growth Differ 41:217–225

Part III
Basic Knowledge of Female Gametes and Sperm–Egg Interaction in Aquatic Animals

Chapter 6
Structure of Mature Oocytes

Oliana Carnevali, Isabel Forner-Piquer, and Giorgia Gioacchini

Abstract In aquaculture, the growth and maturation of oocyte associated with the quality of egg represent a crucial point for reproductive success and therefore, for breeding in aquaculture. Knowing the components that make up an oocyte, their natural distribution and abundance in the various stages of maturation turns out to be of fundamental importance for the understanding of the oogenesis process. This chapter describes the different components present within the oocyte and their origin.

The components have been described in relation to the different phases of oocyte maturation and in several experimental models with different reproductive strategies. A focus was also addressed to the endocrine control of oocyte growth and maturation and to the detrimental effects of contaminants with hormonal activity on oocyte growth and maturation.

Keywords Egg envelope · Vitellogenesis · Yolk · Cortical alveoli

6.1 Introduction

Fish have developed a variety of reproductive strategies, where the predominant pattern is the gonochorism (two separate sexes: male and female, eggs and sperm housed in two separate sexes). However, some species display hermaphroditism during their life span, it means, they may produce mature gametes of both sexes and undergo one or more sex changes.

Basically, there are two types of hermaphroditism:

1. Sequential or successive: The most current, when the organisms are male for the first part of their lifetime and then, sex inversion occurs to become female (pro-

O. Carnevali (✉) · I. Forner-Piquer · G. Gioacchini
Dipartimento di Scienze della Vita e dell'ambiente, Università Politecnica delle Marche, Ancona, Italy
e-mail: o.carnevali@univpm.it

M. Yoshida, J. F. Asturiano (eds.), *Reproduction in Aquatic Animals*,
https://doi.org/10.1007/978-981-15-2290-1_7

tandrous, i.e., *Sparus aurata, Pagrus pagrus*) or vice versa, from female to male (protogynous, i.e., labrids) (Jalabert 2005).

2. Simultaneous/synchronous hermaphroditism: It is characterized by the simultaneous presence of testicular and ovarian tissues in the same gonad.

Teleosts are mostly oviparous with external fertilization; however, viviparism (seahorses, pipefishes) and ovoviviparism (guppies family, *Poecilia reticulata*) are existing even if less frequent.

Teleosts can also be divided among pelagic and demersal spawners. Pelagic spawners are characterized by the release of larger number of eggs with smaller size and lesser yolk. Differently, the demersal/benthonic spawners tend to produce bigger eggs with higher quantity of yolk (Olivotto et al. 2017). Furthermore, the oocyte envelope among pelagic and demersal eggs is different depending on the characteristics of the surrounding environment (Berois et al. 2011; Jiang et al. 2010).

Three modes of ovarian development have been identified in teleost, according to the oocyte growth pattern (Khan and Thomas 1998; Lubzens et al. 2010):

1. Synchronous: All oocytes present in the ovary are in the same stage of development. Generally, this strategy is found in species that spawn once a year or once in their lifetime and then, die (i.e., *Salmo trutta*, *Oncorhynchus kisutch)*.
2. Group—synchronous: Mainly two populations or clutches of oocytes at different developmental stages are simultaneously present during the reproductive season. Basically, several egg layings happen during a spawning season (Asturiano et al. 2000), where the smaller oocytes will be recruited and spawned the following breeding season (Murua and Saborido-Rey 2003) (i.e., *Seriola dumerili, Umbrina cirrosa, Oncorhynchus mykiss*). This is the most common condition among teleost (Wallace and Selman 1981).
3. Asynchronous: All oocyte stages are present within the same ovary, without a dominant population. The eggs are usually spawned along different batches (i.e., *Sparus aurata, Pagrus*, *Danio rerio*, *Fundulus heteroclitus*, seahorses).

The development of an egg is a multifaceted process. Upon ovulation, the egg is self-sufficient to protect and sustain the developing embryo until hatching, with very little take-up from the surrounding environment. Therefore, all the contents of an egg, whether in the form of mRNA, nutrients, or hormones, must be incorporated during its development as an oocyte. The production of a good-quality egg relies upon the correct progression of oogenesis, and this coordinated assembly is controlled by an interplay of endocrine and intra-ovarian factors, both paracrine and autocrine (Tyler and Sumpter 1996). During oogenesis, fish oocyte autonomously makes most of the components of the machinery for DNA and protein synthesis, as well as "maternal" mRNAs, needed immediately after fertilization. However, non-ovarian tissues, mainly the liver, produce typical egg constituents such as yolk proteins, cholesterol, lipoproteins, some zona radiata (ZR) proteins and vitamins. The assembly of the developing egg is therefore a fascinating example of interplay among oocytes and extra-oocyte tissues, finely regulated by endocrine paracrine factors. However, several cultured species exhibit some degrees of reproductive malfunction since that interplay among the reproductive axis is altered, requiring hormonal therapies (Mylonas and Zohar 2007). For that reason, oogenesis as well as spermatogenesis must be fully understood to improve the reproductive processes of cultured species.

This chapter reviews the current knowledge on teleost mature oocyte and the processes involved in oocyte maturation. It briefly describes the different stages of the oocyte growth, the processes involved in the oocyte maturation, the morphological aspects of the mature eggs, as well as the endocrine processes to achieve the complete oogenesis.

6.2 Oocyte Growth

To date, almost all the teleosts studied present the same basic pattern of oocyte growth. Generally, the oocyte growth can be classified into six steps (Tyler and Sumpter 1996) that will be shortly detailed below. However, since the aim of the present chapter is to describe the structure of the mature oocyte, the formation of the primordial germ cells (PGC) and their transformation into oogonia will be omitted.

6.2.1 Oogonia

Each oogonia is a diploid cell, with two copies of each chromosome. For each chromosome, a sister chromatid pair from the mother and from the father are present. Afterward, the granulosa and theca cells, together constituting a real primary follicle, will surround this primary oocyte.

6.2.2 Primary Oocyte

At the beginning of this stage, the oocyte contains a meager amount of cytoplasm and a large centrally located nucleus or germinal vesicle (GV), which will increase in size and where many nucleoli will appear (Wallace and Selman 1981). Another characteristic of this stage is the formation of lampbrush chromosomes (Wallace and Selman 1981) and the intense RNA synthesis (Tyler and Sumpter 1996).

The primary oocyte growth is initiated with the first meiotic division. During the prophase, the nuclear material (paternal and maternal chromatids) experiences several structural changes to finally be arrested in the diplotene stage, where each chromatid pair will be packed in dense complexes (Le Menn et al. 2007).

Additionally, Balbiani body develops from material aggregations and migrates to the periphery to be dispersed into smaller fragments. The Balbiani body will be a key factor for later embryogenesis (Elkouby and Mullins 2017).

At the onset of oocyte primary growth, the oocyte extends microvilli. These microvilli will elongate from the oocyte surface toward the granulosa cells (GC) and will become indispensable for the exchanges of essential macromolecules for oocyte growth. At this stage, the oocyte starts to build in the space between the oolemma and the granulosa cells (GC), an extra-oocyte matrix, the zona radiata (ZR), starting from the base of these microvilli (Le Menn et al. 2007; Selman et al. 1993). The term "zona radiata" is due to the structure of this protective layer, which perfectly describes the radial channels, formed by the deposition of the ZR around the microvilli.

The GC and theca cells (TC) highly proliferate, constituting a regular epithelium around the oocyte. Furthermore, during this stage, the oocytes will gain considerably size concomitant with the reduction of nucleus–cytoplasmic ratio (Wallace and Selman 1981).

6.2.3 *Cortical Alveolus Stage*

The cortical granules are the first structures to appear in the oocyte.

As they increase in size and number, they move forming a peripheral ring. Contemporarily with cortical alveoli formation and ZR deposition, accumulation of lipid droplets occurs.

6.2.4 *Vitellogenesis*

This stage is characterized by the uptake of vitellogenin (VTG) into the oocytes. VTG is synthesized mainly in the liver, under the regulation of estradiol (E_2) (Babin et al. 2007), and circulates as a large phosphoglycolipoprotein dimer. The VTG, once in the plasma, reaches the oocyte by passing from the theca capillaries to the granulosa layer, coming at the oocyte surface through the pore canals of the ZR. Then, it is sequestered by a receptor-mediated endocytosis in specialized areas of the membrane called "coated pits." Coated pits pinch off into the peripheral cytoplasm and give rise to coated vesicles that fuse with lysosome-like multivesicular bodies. In those multivesicular bodies, proteolytic cleavage of VTG into yolk proteins occurs by aspartic endopeptidases named cathepsins (Carnevali et al. 1999a, b). The VTG after its proteolytic cleavage is stored as yolk, constituting an essential element for the embryo development and the major food supply for the embryos. However, the accumulation of yolk for the oncoming embryo is an important metabolic effort for the maternal organism (Jalabert 2005), and depending on the species, the vitellogenesis shows variable durability, from 1 day to several weeks. At the end of the vitellogenic growth, the ZR is fully formed.

During the vitellogenesis, the ovary is reaching an important volume before the ovulation. Generally, the vitellogenesis is the principal factor for the growth of the oocyte. However, in other species, rather than vitellogenesis, the hydration at the end of the egg growth is the main responsible for the enlarged size.

At this point, the oocytes are not fertilizable, and the nucleus is still in diplotene stage and centrally located. However, at the end of the vitellogenesis, the nucleus will move to the periphery, indicating the start of the final maturation.

6.2.5 *Oocyte Maturation*

This stage is accurately described in the next sections.

Briefly, at the end of growth phase, maturation is an essential process before ovulation to render the oocyte fertilizable, which basically consists of a "nuclear maturation" characterized by the resumption of meiosis, the breakdown of the germinal vesicle (GVBD), the chromosome condensation, and the formation of the first polar body (Nagahama and Yamashita 2008); and a "cytoplasmic maturation" characterized by the reduction or stopping of endocytosis, yolk globule coalescence, cytoplasm clarification, and the increase of the oocyte volume due to hydration process (Jalabert 2005) (Fig. 6.1).

6.2.6 *Ovulation*

Once matured, the egg is ovulated surrounded by the ZR (or chorion). In the mature eggs, we can find the micropyle, the structure through which the spermatozoa will enter into the egg for its fertilization.

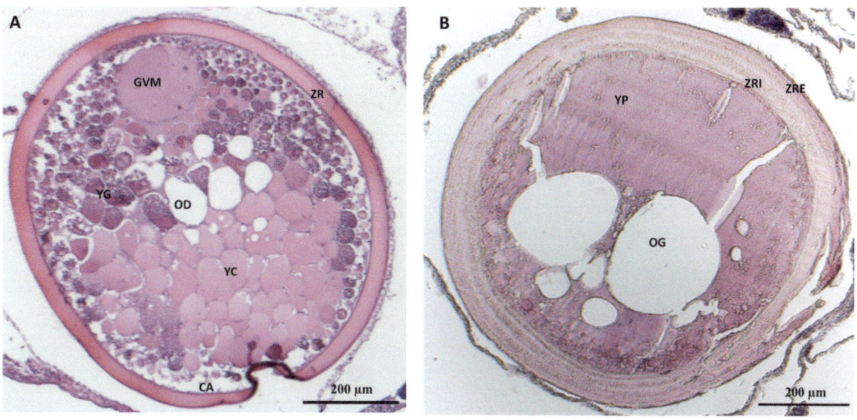

Fig. 6.1 (**a**) Swordfish oocyte in maturation; (**b**) swordfish mature oocyte. *GVM* germinal vesicle migration, *ZR* zona radiata, *ZRI* zona radiata interna, *ZRE* zona radiata externa, *OD* oil droplets, *OG* oil globule, *CA* cortical alveoli, *YG* yolk globules, *YC* yolk coalescence, *YP* yolk platelet

The separation from the follicle layers and the expulsion of the egg are triggered by several signals including prostaglandins, gonadotropins, and proteases, among others, regulating the dissolution of the extracellular matrix of the follicular layers.

6.3 Structure of the Mature Oocyte

When fish oocyte maturation completes the process, it results in a fertilizable female gamete or egg (Guraya 1986).

In this section, structural changes occurring to the cellular component (egg envelope, cortical alveoli, vitellogenin vesicles, oil droplets, and nucleus) during oocyte maturation will be described (Fig. 6.2).

6.3.1 Egg Envelope

Two major cell layers, an outer TC layer and an inner GC layer, and an acellular envelope, the ZR, surround teleost oocytes, as in other vertebrates.

6.3.1.1 Cellular Components

Ovarian follicular cells play a significant role during teleost oogenesis supporting and controlling the process (Nagahama and Yamashita 2008). The ovarian follicles comprise two main somatic cell layers, an inner GC layer and an outer TC layer,

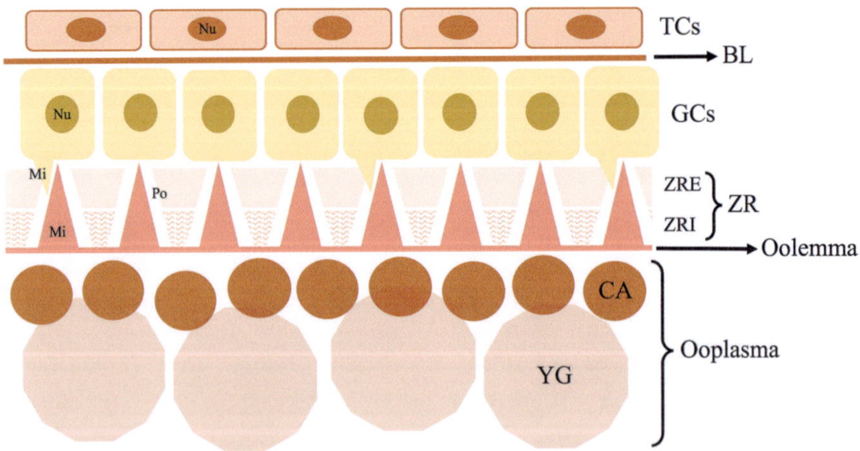

Fig. 6.2 Structure of a mature follicle. *TCs* theca cells, *BL* basal lamina, *GCs* granulosa cells, *ZR* zona radiata, *ZRI* zona radiata interna, *ZRE* zona radiata externa; *Mi* microvilli, *Po* porus, *Nu* nucleus, *CA* cortical alveoli, *YP* yolk platelet

separated by a rather thick basement membrane consisting of a series of membranous layers. Within the oocyte growth, the GCs undergo morphological changes. During the maturation phase, they change from squamous to cuboidal or columnar shape. The TC layer is composed of fibroblast-like cells, collagen fibrils, blood capillaries, and a small number of special thecal cells (ST cells), which are larger than the ordinary TCs and contain a centrally located nucleus.

During early maturation, dramatic structure alterations become apparent in the GC as maturation proceeds, resulting in a drastic reduction of microvilli and in a reduction of macromolecular exchange with the oocyte. During nuclear migration, the GCs become stretched, and some wide intercellular spaces appear between adjacent cells due to the loss of intracellular GAP junction (Matsuyama et al. 1991).

6.3.1.2 Zona Radiata (ZR)

Vertebrate eggs are surrounded by an acellular envelope, known by several synonyms as the egg envelope, vitelline membrane, zona pellucida, or ZR (Litscher and Wassarman 2018).

This envelope consists mainly of 2–4 major proteins, the zona radiata proteins (ZRP) (Modig et al. 2006, 2007). In fish, the synthesis of ZRPs occurs in the liver, or in the ovary or both (Conner and Hughes 2003; Hyllner et al. 1995, 2001). In any case, the ZRP precursors synthesized by the liver are transported to the ovary, where they are incorporated by the oocyte via endocytic-coated pits and modified with oligosaccharides within the Golgi apparatus. Then, they reshuffled via vesicles to the oocyte plasma membrane and deposited in the innermost layer of the thickening ZR (Abraham et al. 1984).

ZR continues differentiation throughout the growth of the oocyte and becomes highly ordered and architecturally complex (Le Menn et al. 2007; Modig et al. 2007). When the ZR is fully formed at the end of vitellogenic growth, it is generally composed of a thin outer layer and a thick inner layer (Hyllner et al. 1995). ZR internal and external architecture differs among fish species and does not seem to be similar even in species of one genus (Kaviani et al. 2013). The outer layer, the ZR externa (ZRE), is the first to be laid down. The inner layer, the ZR interna (ZRI), is deposited in the extra-oocyte matrix, gradually displacing the ZRE toward the GCs. Unlike the amorphous structure of the ZRE, the ZRI has reticulated deposits with twisted arrangements, giving the arched appearance of a polymerized fibrillar secretion deposited in a matrix. After germinal vesicle breakdown (GVBD), the ZRI becomes more compact, and there is a loss of all the microvilli connecting oocyte to GC from the pore canals. Prior to ovulation, the ZR becomes denser, striated, without pore canals, and spaced from GCs.

ZR plays an important role during fertilization. Glycoproteins of ZRE apparently have an affinity for spermatozoa and guide the single spermatozoa into the micropyle to reach the egg cell. After the egg is activated by the spermatozoa, the micropyle closes preventing polyspermy (Amanze and Iyengar 1990; Hart 1990). ZR also provides protection against mechanical disturbances for the developing embryo during the first fragile period (Jiang et al. 2010; Riehl and Patzner 1998). Finally, ZR possesses antimicrobial and antibactericidal functions, protecting the egg (Modig et al. 2007).

In some teleosts, such as sturgeon and medaka, outside the ZRE, follicular cells synthesize an additional layer called extrachorion. This enables females to attach the eggs in a position with optimal environmental conditions (Murray et al. 2013; Zelazowska 2014; Zelazowska 2010).

6.3.2 *Yolk Mass (YM) Components*

Teleost egg contains a substantial yolk mass that serves as a protein- and lipid-rich source of nutrients for embryonic development and larval growth. Two distinct types of yolk components can be found in teleost oocyte: vitellogenin vesicles (or globules) and oil droplets, which are composed by extra-ovarian materials selectively incorporated, modified, and stored by the oocyte during secondary growth.

The changes occurring in the cytoplasm during maturation are collectively referred as "cytoplasmic maturation," which includes fusion of yolk granules (or globules) and clarification of the ooplasm, coalescence of neutral lipid droplets, if present, and hydration of the ooplasm leading to an increase in oocyte volume (Cerdà et al. 2013; Finn 2007). The rise of oocyte volume is small in most freshwater teleosts (~1- to 3-fold) but become more evident in marine species producing pelagic eggs (~3- to 8-fold), and it is due to a significant increase in oocyte water content, typically ranging from 76 to 93% in pelagic eggs but only from 54 to 76% in eggs of benthophil teleosts (LaFleur et al. 2005). The increased water content may serve as an adaptation to the hyperosmotic conditions of seawater, providing a water reservoir in the embryos until osmoregulatory organs develop, improving oxygen exchange by increasing the surface area of the eggs or assisting in dispersal of the eggs due to their higher buoyancy in seawater (Chauvigné 2011).

6.3.2.1 Yolk Vesicles

A large portion of the yolk mass is represented by components derived from vitellogenins (VTGs). VTGs are phospholipoglycoproteins found in the blood of all oviparous vertebrate females during vitellogenesis. A number of studies indicated that teleosts have at least three different VTGs: VTGAa, VTGAb, and VTGC, and all of them are incorporated in the oocyte. Their amino acid sequence can be divided into several domains located in linear fashion: NH_2-heavy chain of lipovitellin (LvH)-phosvitin (polyserine domain)-light chain of lipovitellin (LvL)-β′component (β′c)-COOH. VtgC is similar to VtgAs but lacks the phosvitin domain. Proteolytic cleavage of VTG into yolk proteins occurs by aspartic endopeptidases named Cathepsins (Amano et al. 2010; Carnevali et al. 2006; Selman et al. 1993). Most evidences indicate that protein uptake into oocytes continues during the maturation phase, but abruptly stops at the germinal vesicle breakdown (Wallace 1985), when the oocyte initiates the final step toward a non-sequestering, relatively impermeable egg.

During oocyte maturation, especially in pelagophil species, yolk proteins are further cleaved, generating free amino acids (FAAs) and small peptides used for oocyte hydration. During this process, the yolk globules fuse among them forming a central mass of liquid yolk, and the internal crystalline structures disassemble conferring to the mature oocytes their characteristic transparency (Carnevali et al. 2006). The hydrolysis of VTGAa- and VTGAb-derived yolk proteins occurs concomitantly by the action of cathepsin L and/or cathepsin B that are specifically activated during maturation (Babin et al. 2007; Carnevali et al. 1999a, b, 2001). In most species, LvHb (from VTGB) is dissociated into two monomers, LvHa (from VTGA), phosvitins, and β-component are extensively degraded to produce FAA, and LvLs are partially hydrolyzed. Thus, the degradation of the LvHa fundamentally contributes to the pool of FAAs for oocyte hydration, whereas the LvHb and LvL remain stored in the oocyte as a source of nutrients for further embryonic development (Ohkubo et al. 2006). VTGC-derived yolk proteins are only slightly hydrolyzed during oocyte maturation, and therefore, they do not seem to be major contributors to the hydration of the oocyte (Amano et al. 2010). In benthophil species, however, only the LvHa is partially cleaved, while phosvitins are fully degraded, and altogether results in a more modest increase of the FAA pool (Finn 2007; LaFleur et al. 2005). The increase in the osmotic pressure in the ooplasm of maturing oocytes due to the increase of FAAs is the major driving force for oocyte hydration in marine teleosts (Skoblina 2010). However, a specific role in teleost oocyte hydration was assigned to aquaporins (AQPS), which are members of a superfamily of integral membrane proteins that act as molecular pores or channels to facilitate the rapid and selective flux of water or other small solutes across biological membranes. Aquaporins transport water along an osmotic gradient, where the direction and force of water flow is determined by the orientation and steepness of the gradient, respectively (Cerdà et al. 2013).

In freshwater teleosts spawning demersal eggs, the VTG-derived yolk proteins can represent 80–90% or more of the dry weight of the ovulated eggs (Le Menn et al. 2007). In addition to the proteins source, VTG is an important carrier of lipids into growing oocytes of species that spawn demersal eggs lacking a prominent oil globule (Tyler and Sumpter 1996; Wallace 1985). Teleost VTG contains ~20% lipid by weight, and about 60–80% of this lipid can be phospholipids such as the polar lipids, phosphatidylcholine, or phosphatidyl-ethanolamine, which are typically rich in polyunsaturated fatty acids and are important membrane components in all organisms (Johnson 2009; Silversand and Haux 1995).

6.3.2.2 Oil Globules

In teleosts spawning pelagic eggs, especially marine perciforms, full-grown oocytes contain oil droplets which occupy up to half or more of ooplasm volume (Lønning et al. 1988). A common feature of oocyte "cytoplasmic maturation" in species with prominent oil droplets in their ooplasm is the coalescence and fusion of these droplets to form one or few larger oil globules that are present at the vegetal pole in the ovulated egg (Kagawa et al. 2011, 2013; Lubzens et al. 2010). In contrast with the VTG-associated lipids, oil globules contain mainly neutral lipids such as triglycerides and wax or steryl esters. These neutral lipids are rich in monounsaturated fatty acids that, in fishes, preferentially serve as metabolic energy reserves (Modig et al. 2006). Thus, it appears that VTG mainly transports structural lipids and essential fatty acids into growing oocytes to support embryo tissue growth, whereas neutral lipids are taken up by other means and stored as isolated oil inclusions for meeting the energy demands during embryogenesis until the last stages of endogenous larval nutrition (Wiegand 1996).

6.3.2.3 Other Yolk Components

Although yolk proteins and lipids constitute the bulk of material in fish eggs, there are other molecules that are accumulated during oogenesis and could be produced by the oocyte itself or derived from sources outside of the oocyte, such as vitamins and metals that are equally required for enzyme activities. Their content in fish eggs is related to egg and offspring viability (Emata et al. 2000; Izquierdo et al. 2001; Kjørsvik et al. 1990; Palace et al. 2006).

Among the vast number of maternal RNAs present in the yolk egg are those encoding for hormones, hormone receptors, and binding proteins. Additionally, sex steroid hormones, corticosteroids, and thyroid hormones that may function in very early development, before the zygote is capable of its own synthesis are present (Li et al. 2012; Paitz et al. 2015; Vassallo et al. 2014).

6.3.3 Cortical Alveoli

In addition to the YM, ooplasm is characterized by the presence of a number of cortical alveoli containing materials produced by the oocyte itself during the primary growth phase. Cortical alveoli are specialized membrane-limited vesicles of variable size, originating from Golgi apparatus and later displaced toward the oocyte cortex, containing a complex protein mixture including hyaline, hyosophorins, lectins (i.e., C-type lectin, rhamnose-binding lectin), and proteases including alveolin (Finn 2007; Gallo and Costantini 2012; Shibata et al. 2012; Viana et al. 2018). Their relative depth increases as they grow in size and number and then decreases during later vitellogenic growth as yolk accumulates centripetally in the oocyte. Their content is released to the egg surface after fertilization as part of the "cortical reaction." This release leads to the restructuring of ZRPs forming the chorion preventing polyspermy (Selman et al. 1993).

6.3.4 Nucleus

During maturation phase, nucleus undergoes several dramatic changes including resumption of meiosis (completion of the first meiotic division followed by progression to metaphase II), the germinal vesicle migration (GVM), chromosome condensation, assembly of the meiotic spindle, the germinal vesicle breakdown (GVBD), and the formation of the first polar body.

Oocyte in late vitellogenesis is arrested in prophase I of meiosis with a central, large GV containing many nucleoli and a very high transcriptional activity. In an oocyte in maturation, the centrally positioned GV must be repositioned to the periphery in order to produce the unequal cell division resulting in the first polar body and the secondary oocyte. The exact mechanism of GVM is not still fully understood, but it probably involves changes in the oocyte cytoskeleton and may also use the inherent polarity and density differences in ooplasm to allow "buoyancy" and to propel the GV toward the oocyte periphery (Lessman 2009; Lessman et al. 2007).

Concomitantly to the migration, dissolution of the outer GV membrane (GVBD), which is a prerequisite for fertilization, occurs. Disassembly or dissolution of the nuclear envelope is an integral part of the higher eukaryote cell division process and is generally involving changes in Lamin phosphorylation resulting in vesiculation of the nuclear membranes (Yamaguchi et al. 2006).

During the period of meiotic arrest, the chromatin is in a de-condensed conformation, but during oocyte maturation, it changes to a condensed conformation, which is accompanied by the global transcriptional repression that precedes GVBD.

6.4 Hormonal Control of Maturation Process

The endocrine regulation of oocyte maturation has been extensively investigated in fish (Nagahama et al. 1993, 1995; Nagahama and Yamashita 2008; Ramezani-Fard et al. 2013). Studies using well-characterized in vitro systems as well as in vivo systems have revealed that oocyte maturation in fish is regulated by three mediators: the gonadotropin luteinizing hormone (LH), maturation-inducing hormone (MIH), and maturation promoting factor (MPF) (Fig. 6.3).

In many teleosts, LH blood levels begin to rise at or near the time when the vitellogenic growth is completed (Khan and Thomas 1998). Although LH is the primary endocrine factor responsible for the induction of oocyte maturation, this hormone does not seem to act directly on the oocyte to induce meiotic maturation. LH binds its receptor on follicular cells and stimulates a sequence of events, including increased

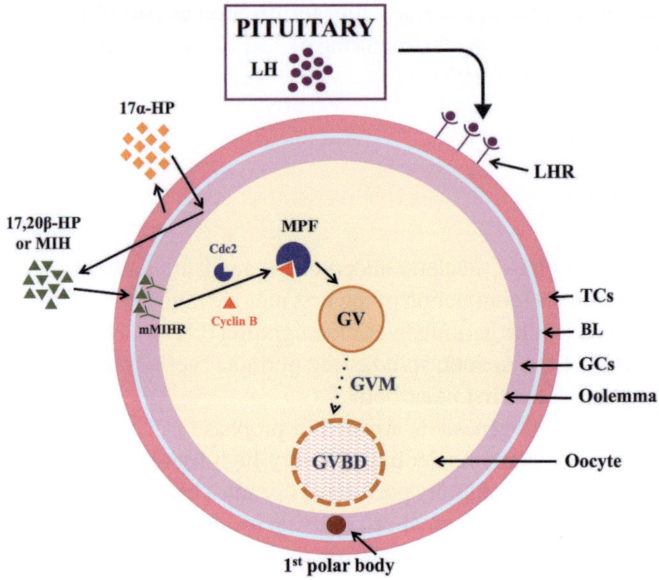

Fig. 6.3 Endocrine paracrine control of follicle maturation in teleosts. A teleost follicle contains an oocyte (O), one layer of granulosa cells (GCs) and one layer of theca cells (TCs) separated by a basement membrane (BL). As a consequence of the LH binding with its receptor (LHR) in theca cells membrane, the synthesis of 17alpha-hydroxy-progesterone (17α-HP) is induced. This hormone diffuses into the GCs where it is converted to 17,20-dihydroxy-4-pregnen-3-one (17,20β-HP), one of the common forms of MIH. MIH binds its membrane receptor (mMIHR) located in oolemma and let the activation of the maturation promoting factor (MPF) composed by cyclin B and Cdc2. Subsequently, MPF induces GV migration (GVM), germinal vesicle breakdown (GVBD), chromosome condensation, and the extrusion of the first polar body. (The model is based on Nagahama et al. 1995 and Nagahama and Yamashita 2008)

lhr transcripts, and an LH-induced switch in the follicular steroidogenic pathway from E_2 to progestin production (Nagahama and Yamashita 2008). In particular, the cellular machinery that induces the resumption of meiosis consists on changes of sex steroids levels. In thecal cells, the synthesis of 17α-hydroxyprogesterone (17α-HP) becomes predominant respect to androgens while in granulosa cells, the 20β- or 20β,21-hydroxylation of 17α-HP leads to the synthesis of maturation-inducing hormone (17α,20β-HP or MIH).

The second step of the hormonal control of oocyte maturation consists on the capacity acquisition to respond to the MIH signal (termed "oocyte maturational competence," OMC) concomitant with the expression of the membrane progesterone receptor (mMIHR) in the oocyte (Patiño et al. 2003). Finally, the last step of this hormonal cascade involves the MIH-induced activation of maturation-promoting factor (MPF) which trigger the meiotic arrest in prophase I (Nagahama and Yamashita 2008). Possibly due to a decline in oocyte cAMP levels, this transduction pathway activates cytosolic MPF, which consists of cyclin B and Cdc2

(Nagahama 1997). Active MPF directly causes the resumption of meiosis (Ramezani-Fard et al. 2013).

Although MIH is generally considered a necessary and sufficient mediator of LH-induced meiotic resumption in teleost ovarian follicles, a number of other factors may also mediate or modulate this process. For instance, insulin-like growth factors (IGF) (Irwin et al. 2012), activins (Ge 2000), epidermal growth factor (EGF), transforming growth factor α (TGFα) (Kohli et al. 2005), among others, have been shown to induce meiotic resumption in maturational competent follicles of teleosts.

In addition to the stimulatory control of oocyte maturation induced by LH, meiotic arrest is maintained by E_2 and other growth factors (such as bone morphogenetic protein [BMP15], growth differentiation factor [GDF9], and TGFβ) (Pang and Thomas 2009; Van Der Kraak and Lister 2011).

6.5 Alterations of the Oocyte Maturation by External Factors

The formation and development of teleost oocyte is a large process that comprises hormonal coordination and environmental stimuli. However, failures of such process are existing in farmed species due to the existence of an artificial environment and the human manipulation, inducing deficiencies on the reproductive performances (Mylonas and Zohar 2007).

In addition to the farmed conditions, the raising of the water temperatures due to climate change and the presence of pollutants pose new threats to fish reproduction.

In fact, a wild variety of endocrine-disrupting chemicals (EDCs) have been detected in aquatic systems (i.e., plasticizers, pesticides, surfactants, pharmaceutical compounds). According to the World Health Organization, an EDC is “an exogenous substance or mixture that alters function(s) of the endocrine system and consequently causes adverse health effects in an impact organism, or its progenies, or (sub)populations” (World Health Organization 2002). Indeed, to date, several studies have been published regarding the effects of different EDCs on the teleost gametes and reproductive system (Carnevali et al. 2018). For instance, it was recently evidenced that the plastic component bisphenol A at environmental concentrations induced morphological alterations in mature follicles of zebrafish and alter the gene expression of oocyte maturation markers (Santangeli et al. 2016). The plasticizer di-isononyl phthalate (DiNP) decreased the frequency of mature oocytes in zebrafish ovaries (Santangeli et al. 2017) and changed the biochemical composition of the vitellogenic oocytes. In female murray rainbowfish (*Melanotaenia fluviatilis*), the plasticizer di-n-butyl phthalate (DnBP) reduced the thickness of the chorion in late vitellogenic oocytes (Bhatia et al. 2013). Another xenobiotic, the medicine tamoxifen used for breast cancer, stimulated oocyte maturation on zebrafish (Tokumoto et al. 2005) as well as the organophosphate flame retardant, TDCPP, which increased the frequency of mature oocyte in zebrafish ovaries (Wang et al. 2015).

Water temperature is an essential regulatory factor for fish reproduction, involved in the control of all reproductive process, from gametogenesis to larval survival. Certainly, temperature alterations have the capacity to affect the hypothalamus—pituitary—gonadal axis (Pankhurst and Munday 2011). For instance, as reviewed by Pankhurst and King, exposure to elevated temperatures may impair reproduction of the Atlantic salmon and the rainbow trout among other species (Pankhurst and King 2010). In the Arctic charr, the rising of the temperatures inhibited the ovulation, reduced the sensitivity to 17,20βP and inhibited LH secretion, impairing the oocyte maturation (Gillet et al. 2011).

To solve reproductive dysfunctionalities in farmed species, the aquaculture industry has been dealing with several external factors, as temperature, photoperiod, or hormonal treatments to improve the reproductive outcomes (Mylonas et al. 2010, 2017; Mylonas and Zohar 2007). Recently, an improvement in broodstock nutrition has been proved to enhance the gamete quality, opening a potential new window for captivity reproduction. For instance, the beneficial use of probiotics on reproduction have been studied in several species (reviewed in Gioacchini et al. 2014). Specifically, probiotics supplements induced responsiveness to MIH toward the oocyte maturation (Gioacchini et al. 2012), increased the expression of genes coding for oocyte maturation markers in zebrafish (Gioacchini et al. 2011, 2012) and varied the biochemical composition of mature oocytes (Giorgini et al. 2010).

6.6 Conclusions and Future Prospective

In the present chapter, the composition and the structure of the different layers surrounding the oocyte together with the cytoplasmic components of fish oocyte were described. Within the eggs of fish, as in other oviparous vertebrates, all nutrients necessary to support further embryo development are stored while on its surface macromolecules are deposited to protect it. A coordinated transport of all these components occurs during oogenesis to guarantee a good egg quality.

Recently, changes in oocyte maturation as well as an increase of apoptotic process were observed when fish were exposed to environmental relevant concentrations of EDCs, providing clear evidences about their effects on epigenetic control. In addition, we are aware that climate changes could also affect the endocrine system and in turn, could lead to alterations in the reproductive system and decrease the fitness of the wild populations and farmed species.

The oocyte structure and components may be used as an early warning for reproduction toxicity in polluted aquatic environment or associated with inappropriate diet or climate changes. The morphological alteration of the oocyte is generally associated with physiological dysfunction that may compromise the reproduction and consequently, the availability of good quality eggs and so of larvae for the aquaculture industry. The identification of morphological markers of egg quality may support the farmers in providing appropriate diet and environmental conditions to improve larval survival.

Acknowledgments The authors would like to thank Luca Marisaldi from the Università Politecnica delle Marche (Italy) for kindly providing Fig. 6.1.

References

Abraham M, Hilge V, Lison S, Tibika H (1984) The cellular envelope of oocyte in teleosts. Cell Tissue Res 235:403–410

Amano H, Mochizuki M, Fujita T, Hiramatsu N, Todo T, Hara A (2010) Purification and characterization of a novel incomplete-type vitellogenin protein (VgC) in Sakhalin taimen (Hucho perryi). Comp Biochem Physiol A Mol Integr Physiol 157:41–48. https://doi.org/10.1016/j.cbpa.2010.05.006

Amanze D, Iyengar A (1990) The micropyle: a sperm guidance system in teleost fertilization. Development 109:495–500

Asturiano JF, Sorbera LA, Ramos J, Kime DE, Carrilo M, Zanuy S (2000) Hormonal regulation of the European sea bass reproductive cycle: an individualized female approach. J Fish Biol 56:1155–1172. https://doi.org/10.1111/j.1095-8649.2000.tb02131.x

Babin PJ, Carnevali O, Lubzens E, Schneider WJ (2007) Molecular aspects of oocyte vitellogenesis in fish. In: The fish oocyte. Springer, Dordrecht, pp 39–76

Berois N, Arezo MJ, Papa NG (2011) Gamete interactions in teleost fish: the egg envelope. Basic studies and perspectives as environmental biomonitor. Biol Res 44:119–124. https://doi.org/10.4067/S0716-97602011000200002

Bhatia H, Kumar A, Du J, Chapman J, McLaughlin MJ (2013) Di-n-butyl phthalate causes antiestrogenic effects in female Murray rainbowfish (*Melanotaenia fluviatilis*). Environ Toxicol Chem 32:2335–2344. https://doi.org/10.1002/etc.2304

Carnevali O, Carletta R, Cambi A, Vita A, Bromage N (1999a) Yolk formation and degradation during oocyte maturation in seabream *Sparus aurata*: involvement of two lysosomal proteinases. Biol Reprod 60:140–146

Carnevali O, Centonze F, Brooks S, Marota I, Sumpter JP (1999b) Molecular cloning and expression of ovarian cathepsin D in seabream, *Sparus aurata*. Biol Reprod 61:785–791

Carnevali O, Mosconi G, Cambi A, Ridolfi S, Zanuy S, Polzonetti-Magni AM (2001) Changes of lysosomal enzyme activities in sea bass (*Dicentrarchus labrax*) eggs and developing embryos. Aquaculture 202:249–256. https://doi.org/10.1016/S0044-8486(01)00775-X

Carnevali O, Cionna C, Tosti L, Lubzens E, Maradonna F (2006) Role of cathepsins in ovarian follicle growth and maturation. Gen Comp Endocrinol 146:195–203. https://doi.org/10.1016/j.ygcen.2005.12.007

Carnevali O, Santangeli S, Forner-Piquer I, Basili D, Maradonna F (2018) Endocrine-disrupting chemicals in aquatic environment: what are the risks for fish gametes? Fish Physiol Biochem 44:1561–1576. https://doi.org/10.1007/s10695-018-0507-z

Cerdà J, Zapater C, Chauvigné F, Finn RN (2013) Water homeostasis in the fish oocyte: new insights into the role and molecular regulation of a teleost-specific aquaporin. Fish Physiol Biochem 39:19–27. https://doi.org/10.1007/s10695-012-9608-2

Chauvigné F (2011) Role of aquaporins during teleost gametogenesis and early embryogenesis. Front Physiol 2:66. https://doi.org/10.3389/fphys.2011.00066

Conner SJ, Hughes DC (2003) Analysis of fish ZP1/ZPB homologous genes—evidence for both genome duplication and species-specific amplification models of evolution. Reproduction 126:347–352

Elkouby YM, Mullins MC (2017) Coordination of cellular differentiation, polarity, mitosis and meiosis—new findings from early vertebrate oogenesis. Dev Biol 430:275–287. https://doi.org/10.1016/J.YDBIO.2017.06.029

Emata AC, Borlongan IG, Damaso JP (2000) Dietary vitamin C and E supplementation and reproduction of milkfish *Chanos chanos* Forsskal. Aquac Res 31:557–564. https://doi.org/10.1046/j.1365-2109.2000.00467.x

Finn RN (2007) The maturational disassembly and differential proteolysis of paralogous vitellogenins in a marine pelagophil yeleost: a conserved mechanism of oocyte hydration. Biol Reprod 76:936–948. https://doi.org/10.1095/biolreprod.106.055772

Gallo A, Costantini M (2012) Glycobiology of reproductive processes in marine animals: the state of the art. Mar Drugs 10:2861–2892. https://doi.org/10.3390/md10122861

Ge W (2000) Roles of the activin regulatory system in fish reproduction. Can J Physiol Pharmacol 78:1077–1085

Gillet C, Breton B, Mikolajczyk T, Bodinier P, Fostier A (2011) Disruption of the secretion and action of 17,20β-dihydroxy-4-pregnen-3-one in response to a rise in temperature in the Arctic charr, *Salvelinus alpinus*. Consequences on oocyte maturation and ovulation. Gen Comp Endocrinol 172:392–399. https://doi.org/10.1016/J.YGCEN.2011.04.002

Gioacchini G, Lombardo F, Merrifield DL, Silvi S, Alberto C, Avella MA, Carnevali O (2011) Effects of probiotic on zebrafish reproduction. J Aquac Res Development s1:1–6. https://doi.org/10.4172/2155-9546.S1-002

Gioacchini G, Giorgini E, Merrifield DL, Hardiman G, Borini A, Vaccari L, Carnevali O (2012) Probiotics can induce follicle maturational competence: the *Danio rerio* case. Biol Reprod 86(3):65. https://doi.org/10.1095/biolreprod.111.094243

Gioacchini G, Giorgini E, Vaccari L, Carnevali O (2014) Can probiotics affect reproductive processes of aquatic animals? In: Aquaculture nutrition. Wiley, Chichester, pp 328–346

Giorgini E, Conti C, Ferraris P, Sabbatini S, Tosi G, Rubini C, Vaccari L, Gioacchini G, Carnevali O (2010) Effects of *Lactobacillus rhamnosus* on zebrafish oocyte maturation: an FTIR imaging and biochemical analysis. Anal Bioanal Chem 398:3063–3072. https://doi.org/10.1007/s00216-010-4234-2

Guraya SS (1986) The cell and molecular biology of fish oogenesis. Monogr Dev Biol 18:1–223

Hart NH (1990) Fertilization in teleost fishes: mechanisms of sperm-egg interactions. Int Rev Cytol 121:1–66

Hyllner SJ, Barber HF-P, Larsson DGJ, Haux C (1995) Amino acid composition and endocrine control of vitelline envelope proteins in European sea bass (*Dicentrarchus labrax*) and gilthead sea bream (*Sparus aurata*). Mol Reprod Dev 41:339–347. https://doi.org/10.1002/mrd.1080410309

Hyllner SJ, Westerlund L, Olsson P-E, Schopen A (2001) Cloning of rainbow trout egg envelope proteins: members of a unique group of structural proteins. Biol Reprod 64:805–811. https://doi.org/10.1095/biolreprod64.3.805

Irwin DA, Van G, Kraak D (2012) Regulation and actions of insulin-like growth factors in the ovary of zebrafish (Danio rerio). Gen Comp Endocrinol 177(1):187–194. https://doi.org/10.1016/j.ygcen.2012.03.006

Izquierdo M, Fernández-Palacios H, Tacon AG (2001) Effect of broodstock nutrition on reproductive performance of fish. Aquaculture 197:25–42. https://doi.org/10.1016/S0044-8486(01)00581-6

Jalabert B (2005) Particularities of reproduction and oogenesis in teleost fish compared to mammals. Reprod Nutr Dev 45:261–279. https://doi.org/10.1051/rnd:2005019

Jiang Y-Q, Zhang T-T, Yang W-X (2010) Formation of zona radiata and ultrastructural analysis of egg envelope during oogenesis of Chinese perch *Siniperca chuatsi*. Micron 41:7–14. https://doi.org/10.1016/j.micron.2009.07.004

Johnson RB (2009) Lipid deposition in oocytes of teleost fish during secondary cocyte growth. Rev Fish Sci 17:78–89. https://doi.org/10.1080/10641260802590004

Kagawa H, Kishi T, Gen K, Kazeto Y, Tosaka R, Matsubara H, Matsubara T, Sawaguchi S (2011) Expression and localization of aquaporin 1b during oocyte development in the Japanese eel (*Anguilla japonica*). Reprod Biol Endocrinol 9:71. https://doi.org/10.1186/1477-7827-9-71

Kagawa H, Sakurai Y, Horiuchi R, Kazeto Y, Gen K, Imaizumi H, Masuda Y (2013) Mechanism of oocyte maturation and ovulation and its application to seed production in the Japanese eel. Fish Physiol Biochem 39:13–17. https://doi.org/10.1007/s10695-012-9607-3

Kaviani EF, Shabanipour N, Mirnategh SB (2013) Light and electron microscope structural study of the zona radiata in the oocyte of zebrafish (*Danio rerio*). J Electron Microsc 62:377–382. https://doi.org/10.1093/jmicro/dfs086

Khan IA, Thomas P (1998) Ovarian cycle, teleost fish. In: Skinner M, Knobil E, Neill J (eds) Encyclopedia of reproduction, 1st edn. Academic, San Diego, pp 551–564

Kjørsvik E, Mangor-Jensen A, Holmefjord I (1990) Egg quality in fishes. Adv Mar Biol 26:71–113. https://doi.org/10.1016/S0065-2881(08)60199-6

Kohli G, Clelland E, Peng C (2005) Potential targets of transforming growth factor-beta1 during inhibition of oocyte maturation in zebrafish. Reprod Biol Endocrinol 3:53. https://doi.org/10.1186/1477-7827-3-53

LaFleur GJ, Raldúa D, Fabra M, Carnevali O, Denslow N, Wallace RA, Cerdà J (2005) Derivation of major yolk proteins from parental vitellogenins and alternative processing during oocyte maturation in Fundulus heteroclitus. Biol Reprod 73:815–824. https://doi.org/10.1095/biolreprod.105.041335

Le Menn F, Cerdà J, Babin PJ (2007) Ultrastructural aspects of the ontogeny and differentiation of ray-finned fish ovarian follicles. In: The fish oocyte. Springer, Dordrecht, pp 1–37

Lessman CA (2009) Oocyte maturation: converting the zebrafish oocyte to the fertilizable egg. Gen Comp Endocrinol 161:53–57. https://doi.org/10.1016/j.ygcen.2008.11.004

Lessman CA, Nathani R, Uddin R, Walker J, Liu J (2007) Computer-aided meiotic maturation assay (CAMMA) of zebrafish (*Danio rerio*) oocytes *in vitro*. Mol Reprod Dev 74:97–107. https://doi.org/10.1002/mrd.20530

Li M, Christie HL, Leatherland JF (2012) The *in vitro* metabolism of cortisol by ovarian follicles of rainbow trout (*Oncorhynchus mykiss*): comparison with ovulated oocytes and pre-hatch embryos. Reproduction 144:713–722. https://doi.org/10.1530/REP-12-0354

Litscher ES, Wassarman PM (2018) The fish egg's zona pellucida. Curr Top Dev Biol 130:275–305. https://doi.org/10.1016/bs.ctdb.2018.01.002

Lønning S, Køjrsvik E, Falk-petersen I (1988) A comparative study of pelagic and demersal eggs from common marine fishes in northern Norway. Sarsia 73:49–60. https://doi.org/10.1080/00364827.1988.10420671

Lubzens E, Young G, Bobe J, Cerdà J (2010) Oogenesis in teleosts: how fish eggs are formed. Gen Comp Endocrinol 165:367–389. https://doi.org/10.1016/J.YGCEN.2009.05.022

Matsuyama M, Nagahama Y, Matsuura S (1991) Observations on ovarian follicle ultrastructure in the marine teleost, *Pagrus major*, during vitellogenesis and oocyte maturation. Aquaculture 92:67–82. https://doi.org/10.1016/0044-8486(91)90009-V

Modig C, Modesto T, Canario A, Cerdà J, von Hofsten J, Olsson P-E (2006) Molecular characterization and expression pattern of zona pellucida proteins in gilthead seabream (*Sparus aurata*). Biol Reprod 75:717–725. https://doi.org/10.1095/biolreprod.106.050757

Modig C, Westerlund L, Olsson P-E (2007) Oocyte zona pellucida proteins. In: The fish oocyte. Springer, Dordrecht, pp 113–139

Murray DS, Bain MM, Adams CE (2013) Adhesion mechanisms in European whitefish *Coregonus lavaretus* eggs: is this a survival mechanism for high-energy spawning grounds? J Fish Biol 83:1221–1233. https://doi.org/10.1111/jfb.12218

Murua H, Saborido-Rey F (2003) Female reproductive strategies of marine fish species of the North Atlantic. J Northw Atl Fish Sci 33:23–31

Mylonas CC, Zohar Y (2007) Promoting oocyte maturation, ovulation and spawning in farmed fish. In: The fish oocyte. Springer, Dordrecht, pp 437–474

Mylonas CC, Fostier A, Zanuy S (2010) Broodstock management and hormonal manipulations of fish reproduction. Gen Comp Endocrinol 165:516–534. https://doi.org/10.1016/J.YGCEN.2009.03.007

Mylonas CC, Duncan NJ, Asturiano JF (2017) Hormonal manipulations for the enhancement of sperm production in cultured fish and evaluation of sperm quality. Aquaculture 472:21–44. https://doi.org/10.1016/J.AQUACULTURE.2016.04.021

Nagahama Y (1997) 17 alpha,20 beta-dihydroxy-4-pregnen-3-one, a maturation-inducing hormone in fish oocytes: mechanisms of synthesis and action. Steroids 62:190–196

Nagahama Y, Yamashita M (2008) Regulation of oocyte maturation in fish. Dev Growth Differ 50:S195–S219. https://doi.org/10.1111/j.1440-169X.2008.01019.x

Nagahama Y, Yoshikuni M, Yamashita M, Sakai N, Tanaka M (1993) Molecular endocrinology of oocyte growth and maturation in fish. Fish Physiol Biochem 11:3–14. https://doi.org/10.1007/BF00004545

Nagahama Y, Yoshikuni M, Yamashita M, Tokumoto T, Katsu Y (1995) Regulation of oocyte growth and maturation in fish. Curr Top Dev Biol 30:103–145

Ohkubo N, Sawaguchi S, Hamatsu T, Matsubara T (2006) Utilization of free amino acids, yolk proteins and lipids in developing eggs and yolk-sac larvae of walleye Pollock *Theragra chalcogramma*. Fish Sci 72:620–630. https://doi.org/10.1111/j.1444-2906.2006.01192.x

Olivotto I, Leu M-Y, Blázquez M (2017) Life cycles in marine ornamental species - fishes as a case study. In: Marine ornamental species aquaculture. Wiley, Chichester, pp 23–49

Paitz RT, Mommer BC, Suhr E, Bell AM (2015) Changes in the concentrations of four maternal steroids during embryonic development in the threespined stickleback (*Gasterosteus aculeatus*). J Exp Zool A Ecol Genet Physiol 323:422–429. https://doi.org/10.1002/jez.1937

Palace VP, Werner J, Werner J (2006) Vitamins A and E in the maternal diet influence egg quality and early life stage development in fish: a review. Sci Mar 70:41–57. https://doi.org/10.3989/scimar.2006.70s241

Pang Y, Thomas P (2009) Involvement of estradiol-17beta and its membrane receptor, G protein coupled receptor 30 (GPR30) in regulation of oocyte maturation in zebrafish, *Danio rario*. Gen Comp Endocrinol 161:58–61. https://doi.org/10.1016/j.ygcen.2008.10.003

Pankhurst NW, King HR (2010) Temperature and salmonid reproduction: implications for aquaculture. J Fish Biol 76:69–85. https://doi.org/10.1111/j.1095-8649.2009.02484.x

Pankhurst NW, Munday PL (2011) Effects of climate change on fish reproduction and early life history stages. Mar Freshw Res 62:1015–1026. https://doi.org/10.1071/MF10269

Patiño R, Thomas P, Yoshizaki G (2003) Ovarian follicle maturation and ovulation: an integrated perspective. Fish Physiol Biochem 28:305–308. https://doi.org/10.1023/B:FISH.0000030565.74702.0a

Ramezani-Fard E, Kamarudin MS, Harmin SA (2013) Endocrine control of oogenesis in teleosts. Asian J Anim Vet Adv 8:205–215. https://doi.org/10.3923/ajava.2013.205.215

Riehl R, Patzner RA (1998) Minireview: the modes of egg attachment in teleost fishes. Ital J Zool 65:415–420. https://doi.org/10.1080/11250009809386857

Santangeli S, Maradonna F, Gioacchini G, Cobellis G, Piccinetti CC, Dalla Valle L, Carnevali O (2016) BPA-induced deregulation of epigenetic patterns: effects on female zebrafish reproduction. Sci Rep 6:21982. https://doi.org/10.1038/srep21982

Santangeli S, Maradonna F, Zanardini M, Notarstefano V, Gioacchini G, Forner-Piquer I, Habibi H, Carnevali O (2017) Effects of diisononyl phthalate on *Danio rerio* reproduction. Environ Pollut 231:1051–1062. https://doi.org/10.1016/j.envpol.2017.08.060

Selman K, Wallace RA, Sarka A, Qi X (1993) Stages of oocyte development in the zebrafish, *Brachydanio rerio*. J Morphol 218:203–224. https://doi.org/10.1002/jmor.1052180209

Shibata Y, Iwamatsu T, Suzuki N, Young G, Naruse K, Nagahama Y, Yoshikuni M (2012) An oocyte-specific astacin family protease, alveolin, is released from cortical granules to trigger egg envelope hardening during fertilization in medaka (*Oryzias latipes*). Dev Biol 372:239–248. https://doi.org/10.1016/j.ydbio.2012.09.016

Silversand C, Haux C (1995) Fatty acid composition of vitellogenin from four teleost species. J Comp Physiol B 164:593–599. https://doi.org/10.1007/BF00389799

Skoblina MN (2010) Hydration of oocytes in teleost fishes. Russ J Dev Biol 41:1–12. https://doi.org/10.1134/S1062360410010017

Tokumoto T, Tokumoto M, Nagahama Y (2005) Induction and inhibition of oocyte maturation by EDCs in zebrafish. Reprod Biol Endocrinol 3:69. https://doi.org/10.1186/1477-7827-3-69

Tyler CR, Sumpter JP (1996) Oocyte growth and development in teleosts. Rev Fish Biol Fish 6:287–318. https://doi.org/10.1007/BF00122584

Van Der Kraak G, Lister AL (2011) The inhibitory control of oocyte maturation in the zebrafish (*Danio rerio*): the role of the G protein-coupled estrogen receptor and epidermal growth factor. Biol Reprod 85:6–8. https://doi.org/10.1095/biolreprod.111.092411

Vassallo BG, Paitz RT, Fasanello VJ, Haussmann MF (2014) Glucocorticoid metabolism in the in ovo environment modulates exposure to maternal corticosterone in Japanese quail embryos (*Coturnix japonica*). Biol Lett 10:20140502. https://doi.org/10.1098/rsbl.2014.0502

Viana IKS, Gonçalves LAB, Ferreira MAP, Mendes YA, Rocha RM (2018) Oocyte growth, follicular complex formation and extracellular-matrix remodeling in ovarian maturation of the imperial zebra pleco fish *Hypancistrus zebra*. Sci Rep 8:13760. https://doi.org/10.1038/s41598-018-32117-7

Wallace RA (1985) Vitellogenesis and oocyte growth in nonmammalian vertebrates. In: Oogenesis. Springer, Boston, pp 127–177

Wallace RA, Selman K (1981) Cellular and dynamic aspects of oocyte growth in teleosts. Am Zool 21:325–343

Wang Q, Lam JCW, Han J, Wang X, Guo Y, Lam PKS, Zhou B (2015) Developmental exposure to the organophosphorus flame retardant tris(1,3-dichloro-2-propyl) phosphate: estrogenic activity, endocrine disruption and reproductive effects on zebrafish. Aquat Toxicol 160:163–171. https://doi.org/10.1016/J.AQUATOX.2015.01.014

Wiegand MD (1996) Composition, accumulation and utilization of yolk lipids in teleost fish. Rev Fish Biol Fish 6:259–286. https://doi.org/10.1007/BF00122583

World Health Organization (2002) International programme on chemical safety: global assessment of the state-of-the-science of endocrine disruptors. World Health Organization, Geneva

Yamaguchi A, Katsu Y, Matsuyama M, Yoshikuni M, Nagahama Y (2006) Phosphorylation of the p34cdc2 target site on goldfish germinal vesicle lamin B3 before oocyte maturation. Eur J Cell Biol 85:501–517. https://doi.org/10.1016/j.ejcb.2006.02.002

Żelazowska M (2010) Formation and structure of egg envelopes in Russian sturgeon *Acipenser gueldenstaedtii* (Acipenseriformes: Acipenseridae). J Fish Biol 76:694–706. https://doi.org/10.1111/j.1095-8649.2009.02527.x

Zelazowska M (2014) Ultrastructure and histochemistry of the periplasm in oocytes of sturgeons during egg envelope formation. Folia Biol (Praha) 62:377–385. https://doi.org/10.3409/fb62_4.377

Chapter 7
Gametogenesis, Spawning, and Fertilization in Bivalves and Other Protostomes

Ryusaku Deguchi and Makoto Osada

Abstract The diversity of protostomes is demonstrated by the classification of these animals into more than 20 phyla and 1,000,000 species. Many species of aquatic protostomes are considered valuable for basic studies of reproduction as well as for fishery resources. The aim of this chapter is to provide a brief overview of the process and mechanism of fertilization in three protostome groups, the mollusks, annelids, and arthropods, in which various modes of sexual reproduction have evolved. Regarding the series of interactions between oocytes and sperm at fertilization, we have described the structural changes in gametes and the regulatory mechanisms of polyspermy block and increases in intracellular Ca^{2+}, which enable the successful fertilization of oocytes. Prior to fertilization, oocytes and sperm are produced and matured in gonads and released at the optimal time. The second half of the article focuses on gonial cell multiplication, oocyte growth, and spawning (oocyte release and sperm release) in bivalve mollusks, in which these processes are precisely regulated by endocrine systems. Although bivalves share many endocrine regulatory molecules with vertebrates, they also employ unique mechanisms such as the use of the neurohormone serotonin (5-hydroxytryptamine, 5-HT), which acts directly on oocytes and sperm to induce oocyte maturation, sperm activation, and spawning.

Keywords Mollusk · Annelid · Arthropod · Bivalve · Polyspermy block · Ca^{2+} rise · Gonial multiplication · Vitellogenesis · Oocyte maturation · Sperm motility · Spawning · Endocrine control

---

R. Deguchi (✉)
Department of Biology, Miyagi University of Education, Sendai, Miyagi, Japan
e-mail: deguchi@staff.miyakyo-u.ac.jp

M. Osada
Graduate School of Agricultural Science, Tohoku University, Sendai, Miyagi, Japan
e-mail: makoto.osada.a8@tohoku.ac.jp

M. Yoshida, J. F. Asturiano (eds.), *Reproduction in Aquatic Animals*,
https://doi.org/10.1007/978-981-15-2290-1_8

7.1 Introduction

Both protostomes and deuterostomes are characterized by the presence of three germ layers and bilateral symmetry. During the embryogenesis of most (but not all) protostomes, the fate of the initial gastrulation site blastopore forms the mouth, whereas the anus is formed later at a different site. The opposite situation is observed in deuterostomes (Hejnol and Martindale 2009). Recent molecular phylogenetic analyses support the concept of two main protostome clades, Spiralia (e.g., mollusks, annelids, and nemerteans) and Ecdysozoa (e.g., arthropods and nematodes), although traditionally, annelids and arthropods have been considered closely related because of their similarly segmented body plans (Dunn et al. 2014). Compared with the approximately 70,000 described species of deuterostomes, protostomes include a more diverse range of more than 1,000,000 described species (Chapman 2009). In particular, mollusks, annelids, and arthropods include many taxa and species that have expanded their habitats from ocean to fresh waters and even to terrestrial environments, leading to the diversification of their reproductive modes.

Various protostomes are captured or cultivated worldwide to benefit humans. Many species of mollusks, such as bivalves (clams, mussels, oysters, and scallops), gastropods (abalones and snails), and cephalopods (squids and octopuses), as well as arthropods such as crustaceans (shrimps, lobsters, and crabs), are important to marine or freshwater fishery resources. Annelids, such as lugworms and earthworms, are used as food for animal cultivation and as bait in the fishing industry. The fruit fly *Drosophila melanogaster* (Markow 2015) and the nematode worm *Caenorhabditis elegans* (Frézal and Félix 2015) are considered excellent "model organisms" for various types of biological studies, given their rapid life cycles, small sizes, ease of laboratory cultivation, and abundant molecular information. However, studies of oocyte maturation and fertilization in protostomes have been mainly conducted in "non-model species" that live in aquatic environments and are fertilized externally, and this research area has only recently expanded to fruit flies and nematodes (e.g., Von Stetina and Orr-Weaver 2011; Singaravelu and Singson 2013; Kaneuchi et al. 2015).

In Sect. 7.2, we have provided a brief overview of the process and mechanism of fertilization in three protostome groups: mollusks, annelids, and arthropods. In Sect. 7.3, we have provided a more detailed description of the endocrine systems that regulate gonial cell multiplication, oocyte growth, and spawning in bivalve mollusks.

7.2 Overview of the Fertilization Process and Mechanism in Protostomes

7.2.1 *Mollusks*

7.2.1.1 Mode of Sexual Reproduction

The phylum Mollusca comprises eight classes (Smith et al. 2011; Kocot et al. 2011), including Bivalvia (Pelecypoda), Gastropoda, and Cephalopoda. Approximately 85,000 mollusk species have been described (Chapman 2009), of which a majority of them are ocean dwellers; however, some bivalves and gastropods inhabit freshwater niches, and some gastropods are terrestrial. Most aquatic mollusks possess gills for respiration, and land snails and land slugs can take up oxygen from the air using the mantle cavity as "lung" (Hsia et al. 2013).

Mollusks such as bivalves, limpets (patellogastropods or "true limpets"), and abalones are generally dioecious; the sexually mature adults release either oocytes or sperm to achieve external fertilization. In these animals, fully grown oocytes are stored in the ovaries in a state of arrest at the first prophase (PI) of meiosis, whereas naturally spawned oocytes either remain arrested at PI or resume meiosis toward the second arrest point, i.e., the first metaphase (MI) (Costello et al. 1957; Longo 1983; Deguchi et al. 2015; Table 7.1). In the oocytes of the former PI-type mollusks, fertilization triggers germinal vesicle breakdown (GVBD) and subsequent meiotic events. In contrast, in the latter MI-type species, GVBD has been initiated or completed by the time of spawning, and the remaining meiotic divisions occur after fertilization.

Most aquatic gastropods, including the mud snail *Ilyanassa* (*Tritia*) *obsoleta* (experimental species for spiral cleavage and morphogenesis) (Costello et al. 1957; Goulding and Lambert 2016) and the river snail *Viviparus viviparus* (viviparous species wherein fertilized oocytes develop into young snails inside the mothers) (Jakubik 2012), are dioecious; they copulate to achieve internal fertilization. Conversely, gastropods such as sea hares (including *Aplysia californica*, which is used in studies of learning and memory), sea slugs, land snails, and land slugs are simultaneously hermaphroditic (Heller 1993), although these animals do not normally self-fertilize; rather, self-sperm are usually transferred to other individuals via copulation. The males of cephalopods such as squids implant spermatophores (i.e., sperm-containing capsules) in the internal mantle cavities or on the external body surfaces of female animals (Iwata et al. 2011, 2015). In females, internal fertilization may occur near these sperm storage sites, although the process has not yet been observed directly.

In contrast to the above-mentioned reproductive patterns wherein fertilized eggs or offspring receive genetic information from both parents, some mollusks also serve as examples of uniparental inheritance (gynogenesis and androgenesis). In the freshwater gastropods *Melanoides tuberculata* (Samadi et al. 1999) and *Tarebia granifera* (Miranda et al. 2011), a female can parthenogenetically produce the

Table 7.1 Information about selected protostome species used for recordings of membrane potentials or intracellular Ca^{2+} in oocytes

Phylum (subphylum)/class	Species[a]	Oocyte diameter	Meiotic arrest before activation	Trigger for activation	Changes in membrane potential	Increase in intracellular Ca^{2+}
Mollusca Bivalvia	*Spisula solidissima*	53–56 μm (Costello et al. 1957)	PI	Sperm	Depolarization (Finkel and Wolf 1980)	
	Mactra chinensis	55–60 μm (Deguchi and Osanai 1994b)	PI	Sperm		Single (cortical flash) (Deguchi and Osanai 1994b; Deguchi and Morisawa 2003)
	Mytilus galloprovincialis[b] *Mytilus edulis*[b]	60–70 μm (Sedano et al. 1995)	MI	Sperm	Depolarization (Togo et al. 1995)	Single (Abdelmajid et al. 1993)[c] Oscillations (cortical flash → waves) (Deguchi and Osanai 1994a; this paper)
	Ruditapes philippinarum	60–70 μm (Park and Choi 2004)	MI	Sperm	Depolarization (Leclerc et al. 2000)[d]	Single (Leclerc et al. 2000)[c] Oscillations (cortical flash → waves) (Deguchi unpublished data)
	Tegillarca granosa	~60 μm (Shi et al. 2017)	MI	Sperm	Depolarization (Guo et al. 2017)[d]	Single (Shi et al. 2017)[c]
	Crassostrea gigas	~50 μm (Zhang et al. 2012b)	MI	Sperm		Oscillations (Deguchi and Osanai 1994a)
	Septifer virgatus (*Mytilisepta virgata*)	120–130 μm (Deguchi 2007)	MI	Sperm		Oscillations (cortical flash → waves) (Deguchi 2007; Kashir et al. 2013)
Mollusca Gastropoda	*Lottia kogamogai*	120–130 μm (Deguchi 2007)	MI	Sperm		Single (cortical flash) (Deguchi 2007)
	Lottia gigantea		MI	Sperm	Depolarization (Gould et al. 2001)	Single (Gould et al. 2001)
	Cellana grata	140–150 μm (Deguchi 2007)	MI	Sperm		Single (cortical flash) (Deguchi 2007)
	Haliotis kamtschatkana		MI?	Sperm	Depolarization (Stephano 1992)	

Phylum (subphylum)/class	Species[a]	Oocyte diameter	Meiotic arrest before activation	Trigger for activation	Changes in membrane potential	Increase in intracellular Ca^{2+}
Annelida Echiura	*Urechis caupo*	~120 µm (Jaffe et al. 1979)	PI	Sperm	Depolarization (Jaffe et al. 1979, Gould-Somero et al. 1979)	Single or oscillations (cortical flash) (Stephano and Gould 1997a)
Annelida Polychaeta	*Chaetopterus pergamentaceus*	~100 µm (Costello et al. 1957)	MI	Sperm		Oscillations (local → waves) (Eckberg and Miller 1995)
	Chaetopterus variopedatus	~80 µm (Jaffe 1983)	MI	Sperm	Depolarization (Jaffe 1983)	
	Pseudopotamilla occelata	~150 µm (Nakano et al. 2008)	MI	Sperm		Single (local + cortical flash) (Nakano et al. 2008)
Arthropoda (Crustacea) Malacostraca	*Sicyonia ingentis*	~225 µm (Lindsay et al. 1992)	MI	Seawater (Pillai and Clark 1987)		Two spikes (waves) (Lindsay et al. 1992)
	Palaemon serratus	~650 µm (Goudeau and Goudeau 1989b)	MI	Seawater (Goudeau and Goudeau 1989b)	Hyperpolarization (sperm-independent) (Goudeau and Goudeau 1989b)[e]	Oscillations (Goudeau and Goudeau 1996)
	Homarus gammarus	~1.6 mm (Goudeau and Goudeau 1989b)	MI	Seawater and sperm (Goudeau and Goudeau 1986a; Talbot and Goudeau 1988)	Hyperpolarization (sperm-dependent) (Goudeau and Goudeau 1986a, 1989b)	
	Maia squinado (*Maja squinado*)	~800 µm (Goudeau and Goudeau 1989b)	MI	Seawater and sperm (Goudeau and Goudeau 1989b)	Hyperpolarization (sperm-dependent) (Goudeau and Goudeau 1989a, b)	

[a]Species names described in the original paper and currently accepted (in parentheses) are indicated
[b]The two species names were sometimes used confusingly. *Mytilus edulis* collected in Japan has turned out to be *Mytilus galloprovincialis*
[c]Ca^{2+} indicators were loaded into oocytes by the application of the acetoxymethyl (AM) ester forms, not by direct injection
[d]Changes in membrane potential were measured using fluorescence probes instead of microelectrodes
[e]An initial transient hyperpolarization was followed by the successive phases of depolarization and hyperpolarization

offspring and brood them to young juvenile states. These characteristics have enabled both species to expand their habitats as biological invaders in various regions of the world. The hermaphrodite freshwater triploid clam *Corbicula leana* provides a rare example of natural androgenesis. This clam simultaneously releases MI-arrested oocytes and non-reductional (triploid) sperm; in an MI-arrested oocyte, the axis of the meiotic apparatus is parallel to the surface, and during the first meiotic division after fertilization, the two sets of female chromosomes and centrosomes are emitted as two polar bodies (Komaru et al. 1997, 1998, 2000). The fertilized oocyte, which comprises only a male pronucleus, skips the second meiosis, undergoes mitotic divisions, and develops into a triploid larva in which chromosomes are all derived from the fertilizing sperm (Komaru et al. 1998).

7.2.1.2 Gamete Structures and Fertilization Process

The changes occurring in gamete structures at fertilization have been investigated in detail in marine bivalve, limpet, and abalone species that conduct external fertilization. In most bivalve species previously investigated, the diameters of unfertilized oocytes range from 50 to 70 μm, whereas *Septifer virgatus* (*Mytilisepta virgata*) produces much larger oocytes (120–130 μm in diameter) (Table 7.1). Among representative bivalves, including the surf clam *Spisula solidissima* (PI type) and the mussels *Mytilus edulis* and *M. galloprovincialis* (MI type), the oocyte plasma membrane is surrounded by a thin vitelline envelope (0.5–0.7 μm in width) over a very tiny perivitelline space (0.2–0.3 μm in width), and numerous microvilli (1 μm in length) extend into this envelope (Dan 1962; Humphreys 1962; Rebhun 1962; Longo and Anderson 1970; Kyozuka and Osanai 1994; Fig. 7.1). The outermost structure of the oocyte is a thin jelly coat (<10 μm in width) (Allen 1953; Dan 1962; Rebhun 1962). The morphological features of limpet and abalone oocytes are basically similar, although they are generally larger (100–230 μm in diameter) and equipped with a thicker jelly coat (Sakai et al. 1982; Guerrier et al. 1986; Vacquier and Lee 1993; Deguchi 2007; Kolbin and Kulikova 2011; Balkhair et al. 2016). In limpet oocytes, the perivitelline space is scarce and almost invisible under normal-light microscopy (Guerrier et al. 1986; Buckland-Nicks and Howley 1997). In contrast, abalone oocytes have a wide perivitelline space; for instance, the oocytes of *Haliotis rufescens* have an approximate diameter of 175 μm, a perivitelline space width of approximately 35 μm, and a jelly coat width of approximately 55 μm (Vacquier and Lee 1993). In abalone oocytes, microvilli do not arrive at the vitelline envelope (Sakai et al. 1982; Mozingo et al. 1995).

The males of most marine bivalves, limpets, and abalones release so-called primitive sperm, which comprise a relatively large, conically or cylindrically shaped acrosome, a nucleus containing highly condensed chromatin, a midpiece comprising 4–7 globular or spheroidal mitochondria, and a flagellum exhibiting a typical 9 + 2 structure (Niijima and Dan 1965; Longo and Dornfeld 1967; Longo 1976; Lewis et al. 1980; Sakai et al. 1982; Hodgson and Chia 1993; Buckland-Nicks and Howley 1997). Bivalve and abalone sperm possess a well-developed axial rod

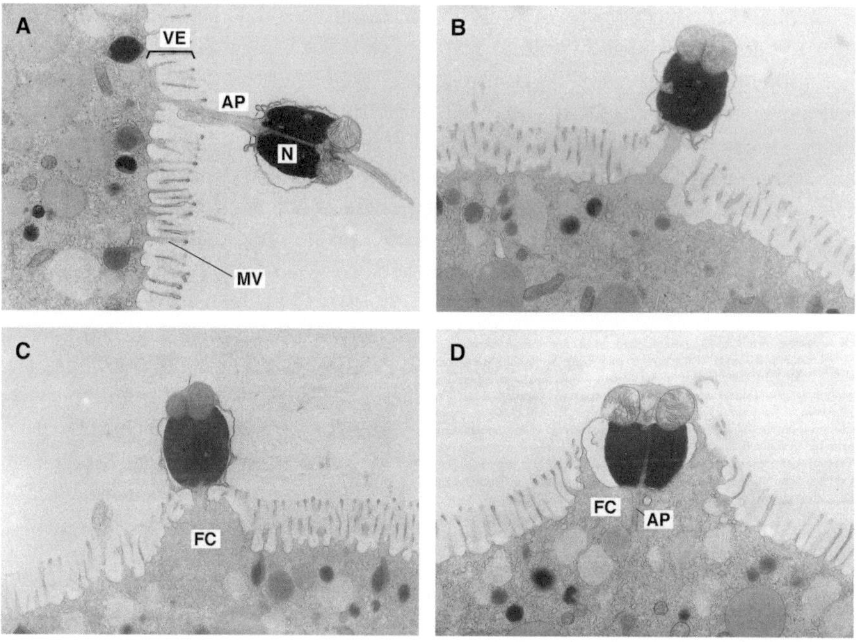

Fig. 7.1 Process of sperm–oocyte fusion in *Mytilus galloprovincialis*. (**a**, **b**) The tip of the acrosomal process of an acrosome-reacted sperm that has fused with microvilli traversing the oocyte vitelline envelope at 30 s post-insemination. (**c**) Formation of the fertilization cone at 1 min post insemination. (**d**) Engulfment of the sperm head into a fertilization cone at 1.5 min post-insemination. *AP* acrosomal process, *FC* fertilization cone, *MV* microvillus, *N* nucleus, *VE* vitelline envelope. (From Kyozuka and Osanai (1994), with permission)

(bundle of actin filaments) even before the acrosome reaction, and this rod sometimes extends from the posterior part of acrosome to the anterior part of the nucleus (Niijima and Dan 1965; Longo and Dornfeld 1967; Lewis et al. 1980; Sakai et al. 1982). Only a few species of limpets produce such a prominent preexisting axial rod in the sperm head (Hodgson and Chia 1993; Buckland-Nicks and Howley 1997).

In mollusk sperm, an acrosome reaction is initiated on or near the vitelline envelope after passage of the jelly coat (Longo 1976; Sakai et al. 1982; Kyozuka and Osanai 1994; Togo et al. 1995; Buckland-Nicks and Howley 1997; Fig. 7.1). Consequent to the acrosome reaction, materials inside the acrosomal vesicle (including the protein lysin in some species) are released outside the sperm, and the anterior end of the sperm is elongated by the actions of actin filaments to form an acrosomal process. In bivalves and limpets, the tip of the acrosome process fuses with microvilli exposed on the outer surface of the vitelline envelope, after which the sperm is engulfed by the oocyte (Longo 1976; Kyozuka and Osanai 1994; Togo et al. 1995; Buckland-Nicks and Howley 1997; Fig. 7.1). In abalones, each acrosome-reacted sperm creates a hole (3 μm in diameter) in the vitelline envelope via a non-enzymatic action of lysin, through which the sperm swims and fuses with the oocyte plasma membrane (Lewis et al. 1982; Sakai et al. 1982; Vacquier and Lee 1993).

In abalones, sperm acrosomal lysin (a non-glycosylated protein comprising 126–138 amino acids) and vitelline envelope receptor for lysin (VERL; a glycosylated protein comprising ~3700 amino acids) have been identified as the molecular partners responsible for the initial recognition between gametes (Vacquier and Lee 1993; Swanson and Vacquier 1997; Kresge et al. 2001; Galindo et al. 2002). Lysin and the initial two domains of VERL (VL1 and VL2) appear to have evolved rapidly under positive Darwinian selection (Vacquier et al. 1997; Metz et al. 1998; Galindo et al. 2003), thus enabling lysin to dissociate VERL in a species-specific manner. Recent structural analyses have revealed the specific properties of lysin that may facilitate rapid evolution (Wilburn et al. 2018), as well as the conformational similarities between the sperm recognition sites of VERL and of ZP2, a mammalian zona pellucida protein (Raj et al. 2017). In *Mytilus*, three proteins (M3, M6, and M7) able to dissolve the vitelline envelope have also been isolated from the sperm acrosome (Takagi et al. 1994). They are considered as Ca^{2+}-dependent carbohydrate-binding proteins and do not share homology with abalone lysin.

Following sperm penetration, bivalve, limpet, and abalone oocytes do not exhibit an obvious cortical reaction such as cortical granule exocytosis (Dan 1962; Humphreys 1967; Longo and Anderson 1970; Longo 1976; Guerrier et al. 1986; Vacquier and Lee 1993; Kyozuka and Osanai 1994), a process observed typically in many other animals, including crustaceans (see later). Nonetheless, the perivitelline space widens slightly in some species, such as *Spisula* (Longo and Anderson 1970).

7.2.1.3 Polyspermy Block

In most animals, polyspermy, defined as the entry of more than two sperm into the oocyte or egg cytoplasm, causes embryonic death during the subsequent developmental process. Therefore, the fertilized oocytes or eggs of many species undergo a rapid change in membrane potential, which acts as a polyspermy block (reviewed by Gould and Stephano 2003; Iwao and Izaki 2018; Jaffe 2018), although it remains unclear whether this mechanism is indeed necessary under physiological conditions (Dale 2016). In all previously examined mollusks, fertilization causes a membrane depolarization event (i.e., positively directed shift in the membrane potential) with a duration of 10 s to 20 min, depending on the species (Finkel and Wolf 1980; Stephano 1992; Togo et al. 1995; Leclerc et al. 2000; Gould et al. 2001; Guo et al. 2017; Table 7.1). This depolarization can be attributed mainly to the influx of Na^+ because the amplitude or duration of depolarization was shown to decrease in low-Na^+ or Na^+-free seawater (Togo et al. 1995; Gould et al. 2001) as the number of polyspermic oocytes increased (Finkel and Wolf 1980; Togo et al. 1995). Ca^{2+} influx also appears to contribute to depolarization (Gould et al. 2001) and affect the rate of polyspermy (Guo et al. 2017). These data support the view that depolarization contributes to the polyspermy block in fertilized mollusk oocytes.

In various other animals, including echinoderms and vertebrates, the vitelline envelope of a fertilized oocyte or egg (sometimes referred to as fertilization membrane) serves as a barrier to prevent additional sperm from entering the perivitelline space; thus, the envelope functions as a late block to polyspermy (reviewed by Wessel et al. 2001; Gould and Stephano 2003; Liu 2011). However, mollusks might not adopt this type of vitelline envelope-mediated polyspermy block. Indeed, after fertilization, the vitelline envelope allows extra sperm to undergo an acrosome reaction and to dissolve it (Lewis et al. 1982; Togo et al. 1995). Nonetheless, acrosome-reacted sperm cannot fuse with the plasma membrane, even after membrane depolarization has ceased (Togo et al. 1995). Currently, the molecular basis of the non-electrical polyspermy block on the plasma membrane observed in mollusks remains completely unknown. In mammals, expression of the oocyte receptor Juno, which binds the sperm protein Izumo1, becomes undetectable on the plasma membrane after fertilization; accordingly, Juno is assumed to play a central role in the non-electrical polyspermy block on the plasma membrane (Bianchi et al. 2014; Bianchi and Wright 2014). Finally, *Mytilus* oocytes appear to possess an additional late polyspermy block mechanism, wherein an aminopeptidase-like protease(s) released from fertilized oocytes reduces the acrosome reaction rate in sperm arriving after fertilization (Togo and Morisawa 1997).

7.2.1.4 Oocyte Activation via Intracellular Ca^{2+} Increase

In all previously studied animals, oocytes or eggs exhibit a transient increase in the intracellular (cytoplasmic) Ca^{2+} concentration at fertilization (reviewed by Stricker 1999; Miyazaki 2006; Horner and Wolfner 2008; Kashir et al. 2013; Swann and Lai 2016). These changes in intracellular Ca^{2+} levels can be measured after directly injecting oocytes with Ca^{2+} indicators or exposing oocytes to acetoxymethyl (AM) ester forms of Ca^{2+} dyes. Both methods work well and produce similar results in mammals. In marine invertebrates (e.g., bivalves), however, the two methods yield different patterns of Ca^{2+} changes even in the same species (Table 7.1).

In oocytes of the PI-type bivalve *Mactra chinensis*, a large increase in Ca^{2+} immediately after fertilization appears as a centripetal Ca^{2+} flow that originates at the whole cortex and spreads to the center (Deguchi and Morisawa 2003). This type of Ca^{2+} increase, which is often described as a cortical flash, is followed by a plateau phase of Ca^{2+} elevation that lasts for approximately 10 min and ceases near the time of GVBD. These intracellular Ca^{2+} changes depend almost entirely on the influx of Ca^{2+}, which likely occurs through voltage-dependent Ca^{2+} channels; the premature cessation of an intracellular increase in Ca^{2+} in the absence of external Ca^{2+} or the presence of Ca^{2+} channel blockers inhibits GVBD or the resumption of meiosis from PI. Several lines of evidence suggest that the resumption of meiosis in other PI-type bivalves also depends on the influx of Ca^{2+} at fertilization (Deguchi and Osanai 1994b and references therein). An increase in the intracellular pH is another important ionic event related to the resumption of meiosis from PI. Indeed, the oocytes of PI-type bivalves, including *Spisula* and *Mactra*, exhibit an increase in intracellular pH or the release of acid at fertilization; however, an intracellular pH above a threshold level is necessary but not sufficient to induce GVBD (Deguchi and Osanai 1994b; Dubé and Eckberg 1997). This situation contrasts strikingly with the oocyte maturation process observed in some MI-type bivalves and limpets, in which an increase in intracellular pH is sufficient to induce the resumption of meiosis from PI without an increase in Ca^{2+} (Guerrier et al. 1986; Deguchi and Osanai 1994b, 1995; Gould et al. 2001).

In contrast to PI-type bivalves, the fertilized oocytes of MI-type bivalves exhibit repeated increases in Ca^{2+} (i.e., Ca^{2+} oscillations) (Deguchi and Osanai 1994a; Deguchi 2007; Kashir et al. 2013; Table 7.1; Fig. 7.2), although AM ester Ca^{2+} dyes can only detect a blunt single increase in Ca^{2+} (Abdelmajid et al. 1993; Leclerc et al. 2000; Shi et al. 2017). During Ca^{2+} oscillations, an initial cortical flash is followed by point-source Ca^{2+} waves that initiate at one cortex (that does not necessarily correspond to the sperm entry site) and propagate to the antipode (Kashir et al. 2013; Fig. 7.2). These Ca^{2+} oscillations may be mediated mainly by the release of Ca^{2+} through inositol 1,4,5-trisphosphate (IP_3)-gated Ca^{2+} channels on the endoplasmic reticulum (Deguchi et al. 1996). Consistent with this process, the oocytes of MI-type bivalves may continue to undergo Ca^{2+} oscillations and resume meiosis from MI even after the removal of external Ca^{2+} (Deguchi and Osanai, 1994a). In contrast, the increase in intracellular Ca^{2+} observed in MI-type limpets exhibits a pattern and external Ca^{2+} dependency similar to those observed in PI-type bivalves, rather than MI-type bivalves (Deguchi 2007; Table 7.1). In limpets, the poorly developed IP_3-dependent Ca^{2+} release mechanism and cortical endoplasmic reticulum (ER) clusters may be partly responsible for the lack of Ca^{2+} oscillations at fertilization (Deguchi 2007). An increase in the intracellular pH is neither observed at fertilization nor required for the resumption of meiosis from MI in bivalve and limpet oocytes (Deguchi and Osanai 1995; Gould et al. 2001).

In mollusks, the molecules that act downstream of a Ca^{2+} increase (and pH increase) at fertilization to regulate the resumption of meiosis from PI or MI remain largely unknown. Experiments involving commonly used agonists and antagonists suggest that protein kinase C (PKC) is a downstream target leading to GVBD in *Spisula* (Eckberg et al. 1987; Colas and Dubé 1998). However, short-lived and constitutively synthesized proteins are likely required for MI arrest and should be degraded or inactivated at fertilization in MI-type bivalves and limpets (Dubé and Dufresne 1990; Colas et al. 1993; Néant and Guerrier 1988; Colas and Dubé 1998). Mitogen-activated protein kinase (MAPK) is one candidate that may be responsible for MI arrest, although this event may also involve other proteins (Portillo-López et al. 2003).

7.2.2 *Annelids*

7.2.2.1 Mode of Sexual Reproduction

Recent molecular analyses suggest that Annelida includes other taxonomic groups that were formerly considered separate phyla (e.g., Echiura and Sipuncula) (Dunn et al. 2008; Weigert et al. 2014; Struck et al. 2015). To date, approximately 17,000

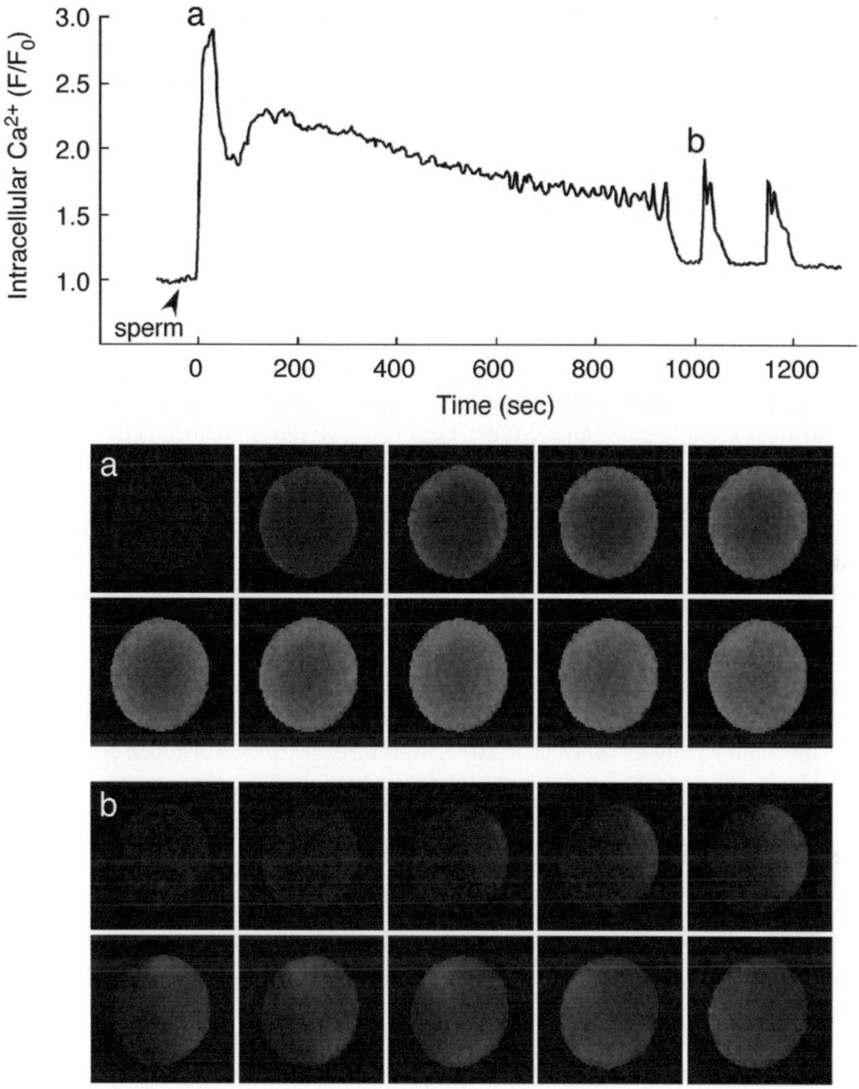

Fig. 7.2 Intracellular Ca^{2+} increases in a *Mytilus galloprovincialis* oocyte at fertilization. Relative changes in fluorescence were measured using the Ca^{2+} indicator Calcium Green-1 dextran (MW 10,000) injected into an oocyte. The fertilized oocyte exhibited an initial Ca^{2+} transient (**a**), followed by a sustained Ca^{2+} elevation and subsequent repetitive Ca^{2+} increases (**b**). Sequential images acquired every 0.4 s demonstrate that the initial Ca^{2+} transient increase takes the form of a cortical flash (**a**), whereas each subsequent Ca^{2+} increase appears as a Ca^{2+} wave propagating from one cortical site to the antipode (**b**)

species belonging to Annelida (including these related phyla) have been described (Chapman 2009). Many annelids, including polychaete (currently thought to be paraphyletic), echiuran, and sipunculid worms, typically live in the ocean, whereas clitellates, such as earthworms and leeches, are abundant in freshwater or moist terrestrial environments. Some polychaete species possess gills for efficient gas exchange, whereas many other annelids respire entirely through their skin (Storch and Alberti 1978; Graham 1988; Chuang et al. 2006).

Similar to mollusks, annelids exhibit a wide range of sexual reproduction modes. Most marine inhabitants are dioecious and external fertilizers (Prevedelli and Simonini 2003; Ram et al. 2008) and are categorized as either PI-type or MI-type species (Costello et al. 1957; Deguchi et al. 2015; Table 7.1). The marine polychaete *Dinophilus gyrociliatus* exhibits a marked sexual dimorphism that is evident even at the egg stage, wherein small "male-type" eggs (40 μm in diameter) develop into dwarf males and larger "female-type" eggs (80–100 μm in diameter) develop into adult females (~1 mm in length) (Prevedelli and Simonini 2003; Simonini et al. 2003). Males copulate with juvenile females (sisters) inside the same capsule and then die. In these females, sperm are thought to have already penetrated the growing oocytes before vitellogenesis (Masui 1985). The females then emerge from the capsule and grow as free-living individuals. Earthworms and leeches are generally hermaphroditic and internal fertilizers; in these animals, sperm are implanted into the partner via copulation (Fernández et al. 1998; Kutschera and Elliott 2010). Some species of earthworms or freshwater worms (*Tubifex* worms) conduct parthenogenetic reproduction, as confirmed by isolation experiments, cytological observations, and molecular analyses (Shen et al. 2012; Marotta et al. 2014; Mustonen et al. 2017).

7.2.2.2 Gamete Structures and Fertilization Process

Studies have mainly explored the fertilization processes of polychaete and echiuran worms that spawn gametes in seawater or brackish water for external fertilization. Previous fertilization experiments have used unfertilized polychaete oocytes ranging from 50 to 300 μm in diameter (Costello et al. 1957; Table 7.1). These oocytes exhibited morphological features similar to those observed in mollusks and are enclosed by a vitelline envelope (0.5–1.5 μm in width) into which microvilli are inserted, regardless of the width of the perivitelline space (Fallon and Austin 1967; Anderson and Eckberg 1983; Sato and Osanai 1983, 1986; Kluge et al. 1995; Shin et al. 2005; Nakano et al. 2008, 2014). Primitive-type polychaete sperm undergo an acrosome reaction on the outermost layer of the vitelline envelope; the subsequent elongation (or not) of an acrosomal process is species-dependent (Anderson and Eckberg 1983; Sato and Osanai 1983, 1986). Subsequently, in some species, the acrosome-reacted sperm fuses with microvilli before penetrating the vitelline envelope whereas, in other species, it fuses with the oocyte plasma membrane after passing through a vitelline envelope via a lysin-mediated process (Colwin and Colwin 1960, 1961; Anderson and Eckberg 1983; Sato and Osanai 1983, 1986). In *Pseudopotamilla occelata* and *Sabellaria vulgaris* oocytes, a wide perivitelline

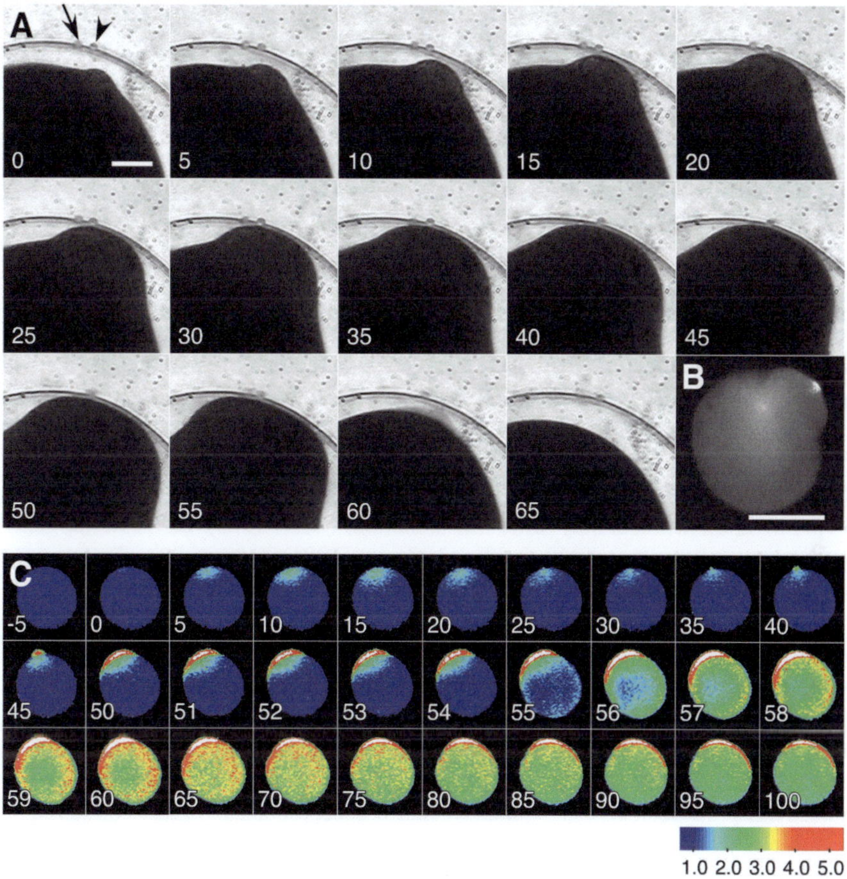

Fig. 7.3 Fertilization-induced morphological and intracellular Ca^{2+} changes in *Pseudopotamilla occelata* oocytes. (**a**) Bright-field images of a fertilized oocyte acquired every 5 s after the formation of a cytoplasmic protrusion was initiated. Of the two sperm bound to the vitelline envelope (arrow and arrowhead), the left (arrow) was engulfed into the tip of the protrusion. Scale bar: 20 μm. (**b**) DAPI-stained sperm nucleus at the tip of the protrusion. Scale bar: 100 μm. (**c**) Fluorescence images of a fertilized oocyte injected with the Ca^{2+} indicator Calcium Green-1 dextran (MW 3,000); the images were acquired every 1 or 5 s and normalized by division in a pixel-to-pixel manner by the resting image obtained immediately before the Ca^{2+} increase (−5 s). An initial local Ca^{2+} increase was followed by the formation of a cytoplasmic protrusion, which was retracted after a subsequent large cortical flash. (From Nakano et al. (2008), with permission)

space (12–15 μm) forms during oocyte maturation (Waterman 1934; Novikoff 1939; Nakano et al. 2008). In these species, the binding of sperm to the vitelline envelope stimulates the oocyte to produce a large cytoplasmic protrusion that extends up to its inner surface and then engulfs the sperm (Fig. 7.3).

Following fertilization, cortical granule exocytosis occurs in some polychaete species such as *Nereis limbata* (*Alitta succinea*) and *Platynereis dumerilii*, resulting in the formation of a jelly coat and perivitelline space (Lillie 1911; Fallon and

Austin 1967; Kluge et al. 1995). In contrast, no obvious post-fertilization changes in cortical granules are observed in *Chaetopterus pergamentaceus*, despite the formation of a perivitelline space (Eckberg 1981). The presence or absence of cortical granule exocytosis may be species-specific even within the genus of *Urechis* (Gould and Stephano 1989; Shin et al. 2005). Unfortunately, little molecular information is known about lysin or the other substances responsible for sperm–oocyte interactions in annelids.

7.2.2.3 Polyspermy Block

In the fertilized oocytes of *Urechis caupo* (PI-type), a large membrane depolarization (from −30 to +50 mV) that depends mainly on Na^+ influx and partly on Ca^{2+} influx is thought to act as a fast electrical polyspermy block during the initial 5–10 min of fertilization (Gould-Somero et al. 1979; Jaffe et al. 1979; Gould and Stephano 1989; Table 7.1). The observed occurrence of a long-lasting membrane depolarization immediately after insemination (Jaffe 1983) and external Na^+-dependent polyspermy (Eckberg and Anderson 1985) suggest that *Chaetopterus* oocytes (MI-type) may employ a similar polyspermy block mechanism.

Polychaete oocytes seem to be equipped with species-specific mechanisms that enable a late but permanent polyspermy block. These mechanisms include a plasma membrane-level block in *Urechis* that prevents the fusion of acrosome-reacted sperm with the oocyte microvilli or plasma membrane independently of the membrane potential (Gould and Stephano 2003), a vitelline envelope-level block in *Chaetopterus* and other polychaetes that may be attributable to the disappearance of microvillus tip vesicles (i.e., putative sperm receptor sites) or the retraction of microvilli from the vitelline envelope (Eckberg and Anderson 1985; Sato and Osanai 1986; Shin et al. 2005), and a jelly coat-level block in *Nereis* wherein the jelly substances released outside the vitelline envelope transport the extra sperm away from the oocyte surface to leave a single fertilizing sperm (Lillie 1911).

7.2.2.4 Oocyte Activation via Intracellular Ca^{2+} Increase

Fertilized *Urechis* oocytes exhibit an increase in intracellular Ca^{2+} that comprises an initial cortical flash and subsequent plateau phase (Stephano and Gould 1997a; Table 7.1). The results of electrophysiology analyses (see above), ^{45}Ca uptake measurements (Johnston and Paul 1977; Jaffe et al. 1979), and experiments using Ca^{2+} channel inhibitors strongly suggest that the increase in Ca^{2+} at fertilization depends mainly on an influx of Ca^{2+}, although Ca^{2+} released from IP_3-dependent internal stores may play a supplementary role (Stephano and Gould 1997a). *Urechis* oocytes also exhibit an increase in intracellular pH at fertilization (Gould and Stephano 1993), and both the increases in Ca^{2+} and pH are required for the resumption of meiosis from PI (Stephano and Gould 1997a), similar to the above-described situations in PI-type bivalves.

In contrast, fertilized *Chaetopterus* oocytes exhibit apparent Ca^{2+} oscillations for several minutes; the initial transient increase in Ca^{2+} always appears as a point-source non-propagated Ca^{2+} wave, which is followed by fully propagating Ca^{2+} waves (Eckberg and Miller 1995; Table 7.1). *Chaetopterus* oocytes can utilize two Ca^{2+}-mobilizing systems, voltage-dependent external Ca^{2+} influx and IP_3-dependent internal Ca^{2+} release (Eckberg and Miller 1995; Thomas et al. 1998; Howell et al. 2003; Yin and Eckberg 2009), although to what extent each mechanism contributes to fertilization-induced Ca^{2+} oscillations remains uncertain.

Pseudopotamilla oocytes exhibit a unique two-step increase in Ca^{2+} at fertilization. In the first step, a non-propagated Ca^{2+} wave that depends mainly on IP_3-dependent Ca^{2+} release occurs in one cortical region and causes the expansion of this region toward the vitelline envelope to engulf a fertilizing sperm, as described above (Nakano et al. 2008, 2014; Fig. 7.3). In the second step, a cortical flash causes a global increase in Ca^{2+} that persists for approximately 10 min. This latter step may be due to a voltage-dependent Ca^{2+} influx and is responsible for the retraction of the cytoplasmic protrusion, disappearance of microvilli in the perivitelline space, and resumption of meiosis from MI in fertilized oocytes (Nakano et al. 2008, 2014; Fig. 7.3).

In *Urechis* sperm, an acrosomal protein containing the active domain VAKKPK binds to the oocyte surface and induces all fertilization-associated events, including membrane depolarization (Gould and Stephano 1991), intracellular increases in Ca^{2+} (Stephano and Gould 1997a) and pH (Stephano and Gould 1997b), and the resumption of meiosis from PI (Gould and Stephano 1991). In *Pseudopotamilla*, the local application of a soluble sperm extract to an unfertilized oocyte mimics a non-propagated increase in Ca^{2+} and, subsequently, the formation of a cytoplasmic protrusion, as seen at fertilization (Nakano and Kyozuka 2015). These externally applicable "sperm factors" in *Urechis* and *Pseudopotamilla* provide a marked contrast to phospholipase C-zeta (PLCζ) in mammalian sperm, which is delivered to the oocyte cytoplasm after the sperm–oocyte fusion step and thus induces Ca^{2+} oscillations and oocyte activation (Kashir et al. 2013; Swann and Lai 2016).

7.2.3 Arthropods

7.2.3.1 Mode of Sexual Reproduction

Arthropoda is the largest phylum in the animal kingdom, with many more than 1,000,000 extant species (Chapman 2009). Most species are insects (~1,000,000 described species) that possess exoskeletal spiracles through which air enters the trachea (Maina 2002; Hsia et al. 2013), and thus these animals thrive in terrestrial but not marine environments (Ikawa et al. 2018). By contrast, crustaceans (~47,000 described species), the second largest group of arthropods, are predominantly found in ocean waters, although freshwater and terrestrial species also exist (Chapman 2009). Most aquatic crustaceans, as well as the horseshoe crab (a marine chelicerate species), use gills for respiration (Farley 2010). In contrast, terrestrial crustaceans absorb oxygen through various mechanisms; for instance, woodlice can breathe using lungs located in their legs (Hsia et al. 2013).

Similar to insects, aquatic crustaceans and horseshoe crabs are typically dioecious; in contrast, many species of barnacles, a type of sedentary marine crustacean, are simultaneous hermaphrodites (Yamaguchi et al. 2013). Crustaceans and horseshoe crabs are either internal or external fertilizers, and their strategies for successful fertilization are highly diverse, even within closely related species. In the American horseshoe crab *Limulus polyphemus*, oocytes released from a female are immediately inseminated by sperm released from a mated male, as well as from "satellite males" that surround the pair (Mattei et al. 2010). In many species of shrimp, external fertilization occurs immediately after a female releases her oocytes together with the sperm that were obtained from copulating male partners and stored in seminal receptacles (Pillai and Clark 1987; Goudeau and Goudeau 1989b). The stalked barnacle *Pollicipes polymerus* conducts "spermcast mating," wherein sperm released from one individual to the surrounding seawater are captured by neighbor individuals and used to fertilize their oocytes in mantle cavities (Barazandeh et al. 2013). Many other barnacles use their penises to transfer self-sperm to partners via "pseudo-copulation." In any case, marine and freshwater arthropods differ from other various aquatic animals because they do not adopt the system of free-spawning broadcast fertilization, which occurs between two types of gametes released simultaneously. In addition to bisexual reproduction, arthropods frequently undergo parthenogenesis; in some species or strains of aquatic crustaceans, such as the water flea *Daphnia* (Hiruta et al. 2010), the brine shrimp *Artemia* (Maccari et al. 2014), and the crayfish *Procambarus* (Vogt et al. 2015; Kato et al. 2016), as well as in various species of insects, the females can produce offspring in the absence of males.

7.2.3.2 Gamete Structures and Fertilization Process

Mature oocytes collected immediately after ovulation or isolation from the ovaries of aquatic arthropods used in fertilization experiments exhibit a remarkable range of diameters from 100 μm (*Pollicipes polymerus*) to 2.5 mm (*Limulus polyphemus*) (Brown and Humphreys 1971; Lewis 1975; Table 7.1). These oocytes commonly contain cortical vesicles inside the plasma membrane, which is covered with a vitelline envelope (e.g., Klepal et al. 1979; Bannon and Brown 1980; Goudeau 1984; Talbot and Goudeau 1988; Pongtippatee-Taweepreda et al. 2004; Dupré and Barros 2011; Niksirat et al. 2015). In barnacles, shrimps, lobsters, and crabs, MI is the arrest point of meiosis before oocyte activation (Walley et al. 1971; Goudeau 1982; Goudeau and Goudeau 1986b, 1989a, b; Lindsay et al. 1992; Table 7.1).

Male horseshoe crabs release primitive and motile sperm that comprise four parts: the apical acrosome, nucleus, mitochondria, and flagellum (9 + 2 or 9 + 0 axoneme pattern) (Brown and Humphreys 1971; Yamamichi and Sekiguchi 1982; Ishijima et al. 1988). Barnacle sperm are also flagellate and motile but are structurally modified to contain an elongated nucleus located alongside three quarters of the axoneme, as well as a single mitochondrion just behind the nucleus (Walley et al. 1971; Kubo et al. 1979; Klepal 1990). Decapods, including shrimp, lobsters, crayfish, hermit crabs, and crabs, produce atypical and immotile sperm for internal or external fertilization. Despite the species-specific morphological characteristics, these decapod sperm share common structures such as a well-developed acrosomal vesicle in the anterior portion and a nucleus composed of uncondensed chromatin in the posterior region. However, these sperm generally lack a flagellate tail containing the axoneme, as well as a mitochondrial midpiece (e.g., Lynn and Clark 1983b; Griffin and Clark 1990; Tudge et al. 2001; Klaus et al. 2009; Simeó et al. 2010; Dupré and Barros 2011; Wang et al. 2015).

Crustacean oocytes undergo cortical modifications and meiotic divisions from MI after spawning or exposure to seawater. In penaeid and caridean shrimps, exposure to seawater triggers the full process of oocyte activation, regardless of the presence or absence of sperm (Goudeau and Goudeau 1989b; Pillai and Clark 1987; Goudeau et al. 1991; Lindsay et al. 1992; Clark and Griffin 1993). Mg^{2+} has been identified as an essential seawater component that triggers oocyte activation in shrimp (Pillai and Clark 1987; Goudeau and Goudeau 1989b; Pongtippatee et al. 2010), although in one exceptional species, normal activation occurs even in the presence of a very low concentration of Mg^{2+} (Rojas and Alfaro 2007).

Oocytes of the penaeid shrimp *Penaeus monodon* (~275 μm in diameter) undergo two-step cortical modifications upon release into seawater (Pongtippatee-Taweepreda et al. 2004). In the first step, cortical rods (10 μm wide and 35–40 μm long, ~400 per oocyte) containing jelly precursor substances are expelled from invaginations of the oocyte cortex (i.e., cortical crypts) within the first minute to form a jelly layer (Fig. 7.4). In the second step, which occurs independently of jelly layer formation and is completed after approximately 15 min, exocytosed cortical vesicles modify the vitelline envelope into a thicker fertilization membrane (i.e., "hatching envelope") and induce the appearance of the perivitelline space (0.5–3 μm in width) (Fig. 7.4). Meiotic divisions from MI, which manifest as extrusions of the first (3–5 min) and second polar bodies (10–15 min), proceed in parallel to these cortical modifications (Fig. 7.4). A similar sperm-independent process of oocyte activation has been observed in other penaeid shrimp species, although the time course is greatly affected by differences in species and experimental conditions (e.g., temperature) (Clark et al. 1980; Pillai and Clark 1987; Lindsay et al. 1992; Clark and Griffin 1993; Rojas and Alfaro 2007; Ikhwanuddin et al. 2015). In fact, emissions of the first and second polar bodies in *Sicyonia ingentis* oocytes incubated at 21°C occurred at intervals of 35–40 and 45–50 min, respectively, after exposure to seawater (Lindsay et al. 1992). *Palaemon serratus* oocytes can also be fully activated in the absence of sperm (Goudeau and Goudeau 1986b, 1989b; Goudeau et al. 1991), although neither the cortical rod-like structure nor initial jelly layer formation is detectable in this or other species of caridean shrimps (Lynn and Clark 1983a; Goudeau et al. 1991; Kim et al. 2007; Dupré and Barros 2011).

The shrimp sperm binds to the oocyte vitelline envelope via an apical appendage, the anterior spike, and undergoes an acrosome reaction that appears to be initiated by molecules in "egg water" (which contains jelly layer components) but not in the

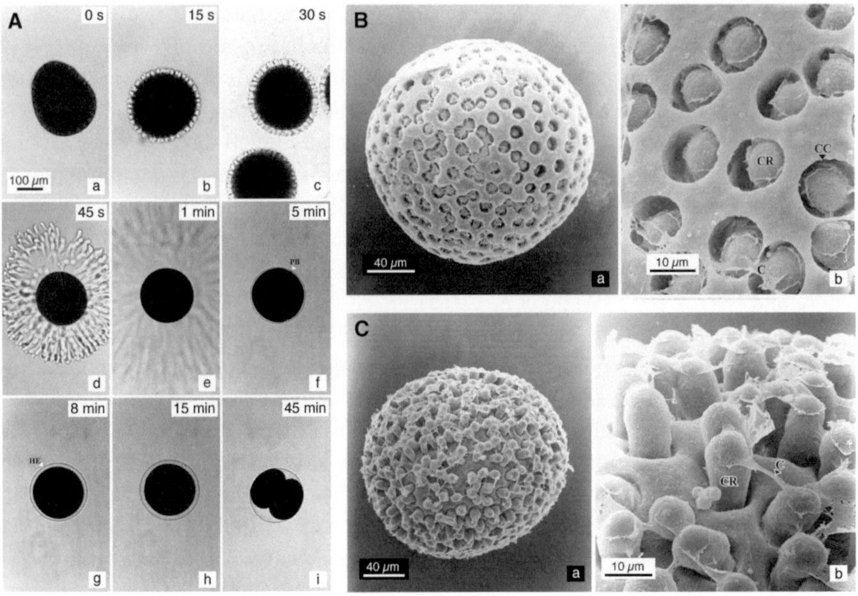

Fig. 7.4 Light and scanning electron micrographs of *Penaeus monodon* oocytes obtained during the activation process. (**a**) Bright-field images revealing cortical modifications and cell cycle progression in oocytes or eggs. Following exposure to seawater (**a**), oocytes extrude cortical rods for the formation of a jelly layer (**b**–**e**), elevation of a hatching envelope (**f**–**h**), and emission of the first and second polar bodies (**f**–**h**) before the first mitotic cleavage (**i**). *PB* first polar body, *HE* hatching envelope. (**b**, **c**) Scanning electron microscopy images of the oocyte surface before and 15 s after exposure to seawater. *CC* cortical crypt, *CR* cortical rod, *C* vitelline envelope (i.e., investment coat). (From Pongtippatee-Taweepreda et al. (2004), with permission)

vitelline envelope (Griffin and Clark 1990; Kruevaisayawan et al. 2008; Ngernsoungnern et al. 2012). Multiple sperm cells can penetrate the vitelline envelope and enter the perivitelline space shortly after binding (Lindsay et al. 1992; Pongtippatee-Taweepreda et al. 2004). However, sperm–oocyte fusion does not occur within the initial 30 min period in *Sicyonia ingentis* (Lindsay et al. 1992). In the caridean shrimp *Rhynchocinetes typus*, an umbrella-like sperm equipped with 9–13 radial arms (each containing uncondensed chromatin extending from the cell body) binds to the vitelline envelope via an anterior spike, enters the perivitelline space and fuses with the oocyte plasma membrane without indicating an apparent acrosome reaction (Dupré and Barros 2011).

In contrast to shrimps, oocyte activation in other crustaceans, such as barnacles (Walley et al. 1971; Lewis 1975; Klepal et al. 1979), lobsters (Goudeau and Goudeau 1986a; Talbot and Goudeau 1988), and crabs (Goudeau and Becker 1982; Goudeau and Goudeau 1989a, b), depends partly or mostly on the presence of sperm. In crabs, for instance, exposure to seawater (or the presence of 30–50 µM Ca^{2+} in Mg^{2+}-free seawater) stimulates the oocytes to release one of two distinct populations of cortical vesicles containing a fine granular material for lifting the vitelline envelope and encourages the resumption of meiosis from MI. Fertilization induces the slow and long-lasting exocytosis of another type of cortical vesicle, which contains ring-shaped elements and is responsible for the formation of a thick fertilization membrane (Goudeau and Becker 1982; Goudeau and Goudeau 1989a, b).

7.2.3.3 Polyspermy Block

Generally, shrimp oocytes that are spawned naturally into normal seawater are mixed with sperm released simultaneously from the same female and are thus monospermic (Lynn and Clark 1983a; Goudeau and Goudeau 1989b; Lindsay et al. 1992). Polyspermy can be induced by incubating spawned oocytes in low-Mg^{2+} or Mg^{2+}-free seawater, as this delays or suppresses both a series of cortical modifications and the resumption of meiosis from MI (Lindsay et al. 1992; Pongtippatee et al. 2010). Immediately after spawning, the oocytes extrude cortical rods that push away most of the potentially binding sperm (Lindsay et al. 1992; Pongtippatee-Taweepreda et al. 2004), and the subsequent formation of a fertilization (hatching) membrane prevents sperm from belatedly entering the perivitelline space (Pongtippatee-Taweepreda et al. 2004). Therefore, cortical modifications may help to eliminate excess sperm around the shrimp oocytes. However, these changes alone are not sufficient to block polyspermy, as multiple sperm can pass through the vitelline envelope immediately after contact, as described above.

Unfertilized *Palaemon* oocytes that have been preincubated in Mg^{2+}-free seawater to avoid activation and inseminated simultaneously with the replacement of external medium by normal seawater were shown to exhibit a series of changes in the membrane potential, namely an initial transient hyperpolarization phase followed by depolarization and final hyperpolarization phases (Goudeau and Goudeau 1986c, 1989b). However, no significant differences in the membrane potential pattern were observed between oocytes that incorporated sperm and those that failed to fertilize, suggesting a lack of "fertilization potential" in oocytes of this species. Therefore, the mechanism by which only one of multiple sperm in the perivitelline space can fuse with the shrimp oocyte plasma membrane remains unclear.

Immediately after artificial insemination in normal seawater, the membrane potential of an oocyte from the crab *Maja squinado* becomes hyperpolarized (from −54 to −80 mV) due to an increase in K^+ permeability (Goudeau and Goudeau 1989a, b; Fig. 7.5). Voltage clamp experiments have shown that sperm can incorporate into oocytes clamped between −65 and −75 mV but not those clamped at ≤ −60 mV or ≥ −80 mV (Fig. 7.5). Naturally spawned *Maja* oocytes are monospermic, and their membranes remain hyperpolarized for several hours (Goudeau and Goudeau 1989a, b). Therefore, membrane hyperpolarization at fertilization likely enables the crab oocyte to incorporate a single fertilizing sperm during an initial short time window and prevents the entry of additional sperm during subsequent long periods. Sperm-induced membrane hyperpolarization has also been detected in lobster oocytes, although its role in polyspermy block is unknown (Goudeau and Goudeau 1986a, 1989b).

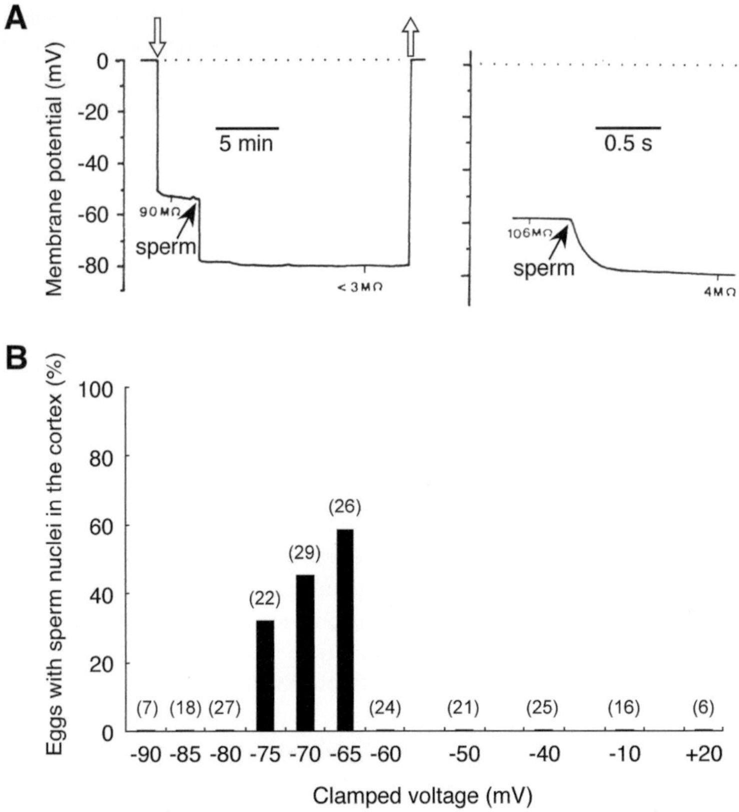

Fig. 7.5 Membrane hyperpolarization in *Maja squinado* oocytes at fertilization. (**a**) Changes in the membrane potential at fertilization were recorded at slow (left) and high speed (right) in two different oocytes. Downward and upward white arrows indicate the times when a microelectrode was inserted into and withdrawn from the oocyte, respectively. (**b**) Sperm entry into voltage-clamped oocytes. Numbers in parentheses indicate the total numbers of oocytes used for the experiments. In the oocytes that were clamped at −75, −70, and −65 mV and permitted sperm entry, an average of 3–6 sperm nuclei were observed in the cortices. (From Goudeau and Goudeau (1989a), with permission)

7.2.3.4 Oocyte Activation via Intracellular Ca^{2+} Increase

In shrimp oocytes, Ca^{2+} ionophores such as A23187 and ionomycin have been shown to trigger cortical modifications and meiosis resumption from MI even in the absence of external Mg^{2+} (Goudeau and Goudeau 1986b; Goudeau et al. 1991), indicating that an intracellular increase in Ca^{2+} plays a role in Mg^{2+}-induced oocyte activation. Indeed, *Sicyonia* oocytes exposed to external Mg^{2+} exhibit a propagating Ca^{2+} wave that initiates from the cortical site (i.e., the side that first comes into contact with Mg^{2+}) (Lindsay et al. 1992; Lindsay and Clark 1994). This initial increase in Ca^{2+} lasts for a few minutes, during which the oocytes undergo cortical contractions associated with the formation of a jelly layer (Lindsay et al. 1992). However, the direct injection of Mg^{2+} into oocytes cannot induce an increase in Ca^{2+} (Lindsay et al. 1992), and several other data suggest the existence of an "Mg^{2+} receptor" on the oocyte surface that stimulates IP_3-dependent Ca^{2+} release, rather than external Ca^{2+} influx (Lindsay and Clark 1994). In activated *Sicyonia* oocytes, the second Ca^{2+} wave occurs at approximately 15 min post-spawning, although the underlying mechanism or function remains unclear (Lindsay et al. 1992). Intracellular changes in Ca^{2+} in response to external Mg^{2+} have also been detected in *Palaemon* oocytes; here, an initial transient increase in Ca^{2+} is followed by Ca^{2+} oscillations lasting for 70–80 min, during which the oocytes undergo the first polar body formation and cortical modification (Goudeau and Goudeau 1996, 1998). These changes in Ca^{2+} can be reversibly interrupted by the absence of external Mg^{2+} and have been proven to be mediated by an IP_3-dependent Ca^{2+} release, although a capacitive (i.e., store-operated) Ca^{2+} entry also appears to contribute to later Ca^{2+} oscillations. Unfortunately, no information about the molecules involved in downstream of Ca^{2+} changes in these shrimp species is available.

Recent studies have demonstrated that *Drosophila* oocytes exhibit an intracellular Ca^{2+} wave at ovulation; the onset and propagation of this wave require an influx of Ca^{2+} through mechanosensitive channels and an IP_3-dependent release of Ca^{2+}, respectively (Kaneuchi et al. 2015). Importantly, in *Drosophila* oocytes, the Ca^{2+} wave proceeds independently of sperm, as observed in shrimp oocytes. The sperm-independent increase in Ca^{2+} and oocyte activation in arthropods provide a striking contrast to the sperm-triggered changes in many other animals. These differences might have facilitated the evolutionary appearance of parthenogenetic species, which are frequently observed in this phylum.

7.3 Overview of the Endocrine System Associated with Reproduction in Bivalves

7.3.1 Preface

In the fishery industry, bivalve species are commercially important organisms harvested as capture and aquaculture products. The reproduction of bivalve species is an essential subject in terms of the biological evaluation of brood stock and the environmental evaluation of aquacultural areas. However, knowledge about the endocrine control associated with gametogenesis and the spawning of bivalve mollusks is limited, and a realization of the stable artificial seed production based on reproductive control is thirsted to improve productivity in aquaculture industry.

The development of oocytes and sperm during gametogenesis is commonly observed in oviparous organisms. During oogenesis, primary germ cells (stem cells) undergo mitotic division, give rise to oogonia (Sastry 1979), and then develop into oocytes upon the accumulation of yolk materials. In the testis, stem cells undergo a series of mitotic divisions associated with decreasing cytoplasmic volume, give rise to spermatogonia (Sastry 1979), and undergo meiosis to form spermatozoa through spermatocytes and spermatids, followed by spermiogenesis. Normally, gametogenesis is an annual cycle, and the changes in gonadal development have been qualitatively classified into several stages; in the scallop, the undifferentiating, early differentiating, growing, mature, spawning, post-spawning, and degenerating stages (Osanai 1975) have been determined based on the characterization of overall gonadal morphology, rather than the quantitative analysis of germ cells.

In this section, the mitotic process during early germ cell development, the oocyte growth process following vitellogenesis, and spawning processes based on oocyte maturation and sperm motility in bivalves are described from the comparative endocrinology point of view.

7.3.2 *Multiplication of Gonial Cells*

The development of gonial cells must be quantified to better understand the endocrine control of early germ cell development in bivalves. The mitotic division of gonial cells has been quantified in some bivalve species. In the blue mussel *Mytilus edulis*, autoradiography was used to demonstrate the incorporation of [^{3}H] thymidine into young germinal cells, particularly spermatogonia, in the mantle tissue (Mathieu 1987). The incorporation of [^{3}H] thymidine into dissociated cell suspensions of mantle tissue was associated with the estimated aspartate transcarbamylase activity. Consequently, this phenomenon was used to quantitatively evaluate the mitogenic influence of an endogenous factor on spermatogonial multiplication (Mathieu 1985, 1987). In a cytologic analysis of the Yesso scallop *Patinopecten (Mizuhopecten) yessoensis*, the incorporation of BrdU into gonial cells during early development revealed that the proliferation of these cells could be divided into phases I and II (Osada et al. 2007). In phase I, the oogonia and spermatogonia underwent slow proliferation. In phase II, oogonia exhibited a termination of proliferation, followed by development into vitellogenic oocytes, whereas spermatogonia underwent rapid proliferation. In vertebrates, developing spermatogonia can be classified into two types: non-proliferated type A spermatogonia or spermatogonial stem cells and type B spermatogonia or differentiated spermatogonia (Miura and Miura 2001). Osada et al. (2007) proposed that in bivalves, proliferation potency could be used to classify spermatogonia as spermatogonial stem cells and a developmental series of differentiated spermatogonia during early spermatogenesis.

7.3.2.1 GnRH Peptides in Mollusks

Mollusks have extremely large nerve cells and are therefore useful to scientists involved in the study of neurophysiological responses. The effects of neural factors on gonial mitosis in mollusks were first reported in *Mytilus* as follows. In an in vitro mussel organ culture experiment, the cerebral ganglia stimulated gonial mitosis as well as the resumption of meiosis in males and previtellogenesis and vitellogenesis in females; however, the ablation of the ganglia did not affect the reproductive function in vivo (Mathieu and Lubet 1980; Lubet and Mathieu 1982). The involvement of neural factors in germ cell multiplication has also been reported in a gastropod, a slipper limpet (Le Gall et al. 1987). Mathieu et al. (1988) reported the occurrence of a neural factor, gonial mitosis-stimulating factor, from the cerebral ganglia of the mussel. This factor had a molecular mass of <5 kDa; it is known to stimulate the incorporation of [^{3}H] thymidine into dissociated cell suspensions of mussel mantle tissue and was detected in the hemolymph and circulatory cells. In bivalve mollusks, this neural factor was thought to be associated with a gonadotropin-releasing hormone (GnRH).

The GnRH superfamily, which includes GnRH, adipokinetic hormone (AKH), corazonin (Crz), and AKH/Crz-related peptides, is almost ubiquitous in bilateral animals (Roch et al. 2011). In vertebrates, GnRH is synthesized in the preoptic area of hypothalamus and transported to the pituitary to promote the release of gonadotropic hormone, which comprises follicle-stimulating hormone and luteinizing hormone (LH). The hypothalamus–pituitary–gonadal axis (HPG axis) connects the brain and gonads via the pituitary and forms the basis for both the neural and endocrine regulation of reproduction in all vertebrates. Interestingly, α-mating factor, a tridecapeptide mating pheromone of yeast, has been identified as a homolog of GnRH and can induce LH release from gonadotrophs at high doses, suggesting that the structural and functional properties of GnRH-related peptides have been conserved during evolution (Loumaye et al. 1982).

Subsequently, the existence and functions of GnRH-like peptides in mollusks have been demonstrated using heterologous GnRH antibodies. In the cephalopod *Octopus vulgaris*, immunopositive GnRH-like peptides were detected in the optic gland, a major endocrine organ (Di Cosmo and Di Cristo 1998). Furthermore, GnRH was detected throughout the central nervous system (CNS) and in both the male and female reproductive ducts (Di Cristo et al. 2002), suggesting that GnRH may have a reproductive function in the octopus. In gastropods, the nervous system of *Hellisoma* (*Planorbella*) *trivolvis* exhibited characteristics consistent with the existence of a GnRH-like peptide that was identified as functionally similar to mammalian GnRH (m-GnRH) (Goldberg et al. 1993). GnRH-like neurons were detected in the nervous systems of freshwater snails, including *Hellisoma*, leading researchers to hypothesize a possible reproductive role (Young et al. 1999). Studies of the marine sea hare, *Aplysia californica*, suggested the presence of multiple forms of a GnRH-like peptide (Zhang et al. 2000), and heterologous GnRH antibodies were used to localize the expression of these peptides to specific sites within the CNS of *Aplysia* (Tsai et al. 2003).

As mentioned above, Mathieu et al. (1988) first reported an in-depth study of the indirect use of mollusk GnRH; here, extracts from the cerebral ganglion and hemolymph of *Mytilus* were shown to promote the incorporation of [^{3}H] thymidine and, consequently, mitosis in gonial cells. Subsequently, the same research group showed that vertebrate GnRHs could also affect the mitosis of gonial cells from the Pacific oyster *Crassostrea gigas*, as well as the mussel, and identified GnRH-like neurons in the CNS of the mussel (Pazos and Mathieu 1999). GnRH signals may be transduced into cells via a membrane receptor, as indicated by the cloning of a GnRH receptor ortholog from the gonads of oysters (Rodet et al. 2005, 2008). Furthermore, mass spectrometry was used to confirm the presence of two GnRH-related peptides (pQNYHFNSNGWQP-NH_2 Cg-GnRH-a; amidated undecapeptide and pQNYHFNSNGWQPG Cg-GnRH-G; non-amidated dodecapeptide) in the CNS of *Crassostrea* (Bigot et al. 2012). However, the specific affinity of the GnRH receptor orthologs for the identified endogenous oyster GnRH-like peptides (Bigot et al. 2012) has not yet been confirmed. In the scallop, Nakamura et al. (2007) detected a GnRH-like peptide in the neurons and fibers of CNS using anti-m-GnRH antibody, whereas no immunopositive GnRH nerve fibers were detected in the gonad. The neural factors extracted from the CNS and the hemocyte lysate, and m-GnRH strongly stimulated the mitosis of spermatogonia in vitro. The reactions of neural factor and m-GnRH were abolished by absorption with an anti-m-GnRH antibody and competition with m-GnRH-specific antagonists, which interfere with the binding of GnRH to the GnRH receptor; these phenomena suggested the presence of an endogenous GnRH-like peptide and GnRH receptor-like receptor. These results suggest that the neural factor and m-GnRH are antigenically similar and that the function of this neural factor in the testis may be mediated through an m-GnRH receptor-like receptor via a neuroendocrine pathway.

The full cDNA sequences of GnRHs in the scallop and the oyster were cloned as py-GnRH and cg-GnRH, respectively. The GnRH-like peptide sequences of both bivalve species were highly similar to oct- and ap-GnRHs (Treen et al. 2012). Similar to other mollusk species, an extra dipeptide insertion after the N-terminal pyro-glutamate residue was detected in both bivalve species (Osada and Treen 2013). The logically predicted pQNFHYSNGWQP-NH_2 (py-GnRH11AA-NH_2), an amidated undecapeptide, was synthesized and shown to stimulate the proliferation of spermatogonia in a tissue culture of the scallop testis, similar to the findings of a previous culture with m-GnRH (Treen et al. 2012). However, this synthesized peptide failed to induce the release of LH release from the quail pituitary, suggesting that although the conserved fundamental molecular structure of mollusk GnRHs might be similar to that observed in other animals, the mollusk forms may not bind to vertebrate receptors (Treen et al. 2012). The predicted peptide pQNFHYSNGWQP-NH_2 (py-GnRH11AA-NH_2) was confirmed via a mass spectrometry analysis of the scallop CNS as an endogenous GnRH-like peptide, together with pQNFHYSNGWQPG-OH (py-GnRH12AA-OH) (Nagasawa et al. 2015c) as two types of GnRHs in the oyster (Bigot et al. 2012).

In mollusks, GnRH has been suggested to play several other physiological functions besides spermatogonial proliferation. oct-GnRH-immunopositive nerve fibers were identified in both the CNS and peripheral organs, suggesting that GnRH may act to modulate higher brain functions in addition to its role as a reproductive factor (Iwakoshi-Ukena et al. 2004; Kanda et al. 2006). oct-GnRH was shown to modulate the contractions of the heart and oviducts (Iwakoshi et al. 2000; Iwakoshi-Ukena et al. 2004). Moreover, oct-GnRH stimulated the production of progesterone, testosterone, and estradiol-17β (E_2) in both the octopus testis and ovaries, suggesting a reproductive role for this factor (Kanda et al. 2006). Ovarian development in the octopus was found to associate with fluctuations in the sex steroid hormones (Di Cosmo et al. 2001; Di Cristo 2013). Ap-GnRH is distributed in the central tissue and modulates behavioral attributes that control parapodia and foot and head movement, but does not act as an acute reproductive trigger for the development of ovotestis or the secretion of egg-laying hormone (Tsai et al. 2010; Sun and Tsai 2011; Sun et al. 2012). Previous findings suggested the presence of multiple forms of GnRH in *Aplysia* (Zhang et al. 2000), and ap-AKH was found to share a common ancestry with AKH/RPCH (red pigment concentrating hormone), as well as ap-GnRH, and to inhibit feeding with consequent reductions in the body and gonadal masses (Johnson et al. 2014). In the scallop, both a py-AKH-like gene and py-AKH receptor gene were found to be expressed predominantly in the CNS (Nagasawa et al. 2017). Bigot et al. (2012) used mass spectrometry to identify two types of GnRH in the oyster CNS, although the functions have not yet been confirmed. In vivo, the scallop py-GnRH accelerated spermatogenesis in the scallop gonad, which was associated with a significant increase in testis mass; conversely, this factor inhibited oocyte development in a process involving apoptosis, suggesting that py-GnRH is associated with an early phenotypic alteration leading to masculinization (Nagasawa et al. 2015a). The py-GnRH signal that induces spermatogenesis and masculinization is thought to be transmitted via the py-GnRH receptor, which is broadly expressed in various tissues, including the gonad and CNS (Nagasawa et al. 2017).

In mollusks without a pituitary, GnRH may be involved in spermatogonial proliferation via the mediation of gonadal steroidogenesis (Osada and Treen 2013). Immunological methods were used to detect estrogen-synthesizing cells along the outside and inside acinar walls of the ovary and testis of the scallop, respectively; notably, this localization pattern was similar to that of Leydig cells and Sertoli cells in the testis, respectively (Matsumoto et al. 1997; Osada et al. 2004b). The observed increases in aromatase activity and E_2 content were synchronous with reproductive progress (Osada et al. 2004b). In vertebrates, E_2 is involved in spermatogenesis (Hess et al. 1997; Pierantoni et al. 2009). Interestingly, oct-GnRH was shown to induce the steroidogenesis of testosterone, progesterone, and estrogen (Kanda et al. 2006), and data mining and PCR cloning have been used to suggest a biosynthetic pathway for sex steroids in the scallop, wherein the mRNA expression of steroidogenesis-related genes was found to associate with gametogenesis (Thitiphuree et al. 2019). Possibly, py-GnRH stimulates estrogen synthesis, which in turn promotes spermatogonial mitosis; both py-GnRH and E_2-induced spermatogonial proliferation could be blocked by an estrogen antagonist (Osada and Treen 2013). Possibly, py-GnRH-stimulated spermatogonial proliferation might be mediated by estrogen, which itself was synthesized via a process induced by py-GnRH (Osada and Treen 2013).

7.3.2.2 Steroid Production

Steroid production and function in the context of mollusk reproduction have been well examined using vertebrate steroids, and these studies have enabled the quantification of steroids via immunological methods and HPLC, the identification of putative steroid production sites, and the induction of sex reversal and oogenesis (see Lafont and Mathieu 2007). However, endogenous steroids and their function in mollusks are recently in contention because convincing evidence for biosynthesis of endogenous steroids by mollusks and the biological effects of endogenous steroids on mollusks are required (Scott 2012, 2013). On the other hand, a potential pathway of steroid synthesis in the scallop has been proposed as stated above (Thitiphuree et al. 2019). Future studies should explore and elucidate the functions of the series of enzymes that compose steroid synthetic pathways and determine the structures of endogenous steroids and their affinities with mollusk-specific receptors.

7.3.3 Oocyte Growth

7.3.3.1 York Protein

The oocyte increases rapidly in size during growth and development. During these phases, the oocytes accumulate large amounts of yolk protein. Vitellin (Vn), a major yolk protein, is stored in the yolk granules of oocytes and used as a nutrient during embryogenesis. In vertebrates, this yolk protein is synthesized from a precursor, vitellogenin (Vtg), which is produced by the liver and transported to the oocytes via the blood circulation system. A few biochemical studies have evaluated yolk proteins in bivalves (Osada et al. 1992b; Suzuki et al. 1992). Yolk protein was purified from *Crassostrea* and *Patinopecten*, and the distribution of this protein in these organisms was detected using a specific antiserum against the purified protein. The detection of this protein in the oocytes confirmed its identity as Vn. In the oyster, the ovarian Vn content increased as the oocyte developed and decreased after spawning. This expression pattern correlated with the profile of the oocyte diameter and the results of histological observations of the ovaries (Li et al. 1998). In the scallop, the Vn content remained low level until the growing stage and reached a peak upon maturation stage, followed by a marked decrease in the degenerating stage after spawning. The level of GI was lowest during the early stage of differentiation and gradually increased to reach a peak immediately before spawning. In the scallop ovary, the Vn content increased in parallel with ovarian development (Osada et al. 2003). These results indicate that oocyte growth depends on the internal accumulation of Vn during vitellogenesis. The finding that no other organ reacted with anti-Vn serum led researchers to predict that Vn is synthesized inside the ovary. Additionally, immunoreactivity against an anti-scallop Vn antibody was also observed in auxiliary cells, which suggested the possibility of Vtg synthesis in these cells (Osada et al. 2003). Low levels of Vn were also detected in the hemolymph of mature females, particularly during the spawning season. This protein appeared to originate from degenerated oocytes. Oocyte degeneration and resorption are not unusual processes in bivalves, and both may be induced by a variety of environmental conditions (Pipe 1987b).

7.3.3.2 Place of Vitellogenesis

The mechanisms associated with vitellogenesis were determined through molecular investigations (Matsumoto et al. 2003; Osada et al. 2004a). The expression of Vtg mRNA peaked during the growing or developing stage, and this level was maintained during the mature stage. In a RT-PCR analysis, Vtg mRNA was detected in the ovary but not the digestive diverticula, consistent with the results of immunohistochemistry analyses. In situ hybridization analyses revealed detectable Vtg mRNA signals in the ovarian follicle cells. Similarly, Vtg mRNA expression was detected in scallop auxiliary cells. In marine bivalves, morphological evidence suggests that autosynthetic yolk formation in the oocytes is the main type of vitellogenesis (Pipe 1987a; Dorange and Le Pennec 1989; Suzuki et al. 1992; Eckelbarger and Davis 1996). In oyster and other bivalve species, the ovarian acini contain only developing oocytes and associated follicle cells within a thin germinal epithelium. Although the functions of follicle cells in the bivalve ovary are not well understood, these cells are thought to play a role in oocyte nutrition. Taken together, the results of RT-PCR, in situ hybridization, and immunocytological analyses indicate that Vtg is synthesized in the auxiliary cells that surround the vitellogenic oocyte via a heterosynthetic pathway and is transferred directly to the oocyte. Similar to bivalves, the Vtg gene is expressed in the follicle cells of the abalone ovary (Matsumoto et al. 2008). In both abalone species and bivalves, follicle cells are present on the stalks of developing oocytes in the maturing ovary. A recent study revealed immunoreactivity to an anti-Vn antibody in abalone ovarian follicle cells and indicated that Vtg gene transcription and translation occur in these cells (Awaji et al. 2011). In addition, immunoreactivity to an anti-Vn antibody was first detected in the stalks of oocytes during the early phase of yolk accumulation; the follicle cells adjacent to the stalks also exhibited positive staining. These observations imply that Vn or Vtg is transported from the follicle cells to the oocyte through the extracellular space around the oocyte stalk. In the scallop, Vn immunoreactivity was detected in the oocytes and auxiliary cells (Osada et al. 2003). Likely, the oyster and scallop employ similar systems of yolk protein transport.

7.3.3.3 Hormonal Control of Vitellogenesis

Previous research in oviparous vertebrates has clearly established the regulation of Vtg gene expression by E_2 through the ER (Polzonetti-Magni et al. 2004). The presence of steroids in marine bivalves has also been reported; for example, the structural identities of endogenous steroids as vertebrate-type sex steroids (progesterone, androstenedione, testosterone, E_2, and estrone) were demonstrated in the mussel using gas chromatography and mass spectrometry (Reis-Henriques et al. 1990). In oysters and scallops, E_2 has been detected in the ovary, and HPLC analyses revealed that the profile of this steroid was synchronous with gametogenesis (Matsumoto et al. 1997). In the scallop ovary, estrogenic cells were identified by observing immunoreactivity against P450 aromatase and E_2, which were found to be distributed along the outside of the acinar wall. Moreover, the levels of aromatase activity and E_2 peaked during the mature stage before spawning (Osada et al. 2004b), while Vtg synthesis was shown to terminate during this stage (Osada et al. 2003, 2004a). These results suggest that E_2, which is synthesized in estrogenic cells via P450 aromatase, may be involved in the induction of Vtg synthesis. The physiological role of estrogen in vitellogenesis has also been reported in various mollusks, although this role remains debated (Scott 2013). Estrogen-induced Vn synthesis has been observed in the scallop (Osada et al. 2003) and oyster (Li et al. 1998). In vitro, E_2 treatment induced an increase in Vn levels in ovarian tissues. Furthermore, an in vitro culture of scallop ovarian tissue that contained cerebral plus pedal ganglion (CPG) extract led to a greater increase in the Vn content. Accordingly, the CPG appears to contain a vitellogenesis-promoting factor (VPF) that regulates Vtg synthesis. However, although E_2 promoted the expression of Vtg mRNA in the ovarian tissue cultured in vitro, VPF had no effect (Osada et al. 2004a). VPF appears to promote vitellogenesis at the level of translation. ER-like immunoreactivity was observed in both growing oocytes and auxiliary cells, which also exhibited Vn immunoreactivity and Vtg mRNA (Osada et al. 2003). These findings suggest the potential involvement of E_2 in the control of ER-mediated vitellogenesis (Fig. 7.6).

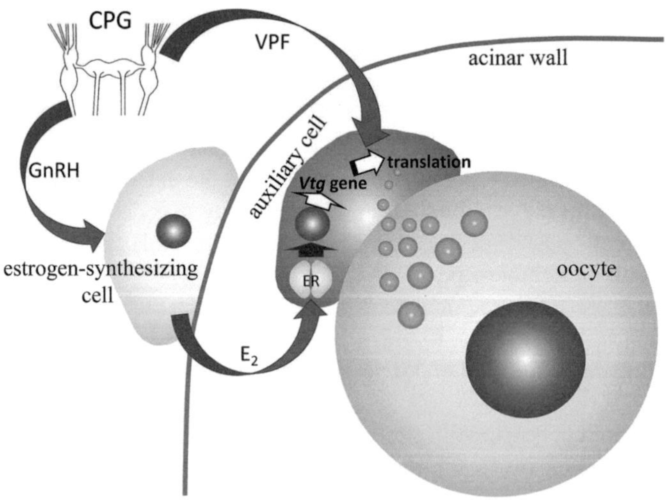

Fig. 7.6 Potential central nervous system–gonadal control of vitellogenesis in scallops. The synthesis of vitellogenin (Vtg) in the auxiliary cell is closely associated with oocyte growth. Estradiol-17β (E_2) promotes the transcription of Vtg mRNA, while vitellogenesis-promoting factor (VPF) from the cerebral plus pedal ganglion (CPG) enhances translation. The GnRH peptide secreted from the CPG is thought to be transported by hemocytes in the circulation system and received by estrogen-synthesizing cells along the outside of the acinar wall to stimulate the secretion of E_2. It remains unknown whether the estrogen receptor (ER) mediates vitellogenesis, although ER immunoreactivity has been detected in auxiliary cells

7.3.3.4 Estrogene Receptors in Mollusk Eggs

ERs are members of the nuclear receptor (NR) superfamily. These receptor proteins share several common features and can be divided into six domains. The DNA-binding domain (DBD) and the ligand-binding domain (LBD) are the most highly conserved domains among species. The DBD recognizes and binds to specific responsive DNA elements. The LBD regulates the hormone-dependent transcription of target genes, such as Vtg.

The oyster ER homolog (Matsumoto et al. 2007), which is highly similar to the ERs of other mollusks such as *Aplysia californica* (Thornton et al. 2003), *Thais* (*Reishia*) *clavigera* (Kajiwara et al. 2006), and *Octopus vulgaris* (Keay et al. 2006), was isolated to facilitate an understanding of estrogen signaling during vitellogenesis in bivalves. A phylogenetic analysis revealed that the isolated oyster ER is an ortholog of *Aplysia* ER, snail ER, and octopus ER. Furthermore, this oyster ER did not activate the expression of luciferase in the presence of E_2 and led to the constitutive activation of reporter transcription. The addition of E_2 did not further enhance this upregulation of reporter gene, consistent with findings from analyses of other mollusk ERs.

The mollusk ER ortholog is described as ER, given its high level of sequence similarity to vertebrate ERs; however, this ER can activate transcription in the absence of a ligand and does not bind to steroid hormones. In the annelid ER, which is responsive to estrogen, the architecture of the ligand pocket has been exceptionally well conserved (Keay and Thornton 2009). The X-ray crystal structure of oyster ER revealed that the ligand pocket is filled with bulky residues that prevent ligand occupancy (Bridgham et al. 2014). The oyster genome possesses 43 putative NR sequences (Vogeler et al. 2014). Specifically, it contains two members of NR3, sex steroid hormone receptor analogs, an ER homolog identified as oyster ER, and an estrogen-related receptor homolog that is constitutively activated and unlikely to bind estrogen. Additional NR3 members that might interact with vertebrate sex steroids were not identified. The results of prior studies of estrogen-induced vitellogenesis suggest that the effects must be mediated by ER activation-independent mechanisms in the presence of estrogens.

Although the functions of mollusk ERs remain uncertain, ER mRNA was detected in the ovary and cerebral ganglia of snails (Kajiwara et al. 2006), and ER expression was observed in both sexes of octopus, although the highest levels were detected in the ovary (Keay et al. 2006). ER mRNA was detected in all tested oyster tissues, and the highest levels were detected in the ovary. Immunohistochemistry with an antiserum against synthetic oyster ER peptide was used to localize oyster ER to the nuclei of follicle cells, the sites of Vtg synthesis, and the oocytes (Matsumoto et al. 2007). In *Mytilus*, two novel forms of ER-like genes were isolated and identified as an ER and estrogen-related receptor (ERR); these proteins were localized to the oocytes and the follicle cells associated with developing oocytes in the ovary, as well as Sertoli cells in the testis and ciliated cells in the gill (Nagasawa et al. 2015b). Furthermore, significant increases in ER and Vtg mRNA expression, but not ERR, were observed when mussels were exposed to estrogens during the early stage of gametogenesis (Ciocan et al. 2010). We cannot rule out the possibility that the ability of E_2 to activate Vtg gene expression in the ovary might be mediated through alternative NRs. Nevertheless, the roles of mollusk ERs remain uncertain.

7.3.4 Spawning

7.3.4.1 Inducion Mechanisms of Spawning

In marine invertebrates, including bivalve mollusks, spawning is considered to correlate with changes in temperature, lunar age, illumination, salinity, phytoplankton abundance, food availability, physical shock, tidal surge, drying, and radical oxygen. These environmental fluctuations reportedly act as natural triggers that induce spawning via endogenous regulation mechanisms (Giese and Kanatani 1987). Blake and Sastry (1979) reported that neurosecretion associated with stage V of the CNS neurosecretory cycle was associated with spawning in the bay scallop *Argopecten irradians*. Spawning-associated histological changes in the ganglion and the induction of spawning following ablation of the ganglion have led to the suggestion that the cerebral ganglion is involved in spawning (see Barber and Blake 2006). The role of the CNS in bivalve spawning has long been predicted, although the endogenous and specific factors that control this process have not yet been identified.

The relationship between specific neurosecretory substances and spawning was initially investigated in *Patinopecten*. Exogenous serotonin (5-hydroxytryptamine, 5-HT) was shown to strongly induce spawning in the scallop in vivo and has been suggested to play an important mechanistic role in bivalve spawning (Matsutani and Nomura 1982). This phenomenon of 5-HT-induced spawning has been reproduced in several other types of marine bivalves (Gibbons and Castagna 1984; Braley 1985; Tanaka and Murakoshi 1985). In the scallop, the 5-HT neuron was identified immunologically in the pedal ganglion, cerebral ganglion, and accessory ganglion adjacent to the visceral ganglion, and 5-HT nerve fibers were also detected around the gonoduct and along the outside of the germinal acini (Matsutani and Nomura 1986a). The localization pattern of the 5-HT neurons and nerve fibers strongly suggests that endogenous 5-HT regulates the spawning process in the gonad. UV ray-irradiated seawater is known to induce spawning in both the scallop and the abalone, *Haliotis discus hannai* (Kikuchi and Uki 1974; Uki and Kikuchi 1974), and the underlying mechanism has been pharmacologically demonstrated. Briefly, UV ray-irradiated seawater stimulates serotonergic mechanisms via dopaminergic mechanisms to induce spawning, and this process is modulated by prostaglandins (Matsutani and Nomura 1986b).

7.3.4.2 Role of Endocrine Systems in Spawning

A quantitative analysis of monoamine and prostaglandin expression during bivalve spawning was conducted to determine the role of each endogenous substance. In *Patinopecten*, Osada et al. (1987) reported a significant decrease in the levels of dopamine in the CNS and gonads of both sexes after UV ray-irradiated seawater-induced spawning, suggesting that the release of dopamine might stimulate the serotonergic mechanisms needed to induce spawning. In another type of scallop,

Argopecten purpuratus, thermal stimulation-induced spawning led to changes of the levels of dopamine, noradrenaline, and serotonin in the CNS, muscle, and gonad (Martínez et al. 1996). Seasonal variations in the gonadal levels of prostaglandin $F_{2\alpha}$($PGF_{2\alpha}$) and prostaglandin E_2 (PGE_2) correlated closely with the reproductive cycle, suggesting that these molecules might be involved in the sexual maturation and spawning of scallops (Osada and Nomura 1990). $PGF_{2\alpha}$ and PGE_2 levels in the gonad were shown to decrease significantly during spawning in females but were found to increase in males. In an in vitro experiment, $PGF_{2\alpha}$ significantly inhibited the 5-HT-induced release of oocytes from ovarian tissue, while PGE_2 enhanced the function of 5-HT; these results suggest that $PGF_{2\alpha}$ and PGE_2 act as suppressive and acceleratory modulators of 5-HT-induced oocyte release, respectively (Matsutani and Nomura 1987). These findings suggest that $PGF_{2\alpha}$ may be a suppressive modulator in the spawning of females and an acceleratory modulator in the spawning of males (Osada et al. 1989).

5-HT can modulate the induction of spawning via steroids, as well as by PGs. The response to artificial stimulation with UV-irradiated seawater tended to increase during sexual maturation (Uki and Kikuchi 1974), suggesting a potential increase in sensitivity to 5-HT upon spawning, depending on maturity. In bivalves, estrogen has been identified as a potential stimulator of gametogenesis, as the seasonal variations associated with gametogenesis and vitellogenesis are accelerated by estrogen (Matsumoto et al. 1997; Li et al. 1998; Osada et al. 2004a, b). Pretreatment with estrogen has promoted 5-HT-induced oocyte release from sections of scallop ovary (Osada et al. 1992a). In bivalves, the 5-HT receptors that transduce 5-HT signals from the surfaces to the interiors of germ cells have been subjected to several pharmacological characterizations. Specifically, pharmacological characterizations have revealed a mixed profile of $5\text{-HT}_1/5\text{-HT}_2$ subtypes and the 5-HT_1 subtype in the oocyte membranes of the scallop and oyster (Osada et al. 1998). A unique type that showed mixed pharmacological properties in the oyster (Kyozuka et al. 1997), a mixed $5\text{-HT}_1/5\text{-HT}_3$ type in the surf clam (Bandivdekar et al. 1991, 1992) or a novel type that was distinct from any mammalian 5-HT receptors in the same surf clam (Krantic et al. 1991, 1993a, b), and an original type in the zebra mussel (Fong et al. 1993) were also reported in the oocyte and sperm. Osada et al. (1998) reported that in the scallop oocyte membrane, 5-HT receptor expression could be induced pharmacologically by E_2 via a genomic mechanism, suggesting that estrogen-induced 5-HT receptors are distributed on the oocyte membrane and increased during oocyte growth. This phenomenon leads to the increased sensitivity to 5-HT associated with spawning.

5-HT exerts neurohormone activity and thus directly mediates the resumption of meiosis in PI-arrested oocytes (Osanai 1985; Hirai et al. 1988; Osanai and Kuraishi 1988; Krantic et al. 1991; Varaksin et al. 1992; Guerrier et al. 1993; Gobet et al. 1994; Fong et al. 1997), as demonstrated by GVBD (Matsutani and Nomura 1987) in oocytes isolated from *Spisula solidissima*, *Spisula sachalinensis*, *Crassostrea gigas*, and *Ruditapes philippinarum*. In the scallop, it is impossible to isolate oocytes from the ovary because of cytolysis after detachment from the germinal epithelium. Here, oocytes exhibiting GVBD were observed in paraffin sections of ovarian tissues treated with 5-HT, and the dose dependency was identical that observed in a variation of 5HT-induced oocyte release (Tanabe et al. 2006). These results suggest that 5-HT plays a primary role in the induction of oocyte maturation during spawning. These facts suggest that the 5-HT receptor is distributed on the surface of the germ cell membrane, as mentioned above.

In mollusks, the primary structures of seven 5-HT receptors were determined through molecular cloning. Six 5-HT receptor cDNAs were cloned from the CNS and reproductive system of the pond snail *Lymnaea stagnalis* and the sea hare *Aplysia californica* (Sugamori et al. 1993; Li et al. 1995; Gerhardt et al. 1996; Angers et al. 1998; Barbas et al. 2002). A full-length cDNA encoding a putative 5-HT receptor was isolated from the ovary of the scallop and identified as a 5-HT_{py} (Tanabe et al. 2010). This 5-HT_{py} was classified as a vertebrate serotonin receptor subtype, 5-HT_1, based on the molecular architecture, homology searches, and a phylogenetic analysis. The 5-HT_{py} was characterized at a high level of probability as an ancestral 5-HT receptor and a member of the 5-HT_1 receptor family coupled with G protein, based on the absence of introns in the coding region of the gene, a relatively long third cytoplasmic loop, and a short fourth inner terminal domain (C-terminal tail) (Albert and Tiberi 2001; Tierney 2001). A positive 5-HT_{py} signal was ubiquitously observed in both the peripheral tissue and the nervous system, as well as the spermatid, oocyte, and ciliated epithelium of the gonoducts of both male and female gonads (Tanabe et al. 2010). These results suggest that 5-HT, possibly via 5-HT_{py}, could mediate a series of spawning-related events comprising the induction of oocyte maturation, sperm motility, and the transport of mature oocytes and sperm through the ciliated epithelium of the gonoducts. In the ovarian tissues, the gene expression of 5-HT_{py} was significantly upregulated by E_2 (Tanabe et al. 2010), consistent with the pharmacological observation that the expression of the 5-HT receptor in the oocyte membrane was induced by E_2 via a genomic action. This expression increased sensitivity to 5-HT in relation to spawning and explained the observed increase in sensitivity to external spawning stimuli associated with maturity (Uki and Kikuchi 1974; Osada et al. 1998).

7.3.4.3 Activation of Sperm Motility and Oocyte Maturation

The 5-HT signal is transmitted into the oocytes and sperm after the binding of 5-HT to 5-HT receptors on the membranes. Sperm cells remain in a quiescent state immediately before ejaculation from the testis. The transduction of a 5-HT signal into the sperm via the 5-HT receptor was suggested to initiate 5-HT-dependent and osmolality-independent sperm motility in marine bivalve mollusks, a process associated with K^+ efflux and Ca^{2+} influx via voltage-dependent ion channels under alkaline conditions (Alavi et al. 2014). Moreover, Na^+ influx was thought to be an important initiator of sperm motility, likely via the regulation of Ca^{2+} exchange (Fig. 7.7) (Alavi et al. 2014). Immediately before ovulation, the oocytes of oviparous animals are generally arrested in the late prophase of meiosis I. At this stage, the oocyte possesses a developmental stage-specific nucleus, the GV, which corre-

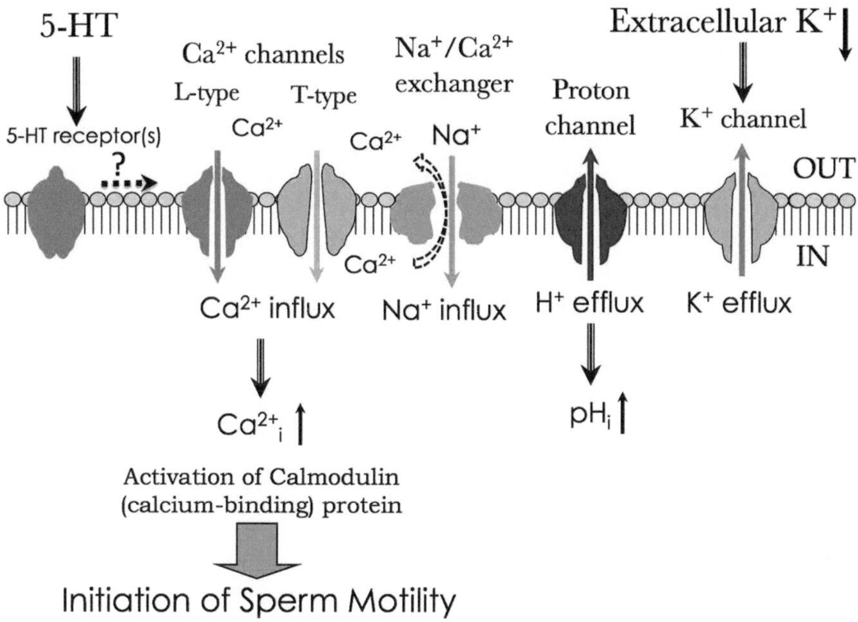

Fig. 7.7 Required ionic flux for the initiation of sperm motility in marine bivalves. The stimulatory effect of serotonin (5-HT) on the initiation of sperm motility is associated with potassium (K^+) and proton (H^+) efflux and calcium (Ca^{2+}) influx through a voltage-dependent K^+ channel, a proton channel and both L-type and T-type voltage-dependent Ca^{2+} channels, respectively. A sodium (Na^+)/Ca^{2+} exchanger regulates Na^+ influx to control intracellular Ca^{2+} levels during the motility period. These steps stimulate Ca^{2+}-dependent or calcium-calmodulin (CaM) protein phosphatase(s) in the flagellum to initiate sperm motility. (From Alavi et al. (2014), with permission)

sponds to the dictyate stage of the oocyte. Meiosis resumption, a process of oocyte maturation, is indicated by GVBD. In the invertebrate species of hydrozoan jellyfish, starfish, and bivalves, oocyte maturation and spawning are initiated by W/RPRPamide, 1-methyladenine, and serotonin, respectively (Kanatani et al. 1969; Hirai et al. 1988; Osanai and Kuraishi 1988; Takeda et al. 2018). In bivalves, 5-HT-induced oocyte maturation was thought to result from a combination of 5-HT signal transduction mechanisms and the cross-talk associated with 5-HT-induced oocyte maturation (Krantic and Rivailler 1996). 5-HT-mediated signaling may reduce cyclic AMP levels in the oocyte cytoplasm, thus inhibiting G_i protein-coupled adenylate cyclase. This 5-HT signal could simultaneously induce the conversion of phosphatidylinositol-4,5-bisphosphate (PIP_2) into IP_3 and diacylglycerol (DAG), which activate G_0 protein-coupled phospholipase C and increase the uptake of Ca^{2+}, which finally activate the G_x protein-coupled, membrane voltage-dependent Ca^{2+} channels. These pathways activate protein kinase A (PKA), PKC, and MAPK, which in turn activate the maturation promoting factor, which comprises cdc2 and cyclin B, and induce oocyte maturation.

Because 5-HT plays roles in the maturation of oocytes and initiation of sperm motility, the in vivo administration of exogenous 5-HT must induce spawning in bivalves. Interestingly, fully grown oocytes in bivalve ovaries are arrested at the dictyate stage; this stage corresponds to the late prophase of meiosis I, which is commonly observed in marine invertebrates (Krantic and Rivailler 1996). Once the highest level of 5-HT-specific binding to the oocyte membrane has been achieved, this state of arrest is maintained for 2 months until spawning (Osada et al. 1998). Furthermore, exogenous 5-HT does not always successfully induce spawning, and the number of released oocytes was found to vary widely among individuals (Matsutani and Nomura 1987). These observations indicate that the modulatory mechanisms regulating 5-HT-induced oocyte/sperm maturation and spawning are driven by maturation-competent extracellular signals. In mammals, the oocyte maturation inhibitor and granulosa cell factor have been identified as heat-stable polypeptides with molecular masses of <2 and 6 kDa, respectively (Sato and Koide 1984; Hillensjo et al. 1985; Tsafriri and Pomerantz 1986; Franchimont et al. 1988). In *Spisula*, the *Spisula* factor, which has a mass <1 kDa (Sato et al. 1985; Kadam and Koide 1990), and an oocyte membrane component with a mass of >18 kDa (Sato et al. 1992) have been identified as oocyte-produced substances that inhibit 5-HT-induced oocyte maturation. However, the bivalve gonad does not contain a vertebrate-like follicle structure, and the endocrine roles of the *Spisula* factor have not been clearly demonstrated.

7.3.4.4 Modulator of 5-HT

As mentioned in Sect. 7.3.4.2, $PGF_{2\alpha}$ has been identified as a potentially suppressive neuromodulator of 5-HT during bivalve spawning. In fact, $PGF_{2\alpha}$ was shown to block the 5-HT-induced release of oocytes from ovarian tissue via the gonoduct but did not inhibit 5-HT-induced oocyte maturation, suggesting that $PGF_{2\alpha}$ might inhibit the cilioexcitatory activity of 5-HT in the gonoducts during the transport of mature oocytes (Tanabe et al. 2006). In addition to $PGF_{2\alpha}$, a novel inhibitor of 5-HT-induced oocyte release from ovarian tissue was identified in the CNS tissues from scallops of both genders. This oocyte maturation arresting factor (OMAF) mainly arrested 5-HT-induced oocyte maturation (Tanabe et al. 2006). OMAF was identified universally in bivalve species of both genders and was thought to be transported from the CNS to the ovary via blood circulation, given the function of OMAF and its identification in the hemolymph. OMAF may prohibit 5-HT-induced oocyte maturation and sperm motility by interfering with the influx of external Ca^{2+} into oocytes and may eventually inhibit spawning (Tanabe et al. 2006; Yuan et al. 2012). The internal amino acid sequences of OMAF, which has a molecular mass of 52 kDa, were determined, and an antibody against a partial peptide of OMAF strongly enhanced the 5-HT-induced release of oocytes and sperm by neutralizing endogenous OMAF and its suppressive activity. These results confirm that OMAF inhibits 5-HT-induced oocyte maturation and sperm motility (Yuan et al. 2012). Taken together, the data identify 5-HT as an essential neurohormone for oocyte maturation and sperm motil-

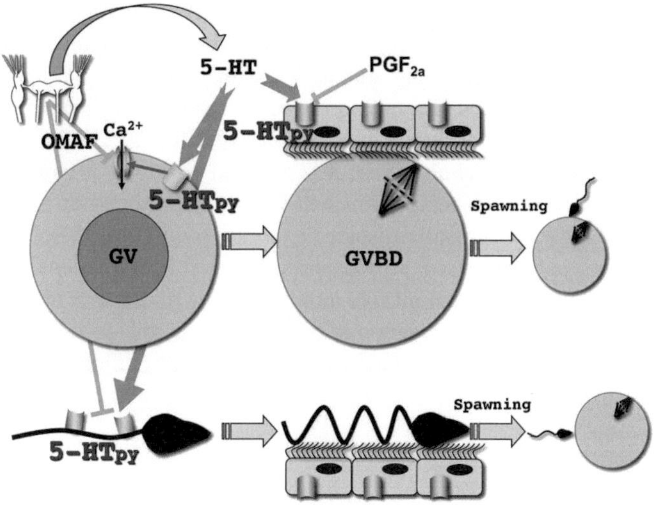

Fig. 7.8 Illustration of the transduction of extracellular signals that regulate oocyte maturation, sperm motility, and spawning in scallops. OMAF, via Ca^{2+} ions, negatively regulates the stimulatory effects of 5-HT on oocyte maturation and sperm motility via the 5-HT receptor. 5-HT and $PGF_{2\alpha}$ regulate the transportation of activated oocytes and sperm through the gonoduct

ity after bivalve spawning. The modes of action of 5-HT on germ cell and gonoduct activation, which are regulated by OMAF and $PGF_{2\alpha}$, are expected to explain the phenomena of arrested oocyte maturation and sperm motility prior to spawning, as well as simultaneous spawning in nature (Fig. 7.8). Moreover, the receptor mechanisms of OMAF and $PGF_{2\alpha}$ and the process by which bivalves can be released from the suppressive effects of OMAF and $PGF_{2\alpha}$ and induced to secrete 5-HT to trigger simultaneous spawning must be elucidated.

7.3.5 *Prospective*

Compared with other animals, including model invertebrates such as fruit flies and nematodes, few genetic studies have focused on mollusks. However, data from genetic analyses of mollusks have begun to accumulate, and molecular studies are contributing to an understanding of the physiological mechanisms in bivalves. The development of next-generation sequencing has accelerated the ability to sequence large amounts of genome transcripts from any type of organism. Accordingly, the genomes of the pearl oyster *Pinctada fucata* (Takeuchi et al. 2012), Pacific oyster (Zhang et al. 2012a), and Yesso scallop (Wang et al. 2017) have been sequenced and used to construct a draft genome database. This database has since been used as a platform for the identification of specific genes related to calcification, an essential phenomenon involved in ornamental pearl production, stress adaptation, shell formation, and larval development. The released pearl oyster draft genome has been further screened for genes associated with reproduction (Matsumoto et al. 2013). De novo transcriptome sequencing and analysis have improved our understanding of sex determination and differentiation, oocyte maturation, growth, and stress responses (Hou et al. 2011; Dheilly et al. 2012; Ghiselli et al. 2012; Pauletto et al. 2014; Teaniniuraitemoana et al. 2014). It is generally difficult to isolate a specific gene based on a conserved region obtained from an alignment of a few mollusk sequences or even taxonomically distant vertebrate sequences; therefore, genome and transcriptome databases will provide strong tools for comprehensive studies of the reproductive mechanisms of bivalve mollusks in the future. The biological functions of genes identified from genomic and transcriptomic resources should be investigated in each animal to clarify their primary physiological functions in bivalve mollusks.

References

Abdelmajid H, Leclerc-David C, Moreau M, Guerrier P, Ryazanov A (1993) Release from the metaphase I block in invertebrate oocytes: possible involvement of Ca^{2+}/calmodulin-dependent kinase III. Int J Dev Biol 37(2):279–290

Alavi SMH, Matsumura N, Shiba K, Itoh N, Takahashi KG, Inaba K, Osada M (2014) Roles of extracellular ions and pH in serotonin-dependent initiation of sperm motility in marine bivalve mollusks. Reproduction 147(3):331–345. https://doi.org/10.1530/REP-13-0418

Albert PR, Tiberi M (2001) Receptor signaling and structure: insights from serotonin-1 receptors. Trends Endocrinol Metab 12(10):453–460. https://doi.org/10.1016/S1043-2760(01)00498-2

Allen RD (1953) Fertilization and artificial activation in the egg of surf clam, *Spisula solidissima*. Biol Bull 105(2):213–239. https://doi.org/10.2307/1538639

Anderson WA, Eckberg WR (1983) A cytological analysis of fertilization in *Chaetopterus pergamentaceus*. Biol Bull 165(1):110–118. https://doi.org/10.2307/1541358

Angers A, Storozhuk MV, Duchaîne T, Castellucci VF, DesGroseillers L (1998) Cloning and functional expression of an *Aplysia* 5-HT receptor negatively coupled to adenylate cyclase. J Neurosci 18(15):5586–5593. https://doi.org/10.1523/JNEUROSCI.18-15-05586.1998

Awaji M, Matsumoto T, Yamano K, Kitamura M, Hara A (2011) Immunohistochemical observations of vitellin synthesis and accumulation processes in ovary of Ezo abalone *Haliotis discus hannai*. Fish Sci 77(2):191–197. https://doi.org/10.1007/s12562-010-0316-5

Balkhair M, Al-Mushikhi A, Rivera R (2016) Embryogenesis and larval development of the Omani abalone (*Haliotis mariae* Wood, 1828). J Shellfish Res 35(3):625–631. https://doi.org/10.2983/035.035.0308

Bandivdekar AH, Segal SJ, Koide SS (1991) Demonstration of serotonin receptors in isolated *Spisula* oocyte membrane. Invert Reprod Dev 19(2):147–150. https://doi.org/10.1080/07924259.1991.9672168

Bandivdekar AH, Segal SJ, Koide SS (1992) Binding of 5-hydroxytryptamine analogs by isolated *Spisula* sperm membrane. Invert Reprod Dev 21(1):43–46. https://doi.org/10.1080/07924259.1992.9672218

Bannon G, Brown GG (1980) Vesicle involvement in the egg cortical reaction of the horseshoe crab, *Limulus polyphemus* L. Dev Biol 76(2):418–427. https://doi.org/10.1016/0012-1606(80)90390-5

Barazandeh M, Davis CS, Neufeld CJ, Coltman DW, Palmer AR (2013) Something Darwin didn't know about barnacles: spermcast mating in a common stalked species. Proc Biol Sci 280(1754):20122919. https://doi.org/10.1098/rspb.2012.2919

Barbas B, Zappulla JP, Angers S, Bouvier M, Castellucci VF, DesGroseillers L (2002) Functional characterization of a novel serotonin receptor (5-HTap$_2$) expressed in the CNS of *Aplysia californica*. J Neurochem 80(2):335–345. https://doi.org/10.1046/j.0022-3042.2001.00703.x
Barber BJ, Blake N (2006) Reproductive physiology (Chapter 6). In: Shumway SE, Parsons GJ (eds) Scallops: biology, ecology and aquaculture. Elsevier, San Diego, CA, pp 357–416
Bianchi E, Wright GJ (2014) Izumo meets Juno: preventing polyspermy in fertilization. Cell Cycle 13(13):2019–2020. https://doi.org/10.4161/cc.29461
Bianchi E, Doe B, Goulding D, Wright GJ (2014) Juno is the egg Izumo receptor and is essential for mammalian fertilization. Nature 508(7497):483–487. https://doi.org/10.1038/nature13203
Bigot L, Zatylny-Gaudin C, Rodet F, Bernay B, Boudry P, Favrel P (2012) Characterization of GnRH-related peptides from the Pacific oyster *Crassostrea gigas*. Peptides 34(2):303–310. https://doi.org/10.1016/j.peptides.2012.01.017
Blake NJ, Sastry AN (1979) Neurosecretory regulation of oögenesis in the bay scallop, *Argopecten irradians irradians* (Lamarck). In: Naylor E, Hortnoll RG (eds) Cyclic phenomena in marine plants and animals. Pergamon Press, New York, pp 181–190
Braley RD (1985) Serotonin-induced spawning in giant clams (Bivalvia: Tridacnidae). Aquaculture 47(4):321–325. https://doi.org/10.1016/0044-8486(85)90217-0
Bridgham JT, Keay J, Ortlund EA, Thornton JW (2014) Vestigialization of an allosteric switch: genetic and structural mechanisms for the evolution of constitutive activity in a steroid hormone receptor. PLoS Genet 10(1):e1004058. https://doi.org/10.1371/journal.pgen.1004058
Brown GG, Humphreys WJ (1971) Sperm–egg interactions of *Limulus polyphemus* with scanning electron microscopy. J Cell Biol 51(3):904–907. https://doi.org/10.1083/jcb.51.3.904
Buckland-Nicks J, Howley B (1997) Spermiogenesis and sperm structure in relation to early events of fertilization in the limpet *Tectura testudinalis* (Müller, 1776). Biol Bull 193(3):306–319. https://doi.org/10.2307/1542933
Chapman AD (2009) Numbers of living species in Australia and the world, 2nd edn. Australian Biodiversity Information Services, Toowoomba
Chuang SC, Lai WS, Chen JH (2006) Influence of ultraviolet radiation on selected physiological responses of earthworms. J Exp Biol 209(21):4304–4312. https://doi.org/10.1242/jeb.02521
Ciocan CM, Cubero-Leon E, Puinean AM, Hill EM, Minier C, Osada M, Fenlon K, Rotchell JM (2010) Effects of estrogen exposure in mussels, *Mytilus edulis*, at different stages of gametogenesis. Environ Pollut 158(9):2977–2984. https://doi.org/10.1016/j.envpol.2010.05.025
Clark WH Jr, Griffin FJ (1993) Acquisition and manipulation of penaeoidean gametes. In: McVey JP (ed) CRC handbook of mariculture: crustacean aquaculture, vol 1, 2nd edn. CRC, London, pp 133–151
Clark WH Jr, Lynn JW, Yudin LA, Persyn HO (1980) Morphology of the cortical reaction in the eggs of *Penaeus aztecus*. Biol Bull 158(2):175–186. https://doi.org/10.2307/1540929
Colas P, Dubé F (1998) Meiotic maturation in mollusc oocytes. Semin Cell Dev Biol 9(5):539–548. https://doi.org/10.1006/scdb.1998.0248
Colas P, Launay C, van Loon AE, Guerrier P (1993) Protein synthesis controls cyclin stability in metaphase I-arrested oocytes of *Patella vulgata*. Exp Cell Res 208(2):518–521. https://doi.org/10.1006/excr.1993.1275
Colwin LH, Colwin AL (1960) Formation of sperm entry holes in the vitelline membrane of *Hydroides hexagonus* (Annelida) and evidence of their lytic origin. J Biophys Biochem Cytol 7(2):315–320. https://doi.org/10.1083/jcb.7.2.315
Colwin LH, Colwin AL (1961) Changes in the spermatozoon during fertilization in *Hydroides hexagonus* (Annelida). I. Passage of the acrosomal region through the vitelline membrane. J Biophys Biochem Cytol 10(2):231–254. https://doi.org/10.1083/jcb.10.2.231
Costello DP, Davidson ME, Eggers A, Fox MH, Henley C (1957) Methods for obtaining and handling marine eggs and embryos. Lancaster Press, Lancaster
Dale B (2016) Achieving monospermy or preventing polyspermy? Res Rep Biol 2016(7):47–57. https://doi.org/10.2147/RRB.S84085
Dan JC (1962) The vitelline coat of the *Mytilus* egg. I. Normal structure and effect of acrosomal lysin. Biol Bull 123(3):531–541. https://doi.org/10.2307/1539574

Deguchi R (2007) Fertilization causes a single Ca^{2+} increase that fully depends on Ca^{2+} influx in oocytes of limpets (Phylum Mollusca, Class Gastropoda). Dev Biol 304(2):652–663. https://doi.org/10.1016/j.ydbio.2007.01.017

Deguchi R, Morisawa M (2003) External Ca^{2+} is predominantly used for cytoplasmic and nuclear Ca^{2+} increases in fertilized oocytes of the marine bivalve *Mactra chinensis*. J Cell Sci 116(2):367–376. https://doi.org/10.1242/jcs.00221

Deguchi R, Osanai K (1994a) Repetitive intracellular Ca^{2+} increases at fertilization and the role of Ca^{2+} in meiosis reinitiation from the first metaphase in oocytes of marine bivalves. Dev Biol 163(1):162–174. https://doi.org/10.1006/dbio.1994.1132

Deguchi R, Osanai K (1994b) Meiosis reinitiation from the first prophase is dependent on the levels of intracellular Ca^{2+} and pH in oocytes of the bivalves *Mactra chinensis* and *Limaria hakodatensis*. Dev Biol 166(2):587–599. https://doi.org/10.1006/dbio.1994.1339

Deguchi R, Osanai K (1995) Serotonin-induced meiosis reinitiation from the first prophase and from the first metaphase in oocytes of the marine bivalve *Hiatella flaccida*: respective changes in intracellular Ca^{2+} and pH. Dev Biol 171(2):483–496. https://doi.org/10.1006/dbio.1995.1298

Deguchi R, Osanai K, Morisawa M (1996) Extracellular Ca^{2+} entry and Ca^{2+} release from inositol 1,4,5-trisphosphate-sensitive stores function at fertilization in oocytes of the marine bivalve *Mytilus edulis*. Development 122(11):3651–3660

Deguchi R, Takeda N, Stricker SA (2015) Calcium signals and oocyte maturation in marine invertebrates. Int J Dev Biol 59(7-9):271–280. https://doi.org/10.1387/ijdb.150239ss

Dheilly NM, Lelong C, Huvet A, Kellner K, Dubos M-P, Riviere G, Boudry P, Favrel P (2012) Gametogenesis in the Pacific oyster *Crassostrea gigas*: a microarray-based analysis identifies sex and stage specific genes. PLoS One 7(5):e36353. https://doi.org/10.1371/journal.pone.0036353

Di Cosmo A, Di Cristo C (1998) Neuropeptidergic control of the optic gland of *Octopus vulgaris*: FMRF-amide and GnRH immunoreactivity. J Comp Neurol 398(1):1–12. https://doi.org/10.1002/(SICI)1096-9861(19980817)398:1<1::AID-CNE1>3.0.CO;2-5

Di Cosmo A, Di Cristo C, Paolucci M (2001) Sex steroid hormone fluctuations and morphological changes of the reproductive system of the female of *Octopus vulgaris* throughout the annual cycle. J Exp Zool 289(1):33–47. https://doi.org/10.1002/1097-010X(20010101/31)289:1<33::AID-JEZ4>3.0.CO;2-A

Di Cristo C (2013) Nervous control of reproduction in *Octopus vulgaris*: a new model. Invert Neurosci 13(1):27–34. https://doi.org/10.1007/s10158-013-0149-x

Di Cristo C, Paolucci M, Iglesias J, Sanchez J, Di Cosmo A (2002) Presence of two neuropeptides in the fusiform ganglion and reproductive ducts of *Octopus vulgaris*: FMRFamide and gonadotropin-releasing hormone (GnRH). J Exp Zool 292(3):267–276. https://doi.org/10.1002/jez.90000

Dorange G, Le Pennec M (1989) Ultrastructural study of oogenesis and oocytic degeneration in *Pecten maximus* from the Bay of St. Brieuc. Mar Biol 103(3):339–348. https://doi.org/10.1007/BF00397268

Dubé F, Dufresne L (1990) Release of metaphase arrest by partial inhibition of protein synthesis in blue mussel oocytes. J Exp Zool 256(3):323–332. https://doi.org/10.1002/jez.1402560312

Dubé F, Eckberg WR (1997) Intracellular pH increase driven by an Na^+/H^+ exchanger upon activation of surf clam oocytes. Dev Biol 190(1):41–54. https://doi.org/10.1006/dbio.1997.8682

Dunn CW, Hejnol A, Matus DQ, Pang K, Browne WE, Smith SA, Seaver E, Rouse GW, Obst M, Edgecombe GD, Sorensen MV, Haddock SHD, Schmidt-Rhaesa A, Okusu A, Kristensen RM, Wheeler WC, Martindale MQ, Giribet G (2008) Broad phylogenomic sampling improves resolution of the animal tree of life. Nature 452(7188):745–749. https://doi.org/10.1038/nature06614

Dunn CW, Giribet G, Edgecombe GD, Hejnol A (2014) Animal phylogeny and its evolutionary implications. Annu Rev Ecol Evol Syst 45:371–395. https://doi.org/10.1146/annurev-ecolsys-120213-091627

Dupré EM, Barros C (2011) In vitro fertilization of the rock shrimp, *Rhynchocinetes typus* (Decapoda, Caridea): a review. Biol Res 44(2):125–133. S0716-97602011000200003

Eckberg WR (1981) An ultrastructural analysis of cytoplasmic localization in *Chaetopterus pergamentaceus*. Biol Bull 160(2):228–239. https://doi.org/10.2307/1540883

Eckberg WR, Anderson WA (1985) Blocks to polyspermy in *Chaetopterus*. J Exp Zool 233(2):253–260. https://doi.org/10.1002/jez.1402330213

Eckberg WR, Miller AL (1995) Propagated and nonpropagated calcium transients during egg activation in the annelid, Chaetopterus. Dev Biol 172(2):654–664. https://doi.org/10.1006/dbio.1995.8043

Eckberg WR, Szuts EZ, Carroll AG (1987) Protein kinase C activity, protein phosphorylation and germinal vesicle breakdown in *Spisula* oocytes. Dev Biol 124(1):57–64. https://doi.org/10.1016/0012-1606(87)90459-3

Eckelbarger KJ, Davis CV (1996) Ultrastructure of the gonad and gametogenesis in the eastern oyster, *Crassostrea virginica*. I. Ovary and oogenesis. Mar Biol 127(1):79–87. https://doi.org/10.1007/BF00993648

Fallon JF, Austin CR (1967) Fine structure of gametes of *Nereis limbata* (Annelida) before and after interaction. J Exp Zool 166(2):225–241. https://doi.org/10.1002/jez.1401660205

Farley RD (2010) Book gill development in embryos and first and second instars of the horseshoe crab *Limulus polyphemus* L. (Chelicerata, Xiphosura). Arthropod Struct Dev 39(5):369–381. https://doi.org/10.1016/j.asd.2010.04.001

Fernández J, Roegiers F, Cantillana V, Sardet C (1998) Formation and localization of cytoplasmic domains in leech and ascidian zygotes. Int J Dev Biol 42(8):1075–1084

Finkel T, Wolf DP (1980) Membrane potential, pH and the activation of surf clam oocytes. Gamete Res 3(3):299–304. https://doi.org/10.1002/mrd.1120030312

Fong PP, Wall DM, Ram JL (1993) Characterization of serotonin receptors in the regulation of spawning in the zebra mussel *Dreissena polymorpha* (Pallas). J Exp Zool 267(5):475–482. https://doi.org/10.1002/jez.1402670502

Fong PP, Deguchi R, Kyozuka K (1997) Characterization of serotonin receptor mediating intracellular calcium increase in meiosis-reinitiated oocytes of the bivalve *Ruditapes philippinarum* from central Japan. J Exp Zool 279(1):89–101. https://doi.org/10.1002/(SICI)1097-010X(19970901)279:1<89::AID-JEZ9>3.0.CO;2-Y

Franchimont P, Demoulin A, Valcke JC (1988) Endocrine, paracrine, and autocrine control of follicle development. Horm Metab Res 20(4):193–203. https://doi.org/10.1055/s-2007-1010793

Frézal L, Félix MA (2015) *C. elegans* outside the Petri dish. Elife 4. https://doi.org/10.7554/eLife.05849

Galindo BE, Moy GW, Swanson WJ, Vacquier VD (2002) Full-length sequence of VERL, the egg vitelline envelope receptor for abalone sperm lysin. Gene 288(1–2):111–117. https://doi.org/10.1016/S0378-1119(02)00459-6

Galindo BE, Vacquier VD, Swanson WJ (2003) Positive selection in the egg receptor for abalone sperm lysin. Proc Natl Acad Sci U S A 100(8):4639–4643. https://doi.org/10.1073/pnas.0830022100

Gerhardt CC, Leysen JE, Planta RJ, Vreugdenhil E, Van-Heerikhuizen H (1996) Functional characterization of a 5-HT$_2$ receptor cDNA cloned from *Lymnaea stagnalis*. Eur J Pharmacol 311(2–3):249–258. https://doi.org/10.1016/0014-2999(96)00410-4

Ghiselli F, Milani L, Chang PL, Hedgecock D, Davis JP, Nuzhdin SV, Passamonti M (2012) De Novo assembly of the Manila clam *Ruditapes philippinarum* transcriptome provides new insights into expression bias, mitochondrial doubly uniparental inheritance and sex determination. Mol Biol Evol 29(2):771–786. https://doi.org/10.1093/molbev/msr248

Gibbons MC, Castagna M (1984) Serotonin as an inducer of spawning in six bivalve species. Aquaculture 40(2):189–191. https://doi.org/10.1016/0044-8486(84)90356-9

Giese A, Kanatani H (1987) Maturation and spawning. In: Giese AC, Pearse JS, Pearse VB (eds) Reproduction of marine invertebrates, Vol. IX—general aspects: seeking unity in diversity. Blackwell Scientific Publication/Boxwood Press, California, CA, pp 251–329

Gobet I, Durocher Y, Leclerc C, Moreau M, Guerrier P (1994) Reception and transduction of the serotonin signal responsible for meiosis reinitiation in oocytes of the Japanese clam *Ruditapes philippinarum*. Dev Biol 164(2):540–549. https://doi.org/10.1006/dbio.1994.1222

Goldberg JI, Garofarlo R, Price CJ, Chang JP (1993) Presence and biological activity of a GnRH-like factor in the nervous system of *Helisoma trivolvis*. J Comp Neurol 336(4):571–582. https://doi.org/10.1002/cne.903360409

Goudeau M (1982) Fertilization in a crab: I. Early events in the ovary, and cytological aspects of the acrosome reaction and gamete contacts. Tissue Cell 14(1):97–111. https://doi.org/10.1016/0040-8166(82)90010-6

Goudeau M (1984) Fertilization in a crab: III. Cytodifferentiation of vesicles enclosing ring-shaped elements involved in the cortical reaction. Gamete Res 9(4):409–424. https://doi.org/10.1002/mrd.1120090406

Goudeau M, Becker J (1982) Fertilization in a crab. II. Cytological aspects of the cortical reaction and fertilization envelope elaboration. Tissue Cell 14(2):273–282. https://doi.org/10.1016/0040-8166(82)90025-8

Goudeau H, Goudeau M (1986a) Electrical and morphological responses of the lobster egg to fertilization. Dev Biol 114(2):325–335. https://doi.org/10.1016/0012-1606(86)90197-1

Goudeau M, Goudeau H (1986b) The resumption of meiotic maturation of the oocyte of the prawn *Palaemon serratus* is regulated by an increase in extracellular Mg^{2+} during spawning. Dev Biol 118(2):361–370. https://doi.org/10.1016/0012-1606(86)90005-9

Goudeau H, Goudeau M (1986c) External Mg^{2+} is required for hyperpolarization to occur in ovulated oocytes of the prawn *Palaemon serratus*. Dev Biol 118(2):371–378. https://doi.org/10.1016/0012-1606(86)90006-0

Goudeau H, Goudeau M (1989a) A long-lasting electrically mediated block, due to the egg membrane hyperpolarization at fertilization, ensures physiological monospermy in eggs of the crab *Maia squinado*. Dev Biol 133(2):348–360. https://doi.org/10.1016/0012-1606(89)90039-0

Goudeau H, Goudeau M (1989b) Electrical responses to fertilization and spontaneous activation in decapod crustacean eggs: characteristics and role. In: Nuccitelli R, Cherr GN, Clark WH Jr (eds) Mechanisms of egg activation. Plenum Press, New York, pp 201–214

Goudeau M, Goudeau H (1996) External Mg^{2+} triggers oscillations and a subsequent sustained level of intracellular free Ca^{2+}, correlated with changes in membrane conductance in the oocyte of the prawn *Palaemon serratus*. Dev Biol 177(1):178–189. https://doi.org/10.1006/dbio.1996.0154

Goudeau H, Goudeau M (1998) Depletion of intracellular Ca^{2+} stores, mediated by Mg^{2+}-stimulated $InsP_3$ liberation or thapsigargin, induces a capacitative Ca^{2+} influx in prawn oocytes. Dev Biol 193(2):225–238. https://doi.org/10.1006/dbio.1997.8799

Goudeau M, Goudeau H, Guillaumin D (1991) Extracellular Mg^{2+} induces a loss of microvilli, membrane retrieval, and the subsequent cortical reaction, in the oocyte of the prawn *Palaemon serratus*. Dev Biol 148(1):31–50. https://doi.org/10.1016/0012-1606(91)90315-T

Gould M, Stephano JL (1989) How do sperm activate eggs in *Urechis* (as well as in polychaetes and molluscs)? In: Nuccitelli R, Cherr GN, Clark WH Jr (eds) Mechanisms of egg activation. Plenum Press, New York, pp 201–214

Gould MC, Stephano JL (1991) Peptides from sperm acrosomal protein that initiate egg development. Dev Biol 146(2):509–518. https://doi.org/10.1016/0012-1606(91)90252-X

Gould MC, Stephano JL (1993) Nuclear and cytoplasmic pH increase at fertilization in *Urechis caupo*. Dev Biol 159(2):608–617. https://doi.org/10.1006/dbio.1993.1268

Gould MC, Stephano JL (2003) Polyspermy prevention in marine invertebrates. Microsc Res Tech 61(4):379–388. https://doi.org/10.1002/jemt.10351

Gould MC, Stephano JL, Ortíz-Barron BD, Pérez-Quezada I (2001) Maturation and fertilization in *Lottia gigantea* oocytes: intracellular pH, Ca^{2+}, and electrophysiology. J Exp Zool 290(4):411–420. https://doi.org/10.1002/jez.1082

Goulding MQ, Lambert JD (2016) Mollusc models I. The snail *Ilyanassa*. Curr Opin Genet Dev 39:168–174. https://doi.org/10.1016/j.gde.2016.07.007

Gould-Somero M, Jaffe LA, Holland LZ (1979) Electrically mediated fast polyspermy block in eggs of the marine worm, *Urechis caupo*. J Cell Biol 82(2):426–440. https://doi.org/10.1083/jcb.82.2.426

Graham JB (1988) Ecological and evolutionary aspects of integumentary respiration: body size, diffusion, and the invertebrata. Am Zool 28(3):1031–1045. https://doi.org/10.1093/icb/28.3.1031

Griffin FJ, Clark WH Jr (1990) Induction of acrosomal filament formation in the sperm of *Sicyonia ingentis*. J Exp Zool 254(3):296–304. https://doi.org/10.1002/jez.1402540308

Guerrier P, Brassart M, David C, Moreau M (1986) Sequential control of meiosis reinitiation by pH and Ca^{2+} in oocytes of the prosobranch mollusk *Patella vulgata*. Dev Biol 114(2):315–324. https://doi.org/10.1016/0012-1606(86)90196-X

Guerrier P, Leclerc-David C, Moreau M (1993) Evidence for the involvement of internal calcium stores during serotonin-induced meiosis reinitiation in oocytes of the bivalve mollusc *Ruditapes philippinarum*. Dev Biol 159(2):474–484. https://doi.org/10.1006/dbio.1993.1257

Guo C, Han Y, Shi W, Zhao X, Teng S, Xiao G, Yan M, Chai X, Liu G (2017) Ca^{2+}-channel and calmodulin play crucial roles in the fast electrical polyspermy blocking of *Tegillarca granosa* (Bivalvia: Arcidae). J Moll Stud 83(3):289–294. https://doi.org/10.1093/mollus/eyx016

Hejnol A, Martindale MQ (2009) The mouth, the anus, and the blastopore–open questions about questionable openings. In: Telford MJ, Littlewood DTJ (eds) Animal evolution. Genomes, fossils, and trees. Oxford University Press, Oxford, pp 33–40

Heller J (1993) Hermaphroditism in molluscs. Biol J Linn Soc 48(1):19–42. https://doi.org/10.1006/bijl.1993.1003

Hess RA, Bunick D, Lee KH, Bahr J, Taylor JA, Korach KS, Lubahn DB (1997) A role for oestrogens in the male reproductive system. Nature 390(6659):509–512. https://doi.org/10.1038/37352

Hillensjo T, Brannstrom M, Chari S, Daume E, Magnusson C, Tornell J (1985) Oocyte maturation as regulated by follicular factors. Ann N Y Acad Sci 442(1):73–79. https://doi.org/10.1111/j.1749-6632.1985.tb37506.x

Hirai S, Kishimoto T, Kadam AL, Kanatani H, Koide SS (1988) Induction of spawning and oocyte maturation by 5-hydroxytryptamine in the surf clam. J Exp Zool 245(3):318–321. https://doi.org/10.1002/jez.1402450312

Hiruta C, Nishida C, Tochinai S (2010) Abortive meiosis in the oogenesis of parthenogenetic *Daphnia pulex*. Chromosome Res 18(7):833–840. https://doi.org/10.1007/s10577-010-9159-2

Hodgson AN, Chia FS (1993) Spermatozoon structure of some North American prosobranchs from the families Lottiidae (Patellogastropoda) and Fissurellidae (Archaeogastropoda). Mar Biol 116(1):97–101. https://doi.org/10.1007/BF00350736

Horner VL, Wolfner MF (2008) Transitioning from egg to embryo: triggers and mechanisms of egg activation. Dev Dyn 237(3):527–544. https://doi.org/10.1002/dvdy.21454

Hou R, Bao Z, Wang S, Su H, Li Y, Du H, Hu J, Wang S, Hu X (2011) Transcriptome sequencing and *de novo* analysis for Yesso scallop (*Patinopecten yessoensis*) using 454 GS FLX. PLoS One 6(6):e21560. https://doi.org/10.1371/journal.pone.0021560

Howell KP, Skipwith A, Galione A, Eckberg WR (2003) Phospholipase C-dependent Ca^{2+} release by worm and mammal sperm factors. Biochem Biophys Res Commun 307(1):47–51. https://doi.org/10.1016/S0006-291X(03)01120-3

Hsia CC, Schmitz A, Lambertz M, Perry SF, Maina JN (2013) Evolution of air breathing: oxygen homeostasis and the transitions from water to land and sky. Compr Physiol 3(2):849–915. https://doi.org/10.1002/cphy.c120003

Humphreys WJ (1962) Electron microscope studies on eggs of *Mytilus edulis*. J Ultrastruct Res 7:467–487. https://doi.org/10.1016/S0022-5320(62)90041-2

Humphreys WJ (1967) The fine structure of cortical granules in eggs and gastrulae of *Mytilus edulis*. J Ultrastruct Res 17(3):314–326. https://doi.org/10.1016/S0022-5320(67)80051-0

Ikawa T, Nozoe Y, Yamashita N, Nishimura N, Ohnoki S, Yusa K, Hoshizaki S, Komaba M, Kawakubo A (2018) A study of the distributions of two endangered sea skaters *Halobates matsumurai* Esaki and *Asclepios shiranui* (Esaki) (Hemiptera: Gerridae: Halobatinae) with special reference to their strategies to cope with tidal currents. Psyche 2018:3464829. https://doi.org/10.1155/2018/3464829

Ikhwanuddin M, Noor-Hidayati AB, Aina-Lyana NMA, Zulaikha H, Muhd-Farouk H, Abol-Munafi AB (2015) *In vitro* fertilization technique in banana shrimp, *Fenneropenaeus merguiensis* (De Man, 1888). J Fish Aquat Sci 10(6):512–522. https://doi.org/10.3923/jfas.2015.512.522

Ishijima S, Sekiguchi K, Hiramoto Y (1988) Comparative study of the beat patterns of American and Asian horseshoe crab sperm: evidence for a role of the central pair complex in forming planar waveforms in flagella. Cell Motil Cytoskeleton 9(3):264–270. https://doi.org/10.1002/cm.970090308

Iwakoshi E, Hisada M, Minakata H (2000) Cardioactive peptides isolated from the brain of a Japanese octopus, *Octopus minor*. Peptides 21(5):623–630. https://doi.org/10.1016/S0196-9781(00)00201-1

Iwakoshi-Ukena E, Ukena K, Takuwa-Kuroda K, Kanda A, Tsutsui K, Minakata H (2004) Expression and distribution of octopus gonadotropin-releasing hormone in the central nervous system and peripheral organs of the octopus (*Octopus vulgaris*) by *in situ* hybridization and immunohistochemistry. J Comp Neurol 477(3):310–323. https://doi.org/10.1002/cne.20260

Iwao Y, Izaki K (2018) Universality and diversity of a fast, electrical block to polyspermy during fertilization in animals. In: Kobayashi K, Kitano T, Iwao Y, Kondo M (eds) Reproductive and developmental strategies. Springer, Japan, pp 499–533. https://doi.org/10.1007/978-4-431-56609-0_24

Iwata Y, Shaw P, Fujiwara E, Shiba K, Kakiuchi Y, Hirohashi N (2011) Why small males have big sperm: dimorphic squid sperm linked to alternative mating behaviours. BMC Evol Biol 11:236. https://doi.org/10.1186/1471-2148-11-236

Iwata Y, Sakurai Y, Shaw P (2015) Dimorphic sperm-transfer strategies and alternative mating tactics in loliginid squid. J Moll Stud 81(1):147–151. https://doi.org/10.1093/mollus/eyu072

Jaffe LA (1983) Fertilization potentials from eggs of the marine worms *Chaetopterus* and *Saccoglossus*. In: Moody WJ, Grinnel AD (eds) The physiology of excitable cells. Alan R. Liss, New York, pp 211–218

Jaffe LA (2018) The fast block to polyspermy: new insight into a century-old problem. J Gen Physiol 150(9):1233–1234. https://doi.org/10.1085/jgp.201812145

Jaffe LA, Gould-Somero M, Holland L (1979) Ionic mechanism of the fertilization potential of the marine worm, *Urechis caupo* (Echiura). J Gen Physiol 73(4):469–492. https://doi.org/10.1085/jgp.73.4.469

Jakubik B (2012) Life strategies of Viviparidae (Gastropoda: Caenogastropoda: Architaenioglossa) in various aquatic habitats: *Viviparus viviparus* (Linnaeus, 1758) and *V. contectus* (Millet, 1813). Folia Malacol 20:145–179. https://doi.org/10.2478/v10125-012-0013-3

Johnson JI, Kavanaugh SI, Nguyen C, Tsai P-S (2014) Localization and functional characterization of a novel adipokinetic hormone in the mollusk, *Aplysia californica*. PLoS One 9(8):e106014. https://doi.org/10.1371/journal.pone.0106014

Johnston RN, Paul M (1977) Calcium influx following fertilization of *Urechis caupo* eggs. Dev Biol 57(2):364–374. https://doi.org/10.1016/0012-1606(77)90221-4

Kadam AL, Koide SS (1990) Inhibition of serotonin-induced oocyte maturation by a *Spisula* factor. J Exp Zool 255(2):239–243. https://doi.org/10.1002/jez.1402550212

Kajiwara M, Kuraku S, Kurokawa T, Kato K, Toda S, Hirose H, Takahashi S, Shibata Y, Iguchi T, Matsumoto T, Miyata T, Miura T, Takahashi Y (2006) Tissue preferential expression of estrogen receptor gene in the marine snail, *Thais clavigera*. Gen Comp Endocrinol 148(3):315–326. https://doi.org/10.1016/j.ygcen.2006.03.016

Kanatani H, Shirai H, Nakanishi K, Kurokawa T (1969) Isolation and identification of meiosis-inducing substance in starfish *Asterias amurensis*. Nature 221(5177):273–274. https://doi.org/10.1038/221273a0

Kanda A, Takahashi T, Satake H, Minakata H (2006) Molecular and functional characterization of a novel gonadotropin-releasing-hormone receptor isolated from the common octopus (*Octopus vulgaris*). Biochem J 395(1):125–135. https://doi.org/10.1042/BJ20051615

Kaneuchi T, Sartain CV, Takeo S, Horner VL, Buehner NA, Aigaki T, Wolfner MF (2015) Calcium waves occur as *Drosophila* oocytes activate. Proc Natl Acad Sci U S A 112(3):791–796. https://doi.org/10.1073/pnas.1420589112

Kashir J, Deguchi R, Jones C, Coward K, Stricker SA (2013) Comparative biology of sperm factors and fertilization-induced calcium signals across the animal kingdom. Mol Reprod Dev 80(10):787–815. https://doi.org/10.1002/mrd.22222

Kato M, Hiruta C, Tochinai S (2016) The behavior of chromosomes during parthenogenetic oogenesis in Marmorkrebs *Procambarus fallax* f. *virginalis*. Zoolog Sci 33(4):426–430. https://doi.org/10.2108/zs160018

Keay J, Thornton JW (2009) Hormone-activated estrogen receptors in annelid invertebrates: implications for evolution and endocrine disruption. Endocrinology 150(4):1731–1738. https://doi.org/10.1210/en.2008-1338

Keay J, Bridgham JT, Thornton JW (2006) The *Octopus vulgaris* estrogen receptor is a constitutive transcriptional activator: evolutionary and functional implications. Endocrinology 147(8):3861–3869. https://doi.org/10.1210/en.2006-0363

Kikuchi S, Uki N (1974) Technical study on artificial spawning of abalone, genus *Haliotis*. II. Effect of irradiated sea water with ultraviolet rays on inducing to spawn. Bull Tohoku Reg Fish Res Lab 33:79–86. (Abstract in English)

Kim YK, Kawazoe I, Jasmani S, Ohira T, Wilder MN, Kaneko T, Aida K (2007) Molecular cloning and characterization of cortical rod protein in the giant freshwater prawn *Macrobrachium rosenbergii*, a species not forming cortical rod structures in the oocytes. Comp Biochem Physiol B Biochem Mol Biol 148(2):184–191. https://doi.org/10.1016/j.cbpb.2007.05.008

Klaus S, Schubart CD, Brandis D (2009) Ultrastructure of spermatozoa and spermatophores of old world freshwater crabs (Brachyura: Potamoidea: Gecarcinucidae, Potamidae, and Potamonautidae). J Morphol 270(2):175–193. https://doi.org/10.1002/jmor.10678

Klepal W (1990) The fundamentals of insemination in cirripedes. Oceanogr Mar Biol Annu Rev 28:353–379

Klepal W, Barnes H, Barnes M (1979) Studies of the reproduction of cirripedes. VII. The formation and fine structure of the fertilization membrane and egg case. J Exp Mar Biol Ecol 36(1):53–78. https://doi.org/10.1016/0022-0981(79)90100-X

Kluge B, Lehmann-Greif M, Fischer A (1995) Long-lasting exocytosis and massive structural reorganisation in the egg periphery during cortical reaction in *Platynereis dumerilii* (Annelida, Polychaeta). Zygote 3(2):141–156. https://doi.org/10.1017/S0967199400002513

Kocot KM, Cannon JT, Todt C, Citarella MR, Kohn AB, Meyer A, Santos SR, Schander C, Moroz LL, Lieb B, Halanych KM (2011) Phylogenomics reveals deep molluscan relationships. Nature 477(7365):452–456. https://doi.org/10.1038/nature10382

Kolbin KG, Kulikova VA (2011) Reproduction and larval development of the limpet *Lottia persona* (Rathke, 1833) (Gastropoda: Lottiidae). Russ J Mar Biol 37(3):239–242. https://doi.org/10.1134/S1063074011030072

Komaru A, Konishi K, Nakayama I, Kobayashi T, Sakai H, Kawamura K (1997) Hermaphroditic freshwater clams in the genus *Corbicula* produce non-reductional spermatozoa with somatic DNA content. Biol Bull 193(3):320–323. https://doi.org/10.2307/1542934

Komaru A, Kawagishi T, Konishi K (1998) Cytological evidence of spontaneous androgenesis in the freshwater clam *Corbicula leana* Prime. Dev Genes Evol 208(1):46–50. https://doi.org/10.1007/s004270050152

Komaru A, Ookubo K, Kiyomoto M (2000) All meiotic chromosomes and both centrosomes at spindle pole in the zygotes discarded as two polar bodies in clam *Corbicula leana*: unusual polar body formation observed by antitubulin immunofluorescence. Dev Genes Evol 210(5):263–269. https://doi.org/10.1007/s004270050313

Krantic S, Rivailler P (1996) Meiosis reinitiation in molluscan oocytes: a model to study the transduction of extracellular signals. Invert Reprod Dev 30(1–3):55–69. https://doi.org/10.1080/07924259.1996.9672532

Krantic S, Dubé F, Quirion R, Guirrier P (1991) Pharmacology of the serotonin-induced meiosis reinitiation in *Spisula solidissima* oocytes. Dev Biol 146(2):491–498. https://doi.org/10.1016/0012-1606(91)90250-7

Krantic S, Dubé F, Guerrier P (1993a) Evidence for a new subtype of serotonin receptor in oocytes of the surf clam *Spisula solidissima*. Gen Comp Endocrinol 90(1):125–131. https://doi.org/10.1006/gcen.1993.1067

Krantic S, Guerrier P, Dubé F (1993b) Meiosis reinitiation in surf clam oocytes is mediated via a 5-hydroxytryptamine$_5$ serotonin membrane receptor and a vitelline envelope-associated high affinity binding site. J Biol Chem 268(11):7983–7989

Kresge N, Vacquier VD, Stout CD (2001) Abalone lysin: the dissolving and evolving sperm protein. Bioessays 23(1):95–103. https://doi.org/10.1002/1521-1878(200101)23:1<95::AID-BIES1012>3.0.CO;2-C

Kruevaisayawan H, Vanichviriyakit R, Weerachatyanukul W, Iamsaard S, Withyachumnarnkul B, Basak A, Tanphaichitr N, Sobhon P (2008) Induction of the acrosome reaction in black tiger shrimp (*Penaeus monodon*) requires sperm trypsin-like enzyme activity. Biol Reprod 79(1):134–141. https://doi.org/10.1095/biolreprod.107.066316

Kubo M, Nakashima S, Tsukahara J, Ishikawa M (1979) Spermiogenesis in barnalces with special reference to organization of the accessory body. Dev Growth Differ 21(5):445–456. https://doi.org/10.1111/j.1440-169X.1979.00445.x

Kutschera U, Elliott JM (2010) Charles Darwin's observations on the behaviour of earthworms and the evolutionary history of a giant endemic species from Germany, *Lumbricus badensis* (Oligochaeta: Lumbricidae). Appl Environ Soil Sci 2010:823047. https://doi.org/10.1155/2010/823047

Kyozuka K, Osanai K (1994) Functions of the egg envelope of *Mytilus edulis* during fertilization. Bull Mar Biol Stn Asamushi Tohoku Univ 19(2):79–92

Kyozuka K, Deguchi R, Yoshida N, Yamashita M (1997) Change in intracellular Ca^{2+} is not involved in serotonin-induced meiosis reinitiation from the first prophase in oocytes of the marine bivalve *Crassostrea gigas*. Dev Biol 182(1):33–41. https://doi.org/10.1006/dbio.1996.8470

Lafont R, Mathieu M (2007) Steroids in aquatic invertebrates. Ecotoxicology 16(1):109–130. https://doi.org/10.1007/s10646-006-0113-1

Le Gall S, Feral C, Lengronne C, Porchet M (1987) Partial purification of the endocrine mitogenic factor in the mollusk *Crepidula fornicata* L. Comp Biochem Physiol B 86(2):393–396. https://doi.org/10.1016/0305-0491(87)90311-7

Leclerc C, Guerrier P, Moreau M (2000) Role of dihydropyridine-sensitive calcium channels in meiosis and fertilization in the bivalve molluscs *Ruditapes philippinarum* and *Crassostrea gigas*. Biol Cell 92(3–4):285–299. https://doi.org/10.1016/S0248-4900(00)01069-8

Lewis CA (1975) Development of the gooseneck barnacle *Pollicipes polymerus* (Cirripedia: Lepadomorpha): fertilization through settlement. Mar Biol 32(2):141–153. https://doi.org/10.1007/BF00388507

Lewis CA, Leighton DL, Vacquier VD (1980) Morphology of abalone spermatozoa before and after the acrosome reaction. J Ultrastruct Res 72(1):39–46. https://doi.org/10.1016/S0022-5320(80)90133-1

Lewis CA, Talbot CF, Vacquier VD (1982) A protein from abalone sperm dissolves the egg vitelline layer by a nonenzymatic mechanism. Dev Biol 92(1):227–239. https://doi.org/10.1016/0012-1606(82)90167-1

Li XC, Giot JF, Kuhl D, Hen R, Kandel ER (1995) Cloning and characterization of two related serotonergic receptors from the brain and the reproductive system of *Aplysia* that activate phospholipase C. J Neurosci 15(11):7585–7591. https://doi.org/10.1523/JNEUROSCI.15-11-07585.1995

Li Q, Osada M, Suzuki T, Mori K (1998) Changes in vitellin during oogenesis and effect of estradiol-17β on vitellogenesis in the Pacific oyster *Crassostrea gigas*. Invert Reprod Dev 33(1):87–93. https://doi.org/10.1080/07924259.1998.9652345

Lillie FR (1911) Studies of fertilization in *Nereis*. I. The cortical changes in the egg. II. Partial fertilization. J Morphol 22(2):361–393. https://doi.org/10.1002/jmor.1050220208

Lindsay LL, Clark WH Jr (1994) Signal transduction during shrimp oocyte activation by extracellular Mg^{2+}: roles of inositol 1,4,5- trisphosphate, tyrosine kinases and G-proteins. Development 120(12):3463–3472

Lindsay LL, Hertzler PL, Clark WH Jr (1992) Extracellular Mg^{2+} induces an intracellular Ca^{2+} wave during oocyte activation in the marine shrimp *Sicyonia ingentis*. Dev Biol 152(1):94–102. https://doi.org/10.1016/0012-1606(92)90159-E

Liu M (2011) The biology and dynamics of mammalian cortical granules. Reprod Biol Endocrinol 9:149. https://doi.org/10.1186/1477-7827-9-149

Longo FJ (1976) Ultrastructural aspects of fertilization in spiralian eggs. Am Zool 16(3):375–394. https://doi.org/10.1093/icb/16.3.375

Longo FJ (1983) Meiotic maturation and fertilization. In: Wilbur KM (ed) The mollusca, vol 3. Academic Press, New York, pp 49–89

Longo FJ, Anderson E (1970) An ultrastructual analysis of fertilization in the surf clam, *Spisula solidissima*. I. Polar body formation and development of the female pronucleus. J Ultrastruct Res 33(5):495–514. https://doi.org/10.1016/S0022-5320(70)90177-2

Longo FJ, Dornfeld EJ (1967) The fine structure of spermatid differentiation in the mussel, *Mytilus edulis*. J Ultrastruct Res 20(5):462–480. https://doi.org/10.1016/S0022-5320(67)80113-8

Loumaye E, Thorner J, Catt KJ (1982) Yeast mating pheromone activates mammalian gonadotrophs: evolutionary conservation of a reproductive hormone? Science 218(4579):1323–1325. https://doi.org/10.1126/science.6293058

Lubet P, Mathieu M (1982) The action of internal factors on gametogenesis in pelecypod molluscs. Malacologia 22(1–2):131–136

Lynn JW, Clark WH Jr (1983a) A morphological examination of sperm-egg interaction in the freshwater prawn, *Macrobrachium rosenbergii*. Biol Bull 164(3):446–458. https://doi.org/10.2307/1541254

Lynn JW, Clark WH Jr (1983b) The fine structure of the mature sperm of the freshwater prawn, *Macrobrachium rosenbergii*. Biol Bull 164(3):459–470. https://doi.org/10.2307/1541255

Maccari M, Amat F, Hontoria F, Gómez A (2014) Laboratory generation of new parthenogenetic lineages supports contagious parthenogenesis in *Artemia*. Peer J 2:e439. https://doi.org/10.7717/peerj.439

Maina JN (2002) Structure, function and evolution of the gas exchangers: comparative perspectives. J Anat 201(4):281–304. https://doi.org/10.1046/j.1469-7580.2002.00099.x

Markow TA (2015) The secret lives of *Drosophila* flies. Elife 4. https://doi.org/10.7554/eLife.06793

Marotta R, Crottini A, Raimondi E, Fondello C, Ferraguti M (2014) Alike but different: the evolution of the *Tubifex tubifex* species complex (Annelida, Clitellata) through polyploidization. BMC Evol Biol 14(1):73. https://doi.org/10.1186/1471-2148-14-73

Martínez G, Saleh FL, Mettifogo L, Campos E, Inestrosa N (1996) Monoamines and the release of gamets by the scallop *Argopecten purpuratus*. J Exp Zool 274(6):365–372. https://doi.org/10.1002/(SICI)1097-010X(19960415)274:6<365::AID-JEZ5>3.0.CO;2-M

Masui Y (1985) Meiotic arrest in animal oocytes. In: Metz CB, Monroy A (eds) Biology of fertilization, vol 1. Academic Press, New York, pp 189–219

Mathieu M (1985) Partial characterization of aspartate transcarbamylase from the mantle of the mussel *Mytilus edulis*. Comp Biochem Physiol B 82(4):667–674. https://doi.org/10.1016/0305-0491(85)90505-X

Mathieu M (1987) Utilization of aspartate transcarbamylase activity in the study of neuroendocrine control of gametogenesis in *Mytilus edulis*. J Exp Zool 241(2):247–252. https://doi.org/10.1002/jez.1402410211

Mathieu M, Lubet P (1980) Analyse expérimentale en cultures d'organes de l'action des ganglions nerveux sur la gonade adulte de la moule. Bull Soc Zool Fr 105:149–153

Mathieu M, Lenoir F, Robbins I (1988) A gonial mitosis-stimulating factor in cerebral ganglia and hemolymph of the marine mussel *Mytilus edulis* L. Gen Comp Endocrinol 72(2):257–263. https://doi.org/10.1016/0016-6480(88)90208-0

Matsumoto T, Osada M, Osawa Y, Mori K (1997) Gonadal estrogen profile and immunohistochemical localization of steroidogenic enzymes in the oyster and scallop during sexual maturation. Comp Biochem Physiol B 118(4):811–817. https://doi.org/10.1016/S0305-0491(97)00233-2

Matsumoto T, Nakamura AM, Mori K, Kayano T (2003) Molecular characterization of a cDNA encoding putative vitellogenin from the Pacific oyster *Crassostrea gigas*. Zoolog Sci 20(1):37–42. https://doi.org/10.2108/zsj.20.37

Matsumoto T, Nakamura AM, Mori K, Akiyama I, Hirose H, Takahashi Y (2007) Oyster estrogen receptor: cDNA cloning and immunolocalization. Gen Comp Endocrinol 151(2):195–201. https://doi.org/10.1016/j.ygcen.2007.01.016

Matsumoto T, Yamano K, Kitamura M, Hara A (2008) Ovarian follicle cells are the site of vitellogenin synthesis in the Pacific abalone *Haliotis discus hannai*. Comp Biochem Physiol A 149(3):293–298. https://doi.org/10.1016/j.cbpa.2008.01.003

Matsumoto T, Masaoka T, Fujiwara A, Nakamura Y, Satoh N, Awaji M (2013) Reproduction-related genes in the pearl oyster genome. Zoolog Sci 30(10):826–850. https://doi.org/10.2108/zsj.30.826

Matsutani T, Nomura T (1982) Induction of spawning by serotonin in the scallop, *Patinopecten yessoensis* (Jay). Mar Biol Lett 3:353–358

Matsutani T, Nomura T (1986a) Serotonin-like immunoreactivity in the central nervous system and gonad of the scallop, *Patinopecten yessoensis*. Cell Tissue Res 244(3):515–517. https://doi.org/10.1007/BF00212528

Matsutani T, Nomura T (1986b) Pharmacological observations on the mechanism of spawning in the scallop *Patinopecten yessoensis*. Bull Jpn Soc Sci Fish 52(9):1589–1594. https://doi.org/10.2331/suisan.52.1589

Matsutani T, Nomura T (1987) *In vitro* effects of serotonin and prostaglandins on release of eggs from the ovary of the scallop, *Patinopecten yessoyensis*. Gen Comp Endocrinol 67(1):111–118. https://doi.org/10.1016/0016-6480(87)90210-3

Mattei JH, Beekey MA, Rudman A, Woronik A (2010) Reproductive behavior in horseshoe crabs: does density matter? Curr Zool 56(5):634–642

Metz EC, Robles-Sikisaka R, Vacquier VD (1998) Nonsynonymous substitution in abalone sperm fertilization genes exceeds substitution in introns and mitochondrial DNA. Proc Natl Acad Sci U S A 95(18):10676–10681

Miranda NA, Perissinotto R, Appleton CC (2011) Population structure of an invasive parthenogenetic gastropod in coastal lakes and estuaries of northern KwaZulu-Natal, South Africa. PLoS One 6(8):e24337. https://doi.org/10.1371/journal.pone.0024337

Miura T, Miura C (2001) Japanese eel: a model for analysis of spermatogenesis. Zoolog Sci 18(8):1055–1063. https://doi.org/10.2108/zsj.18.1055

Miyazaki S (2006) Thirty years of calcium signals at fertilization. Semin Cell Dev Biol 17(2):233–243. https://doi.org/10.1016/j.semcdb.2006.02.007

Mozingo NM, Vacquier VD, Chandler DE (1995) Structural features of the abalone egg extracellular matrix and its role in gamete interaction during fertilization. Mol Reprod Dev 41(4):493–502. https://doi.org/10.1002/mrd.1080410412

Mustonen M, Haimi J, Kesäniemi J, Högmander H, Knott KE (2017) Variation in gene expression within clones of the earthworm *Dendrobaena octaedra*. PLoS One 12(4):e0174960. https://doi.org/10.1371/journal.pone.0174960

Nagasawa K, Oouchi H, Itoh N, Takahashi KG, Osada M (2015a) *In vivo* administration of scallop GnRH-like peptide influences on gonad development in the Yesso scallop, *Patinopecten yessoensis*. PLoS One 10(6):e0129571. https://doi.org/10.1371/journal.pone.0129571

Nagasawa K, Treen N, Kondo R, Otoki Y, Itoh N, Rotchell JM, Osada M (2015b) Molecular characterization of an estrogen receptor and estrogen-related receptor and their autoregulatory capabilities in two *Mytilus* species. Gene 564(2):153–159. https://doi.org/10.1016/j.gene.2015.03.073

Nagasawa K, Osugi T, Suzuki I, Itoh N, Takahashi KG, Satake H, Osada M (2015c) Characterization of GnRH-like peptides from the nerve ganglia of Yesso scallop, *Patinopecten yessoensis*. Peptides 71:202–210. https://doi.org/10.1016/j.peptides.2015.07.022

Nagasawa K, Muroi M, Thitiphuree T, Minegishi Y, Itoh N, Osada M (2017) Cloning of invertebrate gonadotropin-releasing hormone receptor (GnRHR)-like gene in Yesso scallop, *Patinopecten yessoensis*. Agric Gene 3:46–56. https://doi.org/10.1016/j.aggene.2016.11.005

Nakamura S, Osada M, Kijima A (2007) Involvement of GnRH neuron in the spermatogonial proliferation of the scallop, *Patinopecten yessoensiss*. Mol Reprod Dev 74(1):108–115. https://doi.org/10.1002/mrd.20544

Nakano T, Kyozuka K (2015) Soluble sperm extract specifically recapitulates the initial phase of the Ca^{2+} response in the fertilized oocyte of *P. occelata* following a G-protein/PLCβ signaling pathway. Zygote 23(6):821–835. https://doi.org/10.1017/S0967199414000501

Nakano T, Kyozuka K, Deguchi R (2008) Novel two-step Ca^{2+} increase and its mechanisms and functions at fertilization in oocytes of the annelidan worm *Pseudopotamilla occelata*. Dev Growth Differ 50(5):365–379. https://doi.org/10.1111/j.1440-169X.2008.01022.x

Nakano T, Deguchi R, Kyozuka K (2014) Intracellular calcium signaling in the fertilized eggs of Annelida. Biochem Biophys Res Commun 450(3):1188–1194. https://doi.org/10.1016/j.bbrc.2014.06.056

Néant I, Guerrier P (1988) Meiosis reinitiation in the mollusc *Patella vulgata*. Regulation of MPF, CSF, and chromosome condensation activity by intracellular pH, protein synthesis and phosphorylation. Development 102(3):505–516

Ngernsoungnern P, Ngernsoungnern A, Chaiseha Y, Sretarugsa P (2012) Role of vitelline envelope during fertilization in the black tiger shrimp, *Penaeus monodon*. Acta Histochem 114(7):659–664. https://doi.org/10.1016/j.acthis.2011.11.013

Niijima L, Dan J (1965) The acrosome reaction in *Mytilus edulis*. I. Fine structure of the intact acrosome. J Cell Biol 25:243–248. https://doi.org/10.1083/jcb.25.2.243

Niksirat H, Kouba A, Kozák P (2015) Ultrastructure of egg activation and cortical reaction in the noble crayfish *Astacus astacus*. Micron 68:115–121. https://doi.org/10.1016/j.micron.2014.09.010

Novikoff AB (1939) Surface changes in unfertilized and fertilized eggs of *Sabellaria vulgaris*. J Exp Zool 82(2):217–237. https://doi.org/10.1002/jez.1400820204

Osada M, Nomura T (1990) The levels of prostaglandins associated with the reproductive cycle of the scallop, *Patinopecten yessoensis*. Prostaglandins 40(3):229–239. https://doi.org/10.1016/0090-6980(90)90011-J

Osada M, Treen N (2013) Molluscan GnRH associated with reproduction. Gen Comp Endocrinol 181:254–258. https://doi.org/10.1016/j.ygcen.2012.09.002

Osada M, Matsutani T, Nomura T (1987) Implication of catecholamines during spawning in marine bivalve molluscs. Int J Invert Reprod Dev 12(3):241–252. https://doi.org/10.1080/01688170.1987.10510324

Osada M, Nishikawa M, Nomura T (1989) Involvement of prostaglandins in the spawning of the scallop *Patinopecten yessoensis*. Comp Biochem Physiol C 94(2):595–601. https://doi.org/10.1016/0742-8413(89)90119-9

Osada M, Mori K, Nomura T (1992a) *In vitro* effects of estrogen and serotonin on release of eggs from the ovary of the scallop. Nippon Suisan Gakkaishi 58(2):223–227. https://doi.org/10.2331/suisan.58.223

Osada M, Unuma T, Mori K (1992b) Purification and characterization of a yolk protein from the scallop ovary. Nippon Suisan Gakkaishi 58(12):2283–2289. https://doi.org/10.2331/suisan.58.2283

Osada M, Nakata A, Matsumuto T, Mori K (1998) Pharmacological characterization of serotonin receptor in the oocyte membrane of bivalve molluscs and its formation during oogenesis. J Exp Zool 281(2):124–131. https://doi.org/10.1002/(SICI)1097-010X(19980601)281:2<124::AID-JEZ6>3.0.CO;2-Q

Osada M, Takamura T, Sato H, Mori K (2003) Vitellogenin synthesis in the ovary of scallop, *Patinopecten yessoensis*: control by estradiol-17 beta and the central nervous system. J Exp Zool 299A(2):172–179. https://doi.org/10.1002/jez.a.10276

Osada M, Harata M, Kishida M, Kijima A (2004a) Molecular cloning and expression analysis of vitellogenin in scallop, *Patinopecten yessoensis* (Bivalvia, Mollusca). Mol Reprod Dev 67(3):273–281. https://doi.org/10.1002/mrd.20020

Osada M, Tawarayama H, Mori K (2004b) Estrogen synthesis in relation to gonadal development of Japanese scallop, *Patinopecten yessoensis*: gonadal profile and immunolocalization of P450 aromatase and estrogen. Comp Biochem Physiol B 139(1):123–128. https://doi.org/10.1016/j.cbpc.2004.07.002

Osada M, Nakamura S, Kijima A (2007) Quantitative analysis of the pattern of gonial proliferation during sexual maturation in the Japanese scallop *Patinopecten yessoensis*. Fish Sci 73(6):1318–1324. https://doi.org/10.1111/j.1444-2906.2007.01470.x

Osanai K (1975) Seasonal gonad development and sex alteration in the scallop, *Patinopecten yessoensis*. Bull Mar Biol St Asamushi Tohoku Univ 15(2):81–88

Osanai K (1985) *In vitro* induction of germinal vesicle breakdown in oyster oocytes. Bull Mar Biol Stn Asamushi Tohoku Univ 18(1):1–9

Osanai K, Kuraishi R (1988) Response of oocytes to meiosis-inducing agents in pelecypods. Bull Mar Biol Stn Asamushi Tohoku Univ 18(2):45–56

Park KI, Choi KS (2004) Application of enzyme-linked immunosorbent assay for studying of reproduction in the Manila clam *Ruditapes philippinarum* (Mollusca: Bivalvia): I. Quantifying eggs. Aquaculture 241(1–4):667–687. https://doi.org/10.1016/j.aquaculture.2004.08.017

Pauletto M, Milan M, de Sousa JT, Huvet A, Joaquim S, Matias D, Leitão A, Patarnello T, Bargelloni L (2014) Insights into molecular features of *Venerupis decussata* oocytes: a microarray-based study. PLoS One 9(12):e113925. https://doi.org/10.1371/journal.pone.0113925

Pazos AJ, Mathieu M (1999) Effects of five natural gonadotropin-releasing hormones on cell suspensions of marine bivalve gonad: stimulation of gonial DNA synthesis. Gen Comp Endocrinol 113(1):112–120. https://doi.org/10.1006/gcen.1998.7186

Pierantoni R, Cobellis G, Meccariello R, Cacciola G, Chianese R, Chioccarelli T, Fasano S (2009) Testicular gonadotropin-releasing hormone activity, progression of spermatogenesis, and sperm transport in vertebrates. Ann N Y Acad Sci 1163:279–291. https://doi.org/10.1111/j.1749-6632.2008.03617.x

Pillai MC, Clark WH Jr (1987) Oocyte activation in the marine shrimp, *Sicyonia ingentis*. J Exp Zool 244(2):325–329. https://doi.org/10.1002/jez.1402440217

Pipe RK (1987a) Oogenesis in the marine mussel *Mytilus edulis*: an ultrastructural study. Mar Biol 95(3):405–414. https://doi.org/10.1007/BF00409571

Pipe RK (1987b) Ultrastructural and cytochemical study on interactions between nutrient storage cells and gametogenesis in the mussel *Mytilus edulis*. Mar Biol 96(4):519–528. https://doi.org/10.1007/BF00397969

Polzonetti-Magni AM, Mosconi G, Soverchia L, Kikuyama S, Carnevali O (2004) Multihormonal control of vitellogenesis in lower vertebrates. Int Rev Cytol 239:1–45. https://doi.org/10.1016/S0074-7696(04)39001-7

Pongtippatee P, Luppanakane R, Thaweethamsewee P, Kirirat P, Weerachatyanukul W, Withyachumnarnkul B (2010) Delay of the egg activation process in the black tiger shrimp *Penaeus monodon* by manipulation of magnesium levels in spawning water. Aquacult Res 41(2):227–232. https://doi.org/10.1111/j.1365-2109.2009.02322.x

Pongtippatee-Taweepreda P, Chavadej J, Plodpai P, Pratoomchart B, Sobhon P, Weerachatyanukul W, Withyachumnarnkul B (2004) Egg activation in the black tiger shrimp *Penaeus monodon*. Aquaculture 234(1–4):183–198. https://doi.org/10.1016/j.aquaculture.2003.10.036

Portillo-López A, Gould MC, Stephano JL (2003) MAPK is involved in metaphase I arrest in oyster and mussel oocytes. Biol Cell 95(5):275–282. https://doi.org/10.1016/S0248-4900(03)00054-6

Prevedelli D, Simonini R (2003) Life cycles in brackish habitats: adaptive strategies of some polychaetes from the Venice lagoon. Oceanol Acta 26(1):77–84. https://doi.org/10.1016/S0399-1784(02)01232-X

Raj I, Sadat Al Hosseini H, Dioguardi E, Nishimura K, Han L, Villa A, de Sanctis D, Jovine L (2017) Structural basis of egg coat-sperm recognition at fertilization. Cell 169(7):1315–1326. https://doi.org/10.1016/j.cell.2017.05.033

Ram JL, Fei X, Danaher SM, Lu S, Breithaupt T, Hardege JD (2008) Finding females: pheromone-guided reproductive tracking behavior by male *Nereis succinea* in the marine environment. J Exp Biol 211(5):757–765. https://doi.org/10.1242/jeb.012773

Rebhun LI (1962) Electron microscope studies on the vitelline membrane of the surf clam, *Spisula solidissima*. J Ultrastruct Res 6(1):107–122. https://doi.org/10.1016/S0022-5320(62)90064-3
Reis-Henriques MA, Le Guellec D, Remy-Martin JP, Adessi GL (1990) Studies of endogenous steroids from the marine mollusc *Mytilus edulis* L. By gas chromatography and mass spectrometry. Comp Biochem Physiol B 95(2):303–309. https://doi.org/10.1016/0305-0491(90)90080-D
Roch GJ, Busby ER, Sherwood NM (2011) Evolution of GnRH: diving deeper. Gen Comp Endocrinol 171(1):1–16. https://doi.org/10.1016/j.ygcen.2010.12.014
Rodet F, Lelong C, Dubos M-P, Costil K, Favrel P (2005) Molecular cloning of a molluscan gonadotropin-releasing hormone receptor orthologue specifically expressed in the gonad. Biochim Biophys Acta 1730(3):187–195. https://doi.org/10.1016/j.bbaexp.2005.05.012
Rodet F, Lelong C, Dubos M-P (2008) Favrel P (2008) Alternative splicing of a single precursor mRNA generates two subtypes of gonadotropin-releasing hormone receptor orthologues and their variants in the bivalve mollusc *Crassostrea gigas*. Gene 414(1–2):1–9. https://doi.org/10.1016/j.gene.2008.01.022
Rojas E, Alfaro J (2007) *In vitro* manipulation of egg activation in the open thelycum shrimp *Litopenaeus*. Aquaculture 264(1–4):469–474. https://doi.org/10.1016/j.aquaculture.2006.12.025
Sakai Y, Shiroya Y, Haino-Fukushima K (1982) Fine structural changes in the acrosome reaction of the Japanese abalone, *Haliotis disus*. Dev Growth Differ 24(6):531–542. https://doi.org/10.1111/j.1440-169X.1982.00531.x
Samadi S, Mavárez J, Pointier JP, Delay B, Jarne P (1999) Microsatellite and morphological analysis of population structure in the parthenogenetic freshwater snail *Melanoides tuberculata*: insights into the creation of clonal variability. Mol Ecol 8(7):1141–1153. https://doi.org/10.1046/j.1365-294x.1999.00671.x
Sastry AN (1979) Pelecypoda (exclusive ostreidae). In: Giese AC, Pearse JS (eds) Reproduction of marine invertebrates, vol 5. Academic Press, New York, pp 113–292
Sato E, Koide SS (1984) A factor from bovine granulose cells preventing oocyte maturation. Differentiation 26(1):59–62. https://doi.org/10.1111/j.1432-0436.1984.tb01374.x
Sato M, Osanai K (1983) Sperm reception by an egg microvillus in the polychaete, *Tylorrhynchus heterochaetus*. J Exp Zool 227(3):459–469. https://doi.org/10.1002/jez.1402270315
Sato M, Osanai K (1986) Morphological identification of sperm receptors above egg microvilli in the polychaete, *Neanthes japonica*. Dev Biol 113(2):263–270. https://doi.org/10.1016/0012-1606(86)90161-2
Sato E, Wood HN, Lynn DG, Sahni MK, Koide SS (1985) Meiotic arrest in oocytes regulated by a *Spisula* factor. Biol Bull 169(2):334–341. https://doi.org/10.2307/1541486
Sato E, Toyoda Y, Segal SJ, Koide SS (1992) Oocyte membrane components preventing trypsin-induced germinal vesicle breakdown in surf clam oocyte. J Reprod Dev 38(4):309–315. https://doi.org/10.1262/jrd.38.309
Scott AP (2012) Do mollusks use vertebrate sex steroids as reproductive hormones? Part I: critical appraisal of the evidence for the presence, biosynthesis and uptake of steroids. Steroids 77(13):1450–1468. https://doi.org/10.1016/j.steroids.2012.08.009
Scott AP (2013) Do mollusks use vertebrate sex steroids as reproductive hormones? II. Critical review of the evidence that steroids have biological effects. Steroids 78(2):268–281. https://doi.org/10.1016/j.steroids.2012.11.006
Sedano FJ, Rodríguez JL, Ruiz C, García-Martín LO, Sánchez JL (1995) Biochemical composition and fertilization in the eggs of *Mytilus galloprovincialis* (Lamarck). J Exp Mar Biol Ecol 192(1):75–85. https://doi.org/10.1016/0022-0981(95)00062-V
Shen HP, Yu HT, Chen JH (2012) Parthenogenesis in two Taiwanese mountain earthworms *Amynthas catenus* Tsai et al., 2001 and *Amynthas hohuanmontis* Tsai et al., 2002 (Oligochaeta, Megascolecidae) revealed by AFLP. Eur J Soil Biol 51:30–36. https://doi.org/10.1016/j.ejsobi.2012.03.007
Shi W, Han Y, Guo C, Zhao X, Liu S, Su W, Wang Y, Zha S, Chai X, Liu G (2017) Ocean acidification hampers sperm-egg collisions, gamete fusion, and generation of Ca^{2+} oscillations of a broadcast spawning bivalve, *Tegillarca granosa*. Mar Environ Res 130:106–112. https://doi.org/10.1016/j.marenvres.2017.07.016

Shin KS, Kwon HJ, Kim WJ (2005) Attribution of cortical granules to formation of fertilization envelopes and polyspermy block in *Urechis unicinctus*. Integr Biosci 9(2):57–64. https://doi.org/10.1080/17386357.2005.9647252

Simeó CG, Kurtz K, Rotllant G, Chiva M, Ribes E (2010) Sperm ultrastructure of the spider crab *Maja brachydactyla* (Decapoda: Brachyura). J Morphol 271(4):407–417. https://doi.org/10.1002/jmor.10806

Simonini R, Molinari F, Pagliai AM, Ansaloni I, Prevedelli D (2003) Karyotype and sex determination in *Dinophilus gyrociliatus* (Polychaeta: Dinophilidae). Mar Biol 142(3):441–445. https://doi.org/10.1007/s00227-002-0979-2

Singaravelu G, Singson A (2013) Calcium signaling surrounding fertilization in the nematode *Caenorhabditis elegans*. Cell Calcium 53(1):2–9. https://doi.org/10.1016/j.ceca.2012.11.009

Smith SA, Wilson NG, Goetz FE, Feehery C, Andrade SCS, Rouse GW, Giribet G, Dunn CW (2011) Resolving the evolutionary relationships of molluscs with phylogenomic tools. Nature 480(7377):364–367. https://doi.org/10.1038/nature10526

Stephano JL (1992) A study of polyspermy in abalone. In: Shepherd SA, Tegner MJ, Guzman del Proo SA (eds) Abalone of the world: biology, fisheries and culture. Fishing News Books, Oxford, pp 518–526

Stephano JL, Gould MC (1997a) The intracellular calcium increase at fertilization in *Urechis caupo* oocytes: activation without waves. Dev Biol 191(1):53–68. https://doi.org/10.1006/dbio.1997.8709

Stephano JL, Gould MC (1997b) Parthenogenesis in *Urechis caupo* (Echiura). II. Role of intracellular pH in parthenogenesis induction. Dev Growth Differ 39(1):99–104. https://doi.org/10.1046/j.1440-169X.1997.00010.x

Storch V, Alberti G (1978) Ultrastructural observations on the gills of polychaetes. Helgoland Wiss Meer 31(1–2):169–179. https://doi.org/10.1007/BF02296995

Stricker SA (1999) Comparative biology of calcium signalling during fertilization and egg activation in animals. Dev Biol 211(2):157–176. https://doi.org/10.1006/dbio.1999.9340

Struck TH, Golombek A, Weigert A, Franke FA, Westheide W, Purschke G, Bleidorn C, Halanych KM (2015) The evolution of annelids reveals two adaptive routes to the interstitial realm. Curr Biol 25(15):1993–1999. https://doi.org/10.1016/j.cub.2015.06.007

Sugamori KS, Sunahara RK, Guan HC, Bulloch AG, Tensen CP, Seeman P, Niznik HB, Van Tol HH (1993) Serotonin receptor cDNA, cloned from *Lymnaea stagnalis*. Proc Natl Acad Sci U S A 90(1):11–15. https://doi.org/10.1073/pnas.90.1.11

Sun B, Tsai P-S (2011) A gonadotropin-releasing hormone-like molecule modulates the activity of diverse central neurons in a gastropod mollusk, *Aplysia californica*. Front Endocrinol 2:36. https://doi.org/10.3389/fendo.2011.00036

Sun B, Kavanaugh SI, Tsai P-S (2012) Gonadotropin-releasing hormone in protostomes: insights from functional studies on *Aplysia californica*. Gen Comp Endocrinol 176(3):321–326. https://doi.org/10.1016/j.ygcen.2011.11.030

Suzuki T, Hara A, Yamaguchi K, Mori K (1992) Purification and immunolocalization of a vitellin-like protein from the Pacific oyster *Crassostrea gigas*. Mar Biol 113(2):239–245. https://doi.org/10.1007/BF00347277

Swann K, Lai FA (2016) Egg activation at fertilization by a soluble sperm protein. Physiol Rev 96(1):127–149. https://doi.org/10.1152/physrev.00012.2015

Swanson WJ, Vacquier VD (1997) The abalone egg vitelline envelope receptor for sperm lysin is a giant multivalent molecule. Proc Natl Acad Sci U S A 94(13):6724–6729. https://doi.org/10.1073/pnas.94.13.6724

Takagi T, Nakamura A, Deguchi R, Kyozuka K (1994) Isolation, characterization, and primary structure of three major proteins obtained from *Mytilus edulis* sperm. J Biochem 116(3):598–605. https://doi.org/10.1093/oxfordjournals.jbchem.a124566

Takeda N, Kon Y, Quiroga Artigas G, Lapébie P, Barreau C, Koizumi O, Kishimoto T, Tachibana K, Houliston E, Deguchi R (2018) Identification of jellyfish neuropeptides that act directly as oocyte maturation-inducing hormones. Development 145(2):dev156786. https://doi.org/10.1242/dev.156786

Takeuchi T, Kawashima T, Koyanagi R, Gyoja F, Tanaka M, Ikuta T, Shoguchi E, Fujiwara M, Shinzato C, Hisata K, Fujie M, Usami T, Nagai K, Maeyama K, Okamoto K, Aoki H, Ishikawa T, Masaoka T, Fujiwara A, Endo K, Endo H, Nagasawa H, Kinoshita S, Asakawa S, Watabe S, Satoh N (2012) Draft genome of the pearl oyster *Pinctada fucata*: a platform for understanding bivalve biology. DNA Res 19(2):117–130. https://doi.org/10.1093/dnares/dss005

Talbot P, Goudeau M (1988) A complex cortical reaction leads to formation of the fertilization envelope in the lobster, *Homarus*. Gamete Res 19(1):1–18. https://doi.org/10.1002/mrd.1120190102

Tanabe T, Osada M, Kyozuka K, Inaba K, Kijima A (2006) Novel oocyte maturation arresting factor in the central nervous system of scallops inhibits serotonin-induced oocyte maturation and spawning of bivalve mollusks. Gen Comp Endocrinol 147(3):352–361. https://doi.org/10.1016/j.ygcen.2006.02.004

Tanabe T, Yuan Y, Nakamura S, Itoh N, Takahashi KG, Osada M (2010) The role in spawning of a putative serotonin receptor isolated from the germ and ciliary cells of the gonoduct in the gonad of the Japanese scallop, *Patinopecten yessoensis*. Gen Comp Endocrinol 166(3):620–627. https://doi.org/10.1016/j.ygcen.2010.01.014

Tanaka Y, Murakoshi M (1985) Spawning induction of the hermaphroditic scallop, *Pecten albicans*, by injection with serotonin. Bull Natl Res Inst Aquacult 7:9–12

Teaniniuraitemoana V, Huvet A, Levy P, Klopp C, Lhuillier E, Gaertner-Mazouni N, Gueguen Y, Le Moullac G (2014) Gonad transcriptome analysis of pearl oyster *Pinctada margaritifera*: identification of potential sex differentiation and sex determining genes. BMC Genomics 15:491. https://doi.org/10.1186/1471-2164-15-491

Thitiphuree T, Nagasawa K, Osada M (2019) Molecular identification of steroidogenesis-related genes in scallops and their potential roles in gametogenesis. J Steroid Biochem Mol Biol 186:22–33. https://doi.org/10.1016/j.jsbmb.2018.09.004

Thomas TW, Eckberg WR, Dubé F, Galione A (1998) Mechanisms of calcium release and sequestration in eggs of *Chaetopterus pergamentaceus*. Cell Calcium 24(4):285–292. https://doi.org/10.1016/S0143-4160(98)90052-5

Thornton JW, Need E, Crews D (2003) Resurrecting the ancestral steroid receptor: ancient origin of estrogen signaling. Science 301(5640):1714–1717. https://doi.org/10.1126/science.1086185

Tierney AJ (2001) Structure and function of invertebrate 5-HT receptors: a review. Comp Biochem Physiol A 128(4):791–804. https://doi.org/10.1016/S1095-6433(00)00320-2

Togo T, Morisawa M (1997) Aminopeptidase-like protease released from oocytes affects oocyte surfaces and suppresses the acrosome reaction in establishment of polyspermy block in oocytes of the mussel *Mytilus edulis*. Dev Biol 182(2):219–227. https://doi.org/10.1006/dbio.1996.8483

Togo T, Osanai K, Morisawa M (1995) Existence of three mechanisms for blocking polyspermy in oocytes of the mussel *Mytilus edulis*. Biol Bull 189(3):330–339. https://doi.org/10.2307/1542150

Treen N, Itoh N, Miura H, Kikuchi I, Ueda T, Takahashi KG, Ubuka T, Yamamoto K, Sharp PJ, Tsutsui K, Osada M (2012) Mollusc gonadotropin-releasing hormone directly regulates gonadal functions: a primitive endocrine system controlling reproduction. Gen Comp Endocrinol 176(2):167–172. https://doi.org/10.1016/j.ygcen.2012.01.008

Tsafriri A, Pomerantz SH (1986) Oocyte maturation inhibitor. Clin Endocrinol Metab 15(1):157–170. https://doi.org/10.1016/S0300-595X(86)80047-0

Tsai P-S, Maldonado TA, Lunden JB (2003) Localization of gonadotropin-releasing hormone in the central nervous system and a peripheral chemosensory organ of *Aplysia californica*. Gen Comp Endocrinol 130(1):20–28. https://doi.org/10.1016/S0016-6480(02)00519-1

Tsai P-S, Sun B, Rochester JR, Wayne NL (2010) Gonadotropin-releasing hormone-like molecule is not an acute reproductive activator in the gastropod, *Aplysia californica*. Gen Comp Endocrinol 166(2):280–288. https://doi.org/10.1016/j.ygcen.2009.09.009

Tudge CC, Scheltinga DM, Jamieson BGM (2001) Spermatozoal morphology in the "symmetrical" hermit crab, *Pylocheles* (*Bathycheles*) sp. (Crustacea, Decapoda, Anomura, Paguroidea, Pylochelidae). Zoosystema 23(1):117–130

Uki N, Kikuchi S (1974) On the effect of irradiated seawater with ultraviolet rays on inducing spawning of the scallop, *Patinopecten yessoensis* (Jay). Bull Tohoku Reg Fish Res Lab 34:87–92. (Abstract in English)

Vacquier VD, Lee YH (1993) Abalone sperm lysin: unusual mode of evolution of a gamete recognition protein. Zygote 1(3):181–196. https://doi.org/10.1017/S0967199400001465

Vacquier VD, Swanson WJ, Lee YH (1997) Positive Darwinian selection on two homologous fertilization proteins: what is the selective pressure driving their divergence? J Mol Evol 44(Suppl 1):S15–S22. https://doi.org/10.1007/PL00000049

Varaksin AA, Varaksina GS, Reunova OV, Latyshev NA (1992) Effect of serotonin, some fatty acids and their metabolites on reinitiation of meiotic maturation in oocytes of bivalve *Spisula sachalinensis* (Schrenk). Comp Biochem Physiol C 101(3):627–630. https://doi.org/10.1016/0742-8413(92)90097-Q

Vogeler S, Galloway TS, Lyons BP, Bean TP (2014) The nuclear receptor gene family in the Pacific oyster, *Crassostrea gigas*, contains a novel subfamily group. BMC Genomics 15:369. https://doi.org/10.1186/1471-2164-15-369

Vogt G, Falckenhayn C, Schrimpf A, Schmid K, Hanna K, Panteleit J, Helm M, Schulz R, Lyko F (2015) The marbled crayfish as a paradigm for saltational speciation by autopolyploidy and parthenogenesis in animals. Biol Open 4(11):1583–1594. https://doi.org/10.1242/bio.014241

Von Stetina JR, Orr-Weaver TL (2011) Developmental control of oocyte maturation and egg activation in metazoan models. Cold Spring Harb Perspect Biol 3(10):a005553. https://doi.org/10.1101/cshperspect.a005553

Walley LJ, White F, Brander KM (1971) Sperm activation and fertilization in *Balanus balanoides*. J Mar Biol Assoc UK 51(2):489–494. https://doi.org/10.1017/S0025315400031933

Wang YL, Sun WJ, He L, Li Q, Wang Q (2015) Morphological alterations of all stages of spermatogenesis and acrosome reaction in Chinese mitten crab *Eriocheir sinensis*. Cell Tissue Res 360(2):401–412. https://doi.org/10.1007/s00441-014-2092-5

Wang S, Zhang J, Jiao W, Li J, Xun X, Sun Y, Guo X, Huan P, Dong B, Zhang L, Hu X, Sun X, Wang J, Zhao C, Wang Y, Wang D, Huang X, Wang R, Lv J, Li Y, Zhang Z, Liu B, Lu W, Hui Y, Liang J, Zhou Z, Hou R, Li X, Liu Y, Li H, Ning X, Lin Y, Zhao L, Xing Q, Dou J, Li Y, Mao J, Guo H, Dou H, Li T, Mu C, Jiang W, Fu Q, Fu X, Miao Y, Liu J, Yu Q, Li R, Liao H, Li X, Kong Y, Jiang Z, Chourrout D, Li R, Bao Z (2017) Scallop genome provides insights into evolution of bilaterian karyotype and development. Nat Ecol Evol 1(5):0120. https://doi.org/10.1038/s41559-017-0120

Waterman AJ (1934) Observation on reproduction, prematuration and fertilization in *Sabellaria vulgaris*. Biol Bull 67(1):97–114

Weigert A, Helm C, Meyer M, Nickel B, Arendt D, Hausdorf B, Santos SR, Halanych KM, Purschke G, Bleidorn C, Struck TH (2014) Illuminating the base of the annelid tree using transcriptomics. Mol Biol Evol 31(6):1391–1401. https://doi.org/10.1093/molbev/msu080

Wessel GM, Brooks JM, Green E, Haley S, Voronina E, Wong J, Zaydfudim V, Conner S (2001) The biology of cortical granules. Int Rev Cytol 209:117–206

Wilburn DB, Tuttle LM, Klevit RE, Swanson WJ (2018) Solution structure of sperm lysin yields novel insights into molecular dynamics of rapid protein evolution. Proc Natl Acad Sci U S A 115(6):1310–1315. https://doi.org/10.1073/pnas.1709061115

Yamaguchi S, Yusa Y, Sawada K, Takahashi S (2013) Sexual systems and dwarf males in barnacles: integrating life history and sex allocation theories. J Theor Biol 320:1–9. https://doi.org/10.1016/j.jtbi.2012.12.001

Yamamichi Y, Sekiguchi K (1982) Axoneme patterns of spermatozoa of Asian horseshoe crabs. Experientia 38(10):1219–1220. https://doi.org/10.1007/BF01959750

Yin X, Eckberg WR (2009) Characterization of phospholipases C beta and gamma and their possible roles in *Chaetopterus* egg activation. Mol Reprod Dev 76(5):460–470. https://doi.org/10.1002/mrd.20961

Young KG, Chang JP, Goldberg JI (1999) Gonadotropin-releasing hormone neuronal system of the freshwater snails *Helisoma trivolvis* and *Lymnaea stagnalis*: possible involvement in reproduction. J Comp Neurol 404(4):427–437. https://doi.org/10.1002/(SICI)1096-9861(19990222)404:4<427::AID-CNE1>3.0.CO;2-R

Yuan Y, Tanabe T, Maekawa F, Inaba K, Maeda Y, Itoh N, Takahashi KG, Osada M (2012) Isolation and functional characterization for oocyte maturation and sperm motility of the oocyte maturation arresting factor from the Japanese scallop, *Patinopecten yessoensis*. Gen Comp Endocrinol 179(3):350–357. https://doi.org/10.1016/j.ygcen.2012.09.006

Zhang L, Wayne NL, Sherwood NM, Postigo HR, Tsai P-S (2000) Biological and immunological characterization of multiple GnRH in an opisthobranch mollusk, *Aplysia californica*. Gen Comp Endocrinol 118(1):77–89. https://doi.org/10.1006/gcen.2000.7457

Zhang G, Fang X, Guo X, Li L, Luo R, Xu F, Yang P, Zhang L, Wang X, Qi H, Xiong Z, Que H, Xie Y, Holland PW, Paps J, Zhu Y, Wu F, Chen Y, Wang J, Peng C, Meng J, Yang L, Liu J, Wen B, Zhang N, Huang Z, Zhu Q, Feng Y, Mount A, Hedgecock D, Xu Z, Liu Y, Domazet-Lošo T, Du Y, Sun X, Zhang S, Liu B, Cheng P, Jiang X, Li J, Fan D, Wang W, Fu W, Wang T, Wang B, Zhang J, Peng Z, Li Y, Li N, Wang J, Chen M, He Y, Tan F, Song X, Zheng Q, Huang R, Yang H, Du X, Chen L, Yang M, Gaffney PM, Wang S, Luo L, She Z, Ming Y, Huang W, Zhang S, Huang B, Zhang Y, Qu T, Ni P, Miao G, Wang J, Wang Q, Steinberg CE, Wang H, Li N, Qian L, Zhang G, Li Y, Yang H, Liu X, Wang J, Yin Y, Wang J (2012a) The oyster genome reveals stress adaptation and complexity of shell formation. Nature 490(7418):49–54. https://doi.org/10.1038/nature11413

Zhang Y, Wang Z, Yan X, Yu R, Kong J, Liu J, Li X, Li Y, Guo X (2012b) Laboratory hybridization between two oysters: *Crassostrea gigas* and *Crassostrea hongkongensis*. J Shellfish Res 31(3):619–625. https://doi.org/10.2983/035.031.0304

Chapter 8
Reproduction in the Coral *Acropora*

Masaya Morita and Seiya Kitanobo

Abstract The speciose reef coral *Acropora* spp. has synchronous spawning. Many of its congeners spawn at the same time; hence, gametes of the different species mix in the water column. Nevertheless, the different species are generally reproductively isolated, although introgressive hybridization has been reported for several congeners. The complex reproductive isolation mechanisms in *Acropora* species depend on spawning synchronisms and gamete recognition at fertilization. The reticulate evolution that results from introgressive hybridization in *Acropora* is thought to be related to speciation in the genus. In this chapter, we introduce reproductive isolation in *Acropora*, focusing on spawning synchronization, regulation of sperm motility, gamete-species recognition, and the potential for hybridization in nature.

Keywords Coral · Synchronous spawning · Gamete recognition · Hybridization · Sperm limitation

8.1 Introduction

The coral genus *Acropora* comprises more than 150 species. Sympatric species spawn simultaneously (Babcock et al. 1986; Willis et al. 2006; Baird et al. 2009, 2010; Fogarty et al. 2012). On the Great Barrier Reef (GBR), more than 30 species spawn synchronously in an event termed "mass spawning," which only occurs on a portion of the GBR. Multispecific spawning has been reported in many other Indo-Pacific locations (reviewed by Baird et al. 2009). Inter-specific crossings occur in one-third of sympatric synchronously spawning congeners, with a frequency that varies from high to low (Willis et al. 1997, 2006; Hatta et al. 1999; Fukami et al. 2003; Fogarty et al. 2012; Isomura et al. 2013). Introgressive hybridization occurs in the intercrossing species group, but in most cases, an intercrossing barrier is

M. Morita (✉) · S. Kitanobo
Sesoko Station, Tropical Biosphere Research Center, University of the Ryukyus, Motobu, Okinawa, Japan
e-mail: morita@lab.u-ryukyu.ac.jp

M. Yoshida, J. F. Asturiano (eds.), *Reproduction in Aquatic Animals*,
https://doi.org/10.1007/978-981-15-2290-1_9

maintained by pre- and post-zygotic reproductive isolation mechanisms. Pre-zygotic isolation occurs in a suspended mixture of conspecific and heterospecific gametes, and post-zygotic isolation is expressed as reduced fertility in interspecific hybrids. Here, we review pre-zygotic reproductive isolation in *Acropora*, focusing on gamete production, spawning synchronisms, and species-recognition in the gametes.

8.2 Spawning Synchronism

Spawning synchronisms in the many species of *Acropora* have previously been described. Here, we focus on synchronized spawning in congeners, and the potential for encounters among gametes from different species.

8.2.1 Months and Hours of the Day in Which Acropora Spawns

Most *Acropora* species set their gamete bundles in the evening between 20:30 and 21:00, and release the eggs and sperm from the bundles between 21:30 and 22:30. Prior to spawning, gamete bundles appear on the surfaces of the colonies (Fig. 8.1). The colonies start releasing bundles, which float to the surface of the water column (Fig. 8.2). Most species off Sesoko Island (Okinawa, Japan) begin spawning between 21:30 and 22:30 (Table 8.1), but *A. tenuis* and *A. donei* begin spawning 2 h later (Table 8.1). Most species off Aka Island and at other locations also start spawning between 21:30 and 22:30 (Hatta et al. 1999; Fukami et al. 2003). The time of spawning is similar among species. The gamete bundles are released gradually from the colonies, and not all polyps release their bundles. Spawning dates are mostly

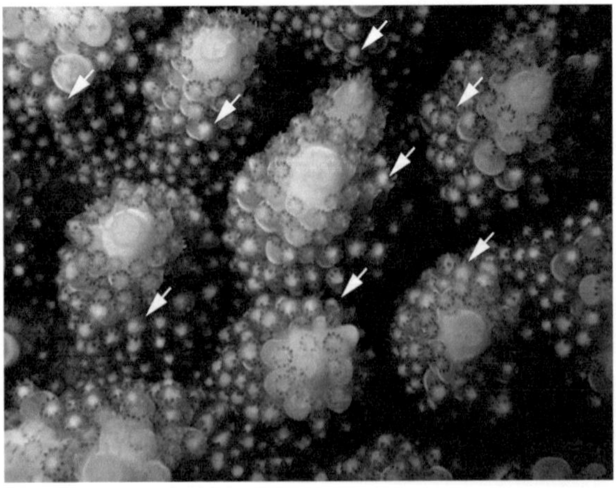

Fig. 8.1 "Setting" prior to spawning in the coral *Acropora digitifera*. Simultaneous hermaphrodite Acropora specimen forming sperm/egg bundles. White arrows point to bundles increasing to the surface of the colony 1.0–1.5 h before spawning

Fig. 8.2 Spawning of the coral *Acropora tenuis*. Single colony of the coral *A. tenuis* beginning to release gamete bundles

Table 8.1 Spawning time of some species of the coral *Acropora* spp. at the Sesoko Island

Species	Spawning month	Spawning time	Specific points in the spawning
Acropora tenuis	End of May–June	19:20–19:50	Releases large amounts of gametes (Fig. 8.2)
Acropora donei	End of May–June	19:50–20:10	Many colonies do not spawn vigorously
Acropora digitifera	End of May–June	21:40–22:30	Spawns vigorously
Acropora florida	End of May–June	21:40–22:20	Spawns vigorously
Acropora intermedia	End of May–June	22:00–22:50	Spawns vigorously
Acropora devaricata	September	21:00–21:15	Spawns vigorously
Acropora sp.1	August	22:00–22:40	Spawns vigorously

coincident among species, although they can span two or three calendar days. Most species primarily spawn on a single "major spawning day" within the range of potential spawning dates. The major spawning days are often different among species. For example, the intercrossing species *A. intermedia* and *A. florida* often have different spawning days around Sesoko Island. Gametogenesis is likely earlier in *A. intermedia* than in *A. florida*. Most spawning in *A. intermedia* occurs when the full moon is in late May, but *A. florida* spawns on the following full moon. Hence, the date of the spawning is likely influenced in part by the timing of gametogenesis.

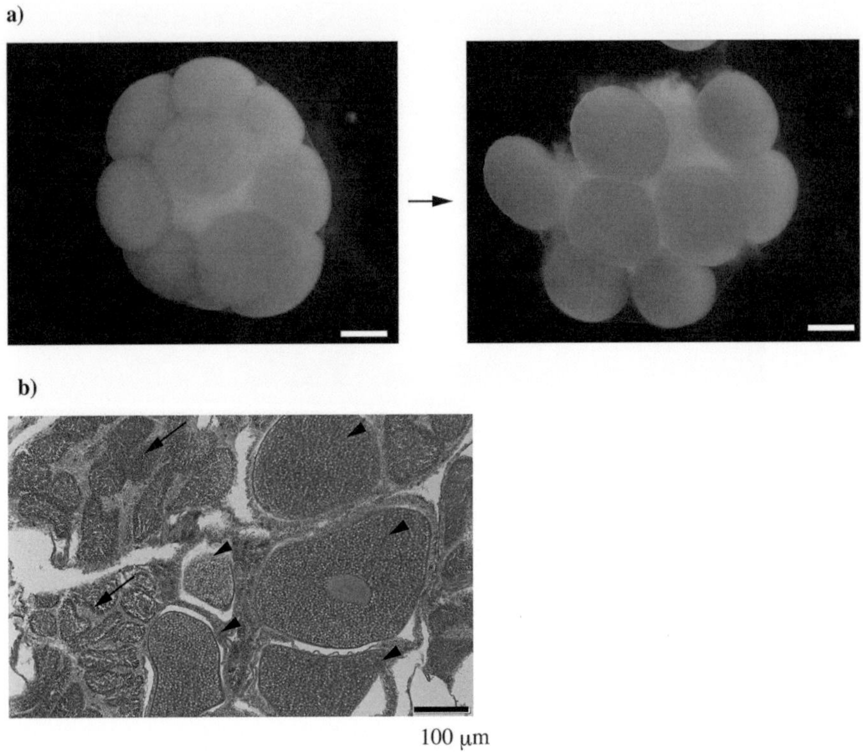

Fig. 8.3 Gamete bundles and gamete production in the coral *Acropora tenuis*. (**a**) Gamete bundles immediately after release during spawning. Each bundle comprises sperm surrounded by eggs. The arrow points to a wrapping membrane-like structure. Sperm and eggs separate after disruption of the membrane-like structure. (**b**) Hematoxylin-eosin stained coral fragment 7 days prior to a full moon in June. Arrowheads indicate eggs; arrows point to mature spermatozoa

8.3 Acropora Gametes

8.3.1 Gametes Bundles

Acropora is a simultaneous hermaphrodite that produces both sperm and eggs. At spawning, the colonies release packages of sperm and eggs in "gamete bundles" (Fig. 8.3a). Histological analyses have shown that sperm and eggs are produced in distinctive tissues lying in close proximity (Fig. 8.3b). *Montipora capitata* is also a member of the family Acroporidae. It forms gamete bundles 1–2 h before spawning (Padilla-Gamino et al. 2011). The egg–sperm bundle is formed by a mucus layer secreted by the oocytes. Little information is available on the wrapping membrane that encloses each bundle.

8.4 The Period Between the Activation of Sperm Motility and Fertilization

There are no precise descriptions of the events extending from egg–sperm interaction through subsequent membrane fusion in *Acropora*. However, numerous studies have examined the potential for fusion between heterospecific gametes. Here, we review gamete interaction with sperm motility regulation, cross-fertilization of heterospecific gametes in vitro, and descriptions of pre-zygotic reproductive isolation observed in sperm choice experiments.

8.4.1 Gamete Interaction with Sperm Motility Regulation

8.4.1.1 Separation of Bundles into Sperm and Eggs After Release

Wrapped gamete bundles separate into sperm and eggs after release (Figs. 8.3a and 8.4). The duration of time between release and separation of the gametes has been reported for several species (Wolstenholme 2004); the durations are highly variable (5 min–4 h, Wolstenholme 2004). Self-fertilization between sperm and eggs is strongly repressed; fertilization between allo-gametes is preferred (see details in Sect. 8.4.2.2).

8.4.1.2 Sperm Dilution and Limitation

After bundle separation into sperm and eggs, the sperm disperses rapidly and fertilization success depends on the degree of dilution; strong dilution is termed "sperm limitation" (e.g., Levitan and Petersen 1995; Yund 2000). Field observations show that an appropriate sperm concentration is maintained for only 1–2 h (Omori et al.

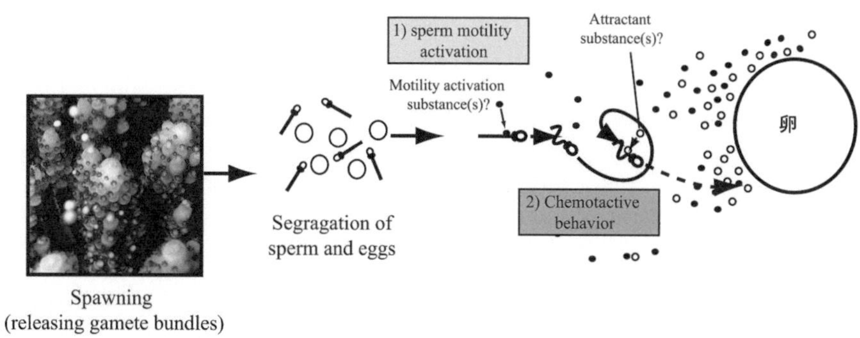

Fig. 8.4 Behavior of sperm in the coral *Acropora tenuis*

2001), and sperm concentration is positively associated with fertilization success (see details in Sect. 8.4.2.1).

8.4.1.3 Egg-Dependent Sperm Motility Regulation

As in most marine sessile and benthic invertebrates, sperm motility regulation in the coral *Acropora* largely depends on substance(s) produced by the eggs. In the acroporid *Montipora digitata*, three highly unsaturated fatty alcohols in the eggs attract sperm (Coll et al. 1994, 1995). The species specificity of sperm attraction is likely a function of the particular mixture combination of these three alcohols. Thus, sperm navigation toward the eggs appears to be mediated by these alcohols, and hybridization is prevented by egg-dependent sperm navigation. A preliminary description of sperm motility regulation in *Acropora* was provided by Morita et al. (2006) (Fig. 8.4). Our ongoing study has shown that the attractant substances were eluted in a distinct fraction containing water-soluble and relatively water-insoluble components. The specificity of sperm activation is still a contentious issue, but extracts from "slicks" (aggregates of eggs from many congeners) activate sperm motility in the most studied species (i.e., *A. florida* and *A. digitifera*). Thus, slicks formed by the eggs of many species may activate sperm motility in multiple congeners, and sperm motility activation in *Acropora* may not be crucial for pre-zygotic reproductive isolation in nature.

8.4.1.4 Activation of Flagellar Movement in Sperm

Activation of sperm flagellar movement is required prior to fertilization. Sperm is mostly quiescent immediately following release into the water. Flagellar motion is activated in response to substance(s) exuded from the eggs (see above). Sperm flagellar motility is activated when the internal pH (pH_i) increases; Ca^{2+}/calmodulin (CaM) cascade(s) are likely involved in activation. In a previous study, we showed that CaM was present in sperm proteins and that inhibition of the Ca^{2+}/CaM-dependent protein kinase suppressed sperm flagellar movement activation. In many marine invertebrates, the influx of Ca^{2+} triggers a chemotactic behavior that changes the direction of sperm movement. For example, sperm of the ascidian *Ciona intestinalis* rapidly changes swimming direction via an increase in the principal band angle (Shiba et al. 2008; Mizuno et al. 2012). This chemotactic behavior is regulated by Ca^{2+} influx, and the reactions occur in response to a chemo-attractant sperm-activating and -attracting factor (Yoshida et al. 2002; Shiba et al. 2008). As in other species, sperm of the corals *A. digitifera*, *A. tenuis*, and *A. gemmifera* respond to an increase in $Ca^{2+}{}_i$, and their trajectories become more circular when they are treated with the Ca^{2+} ionophore A23187 (Morita et al. 2006). Classic experiments have identified a possible role for Ca^{2+} in the chemotactic behavior of *Acropora* (Morita et al. 2006). However, details of the involvement of Ca^{2+}/CaM in chemotactic behavior and motility remain unclear.

8.4.2 Fertilization

8.4.2.1 Self-Fertilization

When gametes are packed in bundles, sperm and eggs from the same parent colony do not fuse (i.e., self-fertilization does not occur). In vitro experiments have shown that self-fertilization rarely occurs after the separation of sperm and eggs (Heyward and Babcock 1986). Gametes of single hermaphrodite *Acropora* individuals have strict allo-recognition that prevents self-fertilization; gametes preferentially fuse with those from other colonies.

8.4.2.2 Sperm Concentration for Fertilization Success

The synchronization of spawning is predicted to promote hybridization. Synchronization benefits reproductive success in broadcast spawning *Acropora*. The rapid dilution of gametes, particularly sperm, reduces fertilization success and is termed "sperm limitation" (Levitan and Petersen 1995; Yund 2000; also see Sect. 8.4.1.2). Sperm concentration is positively associated with fertilization success in *Acropora* (Willis et al. 1997; Nozawa et al. 2015). Previous studies have indicated that optimal concentrations are ca. 10^6 sperm/mL. High concentrations of sperm (termed "polyspermy") do not appear to negatively impact fertilization success. Fertilization rates decrease with declining sperm concentrations below 10^{2-4} sperm/mL. The optimal concentration of ca. 10^6 sperm/mL) persists for approximately 1–2 h following spawning (Omori et al. 2001). The densities of spawning coral colonies likely determine sperm concentrations in the water column. The rare species *A. papillare* is able to successfully form zygotes at low sperm concentrations (Nozawa et al. 2015). Hence, there may be selective pressure to increase gamete adhesion rates at low sperm concentrations. Indeed, some sea urchin genotypes have a gamete adhesion protein, bindin, that promotes high rates of fertilization success at low sperm concentrations (Levitan and Ferrell 2006; Levitan 2012). Surprisingly, the proportions of *bindin* genotypes have not changed with changing sea urchin population numbers over 200 years (Levitan 2012). There is no information on evolutionary changes in the relationship between gamete-recognition proteins and fertilization success in *Acropora*. This relationship may prove crucial during a period of mass bleaching and coral death.

8.4.2.3 Species-Specific Fertilization In Vitro

Species recognition during fertilization is robust, although some exceptions do exist. Robust species recognition is necessary because many congeners spawn synchronously, and gametes need to choose conspecific gametes from other colonies. Gamete species recognition facilitates species-specific fertilization. However, several *Acropora* species are able to cross with heterospecifics. Intercrossing with heterospecifics occurs in about one-third of sympatric and synchronous spawning species (reviewed by Willis et al. 2006), and hybridization is possible. Natural hybrids of Caribbean *A. cervicornis* × *A. palmate* have been reported (Vollmer and Palumbi 2002). Most of the hybrids were F1, and their fertility is doubtful, although one-way introgression has been reported (Vollmer and Palumbi 2002). These hybrids likely propagate through vegetative reproductive mechanisms such as fragmentation. Hybridization between conspecifics and heterospecifics do occur, but post-zygotic reproductive isolation (low fecundity of gametogenesis) prevents "successive" introgressive hybridization between these two species of *Acropora*. Nevertheless, population genetics studies have demonstrated introgression between these species (Vollmer and Palumbi 2002; Palumbi et al. 2012). Spawning synchronization increases introgression between intercrossing species on the GBR (Van Oppen et al. 2000). Hybridization may occur across a range of intercrossing species. However, in vitro fertilization experiments do not provide adequate data for demonstrating that hybridization occurs in wild populations. Population genetics studies do not clearly support the existence of ongoing hybridization.

8.4.2.4 Sperm Choice Experiments

Among intercrossing species, gametes preferentially fuse with conspecific gametes rather than with those of heterospecifics when both conspecific and heterospecific gametes are available. Thus, hybridization in between intercrossing species rarely occurs in nature. Based on in vitro studies on mixtures of isolated gametes from single species pairs (i.e., male gametes from one species, and eggs from another, or vice versa), we can conclude that most distinctive species have the potential to intercross. However, synchronized spawning of sympatric species forms a multispecies gamete mixture. Preferential conspecific mating in this mixture would markedly reduce the chances of hybridization. Sperm choice experiments clearly demonstrate that eggs preferentially fuse with conspecific sperm at optimal sperm concentrations (Kitanobo et al. 2016). *A. intermedia* and *A. florida* are able to produce hybrids, but the eggs of both species preferentially fuse with conspecific sperm when sperm concentrations are optimal (10^6 sperm/mL). Our preliminary study showed that eggs of the intercrossing species *A. tenuis* and *A. donei* also fuse with conspecific sperm when sperm of both species are mixed at optimal sperm concentrations (Morita et al., under review). The frequency of hybridization increases in *A. florida* when sperm are in low concentrations. Thus, the frequency of hybridization in this species is determined by the sperm concentration, which in turn depends in part on the densities of coral colonies. However, although *A. florida* is able to hybridize with *A. intermedia* at low sperm concentrations, the eggs of *A. intermedia* do not fuse with sperm of *A. florida* at any sperm concentration. Hence, the effects of sperm concentrations on hybridization rates are species-specific. The species preference traits of gametes may differ between sperm and eggs prior to membrane fusion. Eggs may fuse with conspecific or heterospecific sperm depending on sperm concentrations. Sperm have a stronger affinity for conspecific eggs. Consequently, gamete fusion is conspecific when sperm concentrations are high, but at low concentrations eggs are able to fuse with both conspecific and heterospecific sperm because the frequency of encounters with conspecific sperm is reduced. Thus, the relative conspecific versus heterospecific preferences of eggs and sperm may relate to hybridization rates in nature. The species-recognition protein of *Acropora* gametes has not been fully characterized. The expression of this protein likely has a role in the non-rigorous species recognition of intercrossing species. Further investigations are required to expand understanding of this topic.

8.5 From Spawning to Fertilization

Acropora is a broadcast spawner. Spawning synchronization is thought to influence the rates of fertilization success in reef-building corals (Van Oppen et al. 2002; Levitan et al. 2004; Fogarty et al. 2012; Teo and Todd 2018). The stages of spawning span gamete release from the bundles, sperm–egg interactions and finally gamete fusion to form zygotes. Numerous studies have identified possible relationships between intercrossing, spawning synchronisms, and introgressive hybridization, but detailed information on these relationships is currently unavailable. In situ observations are essential for a fuller understanding of the reproductive mechanism from spawning to fertilization. Dilution rapidly reduces sperm concentrations in seawater. Optimal concentrations are maintained for only a short period of time. We collected seawater off Sesoko Island immediately after the spawning of *A. tenuis* and *A. donei*. The sperm concentration was very low ($<10^5$ sperm/mL); nevertheless, most gametes fused to form zygotes (Kitanobo et al., unpublished data). The discrepancy in the relationships between fertilization success and sperm concentrations between in situ and in vitro studies should be resolved in further studies.

8.6 Conclusion

Reproductive success in the coral *Acropora* spp. has many determining elements, including spawning synchronism, numbers of released gametes, and fertilization preferences. These elements may relate to the speciation process in the *Acropora* group. Although detailed information is not available for some of the components of reproduction, such as gamete recognition and ongoing hybridization events, it is clear that sexual reproduction in *Acropora* is a crucial life cycle event.

Acknowledgments Preparation of this chapter was supported in part by a JSPS KAKENHI grant (#17K07414) to MM.

References

Babcock RC, Bull GD, Harrison PL, Heyward AJ, Oliver JK, Wallace CC, Willis BL (1986) Synchronous spawnings of 105 scleractinian coral species on the great-barrier-reef. Mar Biol 90:379–394

Baird AH, Guest JR, Willis BL (2009) Systematic and biogeographical patterns in the reproductive biology of scleractinian corals. Annu Rev Ecol Evol Syst 40:551–571

Baird AH, Kospartov MC, Purcell S (2010) Reproductive synchrony in *Acropora* assemblages on reefs of New Caledonia. Pac Sci 64:405–412

Coll JC, Bowden BF, Meehan GV, Konig GM, Carroll AR, Tapiolas DM, Aliño PM, Heaton A, De Nys R, Leone PAea (1994) Chemical aspects of mass spawning in corals. I. Sperm-attractant molecules in the eggs of the scleractinian coral *Montipora digitata*. Mar Biol 118:177–182

Coll JC, Leone PA, Bowden BF, Carroll AR, König GM, Heaton A, De Nys R, Maida M, Alino PM, Willis RH (1995) Chemical aspects of mass spawning in corals. II.(-)-Epi-thunbergol, the sperm attractant in the eggs of the soft coral *Lobophytum crassum* (Cnidaria: Octocorallia). Mar Biol 123:137–143

Fogarty ND, Vollmer SV, Levitan DR (2012) Weak prezygotic isolating mechanisms in threatened Caribbean *Acropora* corals. PLoS One 7:e30486

Fukami H, Omori M, Shimoike K, Hayashibara T, Hatta M (2003) Ecological and genetic aspects of reproductive isolation by different spawning times in *Acropora* corals. Mar Biol 142:679–684

Hatta M, Fukami H, Wang W, Omori M, Shimoike K, Hayashibara T, Ina Y, Sugiyama T (1999) Reproductive and genetic evidence for a reticulate evolutionary history of mass-spawning corals. Mol Biol Evol 16:1607–1613

Heyward AJ, Babcock RC (1986) Self- and cross-fertilization in scleractinian corals. Mar Biol 90:191–195

Isomura N, Iwao K, Fukami H (2013) Possible natural hybridization of two morphologically distinct species of *Acropora* (Cnidaria, Scleractinia) in the Pacific: fertilization and larval survival rates. PLoS One 8:e56701

Kitanobo S, Isomura N, Fukami H, Iwao K, Morita M (2016) The reef-building coral Acropora conditionally hybridize under sperm limitation. Biol Lett 12:20160511

Levitan DR (2012) Contemporary evolution of sea urchin gamete-recognition proteins: experimental evidence of density-dependent gamete performance predicts shifts in allele frequencies over time. Evolution 66:1722–1736

Levitan DR, Ferrell DL (2006) Selection on gamete recognition proteins depends on sex, density, and genotype frequency. Science 312:267–269

Levitan DR, Petersen C (1995) Sperm limitation in the sea. Trends Ecol Evol 10:228–231

Levitan DR, Fukami H, Jara J, Kline D, McGovern TM, McGhee KE, Swanson CA, Knowlton N (2004) Mechanisms of reproductive isolation among sympatric broadcast-spawning corals of the *Montastraea annularis* species complex. Evolution 58:308–323

Mizuno K, Shiba K, Okai M, Takahashi Y, Shitaka Y, Oiwa K, Tanokura M, Inaba K (2012) Calaxin drives sperm chemotaxis by Ca(2)(+)-mediated direct modulation of a dynein motor. Proc Natl Acad Sci U S A 109:20497–20502

Morita M, Nishikawa A, Nakajima A, Iguchi A, Sakai K, Takemura A, Okuno M (2006) Eggs regulate sperm flagellar motility initiation, chemotaxis and inhibition in the coral *Acropora digitifera*, *A. gemmifera* and *A. tenuis*. J Exp Biol 209:4574–4579

Nozawa Y, Isomura N, Fukami H (2015) Influence of sperm dilution and gamete contact time on the fertilization rate of scleractinian corals. Coral Reefs 34:1199–1206

Omori M, Fukami H, Kobinata H, Hatta M (2001) Significant drop of fertilization of *Acropora* corals in 1999: an after-effect of heavy coral bleaching? Limnol Oceanogr 46:704–706

Padilla-Gamino JL, Weatherby TM, Waller RG, Gates RD (2011) Formation and structural organization of the egg-sperm bundle of the scleractinian coral *Montipora capitata*. Coral Reefs 30:371–380

Palumbi SR, Vollmer S, Romano S, Oliver T, Ladner J (2012) The role of genes in understanding the evolutionary ecology of reef building corals. Evol Ecol 26:317–335

Shiba K, Baba SA, Inoue T, Yoshida M (2008) Ca2+ bursts occur around a local minimal concentration of attractant and trigger sperm chemotactic response. Proc Natl Acad Sci U S A 105:19312–19317

Teo A, Todd PA (2018) Simulating the effects of colony density and intercolonial distance on fertilisation success in broadcast spawning scleractinian corals. Coral Reefs 37:891

Van Oppen MJ, Willis BL, Vugt HW, Miller DJ (2000) Examination of species boundaries in the Acropora cervicornis group (Scleractinia, cnidaria) using nuclear DNA sequence analyses. Mol Ecol 9:1363–1373

Van Oppen MJ, Willis BL, Van Rheede T, Miller DJ (2002) Spawning times, reproductive compatibilities and genetic structuring in the *Acropora aspera* group: evidence for natural hybridization and semi-permeable species boundaries in corals. Mol Ecol 11:1363–1376

Vollmer SV, Palumbi SR (2002) Hybridization and the evolution of reef coral diversity. Science 296:2023–2025

Willis BL, Babcock RC, Harrison PL, Wallace CC (1997) Experimental hybridization and breeding incompatibilities within the mating systems of mass spawning reef corals. Coral Reefs 16:S53–S65

Willis BL, van Oppen MJH, Miller DJ, Vollmer SV, Ayre DJ (2006) The role of hybridization in the evolution of reef corals. Annu Rev Ecol Evol Syst 37:489–517

Wolstenholme JK (2004) Temporal reproductive isolation and gametic compatibility are evolutionary mechanisms in the *Acropora humilis* species group (Cnidaria; Scleractinia). Mar Biol 144:567–582

Yoshida M, Murata M, Inaba K, Morisawa M (2002) A chemoattractant for ascidian spermatozoa is a sulfated steroid. Proc Natl Acad Sci U S A 99:14831–14836

Yund PO (2000) How severe is sperm limitation in natural populations of marine free-spawners? Trends Ecol Evol 15:10–13

Chapter 9
Self- and Nonself-Recognition of Gametes in Ascidians

Hitoshi Sawada and Maki Shirae-Kurabayashi

Abstract Ascidians (Tunicata) are hermaphroditic sessile marine invertebrates, which release sperm and eggs nearly simultaneously to the surrounding seawater during the spawning season. To avoid inbreeding, several species, including *Halocynthia roretzi* (Stolidobranchia) and *Ciona intestinalis* type A *(Ciona robusta)* (Phlebobranchia) possess a self-sterility system. In *H. roretzi*, a 70-kDa vitelline coat (VC) protein consisting of 12 EGF-like repeats with polymorphisms, designated as VC70, appears to be involved in gamete interaction and also in self/nonself-recognition. A cysteine-rich secretory protein, designated as Urabin and a type II transmembrane serine protease-1, called TTSP-1, are candidate sperm-borne binding partners for VC70. In *C. intestinalis* type A, on the other hand, a fibrinogen-like VC protein, v-Themis-A and v-Themis-B, and sperm PKDREJ-like protein, s-Themis-A and s-Themis-B, are highly polymorphic among individuals and these proteins appear to play a pivotal role in self/nonself-recognition during gamete interaction. It was recently suggested that three pairs of *v-Themis* and *s-Themis* genes (*s/v-Themis-A, s/v-Themis-B,* and *s/v-Themis-B2)* are responsible for this system. After sperm attachment to the VC of self-eggs, drastic Ca^{2+} influx is elicited, resulting in sperm vigorous movement on the VC followed by sperm detachment from the VC, or by cessation in sperm motility. The C-terminal Ca^{2+}-permeable cation channel domain in s-Themis-B/B2 may be involved in Ca^{2+} influx. Although *s/v-Themis* homologous genes with polymorphisms were detected in the genome of *H. roretzi*, it is not known whether *s/v-Themis* is involved in self/nonself-recognition of *H. roretzi.* Since flowering plants utilize family specific *S*-determinant proteins for self-incompatibility, the mechanism of self/nonself-recognition may be different between stolidobranch and phlebobranch ascidians.

Keywords Sperm lysin · Proteasome · Fertilization · Self/nonself · Self-incompatibility · Ascidian

---

H. Sawada (✉) · M. Shirae-Kurabayashi
Sugashima Marine Biological Laboratory, Graduate School of Science, Nagoya University, Toba, Japan
e-mail: hsawada@bio.nagoya-u.ac.jp

M. Yoshida, J. F. Asturiano (eds.), *Reproduction in Aquatic Animals*,
https://doi.org/10.1007/978-981-15-2290-1_10

9.1 Introduction

Most living organisms adopt sexual reproduction as a reproductive strategy to create genetic diversity in the next generation. Whereas most animals are dioecious, most flowering plants are hermaphrodites. It is known that approximately 60% of angiosperms show self-incompatibility to avoid inbreeding (De Nettancourt 1977), but the others are self-compatible species. Although the mechanisms of self-incompatibility in flowering plants have been studied well, little is known about the mechanism of self-sterility in hermaphroditic animals. Animal self-sterility is achieved by gamete recognition, whereas plant self-incompatibility is achieved by the interaction between non-gamete cells such as pollen and stigmatic cells. In this chapter, therefore, we distinctly utilize the terms "self-sterility" for gamete interaction in animals and "self-incompatibility" for non-gamete interaction in plants. Before focusing on the self/nonself-recognition mechanisms in ascidians, various flowering plant self-incompatibility systems are summarized as follows.

In Brassicaceae, a small-molecular-mass protein with allelic polymorphism, called SP11/SCR, from anther tapetum is attached to the surface of pollen. After attachment of pollen to stigmatic papilla cells, SP11/SCR is recognized by *S*-locus receptor kinase (SRK) in the stigma. If it is recognized as self, pollen germination and tube growth are inhibited, resulting in the block to double fertilization (Sawada et al. 2014; Iwano and Takayama 2012; Takayama and Isogai 2005). In Papaveraceae, a gene pair of pollen *PrpS* and stigmatic *PrsS* is adjoining and highly polymorphic. If PrpS recognizes the PrsS as self, Ca^{2+} influx takes place in pollen, resulting in execution of caspase-like protease-mediated apoptosis in pollen to prohibit self-fertilization. In Solanaceae, it is thought that *S*-RNase in the pistil penetrates into pollen tube and degrades RNA, where multiple *S*-locus F-box proteins (SLF/SBP), a ubiquitin ligase E3, expressed in the pollen tube covalently label ubiquitin to non-self *S*-RNase, resulting in proteasome-mediated degradation of nonself *S*-RNase. As a result, self *S*-RNase can degrade RNAs in self-pollen tube, leading to the prevention of self-fertilization (Iwano and Takayama 2012; Takayama and Isogai 2005).

In contrast to angiosperms, mechanisms of self/nonself recognition during fertilization of hermaphroditic marine animals are not well known. Ascidians (Urochordata or Tunicata) are hermaphroditic sessile marine invertebrates and release sperm and eggs almost simultaneously to the surrounding seawater during the spawning seasons. Several ascidians, including *Ciona intestinalis* type A (recently proposed to be renamed *Ciona robusta* (Brunetti et al. 2015)) and *Halocynthia roretzi,* show self-sterility, although several species such as *Phallusia mammillata* are self-fertile. Whereas these phenomena are well known, the molecular mechanism of self/nonself recognition is poorly understood. Ascidian eggs are covered with glycoproteinaceous investment called the vitelline coat (VC), cellular investments called test cells within the perivitelline space, and a single layer of follicle cells attached to the outer surface of the VC. Since the VC-deprived eggs are self-fertile, it is thought that the self/nonself recognition is achieved at the process of sperm attachment to the VC of the eggs.

Halocynthia roretzi (Order Stolidobranchia) is a useful ascidian for biochemical studies of fertilization (Numakunai and Hoshino 1980; Sawada 2002), since a large quantity of readily fertilizable sperm and eggs can be easily obtained from sexually mature individuals, which are cultured in Onagawa Bay and Mutsu Bay, Japan.

Ciona intestinalis type A (Order Phlebobranchia), a cosmopolitan species, has many advantages for molecular biological and genetic studies, because the genome database is available since 2002 (Dehal et al. 2002), and also because genetic analysis and genome editing experiments are possible in this species (Satou et al. 2005; Kawai et al. 2012; Sakuma et al. 2013).

In the present chapter we mainly focus on the mechanisms of self-sterility in *Halocynthia roretzi* and *Ciona robusta* as a representative of stolidobanch and phlebobranch ascidians, respectively.

9.2 Self/Nonself-Recognition During Fertilization in Stolidobranch Ascidians: VC70 and Its Binding Partners

Halocynthia roretzi is a strictly self-sterile solitary ascidian, inhabiting the rocky subtidal of Japan and Korea. It is known that the attachment of follicle cells to the VC is needed for fertilization in this species (Fuke 1983). Self-sterility in *H. roretzi* is achieved in the interaction of sperm with the VC but not with the follicle cells, since the VC-deprived eggs are self-fertile and also since the eggs having allogeneic follicle cells on the VC are still self-sterile (Fuke 1983).

Immature oocytes in the ovary and mature oocytes treated with acidic seawater (pH 2–3) for 1 min are self-fertile (Fuke and Numakunai 1996). In connection with this, it is interesting to note that the amount of a 70-kDa main component of the VC, referred to as VC70, significantly increased during oocyte maturation, a process of acquisition of self-sterility, and the VC70 is solubilized by 1–5 mM HCl from the isolated VC, which coincided with the fact that the self-sterile eggs become self-fertile by the treatment with pH 2–3 seawater. Furthermore, nonself-sperm, rather than self-sperm, efficiently bound to VC70-immobilized agarose beads. Fertilization ratio decreased when sperm were pretreated with VC70 of nonself eggs rather than HrVC70 of self eggs. VC70 consists of 12 EGF-like repeats, which seem to be generated by the processing of follicle cell trypsin-like protease, ovochymase (Mino and Sawada 2016) from a precursor protein VC120, which is expressed in the oocytes (Harada and Sawada 2007), having an N-terminal signal peptide, 13 EGF-like repeats and C-terminal zona pellucida (ZP) domain (Sawada 2002; Sawada et al. 2002). *VC120* is expressed in the gonad but not in the other organs or cells. VC70 shows significant but not high polymorphisms in amino acids among individuals (probably among alleles), mostly at the regions between the third and fourth cysteine residues in each EGF domain and a region between respective EGF domains. No nucleotide substitution was observed in 3rd, 4th, 6th, and 12th EGF domains among 12 EGF domains so far examined. Each EGF domain corresponds to a single exon. Since a single amino acid substitution impaired the functions of EGF-like repeats resulting in Notch signaling disease (Artavanis-Tsakonas et al. 1995), individually variable VC70 protein appears to be a promising candidate for a self/nonself-recognition protein during fertilization in *H. roretzi* (Sawada et al. 2004) (Table 9.1).

Processing and maturation of a certain allorecognition protein, which is produced in follicle cells, have been proposed in *H. roretzi*, since trypsin inhibitors suppressed the acquisition of self-sterility during oocyte maturation and also since exogenous trypsin was capable of inducing self-sterility in defolliculated immature oocytes (Fuke and Numakunai 1996, 1999). Their finding well coincided with our hypothesis that follicle cell trypsin-like protease ovochymase may produce VC70 from VC120 by limited proteolysis (Mino and Sawada 2016).

In another species in the same genus, *Halocynthia aurantium,* a similar protein VC80, whose precursor is a 130-kDa VC130, has been identified (Ban et al. 2005). VC130 in *H. aurantium* is highly homologous to VC120 in *H. roretzi* (amino acid identity, 83.4%) and polymorphic regions were also similar to those in VC70. Since two alleles of VC80 cDNA were detected in each individual, VC80 appears to be in a diploid expression. One distinct difference between VC70 and VC80 in each species is observed in the eighth EGF domain: eighth domain of VC70 seems to be duplicated in VC80 during *Halocynthia* evolution. Taking into account that a single domain duplication in the sperm receptor would cause a disruption in gamete interaction, these changes in VC80/VC70 could trigger speciation.

Sperm binding proteins toward VC70 were identified by Far Western blotting: the sperm membrane fraction was subjected to SDS-PAGE followed by blotting onto a nitrocellulose membrane and overlaid with VC70 (Urayama et al. 2008). Then VC70-interacting protein was identified by Western blotting using anti-VC70 antibody. A 35-kDa protein, referred to as Urabin, located in the sperm RAFT (LD-DIM) fraction, showed a binding ability to VC70. This protein is a member of the cysteine-rich secretory protein (CRISP) family, containing a C-terminal GPI-anchor signal. In fact, Urabin seems to be a GPI-anchor protein since it was solubilized from the membrane fraction by PI-PLC (Urayama et al. 2008). Deglycosylation of Urabin with PNGase F reduces the molecular weight, and this deglycosylated form failed to bind to VC70, suggesting the importance of N-linked carbohydrate chain in Urabin activity. Some, but not all individuals, express 50-kDa Urabin-L, which was detected by anti-Urabin antibody. Interestingly, however, Urabin-L is unable to bind to VC70 regardless of almost identical sequence to Urabin except for C-terminal $(EXADGD)_6$ repeat. Urabin showed little or no polymorphism and no self/nonself recognition ability in the binding to VC70. However, since anti-Urabin antibody inhibited the allorecognizable sperm binding to VC70-Sepharose and fertilization, Urabin appears to participate in self/nonself-recognition during gamete interaction (Urayama et al. 2008).

Another VC70-interacting protein, referred to as TTSP-1 (type II transmembrane serine protease-1), was identified by yeast two-hybrid screening. This is a 337-kDa protein, which is made up of 3087 amino acids, including an N-terminal signal peptide, a transmembrane domain, 23 sushi domains, 2 ricin domains, a CUB

Table 9.1 Sperm and egg proteins involved in self/nonself-recognition and gamete interaction in ascidians

Species and events	Sperm proteins	Egg or VC proteins	References or URL
Phlebobranch ascidians			
Ciona intestinalis type A			
Self/nonself-recognition	s-Themis-A, B and B2	v-Themis-A, B and B2	Harada et al. (2008) Sawada et al. (2020)
	Trypsin-like protease?	v-Themis-like	Otsuka et al. (2013)
Gamete interaction	70-kDa protein	Apolipoprotein B-like	Yamada et al. (2009)
	Urabin	VC57	Yamaguchi et al. (2011)
	α-L-Fucosidase	Fucose (VC glycoprotein?)	Rosati and De Santis (1980) Hoshi (1986) Lambert and Koch (1988)
	ND	Gln-enriched VC peptides	Kawamura et al. (1991)
		hsp70	Marino et al. (1998, 1999)
Ciona savignyi			
Self/nonself-recognition	s-Themis-A and -B?	v-Themis-A and -B?	ANISEED database
Stolidobranch ascidians			
Halocynthia roretzi			
Self/nonself-recognition	Urabin	VC70	Sawada et al. (2002, 2004) Urayama et al. (2008)
	TTSP-1	VC70	Harada and Sawada (2007)
	s-Themis?	v-Themis?	ANISEED database
Gamete interaction	Proacrosin, Spermosin	Vitellogenin (VC protein)	Akasaka et al. (2010, 2013)
	α-L-Fucosidase	Fucose (VC glycoprotein?)	Matsumoto et al. (2002)
Halocynthia aurantium			
Self/nonself-recognition	ND	VC80	Ban et al. (2005)
	s-Themis?	v-Themis?	ANISEED database

ND Not determined

domain, and a C-terminal serine protease domain (Harada and Sawada 2007). Recombinant TTSP-1 is capable of interacting with VC70, but it is not known whether TTSP-1 can distinguish self and nonself VC70. Further studies are necessary to identify sperm protein(s), other than Urabin or TTSP-1, which are involved in self/nonself-recognition during fertilization. One of the candidates is s-Themis as described in Sect. 9.3. In *Ciona intestinalis* type A, three pairs (A, B and B2) of s(sperm)-Themis and v(vitelline-coat)-Themis proteins seem to be an allele-recognition partner. These proteins are highly polymorphic and appear to recognize the same haplotypic alleles. We have identified four alleles (haplotypes) of *s/v-Themis* genes in the genome of *H. roretzi* by using ANISEED genome database (Brozovic et al. 2016). Furthermore, we detected *s/v-Themis* mRNA expression in the gonad and polymorphism of *v-Themis* cDNA among individuals (Shirae-Kurabayashi et al. unpublished data). Although the polymorphism of *H. roretzi v-Themis* is not so prominent compared to *C. intestinalis* type A *v-Themis* the amino acid substitutions of *H. roretzi* v-Themis are much more variable than that of *H. roretzi* VC70.

In addition to the allorecognition proteins, protein interactions between sperm α-L-fucosidase and the VC glycoprotein containing fucose residue (Matsumoto et al. 2002), and also between sperm trypsin-like proteases, including proacrosin and spermosin, and the vitellogenin C-terminal fragment attached to the VC (Akasaka et al. 2010, 2013) may also support the allorecognition system.

Functional analyses of the genes involved in self-sterility are very difficult in *H. roretzi* because this species takes 3 years to reach sexual maturity in contrast to 3 months in *C. intestinalis* type A. Thus genetic analysis or knockout analysis using genome editing is not easy. Neutralizing antibodies against extracellular domain of the candidate proteins would reveal the function of the proteins involved in self-sterility in this species.

9.3 Self/Nonself-Recognition During Fertilization in Phlebobranch Ascidians: s-Themis, v-Themis, and Other Proteins

In the early part of the twentieth century, Thomas Hunt Morgan reported the mode of self-sterility in the ascidian *Ciona* (Morgan 1910, 1923, 1939, 1942, 1944). As described in our previous review (Harada and Sawada 2008; Sawada et al. 2017), Morgan first discovered that self-sterile *Ciona* eggs become self-fertile after treatment with acidic seawater (pH 2.6) for 2–5 min or after removal of the VC (Morgan 1910, 1923). He also found that nonself-sterile combination was observed in self-fertilized siblings, which was scarcely observed in natural crossing. From these results, he concluded that self-sterility in *Ciona* must be genetically controlled (Morgan 1939, 1942, 1944), which was later re-examined and reconfirmed (Murabe and Hoshi 2002).

Furthermore, Morgan discovered a one-way sterile combination between self-fertilized F1 siblings, i.e., sperm of individual A can fertilize the eggs of individual B, but sperm of individual B cannot fertilize the eggs of individual A. In order to explain these phenomena, he proposed a "haploid sperm hypothesis," in which haploid expression of sperm *S*(self-sterility)-determinant and diploid expression of egg *S*-determinant were assumed (Morgan 1944). If a single locus (locus A) is involved in self-sterility, there are two populations of sperm in A/a heterozygous individual ("A"-expressing sperm and "a"-expressing sperm) but one population of oocytes expressing both alleles (A/a-expressing egg) in heterozygote. (Since ascidian oocytes are arrested at the first metaphase, the diploid expression is reasonable.) In this case, homozygous eggs (A/A-egg or a/a-egg) can be fertilized by heterozygous sperm (a mixture of A-sperm and a-sperm) because either A-sperm ligand or a-sperm ligand is absent in the VC of homozygous eggs, which enables fertilization. In contrast, heterozygous eggs (A/a-egg) cannot be fertilized by homozygous sperm because both A-sperm ligand and a-sperm ligand are present in the VC, which prevents fertilization of A-sperm and a-sperm. In this case, when one-way sterile combination is observed, putative locus A should be homozygous in "male" and heterozygous in "female." Based on these criteria, 70 gene markers were tested whether each gene is homozygous or heterozygous by PCR analysis. By this positional cloning, two candidate loci (locus A in chromosome 2q and locus B in chromosome 7q) have been identified as *S*-determinant loci (Harada et al. 2008).

Among 20 genes in locus A, only one gene product similar to fibrinogen-C-terminal was identified in the VC by LC/MS/MS analysis, and this protein was designated as v(vitelline coat)-Themis-A. On the other hand, among 20 genes, four genes were expressed in the testis, among which only one gene product (PKD1) showed a polymorphism among individuals. Therefore, this protein was thought to be a promising candidate for sperm allorecognition protein, and this was designated as s(sperm)-Themis-A. v-Themis-A sequence showed an allelic variety in the entire region, whereas s-Themis-A showed a hypervariable region (HVR) among alleles at an N-terminal side. Similar gene pairs, referred to as v-Themis-B and s-Themis-B, were identified in locus B regardless of the absence of overall genomic synthenty. Whereas both v-Themis-A and B are highly variable proteins in their entire regions, s-Themis-A and B possess an N-terminal signal peptide, hypervariable region (HVR), GPS (G-protein-coupled proteolytic site) domain, and LH2 (lipoxygenase homology) domain, followed by C-terminal five transmembrane domains in s-Themis-A or by 11 transmembrane domains in s-Themis-B (Fig. 9.1). Interestingly, s-Themis-B possesses a Ca^{2+}-permeable cation channel (PKD channel) domain in its C-terminus. *v-Themis* genes are encoded in the first intron of *s-Themis* genes, but transcribed in the opposite direction (Harada et al. 2008, Fig. 9.1). We recently examined several allelic genes of *s-Themis-B* and *v-Themis-B* to investigate the exon/intron organization of five alleles. Our preliminary data showed that exon/intron organization of *v-Themis* and *s-Themis* is similar but not identical: *v-Themis* genes were always encoded in the first intron of respective *s-Themis* genes (Yamada et al. unpublished data).

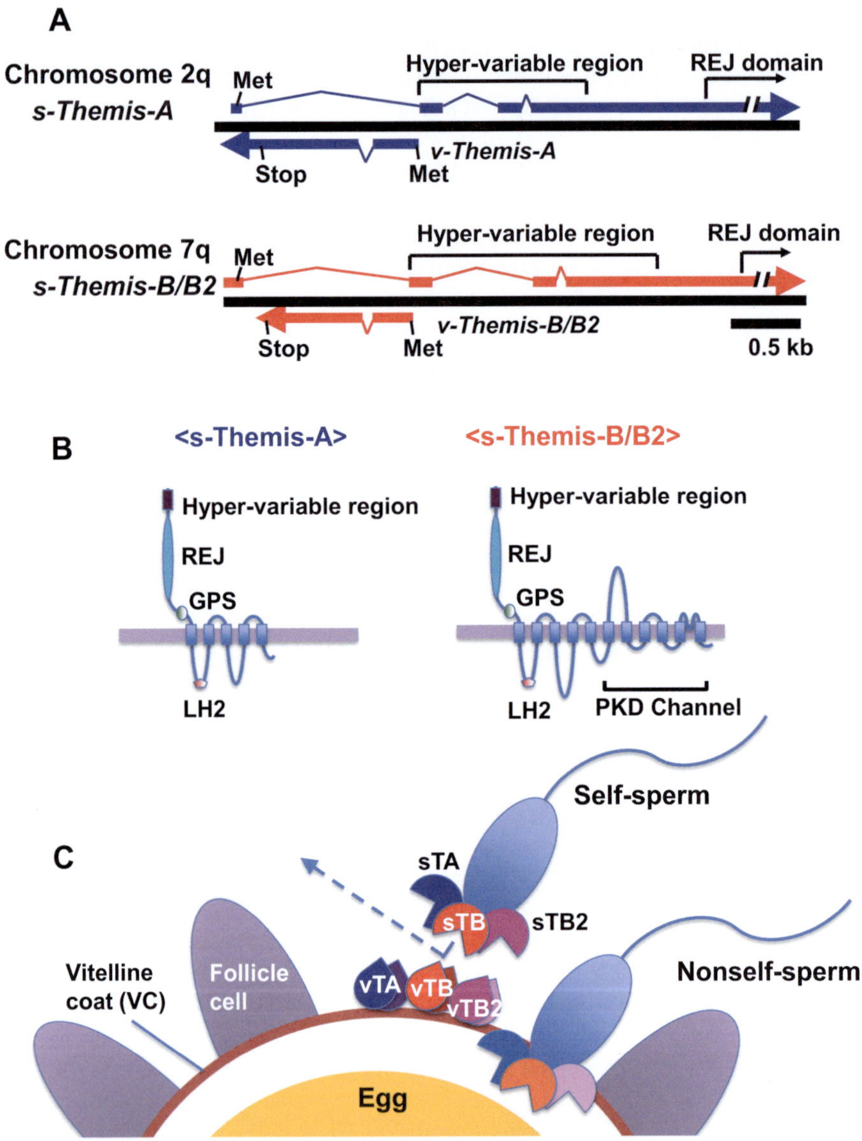

Fig. 9.1 Working hypothesis of the self-sterility mechanism in *Ciona robusta*. (**a**, **b**) In *C. robusta*, it is proposed that sperm PKDREJ-like proteins having an N-terminal hypervariable region, s-Themis-A, B, and B2 interact with fibrinogen-like polymorphic proteins, v-Themis-A, B, and B2 on the VC, respectively. (**c**) If the alleles of s-Themis and v-Themis are from the same haplotypes in A, B and B2, sperm recognizes the VC as self-egg, which allows Ca^{2+} influx, probably via the Ca^{2+}-conducting cation channel (PKD channel) at the C-terminus of s-Themis-B/B2 (**b**). This causes the increase in sperm motility resulting in sperm detachment from the VC or in the eventual decrease in sperm motility (**c**). The s- and v-Themis-mediated self/nonself-recognition process may be supported by the interaction between sperm α-L-fucosidase and fucose residue in the VC and also between sperm Urabin and VC57. v-Themis-like may be involved in the formation of a complex of v-Themis-A and -B

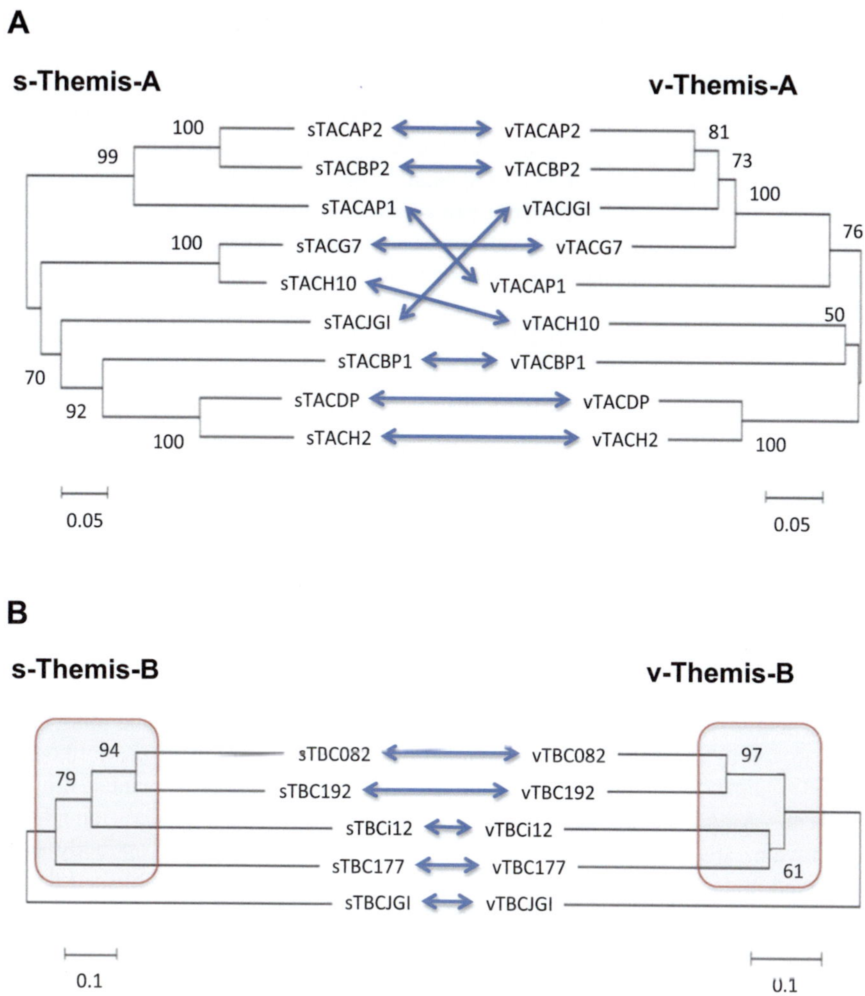

Fig. 9.2 Phylogenetic trees of s/v-Themis-A and s/v-Themis-B. Comparison of phylogenetic trees of s-Themis-A and v-Themis-A (**a**), and those of s-Themis-B and v-Themis-B (**b**). Phylogenetic tree pattern of s-Themis-A well coincided with that of v-Themis-A, except three alleles, such as JGI, AP1, and H10. On the other hand, s-Themis-B and v-Themis-B showed a similar pattern in respective phylogenetic trees (Yamaguchi et al. unpublished data)

Molecular phylogenetic patterns between *s-Themis-A* and *v-Themis-A*, and also between *s-Themis-B* and *v-Themis-B* are very similar except several *s/v-Themis-A* alleles such as AP1, GJI, and H-10 haplotypes (Fig. 9.2). These results imply the coevolutional relationship between *s-Themis-A* and *v-Themis-A*, and also between s-*Themis-B* and *v-Themis-B* (Yamaguchi et al. unpublished data).

Respective gene pairs of *s-Themis* and *v-Themis* always make a single haplotype. There is no exception in the fixed allelic partner so far studied, suggesting the strong partnership between s-Themis-A and v-Themis-A, and also between s-Themis-B and v-Themis-B (Sawada et al. 2020). We cannot rule out a possibility that a certain haplotype, showing no coevolutional relationship, may not function as a self-recognition partner. In fact, we recently identified a novel *s/v-Themis-A*-like gene pair, tentatively designated as *s/v-Themis-A2,* located at 2200 kbp apart from *s/v-Themis-A* gene pair. However, this gene pair appears to be a pseudogene, since several termination codons exist in their ORFs (Yamamoto et al. unpublished data). Therefore, the alleles showing the different patterns of phylogenetic trees may not function as current self-recognition partners but as an intermediate making a new haplotype during evolution.

Haploid expression of *s-Themis* must be regulated by an unknown mechanism, since spermatids are mutually connected with a cytoplasmic bridge forming a multinuclear cell. s-Themis proteins localized on the sperm head membrane may be anchored to a certain region during spermiogenesis. Further studies are necessary to prove the “haploid sperm hypothesis” proposed by Morgan.

After sperm binding to the VC of self-eggs, but not nonself-eggs, drastic Ca^{2+} influx in sperm takes place, which allows vigorous movement on the VC resulting in the sperm detachment from the VC or cessation of the sperm motility (Saito et al. 2012). Since s-Themis-B possesses a cation channel domain in its C-terminus, it is plausible that sperm s-Themis-A and B recognize the same haplotypic v-Themis-A and B, respectively, on the VC, which allows Ca^{2+} influx, probably via the Ca^{2+}-permeable cation channel (PKD channel) in s-Themis-B.

s-Themis-A and B appear to be localized on the tip of sperm head and mitochondrial region in addition to tail region by immunocytochemistry using the antibodies raised against REJ domain (Saito et al. unpublished data). Very interestingly, s-Themis-A and B may be degraded after Ca^{2+} influx elicited by Ca^{2+} ionophore ionomycin, since little or no fluorescence of anti-s-Themis-REJ antibody was observed after treatment of sperm with ionomycin (Saito et al. unpublished data). These results imply that allele-specific strong interaction between s-Themis and v-Themis allows Ca^{2+} influx, which trigger detachment of sperm from the vitelline coat by cleaving the s/v-Themis interaction. Such an allorecognition process may be faster than, or a preceding process of, the activation of lysin, because sperm penetration of the VC is not allowed in self-gamete combination. Our preliminary data showed that low Ca^{2+} ASW allows self-fertilization (Hashimoto et al. unpublished data). These results also support the above hypothesis.

Furthermore, we also found that one additional gene pair, designated as *s/v--Themis-B2*, resides approximately 70 kbp downstream from original *s/v-Themis-B* gene (Sawada et al. 2020). Although nucleotide sequences of *s/v-Themis-B2* genes were very similar to those of *s/v-Themis-B*, a common haplotype has not been observed in *s-Themis-B* and *B2*. Classical genetic analysis suggested the participation of *s/v-Themis-B2* in addition to *s/v-Themis-A* and *B* in self-sterility (Sawada et al. 2020).

Recently, Satou et al. (2014) reported that inbred strain of *Ciona robusta* showed heterozygosity only in *s/v-Themis-B* HVR but not in *s/v-Themis-A* or other regions, suggesting the possibility that the copy number of HVR of *Themis-B, B2* is variable for each individual. Independently, we also found that several alleles of *s/v-Themis-B* were detected in locus B. Therefore, we are now analyzing mRNA expression of each allele for getting more information about the functional alleles (Shirae-Kurabayashi and Sawada unpublished data).

Before our discovery of s/v-Themis, several models had been proposed about self-sterility in *Ciona*. De Santis and her colleagues reported that only nonself-sperm can tightly bind to the VC of glycerinated eggs (Rosati and De Santis 1978) and also that heat shock protein 70 (hsp70) attached to the VC may be responsible for self/nonself recognition (Marino et al. 1998). They also proposed that self-derived peptides, which are produced by the follicle cell proteasome, may be trapped with hsp70 on the VC (Pinto et al. 1995; Marino et al. 1999). However, the putative trapped peptides have not yet been identified. On the other hand, Kawamura and his colleagues found that acid-extract of the egg is able to inhibit the fertilization of eggs inseminated with nonself-spermatozoa. This factor was partially purified but not yet identified (Kawamura et al. 1991).

In addition to s/v-Themis-A, B and B2, the following protein interactions have also been proposed in gamete interaction in *Ciona*, i.e., the interactions between sperm Urabin and the VC57 (*C. intestinalis* type A homolog of *H. roretzi* VC70) (Yamaguchi et al. 2011), between sperm α-L-fucosidase and fucose-containing VC glycoprotein (De Santis et al. 1983; Hoshi 1986) and between sperm 70-kDa ApoBL-interacting protein and ApoBL (apolipoprotein B-like protein) (Yamada et al. 2009). Furthermore, a novel v-Themis-like VC protein, referred to as "v-Themis-like," has been identified as a main component of the acid-extractable VC proteins (Otsuka et al. 2013). This protein, possessing fibrinogen-like domain, showed little or no polymorphism but interacted with s-Themis-B. The binding partner of v-Themis-like on spermatozoa has not been identified, but yeast two hybrid screening suggested that sperm trypsin-like protease is a candidate binding partner for v-Themis-like (Otsuka et al. unpublished data).

9.4 Conclusions

Molecular mechanisms of self/nonself-recognition during gamete interaction have been gradually unveiled. *Halocynthia roretzi* (Order Stolidobranchia) utilizes the VC protein VC70 and sperm proteins urabin and TTSP1 as promising self/nonself-recognition proteins. Although it is still unknown whether urabin and TTSP-1 are directly involved in self/nonself-recognition, they are capable of binding to VC70, a polymorphic EGF-like repeat protein involved in self/nonself-recognition. Although polymorphic *s/v-Themis* homologous genes and their expression were detected in the gonad of *H. roretzi,* it remains elusive whether *s/v-Themis* genes are responsible for self/nonself-recognition during fertilization in *H. roretzi.*

In contrast, *Ciona robusta* (Order Phlebobranchia) has the self-recognition system, which is mediated by sperm receptors s-Themis-A, B, B2 and their cognate VC ligands v-Themis-A, B, B2, respectively. v-Themis proteins are highly polymorphic, whereas s-Themis proteins contain an N-terminal hyper-variable region. Since s-Themis-B and B2 contain a C-terminal Ca^{2+}-permeable cation channel domain, Ca^{2+} influx after sperm binding to the VC of self-eggs may be mediated by this cation channel. v-Themis-like and other gamete proteins may also be indirectly involved in allorecognition during gamete interaction. Further studies on the protein-protein interaction and three-dimensional structure are necessary for elucidation of the molecular mechanisms of self/nonself-recognition during ascidian fertilization.

Acknowledgements This work was supported in part by Grant-in-aid for Scientific Research (B) (JP17H03672) from JSPS and by Grant-in-aid for Scientific Research on Innovative Areas (JP21112001, 21112002) from MEXT, Japan. We are grateful to Gretchen Lambert of University of Washington Friday Harbor Labs for her critical reading of this mauscript.

References

Akasaka M, Harada Y, Sawada H (2010) Vitellogenin C-terminal fragments participate in fertilization as egg-coat binding partners of sperm trypsin-like proteases in the ascidian *Halocynthia roretzi*. Biochem Biophys Res Commun 392:479–484

Akasaka M, Kato KH, Kitajima K, Sawada H (2013) Identification of novel isoforms of vitellogenin expressed in ascidian eggs. J Exp Zool B Mol Dev Evol 320:118–128

Artavanis-Tsakonas S, Matsumoto K, Fortini ME (1995) Notch signaling. Science 268:225–232

Ban S, Harada Y, Yokosawa H, Sawada H (2005) Highly polymorphic vitelline-coat protein HaVC80 from the ascidian, *Halocynthia aurantium*: structural analysis and involvement in self/nonself recognition during fertilization. Dev Biol 286:440–451

Brozovic M, Martin C, Dantec C, Dauga D, Mendez M, Simion P, Percher M, Laporte B, Scornavacca C, Di Gregorio A, Fujiwara S, Gineste M, Lowe EK, Piette J, Racioppi C, Ristoratore F, Sasakura Y, Takatori N, Brown TC, Delsuc F, Douzery E, Gissi C, McDougall A, Nishida H, Sawada H, Swalla BJ, Yasuo H, Lemaire P (2016) Aniseed 2015: a digital framework for the comparative developmental biology of ascidians. Nucleic Acid Res 44(D1):D808–D818

Brunetti R, Gissi C, Pennati R, Caicci F, Gasparini F, Manni L (2015) Morphological evidence that the molecularly determined *Ciona intestinalis* type A and type B are different species: *Ciona robusta* and *Ciona intestinalis*. J Zool Syst Evol Res 53:186–193

De Nettancourt D (1977) Incompatibility in angiosperms. In: Frankel R, Gall GAE, Grossman M, Linskens HF, De Zeeuw D (eds) Monographs on theoretical and applied genetics 3. Springer, Berlin, pp 1–230. https://doi.org/10.1007/978-3-662-12051-4

De Santis R, Pinto MR, Cotellin F, Rosati F, Monroy A, D'Alessio G (1983) A fucosyl glycoprotein component with sperm receptor and sperm-activating activity from the vitelline coat of *Ciona intestinalis* eggs. Exp Cell Res 148:508–513

Dehal P, Satou Y, Campbell RK, Chapman J, Degnan B, De Tomaso A, Davidson B, Di Gregorio A, Gelpke M, Goodstein DM, Harafuji N, Hastings KE, Ho I, Hotta K, Huang W, Kawashima T, Lemaire P, Martinez D, Meinertzhagen IA, Necula S, Nonaka M, Putnam N, Rash S, Saiga H, Satake M, Terry A, Yamada L, Wang HG, Awazu S, Azumi K, Boore J, Branno M, Chin-Bow S, DeSantis R, Doyle S, Francino P, Keys DN, Haga S, Hayashi H, Hino K, Imai KS, Inaba K, Kano S, Kobayashi K, Kobayashi M, Lee BI, Makabe KW, Manohar C, Matassi G, Medina M, Mochizuki Y, Mount S, Morishita T, Miura S, Nakayama A, Nishizaka S, Nomoto H, Ohta F, Oishi K, Rigoutsos I, Sano M, Sasaki A, Sasakura Y, Shoguchi E, Shin-i T, Spagnuolo A, Stainier D, Suzuki MM, Tassy O, Takatori N, Tokuoka M, Yagi K, Yoshizaki F, Wada S, Zhang C, Hyatt PD, Larimer F, Detter C, Doggett N, Glavina T, Hawkins T, Richardson P, Lucas S,

Kohara Y, Levine M, Satoh N, Rokhsar DS (2002) The draft genome of *Ciona intestinalis*: insights into chordate and vertebrate origins. Science 298(5601):2157–2167

Fuke TM (1983) Self and nonself recognition between gametes of the ascidian, *Halocynthia roretzi*. Rouxs Arch Dev Biol 192:347–352

Fuke M, Numakunai M (1996) Establishment of self-sterility of eggs in the ovary of the solitary ascidian, *Halocynthia roretzi*. Rouxs Arch Dev Biol 205:391–400

Fuke M, Numakunai T (1999) Self-sterility of eggs induced by exogenous and endogenous protease in the solitary ascidian *Halacynthia roretzi*. Mol Reprod Dev 52:99–106

Harada Y, Sawada H (2007) Proteins interacting with the ascian vitelline-coat sperm receptor HrVC70 as revealed by yeast two-hybrid screening. Mol Reprod Dev 74:1178–1187

Harada Y, Sawada H (2008) Allorecognition mechanisms during ascidian fertilization. Int J Dev Biol 52:637–645

Harada Y, Takagaki Y, Sugnagawa M, Saito T, Yamada L, Taniguchi H, Shobuchi E, Sawada H (2008) Mechanism of self-sterility in a hermaphroditic chordate. Science 320:548–550

Hoshi M (1986) Sperm glycosidase as a plausible mediator of sperm binding to the vitelline envelope in ascidians. Adv Exp Med Biol 207:251–260

Iwano M, Takayama S (2012) Self/non-self discrimination in angiosperm self-incompatibility. Curr Opin Plant Biol 15:78–83

Kawai N, Ochiai H, Sakuma T, Yamada L, Sawada H, Yamamoto T, Sasakura Y (2012) Efficient targeted mutagenesis of the chordate *Ciona intestinalis* genome with zinc-finger nucleases. Dev Growth Differ 54:535–545

Kawamura K, Nomura M, Kameda T, Shimamoto H, Nakauchi M (1991) Self-nonself recognition activity extracted from self-sterile eggs of the ascidian, *Ciona intestinalis*. Dev Growth Differ 33:139–148

Lambert CC, Koch R (1988) Sperm binding and penetration during ascidian fertilization. Dev Growth Differ 30:325–336

Marino R, Pinto MR, Cotelli F, Lamia CL, De Santis R (1998) The hsp70 protein is involved in the acquisition of gamete self-sterility in the ascidian *Ciona intestinalis*. Development 125:899–907

Marino R, De Santis R, Giuliano P, Pinto MR (1999) Follicle cell proteasome activity and acid extract from the egg vitelline coat prompt the onset of self-sterility in *Ciona intestinalis* oocytes. Proc Natl Acad Sci U S A 96:9633–9636

Matsumoto M, Hirata J, Hirohashi N, Hoshi M (2002) Sperm-egg binding mediated by sperm alpha-L-fucosidase in the ascidian, *Halocynthia roretzi*. Zool Sci 19:43–48

Mino M, Sawada H (2016) Follicle cell trypsin-like protease HrOvochymase: its cDNA cloning, localization, and involvement in the late stage of oogenesis and in the ascidian *Halocynthia roretzi*. Mol Reprod Dev 83:347–358

Morgan TH (1910) Cross- and self-fertilization in *Ciona intestinalis*. Wilhelm Roux Arch Entwick Mech Org 30:206–235

Morgan TH (1923) Removal of the block to self-fertilization in the ascidian *Ciona*. Proc Natl Acad Sci U S A 9:170–171

Morgan TH (1939) The genetic and the physiological problems of self-sterility in *Ciona*. III. Induced self-fertilization. J Exp Zool 80:19–54

Morgan TH (1942) The genetic and the physiological problems of self-sterility in *Ciona*. V. The genetic problem. J Exp Zool 90:199–228

Morgan TH (1944) The genetic and the physiological problems of self-sterility in *Ciona*. VI. Theoretical discussion of genetic data. J Exp Zool 95:37–59

Murabe N, Hoshi M (2002) Re-examination of sibling cross-sterility in the ascidian, *Ciona intestinalis*: genetic background of the self-sterility. Zool Sci 19:527–538

Numakunai T, Hoshino Z (1980) Periodic spawning of three types of the ascidian, *Halocynthia roretzi* (Drasche), under continuous light conditions. J Exp Zool 212:381–387

Otsuka K, Yamada L, Sawada H (2013) cDNA cloning, localization, and candidate binding partners of acid-extractable vitelline-coat protein Ci-v-Themis-like in the ascidian *Ciona intestinalis*. Mol Reprod Dev 80:840–848

Pinto MR, De Stantis R, Marino R, Usui N (1995) Specific induction of self-discrimination by follicle cells in Ciona intestinalis. Dev Growth Differ 37:287–291

Rosati F, De Santis R (1978) Studies on fertilization in the ascidains I. Self-sterility and specific recognition between gametes of *Ciona intestinalis*. Exp Cell Res 112:111–119

Rosati F, De Santis R (1980) Role of the surface carbohydrates in sperm-egg interaction in *Ciona intestinalis*. Nature 283:762–764

Saito T, Shiba K, Inaba K, Yamada L, Sawada H (2012) Self-incompatibility response induced by calcium increase in sperm of the ascidian *Ciona intestinalis*. Proc Natl Acad Sci U S A 109:4158–4162

Sakuma T, Hosoi S, Woltjen K, Suzuki K, Kashiwagi K, Wada H, Ochiai H, Miyamoto T, Kawai N, Sasakura Y, Matsuura S, Okada Y, Kawahara A, Hayashi S, Yamamoto T (2013) Efficient TALEN construction and evaluation methods for human cell and animal applications. Genes Cells 18:315–326

Satou Y, Kawashima T, Shoguchi E, Nakayama A, Satoh N (2005) An integrated database of the ascidian, *Ciona intestinal:* towards functional genomics. Zool Sci 22:837–843

Satou Y, Hirayama K, Mita K, Fujie M, Chiba S, Yoshida R, Endo T, Sasakura Y, Inaba K, Satoh N (2014) Sustained heterozygosity across a self-incompatibility locus in a inbred ascidian. Mol Biol Evol 32:81–90

Sawada H (2002) Ascidian sperm lysin system. Zool Sci 19:139–151

Sawada H, Sakai N, Abe Y, Tanaka E, Takahashi Y, Fujino J, Kodama E, Takizawa S, Yokosawa H (2002) Extracellular ubiquitination and proteasome-mediated degradation of the ascidian sperm receptor. Proc Natl Acad Sci U S A 99:1223–1228

Sawada H, Tanaka E, Ban S, Yamasaki C, Fujino J, Ooura K, Abe Y, Matsumoto K, Yokosawa H (2004) Self/nonself recognition in ascidian fertilization: vitelline coat protein HrVC70 is a candidate allorecognition molecule. Proc Natl Acad Sci U S A 101:15615–15620

Sawada H, Morita M, Iwano M (2014) Self/non-self recognition mechanisms in sexual reproduction: new insight into the self-incompatibility system shared by flowering plants and hermaphroditic animals. Biochem Biophys Res Commun 450:1142–1148

Sawada H, Nakazawa S, Shirae-Kurabayashi M (2017) Ascidian sexual reproductive strategies: mechanisms of sperm-egg interaction and self-sterility. In: Kobayashi K et al (eds) Reproductive and developmental strategies, diversity and community in animals. Springer, Tokyo, pp 479–497

Sawada H, Yamamoto K, Yamaguchi A, Yamada L, Higuchi A, Nukaya H, Fukuoka M, Sakuma T, Yamamoto T, Sasakura Y, Shirae-Kurabayashi M (2020) Three multi-allelic gene pairs are responsible for selfsterility in the ascidian *Ciona intestinalis*. Sci Rep 10:2514

Takayama S, Isogai A (2005) Self-incompatibility in plants. Annu Rev Plant Biol 56:467–489

Urayama S, Harada Y, Nakagawa Y, Ban S, Akasaka M, Kawasaki N, Sawada H (2008) Ascidian sperm glycosylphosphatidylinositol-anchored CRISP-like protein as a binding partner for an allorecognizable sperm receptor on the vitelline coat. J Biol Chem 283:21725–21733

Yamada L, Saito T, Taniguchi H, Sawada H, Harada Y (2009) Comprehensive egg coat proteome of the ascidian *Ciona intestinalis* reveals gamete recognition molecules involved in self-sterility. J Biol Chem 284:9402–9410

Yamaguchi A, Saito T, Yamada L, Taniguchi H, Harada Y, Sawada H (2011) Identification and localization of the sperm CRISP family protein CiUrabin involved in gamete interaction in the ascidian *Ciona intestinalis*. Mol Reprod Dev 78:488–497

Chapter 10
Reproduction of Chondrichthyans

Terence I. Walker

Abstract Species of the class Chondrichthyes (sharks, rays, and chimaeras) exhibit diverse reproductive modes ranging from strictly lecithotrophic oviparity to viviparity with varying matrotrophic nutritional supplement (histotrophy, ovatrophy, adelphotrophy, and placentatrophy). Compared with invertebrate and bony fish taxa, chondrichthyan fishes are long-lived and produce small numbers of large offspring. These characteristics underlie their high vulnerability to harvest as only a small proportion of their populations can be taken annually to ensure sustainable catches and to avoid population depletion. These characteristics also produce challenges for their conservation in the wild and their husbandry associated with commercial fish farming in sea cages and display in aquaria. There is a growing need to address these issues through improved facilities for scientific study and breeding of animals in captivity. Improved conservation of wild stocks and husbandry of captive chondrichthyan animals require a quantitative approach to investigating their maturity, maternity, fecundity, and productivity.

Keywords Oviparity · Viviparity · Maternity · Demography

10.1 Introduction

The class Chondrichthyes comprises squalomorph sharks, galeomorph sharks, batoids (rays), and holocephalans (chimaeras) in 14 taxonomic orders of extant species occurring throughout the world at all depths in the oceans, coastal seas, and some inland freshwater systems. Chondrichthyan fishes form a diverse group of 1250 presently recognised species (Dulvy et al. 2017) sharing the distinguishing features of a cartilaginous skeleton and jaws. The sharks and rays classed as elasmobranchs have common ancestors living about 200 million years ago whereas elasmobranchs and holocephalans originated about 420 million years ago.

T. I. Walker (✉)
School of BioSciences, The University of Melbourne, Parkville, VIC, Australia

School of Biological Sciences, Monash University, Clayton, VIC, Australia

M. Yoshida, J. F. Asturiano (eds.), *Reproduction in Aquatic Animals*,
https://doi.org/10.1007/978-981-15-2290-1_11

Onshore aquaculture and at-sea fish farming facilities and husbandry techniques have in recent years developed rapidly for invertebrate and teleost species, but slowly for chondrichthyan species. The low value of the products from sharks, rays, and chimaeras in the past compared with those from invertebrate and teleost species partly explains this slow development. Life history characteristics such as comparatively large body size, high longevity and age at maturity, long reproductive cycles, and a small number of offspring have discouraged captive breeding programs. Most current understanding of the husbandry of chondrichthyan species is from commercial aquaria facilities holding fishes for public display and from research facilities undertaking studies on captive fish. Nevertheless, there are growing needs for improved technologies for aquaculture and sea-cage farming (ranching) specific to chondrichthyan species.

Demand is increasing for the products, such as meat, fins, liver oil, cartilage, corneas, and leather, from chondrichthyan species in response to severe depletion of many stocks from past unsustainable harvesting practices. However, this demand alone is an insufficient economic stimulus for new aquaculture and fish farming industries. Captive breeding programs for supplying these products are not economically feasible at this time, although the retention of full-term embryos and artificial feeding of captured animals in sea cages for further growth and body-mass gain provide opportunities for adding value to existing fully developed fisheries.

Other drivers for such facilities arise from the need for improved information for better management of the wild stocks best obtained by studying the animals under captive conditions. Such information includes determining levels of cryptic fishing mortality on escaped and released chondrichthyan animals following capture in fishing gear. Other required information includes assessing the effects of other anthropogenic hazards on habitats associated with mineral and hydrocarbon extraction, shipping constructions inshore, and climate change (notably changed currents, water temperature, and pH). Proper evaluation of all these effects requires appropriate facilities with highly controlled conditions for experimentation testing physiological (Bouyoucos et al. 2018; Frick et al. 2009; Frick et al. 2010a; Frick et al. 2010b; Frick et al. 2012; Guida et al. 2016; Heard et al. 2014; Martins et al. 2018; Skomal and Mandelman 2012; Van Rijn and Reina 2010), behavioural (Martins 2017), and transgenerational (Guida et al. 2017) changes in response to specific anthropogenic stressors.

Successful captive breeding programs for chondrichthyan species will have to account for a wide range of reproductive modes associated with diverse morphological and physiological mechanisms and behaviours for ensuring successful copulation, internal fertilisation, embryogenesis, and parturition conditions suitable for the survival of the neonates. Present understanding of behaviour comes mostly from observations of animals held in captivity. Mating involves a range of complex behaviours during precopulatory courtship to ensure the males gain an adequate grip on the female and insert one or both claspers to ensure successful copulation. Observations include females resisting particular males, male co-operation during copulation, and multiple paternity within litters for numerous species (Carrier et al. 2004).

10.2 Chondrichthyan Reproductive Biology

The study of the reproductive biology of chondrichthyan species draws on the diverse fields of anatomy, taxonomy, genetics, physiology, immunology, endocrinology, and demography, and can be related to the behaviour, zoogeography, phylogeny, and ecology. These fields underpin better information to improve sustainable use and to rehabilitate depleted and fragmented populations of these species. More quantitative approaches to reproduction studies enable ecological vulnerability assessment to the stressors of climate change (Chin et al. 2010), fishing (Hobday et al. 2011; Walker 2005a), and other anthropogenic hazards; fisheries assessment (Punt and Walker 1998); species extinction risk assessment (Dulvy et al. 2014); and development of captive breeding programs. Quantitative approaches require applying models that represent the population dynamics of species, where the models range from simple demographic models to highly complex synthetic models combining not only elements of the biology of a species but characteristics of the fishing gear, vessels and fishers (Pribac et al. 2005; Punt et al. 2000). Application of such models to the populations of chondrichthyan species leads to bias without careful distinction of the conditions of maturity, pregnancy, and maternity of the females. These three conditions require determination of the timing and periodicity of reproductive events such as ovulation, gestation, and egg laying or parturition (Walker 2005b).

Maturation and seasonal changes in reproductive condition are under the control of the endocrine system. Environmental cues initiate secretion of gonadotropin-releasing hormone from neurons in the hypothalamus and other parts of the brain. This hormone stimulates the pituitary gland to produce gonadotropins, which in turn promote gametogenesis (oogenesis in females and spermatogenesis in males) and secretion of reproductive steroids in the ovaries or testes. The gonadal steroids—predominantly testosterone, dihydrotestosterone, estradiol, and progesterone—orchestrate sex differentiation and the timing of maturation and reproductive cycles. Thyroid hormones and other hormones such as relaxin and calcitonin appear to have accessory roles (Awruch 2013; Gelsleichter 2004).

10.2.1 Female Reproductive System

Female chondrichthyan fishes have a pair of reproductive tracts and either paired ovaries or a predominant single ovary where the second is vestigial or absent. Each reproductive tract typically consists of a funnel-shaped ostium, anterior oviduct, oviducal gland, isthmus for some species, uterus, and cervix, leading posteriorly to the urogenital sinus common to both tracts and opening into the cloaca (Fig. 10.1).

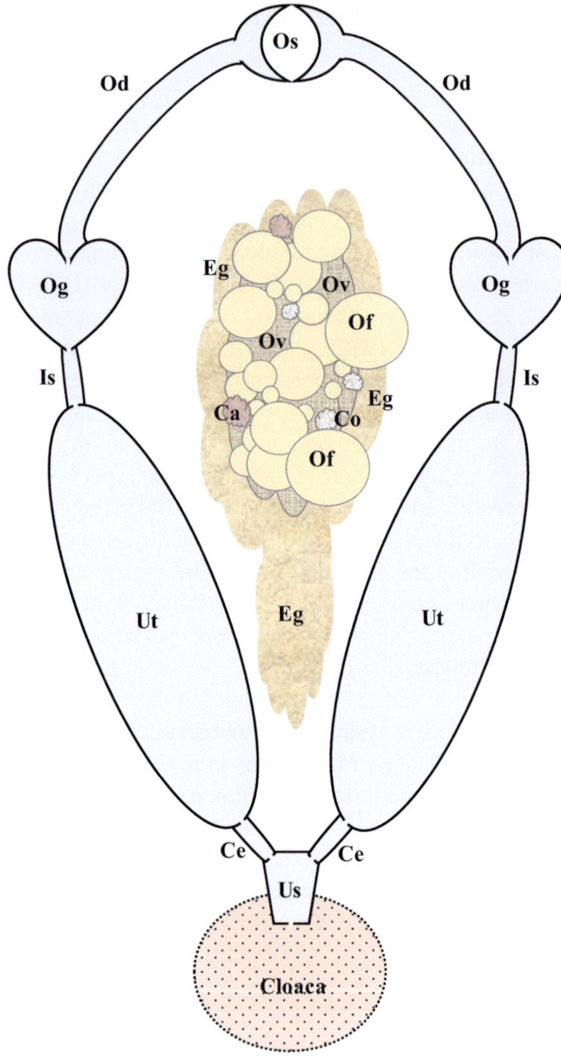

Fig. 10.1 Female chondrichthyan reproductive system. The diagram showing the reproductive tract (pale grey shading) and other structures (textured colour) is a generalised representation and not of a particular species. *Ca* corpora atretica, *Ce* cervix, *Co* corpora luteum, *Eg* epigonal gland, *Is* Isthmus, *Od* oviduct, *Of* ovarian follicle, *Og* oviducal gland, *Os* ostium, *Ov* ovary, *Us* urogenital sinus, *Ut* uterus

10.2.1.1 Ovary and Oogenesis

The chondrichthyan ovary, embedded in the lymphomyleloid tissue of the epigonal gland, is located in the anterior region of the body cavity dorsal to the liver and attached to the dorsal wall by the mesovarium mesentery. The ovary typically consists of follicles of various sizes, atretic follicles, and corpora lutea bound in connective tissue stroma and produces oocytes (germ cells) by oogenesis. As for vertebrates in general, oogenesis involves several stages of transition from female germ cells to mature ova. Primordial germ cells in the germinal epithelium of the ovary proliferate into oogonia and then primary oocytes by several mitotic divisions.

Diploid primary oocytes then divide by the first meiotic division to form haploid secondary oocytes, but the second meiotic division of a secondary oocyte pauses until reactivation with the formation of an ovum at the time of fertilisation (Waltrick et al. 2017).

Each follicle is an oocyte enveloped by proliferating granulosa cells delimited by a basal lamina. In immature animals, the oocytes are small with little or no yolk (Hamlett and Koob 1999). As the animals mature, theca and granulosa cells synthesise and secrete steroid hormones, and the oocytes enlarge through vitellogenesis, whereby phosvitin and lipovitellin of hepatic origin accumulate in the oocytes. Ovulation occurs seasonally when the ovary serially releases the largest oocytes, and the wall of each follicle collapses to form a corpora luteum consisting of a lipid-filled cell derived from the granulosa cells (Storrie 2004). As the period of ovulation ends, enlarged preovulatory follicles undergo resorption of yolk from the oocytes, to form corpora atretica through the process of atresia.

10.2.1.2 Female Reproductive Tracts

Following ovulation, each ovulated oocyte (ovum) is released into the body cavity and propelled anteriorly by cilia into the ostium that bifurcates into the left and right oviducts. The ovum then passes through one of the oviducts and into the oviducal gland. Fertilisation of the ovum and the formation of a protective egg membrane or egg case occur as it passes through the oviducal gland and then the isthmus into the uterus. Depending on the species, the encapsulated egg either undergoes embryogenesis in utero or in an egg case following oviposition to a suitable external habitat. Eggs and embryos in utero are sealed between two uterine sphincters (the isthmus and cervix).

Each of the paired oviducal glands has four distinct zones (proximal club zone, papillary zone, baffle zone, and terminal zone) for several functions. The oviducal glands of egg-laying species are more prominent than of species retaining the fertilised eggs in utero during embryogenesis (Hamlett et al. 2005b). Although observed only in a small number of shark, batoid, and holocephalan species, sperm storage in the oviducal gland most likely occurs for all chondrichthyan species to ensure fertilisation immediately after ovulation and before egg encapsulation. The oviducal gland produces a tertiary egg envelope and egg jelly to surround the zygote and yolk. The tertiary egg envelope can be the egg case of oviparous species, the thin transient egg candle capsule of yolk-sac species, and the thin egg envelope of most placental sharks.

10.2.1.3 Mode of Reproduction, Lecithotrophy, and Matrotrophy

Broadly classed as oviparity (egg-laying) and viviparity (live-bearing), chondrichthyan species have eight modes of reproduction depending on lecithotrophy (nutrition of embryos from egg yolk) and matrotrophy (nutrition of embryos from egg yolk supplemented maternally from other sources) (Hamlett et al. 2005c). Oviparity characterised by highly lecithotrophic fertilised eggs protected in tough leathery egg cases occurs as two reproductive modes: single oviparity and retained oviparity. Single oviparity is where pairs of eggs, one from each uterus, are oviposited sequentially over an extended egg-laying season for external embryonic development. Retained oviparity is where the egg cases remain in the reproductive tract for most of the period for embryonic development before oviposited in habitat suitable for neonates. Viviparity includes six modes of reproduction characterised by fertilised eggs contained in an egg envelope in utero for the duration of gestation. The viviparity modes relate to the absence or presence of several types of nutritional supplementation of egg yolk in the uterus: histotroph (histotrophy), ova (ovatrophy), siblings (intrauterine cannibalism or adelphotrophy), and placental transfer (placentatrophy) (Hamlett et al. 2005c). Yolk-sac viviparity occurs as three recognised modes: yolk-sac viviparity without nutrient supplement, yolk-sac viviparity with limited histotrophy, and yolk-sac viviparity with lipid histotrophy. Other recognised viviparity modes are carcharhinid oophagy, lamnid oophagy, and placental viviparity. For the carcharhinid oophagy mode, large numbers of unfertilised ova are packaged together with the developing embryo in the same egg capsule for ingestion by the embryo. For the lamnid oophagy mode, the developing embryo hatches in utero following depletion of its yolk reserves, and then egg capsules containing small ova are packaged together in the oviducal gland and provided to the embryo for ingestion. In *Carcharias taurus*, the first embryo hatched in each uterus consumes all the other embryos and continues to feed on ova (Gilmore et al. 2005). For the placental viviparity mode in a small number of shark species, the yolk sac of the embryo forms an attachment with the uterine epithelium to form a placenta, and the yolk stalk elongates to form an umbilical cord. Following initial depletion of the yolk stores in the yolk sac, embryonic development depends next on mucoid histotroph and finally on maternal secretions at the placental site. A blood transfer of nutrients via the placenta is unlikely for any species of shark (Hamlett et al. 2005a).

The eight modes of reproduction occur widely among species of Chondrichthyes in five orders of the super-order Squalomorphii, four orders of the Galeomorphii, four orders of the Batoidea, and one order of the Holocephali (Table 10.1). The presence of oviparity in all living species of the Holocephali and a broad range of species across the other super-orders with the one exception of the Squalomorphii supports a long history of the argument for oviparity as the ancestral mode. However, the frequent occurrence of viviparity among extinct holocephalans and the extant Squalomorphii, Galeomorphii, and Batoidea provide evidence for yolk-sac viviparity as the original mode (Musick and Ellis 2005).

10.2.1.4 Female Maturity

The period from the onset of maturity (start of vitellogenesis) to the first ovulation and start of the first pregnancy in most vertebrate species is comparatively short, and the ovarian cycle is annual or less, with only a few exceptions such as among large

Table 10.1 Modes of reproduction by chondrichthyan taxonomic group

Super-order/order	Sub-order or family (where required to subdivide order)	Mode of reproduction
Squalomorphii (squalomorph sharks)		
Echinorhiniformes (bramble sharks)		Yolk-sac viviparity with no histotroph
Hexanchiformes (sixgill and sevengill sharks)		Yolk-sac viviparity with no histotroph
Pristiophoriformes (sawsharks)		Yolk-sac viviparity with no histotroph
Squaliformes (dogfishes)		Yolk-sac viviparity with limited or no histotroph
Squatiniformes (angelsharks)		Yolk-sac viviparity with no histotroph
Galeomorphii (galeomorph sharks)		
Carcharhiniformes (ground sharks)	Scyliorhinidae (catsharks) excluding six species	Single oviparity
	Scyliorhinidae including five species of genus *Halaelurus*	Retained oviparity
	Scyliorhinidae including only *Galeus melastomus*	Retained oviparity
	Proscylliidae (finback catsharks) excluding one species	Yolk-sac viviparity with no histotroph
	Proscylliidae including only *Proscyllium habereri*	Single oviparity
	Pseudotriakidae (false catsharks)	Carcharhinid oophagy
	Leptochariidae (barbelled catsharks)	Placental viviparity
	Sphyrnidae (hammerhead sharks)	Placental viviparity
	Triakidae (hound sharks)	Yolk-sac viviparity with limited or no histotroph
	Hemigaleidae (weasel sharks)	Placental viviparity
	Carcharhinidae (requiem sharks) excluding one species	Placental viviparity
	Carcharhinidae including only *Galeocerdo cuvier*	Yolk-sac viviparity with no histotroph
Heterodontiformes (horn sharks)		Single oviparity
Lamniformes (mackerel sharks)		Lamnid oophagy
Orectolobiformes	Brachaeluridae (blind sharks)	Yolk-sac viviparity with no histotroph
	Hemiscylliidae (bamboo sharks)	Single oviparity
	Ginglymostomatidae (nurse sharks)	Yolk-sac viviparity with no histotroph
	Orectolobidae (wobbegongs)	Yolk-sac viviparity with no histotroph

(continued)

Table 10.1 (continued)

Super-order/order	Sub-order or family (where required to subdivide order)	Mode of reproduction
	Parascylliidae (collar carpet sharks)	Single oviparity
	Rhincodontidae (whale sharks)	Yolk-sac viviparity with no histotroph
	Stegostomatidae (zebra sharks) excluding one species	Single oviparity
	Stegostomatidae including only *Stegostoma fasciatum*	Retained oviparity
Batoidea (rays)		
Myliobatiformes (stingrays and relatives)	Myliobatoidei (stingrays)	Yolk-sac viviparity with lipid histotrophs
	Platyrhinoidei (thornback rays)	Yolk-sac viviparity with no histotroph
	Zanobatoidei (panrays)	Yolk-sac viviparity with no histotroph
Rajiformes (skates)		Single oviparity
Rhinopristiformes (guitarfishes & sawfishes)		Yolk-sac viviparity with no histotroph
Torpediniformes (electric rays)		Yolk-sac viviparity with no histotroph
Holocephali (chimaeras)		
Chimaeriformes (chimaeras)		Single oviparity

Sourced from Musick and Ellis (2005)

mammals. For chondrichthyan species, this period can also be annual or less, but in other species, the period between successive ovulation cycles can be two, three, or for some species possibly more years. For the species with long ovarian cycles (i.e., long period of vitellogenesis), the period from fertilisation to parturition for viviparous species can be, but not necessarily, more than 1 year. Fertilisation occurs during the relatively short period following ovulation when the ovum passes through the oviduct and oviducal gland before encapsulation and entry to the uterus. Storage of sperm in the terminal zone of the oviducal gland (Hamlett et al. 2002a; Smith et al. 2004; Storrie et al. 2008) ensures a supply of sperm for successive fertilisation of ova released by ovulation over a period of several weeks or months. Sperm storage ensures that fertilisation occurs before egg encapsulation and it avoids accumulating eggs in utero obstructing or retarding fresh sperm transiting the uterus to the oviducal gland.

For a mature female's first pregnancy, the period of the first ovarian cycle, together with the subsequent period of pregnancy (i.e., the period of eggs or embryos in utero), is more than 1 year for most species. However, for subsequent pregnancies, the follicles can enlarge through vitellogenesis during part of the ovarian cycle concurrently with pregnancy so that parturition can be annual or, for some species, more frequently. For species producing large-sized follicles, such as *Galeorhinus galeus* (Triakidae) (Walker 2005b), orectolobids (Huveneers et al. 2007), and squaliforms (Braccini et al. 2006; Collonello et al. 2016; Hanchet 1988; Rochowski et al. 2015a) from coastal or upper-slope waters (~40 mm diameter) and mid-slope (60–87 mm diameter) (Girard and Du Buit 1999; Guallart and Vincent 2001; Rochowski et al. 2015b; Yano and Tanaka 1988) the period for vitellogenesis is two, three, or possibly more years.

Female maturity in chondrichthyan species assumed by the presence of enlarged follicles is ambiguous without specification of a specific follicle size. This assumption is arbitrary, subjective, and likely to provide unrepeatable results for a species that has a long ovarian cycle. Less ambiguous criteria for classing an observed female as mature or immature might include a minimum level of selected steroid hormones in the blood plasma, the presence of vitellogenic oocytes, the presence of sperm in the oviducal gland, or the presence of eggs or embryos in utero. However, whether a female is designated mature or immature inevitably varies depending on the criterion adopted. Furthermore, applying a mix of these criteria further confuses the definition of maturity.

The procedure most readily applied and independent of observer bias when undertaking a dissection is to record the colour and diameter of the largest follicle or diameters of a group of the largest follicles. Selecting a follicle diameter that coincides with the size when the oocytes begin vitellogenesis provides a consistent criterion. The criterion of follicle diameter of 0–3 mm as pre-vitellogenic (generally white) (immature) and oocytes >3 mm diameter as vitellogenic (yellowish) (mature) applies to *Galeorhinus galeus* (Walker 2005b), *Mustelus antarcticus* (Triakidae) (Walker 2007), and *Heterodontus portusjacksoni* (Heterodontiformes) (Tovar-Ávila et al. 2007). The criterion of follicles >1 mm as vitellogenic applies to *Urolophus bucculentus*, *U. cruciatus*, *U. paucimaculatus*, and *U. viridis*, and >3 mm as vitellogenic applies to *U. gigas* and *Trygonoptera imitata* (Myliobatoidei) (Trinnie et al. 2014). The criterion of oocytes >11 mm as vitellogenic (mature) applies to *Squalus chloroculus* (Rochowski et al. 2015a) and *Deania calcea* (Squaliformes) (Rochowski et al. 2015b), of yellow oocytes as vitellogenic applies to *Orectolobus* spp. (Huveneers et al. 2007), and yellow and >14 mm to *Squalus megalops* (Braccini et al. 2006).

In fisheries science, the assumption that the number of mature animals in a population (or mature biomass) equates to the number (or biomass) contributing to recruitment is valid for teleost and invertebrate species but not necessarily for all chondrichthyan species. Because the reproductive cycle of many chondrichthyan species exceeds 1 year, it is necessary to invoke the concept of maternity to provide a more accurate predictor of recruitment.

10.2.1.5 Maternity and Pregnancy

The concepts of the periods of pregnancy and maternity underlie the development of one of several equations required for estimating the number of births for viviparous species or the number of egg cases oviposited for oviparous species. The required equation expresses the proportion of the female population producing offspring by the end of the period of parturition or oviposition and thereby contributing to annual recruitment (0+-year-old cohort) at the beginning of the following year referred to as the recruitment year.

The period of pregnancy for a female population is often seasonal and differs between viviparous and oviparous species. For viviparous species, the period of pregnancy is from the start of the period of eggs in utero when the first ovulated egg enters the uterus until the end of the period of macroscopically visible embryos in utero with birth (parturition) of the last embryo. For oviparous species, the period of pregnancy is from the start of the period when the first ovulated egg encapsulated enters the uterus until the end of the period with oviposition of the last egg case. A date to demarcate the end of parturition or oviposition is required to set the last day of the period of maternity and to define the following day as the recruitment date (first day of the recruitment year).

For viviparous species, the period of maternity for a female population has a duration of precisely 1 year immediately preceding the recruitment date and does not necessarily cover the entire period of pregnancy. Establishing the recruitment date and period of maternity provides a basis for classing all females observed in the population as either in maternal condition or non-maternal condition. Any female observed in a pregnant condition before or after the period of maternity is non-maternal. Also, a female in a postpartum condition shortly before the end of the maternal period is probably in maternal condition. The pattern of distribution of the proportion of the females in maternal condition against maternal size provides a basis for determining whether individual females have a period of parturition every 1, 2, or 3 years. Parturition annually, biennially, triennially or longer implies one, two, three, or more years, respectively. Figure 10.2 shows the patterns of the periods of various reproductive conditions with the recruitment date set arbitrarily to 1 January for *Galeorhinus galeus* and *Mustelus antarcticus*.

For oviparous species, the reproductive cycle is annual, and each female with eggs in utero or distended uteri from egg-laying is in maternal condition. The recruitment date immediately follows the egg-hatching period.

10.2.2 Male Reproductive System

Male chondrichthyan fishes have external paired claspers at the bases of the pelvic fins used as intromittent organs for internal fertilisation, although holocephalan species do have additional copulatory appendages. The internal system consists of paired testes and paired reproductive tracts where each consists of genital ducts and a Leydig gland. The genital ducts cover the elongate kidneys embedded in the dorsal abdominal wall and consist of the testis efferent ductules, epididymis, ductus deferens (also vas deferens or Wolffian duct), and seminal vesicle (also ductus deferens ampulla), enclosed by the peritoneum (Fig. 10.3). Also, sharks possess paired siphon sacs whereas batoids have alkaline glands and clasper glands.

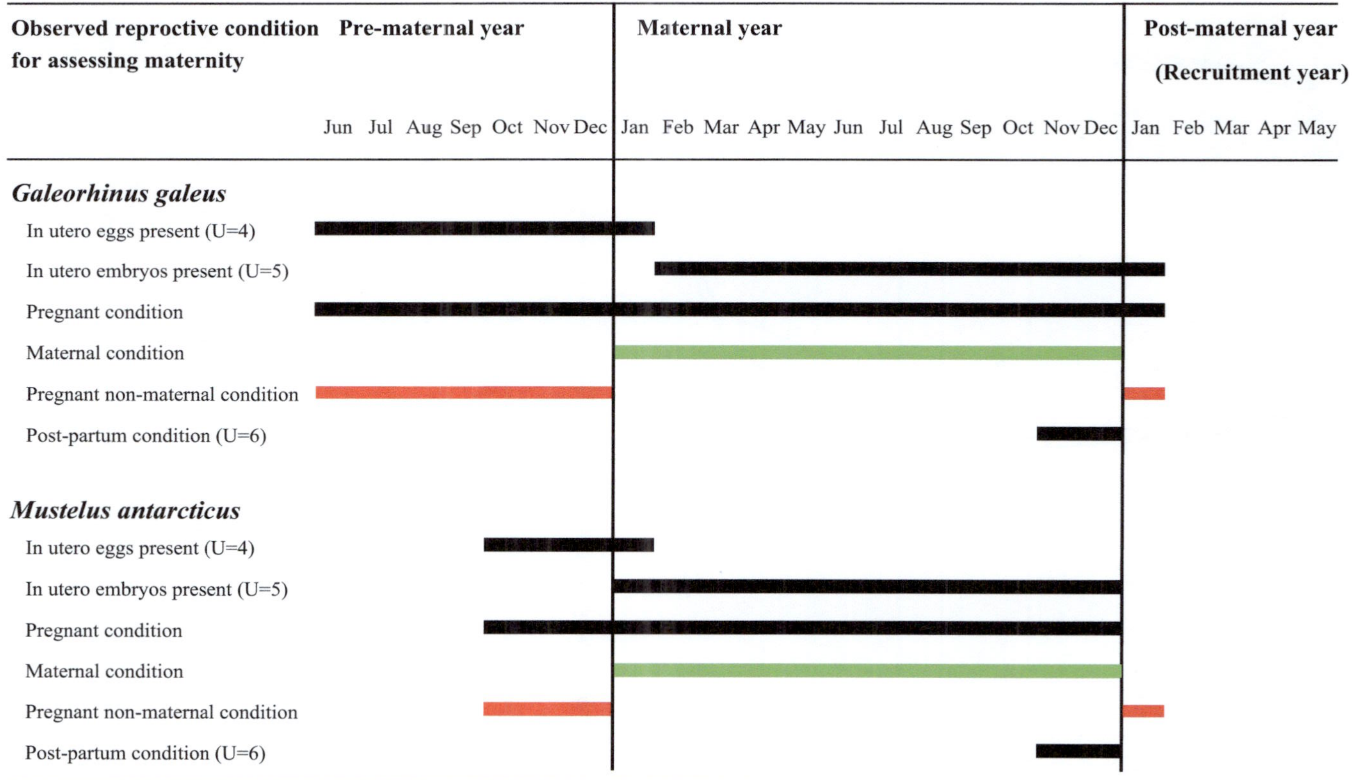

Fig. 10.2 Timing of observed breeding conditions for determining maternity year in two examples of viviparous species. The $U = 6$ condition can occur at any time but only November–December applies for when they are in maternal condition. The displayed information is sourced for *Galeorhinus galeus* from Walker (2005b) and *Mustelus antarcticus* from Walker (2007); the selected recruitment date is 1 January for both species

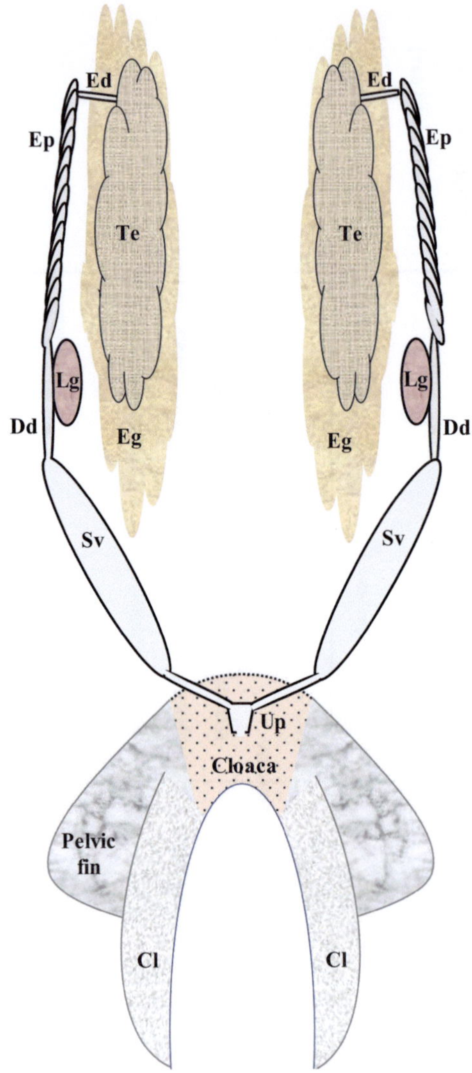

Fig. 10.3 Male chondrichthyan reproductive system. The diagram showing the paired reproductive tracts (pale grey shading) and associated structures (textured colour) is a generalised representation and not of a particular species. *Cl* clasper, *Dd* ductus deferens, *Ed* efferent ductules, *Eg* epigonal gland, *Ep* epididymis, *Lg* Leydig's gland, *Sv* seminal vesicle, *Te* testis, *Up* Urogenital papilla

10.2.2.1 Testis and Spermatogenesis

Paired testes perform the functions of spermatogenesis (germ cell generation) and steroidogenesis (secretion of steroid hormones). Each testis is associated with lymphomyleloid tissue in the anterior portion of the epigonal glands and attached to the dorsal wall by a peritoneal fold forming a mesorchium in the anterior region of the body cavity dorsal to the liver. The right testis is often slightly smaller than the right, and dorsoventral flattening of the testes is more pronounced in batoid species than in the species of other chondrichthyan groups. The testis of immature animals consists of germ cells and somatic cells of separate embryonic origin. In maturing and mature animals, germ cells located in a gonadal zone derived from dividing and differentiating stem cells originating from the yolk stalk migrate to the testis (Engel and Callard 2005; Hamlett 1999).

Spermatogenesis is the transition of a male germ cell through the stages from gonocyte through spermatogonium, spermatocyte, and spermatid to spermatozoon. The process begins in the germinal zone with a spermatogonium (germ cell) and single pre-Sertoli cell (somatic cell) forming a spermatocyst supported in a testicular matrix of connective tissue. Each of the two cells undergoes a series of diploid mitotic divisions where the number of divisions (e.g., 13 for *Squalus acanthias*) and the ratio of daughter spermatogonia to Sertoli cells in a spermatocyst at any step of the divisions (e.g., 1:1 during the first nine divisions for *S. acanthias*) are predictable (Engel and Callard 2005). The germ cells develop not as individual cells, but as isometric clones where the daughter cells, formed by a succession of synchronous divisions of a primary spermatogonium, remain structurally and functionally connected. At the spermatocyte stage with the onset of meiosis, the spermatocysts enlarge markedly in response to the growth of the germ cells and cytoplasm in Sertoli cells. Each of two meiotic haploid divisions doubles the number of spermatocytes, and the Sertoli cells relocate to the periphery of the spermatocyst. During the spermatid stage, the spermatocyst further enlarges as each Sertoli cell forms a large space containing a group of elongating spermatids orientated with the heads peripherally embedded in the basal lamina and tails orientated towards the lumen. The spermatids with emerging flagella are then tightly bundled into mature spermatozoa, which are shed into the interstitial spaces of the testis as the spermatocyst disintegrates before passing through the efferent ductules of the testis into the epididymis. Depending on the time of the year, some of the spermatocysts contain germ cells at various stages of degeneration associated with apoptotic cell death (Engel and Callard 2005; Hamlett 1999). The proportions of spermatocysts at different stages of spermatogenesis vary seasonally for *Sphyrna tiburo* (Sphyrnidae) (Parsons and Grier 1992).

During spermatogenesis, the spermatocysts contain germ cells at different stages of maturation radiating spatially from the germinal zone with the most mature furthest away according to one of three patterns: compound (notably Batoidea), radial (Lamniformes), or linear (mostly Squalomorphii, Galeomorphii, and Holocephali). The compound pattern has spermatocysts formed at each of multiple germinal zones on the dorsal surface radiating in columns across the testis in separate lobes of the testis. The radial pattern is similar to the compound pattern, with the exception that germinal zones occur both on the dorsal surface and at various depths within the testis. The linear pattern has a single germinal zone along the length of the testis opposite the epigonal gland and the sequence of maturing spermatocysts traverses the diameter of the testis (Pratt 1988). Steroid hormones appear to be synthesised by Sertoli cells distributed between spermatocysts (Gelsleichter 2004).

10.2.2.2 Male Reproductive Tracts

Spermatozoa occur throughout the lumena of the paired epididymis, ductus deferens, and seminal vesicles, with secretions from the genital ducts. The process of spermiogenesis, where spermatids mature into motile spermatozoa, occurs in these genital ducts, and sperm bundles form in a matrix as spermatophores (sperm encapsulated) or spermatozeugmata (sperm not encapsulated but tails of peripheral sperm protruding) (Hamlett 1999).

In the holocephalan *Callorhinchus milii* (Chimaeriformes), the epididymis lying laterally to the testis is highly convoluted in mature males with a small luminal diameter continuous with the widening ductus deferens. The ductus deferens then narrows to an isthmus opening into the seminal vesicle, which enlarges during spring when filled with spermatophores and seminal fluid. The ductus deferens lies against the Leydig gland and the seminal vesicle against the paired kidneys (Reardon et al. 2002). The Leydig gland, formed from the anterior portion of the mesonephros, is a branched tubular structure. The epithelium is ciliated, and its secretions contribute to seminal fluid and matrix for spermatophores (Hamlett et al. 2002b). This description is similar for the heterodontid shark *Heterodontus portusjacksoni* and the batoid *Raja eglanteria* (Rajiformes) (Jones et al. 2005).

Ciliated epithelial columnar cells lining the lumen convey the spermatozoa through the genital ducts; only the seminal vesicles have a muscular wall. During copulation and ejaculation, sperm transfer from the seminal vesicle narrowing through the urogenital sinus and urogenital papilla to the dorsal groove on each clasper. Spermatozoa acquire the potential for modest motility while in the terminal regions of the genital ducts, but acquire active, robust motility at ejaculation (Hamlett 1999).

10.2.2.3 Male Copulatory Structures

Male chondrichthyans have external paired claspers extending from the posterior bases of the pelvic fins. As they mature, the claspers calcify, and each articulates freely at the base (basipterygium).

Male sharks possess paired siphon sacs or siphons, expanded epithelia-lined tubular structures surrounded by firm muscle located ventrally under the skin, that open posteriorly through the single apopyle positioned proximally near the clasper grooves. The presence in siphon secretions of serotonin, known to stimulate muscular contraction, suggests it may play a role in the ejaculation of males and uterine contractions in females to move spermatozoa or spermatozeugmata anteriorly, perhaps to the oviducal gland for sperm storage. The siphons are ~12% of body length in *Squalus acanthias* and ~30% in *Mustelus canis* (Jones et al. 2005).

Male batoids lack siphons but possess a nonhomologous structure referred to as the alkaline gland. The gland, located in the posterior region of the body cavity, consists of two separate sacs, each with an efferent duct leading to a genital opening on or near the urinary papilla in the cloaca. In stingrays (Myliobatoidei), but not skates (Rajiformes), each efferent duct merges with the ductus deferens to form a common duct opening on the urinary papilla. Both genital openings and the urinary opening are separate on the urinary papilla. The gland is thought to increase the motility and possibly longevity of spermatozoa (Lacy 2005).

Male skates and stingrays also possess a pair of sub-dermal clasper glands at the base of each pelvic fin and clasper either side of the cloaca. Each clasper gland connects with the apopyle or proximal region of the groove (or in some species the distal end of the groove known as the hypopyle). The function of the clasper gland is uncertain, but the presence of secretory fluids in the tissue mass suggests a potential role in facilitating the transfer of spermatozoa during copulation (Lacy 2005).

In holocephalans, males possess paired pre-pelvic claspers, a tenaculum positioned on the rostrum of the head, and gland beneath the pre-pelvic claspers with a protruding cartilaginous tube thought to produce secretions to facilitate copulation. The females have a pair of pre-pelvic abdominal slits that probably receive the male pre-pelvic claspers. The females also possess a sperm pouch posterior to the cloaca to receive sperm (Jones et al. 2005).

10.2.2.4 Male Maturity

Potential indicators of male maturation include minimum levels of selected steroid hormones in the blood plasma, the condition of the testis, contents of the reproductive tract, and calcification of claspers. The most commonly used indicators for classing males as mature or immature are a macroscopic inspection of claspers (indices $C = 1–3$), testes (indices $T = 1–3$), and seminal vesicles (indices $V = 1–3$) (Table 10.2). Assessment of maturity from the degree of calcification of the claspers is feasible for live animals, whereas assessment from testes development or presence-absence of seminal fluids requires dissection of the animals.

Another indicator of maturity is the stage of spermatogenesis as described in 18 and 7 stages for *Scyliorhinus canicula* (Scyliorhinidae) (Mellinger 1965) and *Sphyrna tiburo* (Parsons and Grier 1992), respectively. Applied to *Mustelus antarcticus* (Walker 2007), the assumed mature condition is when in spermatocysts tightly packed spermatid heads are in contact with Sertoli cells (Stage 17), and mature spermatocytes enter the interstitial spaces of the testis as the spermatocyst disintegrates before passing through the efferent ductules of the testis (Stage 18). The disadvantage of this approach is sectioning testis tissue and histological preparation.

Table 10.2 Indices or staging reproductive condition

Organ or tissue	Index	Description	Maturity assumption
Female			
Ovary	$O = 1$	Largest follicles white and of diameter <2 mm	Immature
	$O = 2$	Largest oocytes yolking and of diameter 2–3 mm	Immature
	$O = 3$	Largest oocytes with yellowish yolk and of diameter >3 mm[a]	Mature
	$O = 4$[b]	Yolked oocytes of diameter >3 mm and corpora atretica readily visible	Mature
Uterus	$U = 1$	Uniformly thin tubular structure	Immature
	$U = 2$	Thin tubular structure partly enlarged posteriorly	Immature
	$U = 3$	Uniformly enlarged tubular structure	Uncertain
	$U = 4$	In utero eggs present without macroscopically visible embryos present	Mature
	$U = 5$	In utero embryos macroscopically visible[a]	Mature
	$U = 6$	Enlarged tubular structure distended[c]	Mature
Male			
Seminal vescicle	$V = 1$	Thin translucent walls and seminal fluids absent	Immature
	$V = 2$	Thickened opaque walls and seminal fluids present	Mature
	$V = 3$	Thickened opaque walls and seminal fluids absent	Mature
Testis	$T = 1$	Thin tissue strip with epigonal gland predominant	Immature
	$T = 2$	Thickened strip with epigonal gland tissue extensive	Immature
	$T = 3$	Enlarged and predominant with epigonal gland tissue negligible	Mature
Clasper[b]	$C = 1$	Pliable with no calcification	Immature
	$C = 2$	Partly calcified	Immature
	$C = 3$	Rigid and fully calcified	Mature
Spermatogenesis	$S = 1$	Microscopically visible spermatocytes at Stage <17 (Mellinger 1965)	Immature
	$S = 2$	Microscopically visible spermatocytes at Stages 17–18 (Mellinger 1965)	Mature

Criteria for classing mature and immature condition and uterus condition used together with time of year for classing maternal and non-maternal condition; adapted from Walker (2005b; 2007) and Trinnie et al. (2016)

[a]Oocyte diameters adopted for *Galeorhinus galeus* (Walker 2005b) and *Mustelus antarcticus* (2007)

[b]Not applicable to oviparous species (see Table 10.1)

[c]Flaccid with extensive trophonemata and histotroph for the sub-order Myliobatoidei (see Table 10.1; Trinnie et al. 2016)

10.3 Quantitative Reproductive Biology

The stability of the population of a species depends on its birth and death rates, while immigration and emigration rates among subpopulations affect populations in separate regions. Birth rate relates to the number of live births for viviparous species and the number of eggs encapsulated in egg cases oviposited for oviparous species.

10.3.1 Sampling for Reproduction Studies

Quantitative studies of reproduction for population biology require the collection of specimens of the species of interest for dissection. It is best to undertake the dissections and record the mass and sizes of various parts of each animal when fresh to avoid measuring error and loss of mass associated with dehydration, freezing, and rigor mortis.

Essential data required from animals of both sexes include the total length, total body mass, and mass of gonads, and for species landed beheaded and eviscerated in fisheries, partial length and landed carcass mass. Data for females include the diameter of one or more of the largest ovarian follicles, the presence-absence (preferably number) of corpora atretica and corpora lutea in the ovary and, for pregnant animals, the number of eggs and embryos in each uterus. Other data include the length, sex, uterus (left or right), and mass (with and without yolk sac) of each embryo and the mass of each in utero egg, and whether the egg is addled (infertile, discoloured, and crenated surface). Appropriate indices for recording the condition of the ovary, oviducal gland, and uteri from the rapid visual inspection are in Table 10.2. Ovary index (O) represents the size and colour of the follicles (O = 1–4), oviducal gland index (G) the size and shape of the gland (G = 1–3), and uterus index (U) appearance, size, and contents of the uteri (U = 1–6).

Data required for males include the length of one or both claspers (from the basipterygium to the distal end) and indices for recording condition of the testes, seminal vesicles, and claspers. Testis index (T) represents shape, size, and relative predominance of testis tissue to epigonal gland tissue (T = 1–3), seminal vesicle index (V) represents the appearance, width, and presence-absence of seminal fluid (V = 1–3), and clasper index (C) represents appearance and rigidity (C = 1–3) (Table 10.2). Collection and fixation of two or three pieces of testis tissue (4–8 mm thick) removed by transverse section from a testis enable subsequent histological preparation and analysis of spermatogenesis.

10.3.2 Quantitative Methods

Simple mathematical equations with appropriate parameters can represent the relationships between pairs of biological variables in reproductive studies for subsequent application in population models. Fitting an implied equation to available data for a pair of selected variables by an appropriate regression method provides estimates of the parameters.

Determination of appropriate equations relating various reproductive variables depends on whether reproduction is continuous or seasonal and whether the reproductive cycle is synchronous or asynchronous among the animals in the population. A discrete reproductive cycle synchronous among the animals might be clear from the timing of ovulation or parturition for a species with an annual cycle. However, to ensure the cycles are not other than annual, it is also necessary to determine the synchronicity and periodicity of oogenesis and gestation among the females and the seasonality of male reproductive condition.

10.3.2.1 Ovarian Cycle and Growth of Follicles

The period of the ovarian cycle is the time from the start of vitellogenesis of a previtellogenic follicle to enlargement ready for ovulation and defined as the period from completion of ovulation commencing one pregnancy to completion of the next ovulation commencing the subsequent pregnancy. For species with synchronous cycles, scattergrams of the size of the largest follicle against the time of year for each of uterus conditions $U = 3$ and $U = 5$ (Table 10.2) indicate the period of the ovarian cycle.

A scattergram for the $U = 5$ condition females (pregnant with macroscopically visible embryos) can indicate the rate of follicular growth. For $U = 3$ females at any time of the year, the largest follicles cluster about one, two, or three sizes, which provide evidence for periods of ovarian cycles of 1, 2, or 3 years, respectively. Verification of the period requires superimposing a straight line, determined by regression of follicle diameter against the day of the year for $U = 5$ females, onto the scattergram for $U = 3$ females. If the superimposed line passes through a single cluster of the points, then the ovarian cycle is likely to be annual. Alternatively, if a wide scatter of data points occur above the line, then extrapolating the regression line for 2 or 3 years indicates whether the ovarian cycle is 2 or 3 years. The linear relationship between the largest follicle diameter, f, against the day of the year, t, is given by

$$f = a + bt,$$

where a and b are parameters estimated by linear regression from a sample of $U = 5$ females.

Application of this method to viviparous species demonstrates ovarian cycles that are annual for *Trygonoptera imitata* (Trinnie et al. 2009) and *Urolophus paucimaculatus* (Trinnie et al. 2014). Examples of biennial ovarian cycles include *U. bucculentus* (Trinnie et al. 2012) and *Squalus megalops* (Braccini et al. 2006) and of triennial cycles include *Squalus chloroculus* (Rochowski et al. 2015a) and *Galeorhinus galeus* (Walker 2005b). For three *Orectolobus* spp. with triennial reproductive cycles, the method shows that most vitellogenic growth occurs during the third year of the ovarian cycle (Huveneers et al. 2007). For the oviparous species *Heterodontus portusjacksoni*, vitellogenesis of an oocyte is 18 months, and although the ovarian cycle is biennial, multiple cohorts of enlarging follicles enable annual ovulation and hence an annual reproductive cycle (Tovar-Ávila et al. 2007). For

Mustelus antarcticus, the method indicates an annual ovarian cycle in the western region of southern Australia and a biennial cycle in the eastern region (Walker 2007).

Several other species also exhibit different ovarian cycles in separate regions. The ovarian cycle of *M. asterias* is annual in the Mediterranean Sea and biennial in the North Atlantic Ocean (Farrell et al. 2010). The ovarian cycle of *Carcharhinus acronotus* in the West Atlantic Ocean is annual off northern Brazil (Hazin et al. 2002) and biennial off the southeastern USA (Driggers et al. 2004). *Carcharhinus isodon* in the central northern Gulf of Mexico has animals with either annual or biennial ovarian cycles in the one discrete locality (Driggers and Hoffmayer 2009). Similarly, separate animals of *Manta alfredi* (Myliobatoidei) with annual and biennial ovarian cycles occur together off southern Mozambique (Marshall and Bennett 2010).

10.3.2.2 Period of Gestation and Growth of Embryos

The period of gestation for an embryo is the time from ovulation when fertilisation of the ovum occurs in the oviducal gland until parturition for a viviparous species or until hatching for oviparous species. Although there are many stages of embryogenesis, the two periods of macroscopically visible eggs versus embryos need distinguishing for the mothers. This distinction helps to categorise females as maternal or non-maternal, and where the growth of the embryos is synchronous among pregnant females, to represent the growth of embryos graphically. Plots of the mean length of embryos (with standard deviation bars) and values of zero for eggs observed for a sample of pregnant females against the day of the year indicate whether the growth of embryos is seasonal and synchronous among the mothers. Plots of mean embryo wet mass and mean yolk-sac wet mass against mean embryo length indicate ~50, ~100, ~1000, and ~6000–7000% wet mass gain from egg to full-term embryo, respectively, for *Squalus megalops* (Braccini et al. 2007), *Galeorhinus galeus* (Walker 2005b), *Mustelus antarcticus* (Walker 2007), and *Urolophus bucculentus* (Trinnic et al. 2012).

Among oviparous species, incubation periods range 9–12 months for *Hydrolagous colliei* (Chimaeriformes) and three species of *Heterodontus* spp., 5–8 months for five species of *Raja* spp. and ~15 months for *R. marginata* (Rajiformes) (Wourms 1977). From measurements of ovarian follicle diameter and presence of egg capsules in utero for *Heterodontus portusjacksoni* indicate an egg-laying period of ~6.5 months (mid-August–February), hatching period of ~5.5 months (June–mid-October), and an incubation period of ~10 months (Tovar-Ávila et al. 2007). For *Cephaloscyllium laticeps*, mature females in captivity oviposit eggs all year but in the wild most eggs are laid over a period of ~6 months (January–June) coinciding with changes in steroid hormone levels in blood plasma (Awruch et al. 2009). Egg-laying, hatching, and incubation periods for *Callorhinchus milii* vary markedly with temperature in the wild among separated egg capsule sampling sites (Lyon et al. 2011).

Among viviparous species, depending on species, an extended period of eggs in utero, preceding the period of macroscopically visible embryos, is diapause (Simpfendorfer 1992; Waltrick et al. 2012; Waltrick et al. 2014). For chondrichthyan species, diapause is mostly obligate diapause where embryogenesis is slowed or delayed ensuring that parturition occurs at the time when seasonal conditions are favourable for the survival of the neonates. Alternatively, facultative diapause is where embryogenesis can be delayed for more than a year until conditions are suitable, as suggested for *Urolophus cruciatus* (Trinnie et al. 2016). *Urolophus cruciatus* has a short (4–6 months) and highly synchronous period of embryonic growth where only a small proportion of the pregnant females observed have macroscopically visible embryos while most have eggs in utero all year. For *U. cruciatus*, it appears that the period of diapause can be several years depending on favourable conditions at an interannual timescale.

For most viviparous species, the period of macroscopically visible embryos is less than 12 months, even for species exhibiting biennial or triennial reproductive cycles. For example, *Galeorhinus galeus* (Walker 2005b) and *Orectolobus* spp. (Huveneers et al. 2007) have triennial reproductive cycles but periods of 10–11 months for the growth of embryos. Similarly, for *Mustelus antarcticus* (Walker 2007) and *M. manazo* (Yamaguchi et al. 2000), which have either annual or biennial reproductive cycles in separate regions, have periods of growth of embryos of 12 or fewer months. Species with unusually short periods of growth of embryos include *Trygonoptera imitata* (5–7 months) and *T. personata* (5 months) (Trinnie et al. 2016). Examples exceeding 12 months are *Urolophus bucculentus* (14–19 months) (Trinnie et al. 2012) and *Squalus acanthias* (Hanchet 1988), *S. choroculus* (Rochowski et al. 2015a), and *S. megalops* (Braccini et al. 2006) at ~2 years.

10.3.2.3 Synchronicity and Periodicity of Reproductive Cycle

The timing of specific reproductive events in a maturing or mature animal relates to its overall reproductive cycle. The female cycle starts with the growth of follicles by vitellogenesis followed by ovulation, fertilisation, and encapsulation for each of a series of sequentially ovulated oocytes as ova. For species of single oviparity, paired fertilised eggs (zygotes) encapsulated in tough egg cases and oviposited over a protracted egg-laying season undergo embryogenesis before hatching from the egg cases in habitat suitable for neonates. For species of retained oviparity, embryogenesis begins in utero and finishes in the egg cases during the period between oviposition and hatching. For viviparous species, the zygotes encapsulated in egg membranes pass through a period of eggs in utero, then a period of embryonic growth (referred to here as the period of macroscopically visible embryos), and finally a short period of parturition. The male reproductive cycle is annual for most species and relates to the female cycle to ensure adequate reserves of spermatozoa in preparation for mating before female ovulation begins. The period of female reproductive cycles varies widely among species from 3 times a year to 3 years or possibly longer.

The reproductive cycles are either seasonal and highly synchronised among the females of a population or nonseasonal and asynchronous depending on species. Determination of the period of the cycle and timing of specific events from year-round observations of animals in the wild population is more straightforward for species with synchronous cycles than for species with asynchronous cycles. Seasonal synchrony is evident for most species with demersal or pelagic lifestyles in waters of depth <200 m, but for many of the 46% of the world's chondrichthyan species inhabiting the waters of the continental slopes and abyss (i.e., ≥200 m depth), the cycles are unknown (Kyne and Simpfendorfer 2010) or appear asynchronous. Asynchronous cycles are evident for squaliform sharks in both deep and waters <200 m, e.g., *Squalus acanthias* (Collonello et al. 2016; Hanchet 1988), *S. chloroculus* (Rochowski et al. 2015a), *S. megalops* (Braccini et al. 2006), *Deania calcea* (Irvine et al. 2012; Rochowski et al. 2015b), and *D. quadrispinosa* (Irvine et al. 2012).

Synchronous reproductive cycles are evident in oviparous species from observations of captive animals and animals sampled in waters <200 m for reproductive condition based on reproductive indices (Table 10.2) or reproductive hormone levels in the blood plasma (Awruch et al. 2009; Barnett et al. 2009). Many of these species oviposit egg cases throughout the year, whereas others have distinct annual egg-laying periods. For example, *Callorhinchus milii* and *Hydrolagous colliei* (Chimaeriformes) have egg-laying seasons of ~4 months (Bell 2012) and 6–8 months (Barnett et al. 2009), respectively. *Heterodontus portusjacksoni* has a ~6-month season (Tovar-Ávila et al. 2007). Several species of the family Scyliorhinidae exhibit discrete seasons, whereas others produce egg cases throughout the year. *Cephaloscyllium laticeps* (Scyliorhinidae) oviposits egg cases at all times of the year but produces most within a 6-month period (Awruch et al. 2009).

The reproductive cycles in viviparous species inhabiting waters of depth <200 m are mostly synchronous among maturing and mature animals in a population, but the period of the cycle is highly variable among species. Differences in the duration of the female reproductive cycle are particularly evident among species of the family Urolophidae (Myliobatoidei) which all exhibit yolk-sac viviparity with lipid histotroph. The shortest reported duration of a cycle is triannual for *Urotrygon rogersi* with three short periods of peak parturition each year (Mejía-Falla et al. 2012). An example of a biannual reproductive cycle is *Urobatis jamaicensis*, which has two seasonal ovulation and parturition periods each year (Fahey et al. 2007). Annual cycles occur for most investigated species of the genera *Trigonoptera* and *Urolophus*, but a biennial cycle is evident for *U. bucculentus* and possibly a lengthier and variable duration for *U. cruciatus* (Trinnie et al. 2016). Similarly, triannual, biannual, and annual reproductive cycles occur for the family Dasyatidae (Myliobatoidei) (Ramírez-Mosqueda et al. 2012). Two periods of seasonal ovulation in *Rhinobatis leucorhynchus* (Rhinopristiformes) is consistent with a biannual cycle (Romero-Caicedo and Carrera-Fernández 2015).

The period of the reproductive cycle is the period from completion of ovulation in one season to completion of ovulation in the following season. Alternatively, it can be defined as the period from completion of parturition (or oviposition) in one season to completion of parturition (or oviposition) the following season. Determination of the period of the reproductive cycle is supported by determining the period of the ovarian cycle and the period of the pregnancy cycle. Although the reproductive cycle for species where the cycle is nonsynchronous among individuals and hypothesised, there are advantages supporting field observations of reproductive condition with analyses of blood plasma levels of steroid hormones. Figure 10.4 illustrates an example of each of an annual and biennial reproductive cycle.

10.3.2.4 Maturity-at-Size and Maternity-at-Size

The form of the equation is the same for expressing three relationships between (1) the proportion of animals mature in the female population and length, (2) the proportion of animals mature in the male population and length, and (3) the proportion of maternal animals in the female population and length. Each relationship between the proportion in a population, P, and length, l, is given by the probit curve

$$P = \frac{c}{\left(1 + e^{-(a+bl)}\right)}$$

where each of maturity and maternity is a dichotomous variable with a value of 0 or 1, and a, b, and c are parameters. In preparation for estimating the three parameters by probit regression for each of the three relationships, each animal in mature female condition (or mature male condition or maternal condition) has a value of 1, whereas each animal in immature condition (or non-maternal condition) has a value of 0 (Walker 2005b).

For a species with annual parturition or oviposition (i.e., maternity is annual), then $c = 1.00$. If parturition or oviposition is biennial or triennial then $c = 0.50$ or $c = 0.33$, respectively. For probit regression analysis, c has a value of 1.00, which is appropriate for the annual maternity and maturity equations. However, if $c < 1.00$, then determination of the parameters for the maternity equation requires additional steps. (1) Calculate the proportion of animals in maternal condition for each length class (e.g., 100 mm). (2) Divide the proportion in each length class by c but if the value of the proportion divided by c exceeds 1.00, then adjust the value to 1.00. (3) Undertake the probit regression. (4) Multiply by c the value of the proportion in each length class determined by probit regression (Walker 2005b). Species with biannual or triannual maternity, respectively, have two or three recruitment events each year where $c = 1.00$ for each event.

Reformulation of the equations determined by probit regression provides more biologically meaningful parameters as

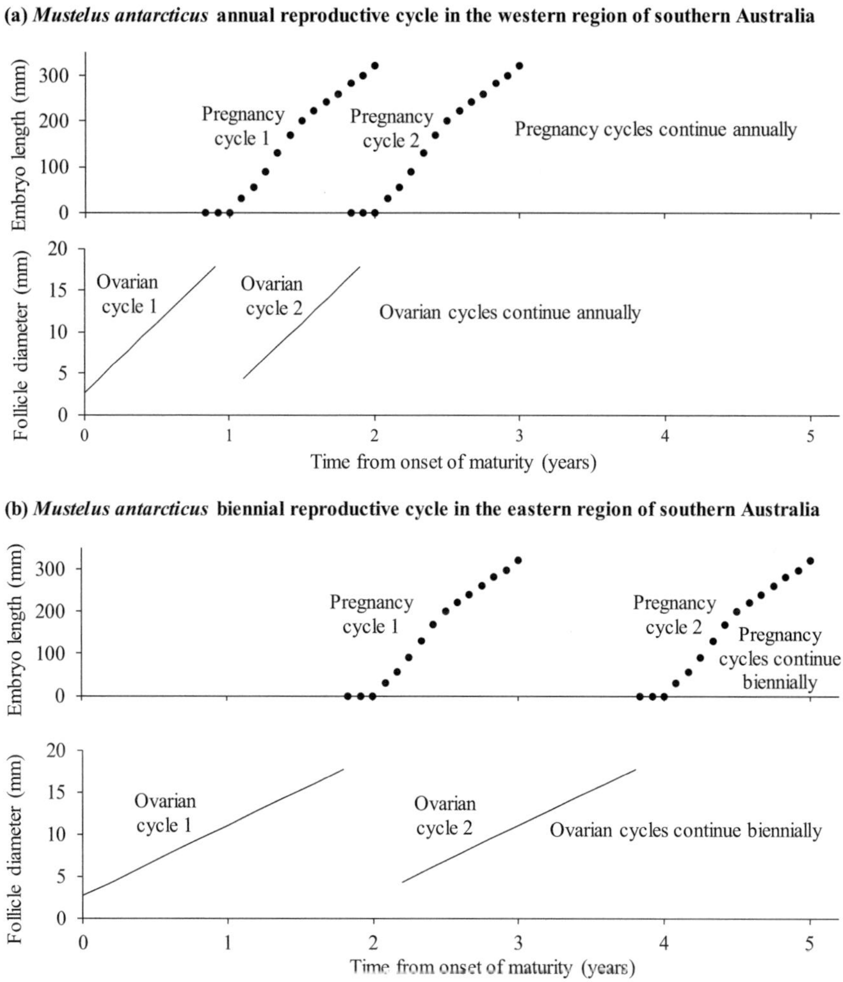

Fig. 10.4 Examples of annual (**a**) and biennial (**b**) reproductive cycles. The reproductive cycle is the period from parturition at the end of one pregnancy cycle to the end of the following pregnancy cycle or the period from ovulation at the end of one ovarian cycle to the end of the following ovarian cycle. The displayed information for the first two reproductive cycles is sourced for *Mustelus antarcticus* from Walker (2007). (**a**) *Mustelus antarcticus* annual reproductive cycle in the western region of southern Australia. (**b**) *Mustelus antarcticus* biennial reproductive cycle in the eastern region of southern Australia

$$P = P_{\max}\left(1+e^{-\ln(19)\left(\frac{l-l_{50}}{l_{95}-l_{50}}\right)}\right)^{-1},$$

where P_{max} is the maximum proportion of animals for the mature female condition, mature male condition, or maternal condition (equivalent to c), and l_{50} and l_{95} are the lengths with the proportions of 50 and 95% of P_{max} at those lengths, respectively (Punt and Walker 1998; Walker 2005b).

See Fig. 10.5 for examples of plots of maturity-at-length and maternity-at-length relationships (ogives). Differences occur in the female maturity and maternity ogives between the western (Fig. 10.5a) and eastern (Fig. 10.5b) regions of southern Australia for *M. antarcticus*. Maturity and maternity are reached at smaller sizes in the western than the cooler eastern region, and maternity is annual in the western region and biennial in the eastern region (Walker 2007). The maternity cycle for *Galeorhinus galeus* in southern Australia is triennial (Walker 2005b) (Fig. 10.5c).

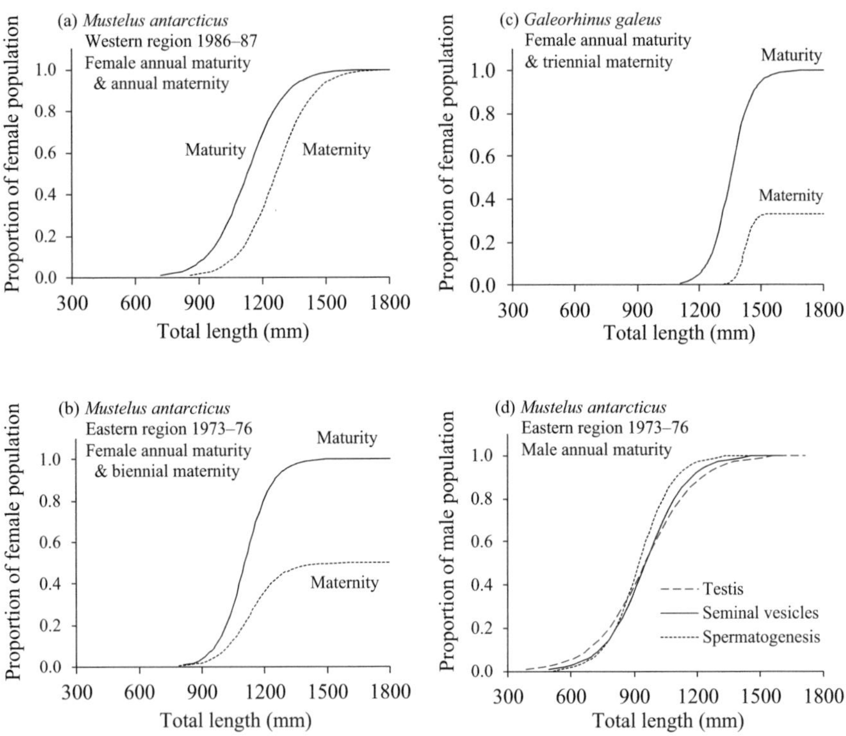

Fig. 10.5 Examples of female (**a**, **b**, and **c**) and male (**d**) annual maturity ogives, annual (**b**) and biennial (**c**) maternity ogives for *Mustelus antarcticus*, and a triennial (**d**) maternity ogive for *Galeorhinus galeus*. The displayed information for *M. antarcticus* and *G. galeus* is sourced from Walker (2007) and Walker (2005b), respectively. In southern Australia, the Eastern Region is east of Kangaroo I and the Western Region is to the west. Years 1973–1976 in the east and 1986–1987 are presented because targeting of *M. antarcticus* increased after those periods to the extent of biasing of maturity and maternity ogives by the effects of length-selective fishing mortality associated with the use of gillnets by the fishing industry. (**a**) *Mustelus antarcticus*: Western region 1986–1987, female annual maturity and annual maternity. (**b**) *Mustelus antarcticus*: Eastern region 1973–1976, female annual maturity and biennial maternity. (**c**) *Galeorhinus galeus*: Female annual maturity and triennial maternity. (**d**) *Mustelus antarcticus*: Eastern region 1973–1976, male annual maturity

Male maturity based on different methods produces slightly different ogives and parameter estimates for *Mustelus antarcticus* (Fig. 10.5d).

10.3.2.5 Litter Size

The relationship between the number of macroscopically visible embryos in utero for viviparous species or number egg cases oviposited for oviparous species in a season, p, and maternal length, l, is often linear given by

$$p = a + bl,$$

where parameters a and b are estimated by linear regression. This relationship occurs for the viviparous species of *Squalus acanthias* (Jones and Ugland 2001), *Mustelus canis* (Conrath and Musick 2002), and *Galeorhinus galeus* (Walker 2005b) and batoids *Trygonoptera imitata* (Trinnie et al. 2009) and *Urolophus bucculentus* (Trinnie et al. 2012). As presented for *M. antarcticus* (Lenanton et al. 1990; Walker 2007), an example of a curvilinear relationship between p and maternal l is given by

$$p = ce^{a+bl},$$

where a and b are parameters estimated by linear regression after reformulation to $\ln(p) = a + bl$, and c is a constant to correct for bias caused by the natural logarithmic transformation of p for the regression (Beauchamp and Olson 1973).

For oviparous chimaeriform species, examples of estimates of the number of egg cases laid annually range 19.5–28.9 for *Hydrolagus colliei* based on the number of eggs posited weekly over 6–8 months by two captive females (Barnett et al. 2009), and 27 for *Harriotta raleighana* and 31 for *Rhinochimaera pacifica* based on the number of enlarged ovarian follicles observed in two animals (Finucci et al. 2017). A study of *Callorhinchus milii* with observations of 92 egg capsules posited by the equivalent of 4.7 captive mature females for the egg-laying season gives an annual estimate of 19.5 eggs oviposited per mature female. This estimate is consistent with the number of observed enlarged ovarian follicles ranging 16–24 early during the egg-laying season, where the number of enlarged follicles correlates positively with maternal length. Diver field surveys indicate that egg cases with dead embryos and holes bored by invertebrates vary from ~30 to >95% in different years (Bell 2012). The number of pre-ovulatory follicles as an indicator of the number of eggs laid annually for *Heterodontus portusjacksoni* range 6–20 (mean = 14, standard deviation = 3.71, n = 29) with a positive correlation with maternal length (Tovar-Ávila et al. 2007).

For viviparous species, reports of the number of neonates produced at each pregnancy range from 2 (one in each uterus) for the adelophagous species *Carcharias taurus* (Lamniformes) to >300 for *Rhincodon typus* (Rhincodontidae) (Joung et al. 1996). Typical litter-size ranges widely for *Mustelus antarcticus* (1–57) (Walker 2007), *Alopias* spp. (1–7), *Prionace glauca* (1–135), *Carcharhinus falciformis* (2–16) (Carcharhiniformes), *Pteroplatytrygon violacea* (2–13) (Snelson et al.

2008), and for *Urolophus paucimaculatus* (1–6) (Myliobatoidei) (Trinnie et al. 2014).

10.3.2.6 Embryo Sex Ratio and Distribution Between Uteri

For viviparous pregnant females with macroscopically visible embryos, it is usual for the sex ratio and the ratio of the number of embryos in the left uterus to the number in the right uterus not to differ significantly from 1:1. Examples of this pattern include *Deania calcea* (Rochowski et al. 2015b), *Squalus chloroculus* (Rochowski et al. 2015a), *Orectolobus ornatus*, *O. maculatus* (Huveneers et al. 2007), *Triakis semifasciata* (Ebert and Ebert 2005), *Galeorhinus galeus* (Walker 2005b), and *Mustelus antarcticus* (Walker 2007). However, for *M. antarcticus*, the number of addled eggs in the left uterus is significantly higher than the number in the right uterus. For pregnant females with eggs only in utero, the ratio of the number of eggs in the left to right uteri does not differ significantly from 1:1 for *G. galeus* or *M. antarcticus*. A significant departure from the 1:1 ratio in the number of embryos and eggs combined between the uteri occurs, but not in the sex ratio, for *Squalus megalops* (Braccini et al. 2006). Several studies report the presence of addled in utero eggs, which represent a minimum occurrence of infertile eggs; e.g., 4.1% for *Squalus acanthias* (Hanchet 1988), 3.4% for *G. galeus* (Walker 2005b), and 6.8% for *M. antarcticus* (Walker 2007).

10.3.2.7 Total Body Mass-at-Size

The relationship between total body mass, w, and length, l, is commonly given by the power curve

$$w = acl^b,$$

for sharks (Olsen 1954) and bony fishes (Ricker 1958) without the constant c, where a and b are parameters determined by linear regression of $\ln(w)$ against $\ln(l)$, and c is a factor correcting for biases caused by natural logarithmic transformation (Beauchamp and Olson 1973; Walker 2005b).

Sexual dimorphism is common for sharks, rays, and chimaeras where females attain a longer maximum length than males and where for some species the mean body mass of the females exceeds that of males with increasing length. For some species, this difference in body mass between the sexes can be explained by additional body mass from pregnancy. For example, four species of Myliobatoidei of the genus *Urolophus*—*U. bucculentus* (Trinnie et al. 2012), *U. paucimaculatus* (Trinnie et al. 2014), *U. viridis* (Trinnie et al. 2015), and *U. cruciatus* (Trinnie et al. 2016)—exhibit no differences in body mass at any length between females and males yet a fifth species *Trygonoptera imitata* does exhibit a difference (Trinnie et al. 2009). However, for some of these species, body mass is higher for pregnant females than for non-pregnant females and males. For *Mustelus antarcticus* (Walker 2007), *Galeorhinus galeus* (Walker 2005b), *Squalus chloroculus* (Rochowski et al. 2015a), and *Heterodontus portusjacksoni* (Tovar-Ávila et al. 2007) body mass is higher in females than males at any length, but body mass is not affected by pregnancy. Body mass of *H. portusjacksoni*, however, is affected by region.

References

Awruch CA (2013) Reproductive endocrinology in chondrichthyans: the present and the future. Gen Comp Endocrinol 192:60–70

Awruch CA, Pankhurst NW, Frusher SD, Stevens JD (2009) Reproductive seasonality and embryo development in the draughtboard shark *Cephaloscyllium laticeps*. Mar Freshw Res 60:1265–1272

Barnett LAK, Earley RL, Ebert DA, Cailliet GM (2009) Maturity, fecundity, and reproductive cycle of the spotted ratfish, *Hydrolagus colliei*. Mar Biol 156:301–316

Beauchamp JJ, Olson JS (1973) Corrections for bias in regression estimates after logarithmic transformation. Ecology 54:1403–1407

Bell JD (2012) Reproduction and ageing of Australian holocephalans and white-fin swell shark. Ph.D. Thesis, Deakin University, Warrnambool, Victoria, Australia

Bouyoucos IA, Weideli OC, Planes S, Simpfendorfer CA, Rummer JL (2018) Dead tired: evaluating the physiological status and survival of neonatal reef sharks under stress. Conser Physiol 6:coy053. https://doi.org/10.1093/conphy/coy053

Braccini JM, Gillanders BM, Walker TI (2006) Determining reproductive parameters for population assessments of chondrichthyan species with asynchronous ovulation and parturition: piked spurdog (*Squalus megalops*) as a case study. Mar Freshw Res 57:105–119

Braccini JM, Hamlett WC, Gillanders BM, Walker TI (2007) Embryo development and maternal–embryo nutritional relationships of piked spurdog (*Squalus megalops*): maternal contribution or embryonic independence? Mar Biol 150:727–739

Carrier JC, Pratt HL, Castro JL (2004) Reproductive biology of elasmobranchs. In: Carrier JC, Musick JA, Heithaus MR (eds) Biology of sharks and their relatives. CRC Press, Boca Raton, pp 269–286

Chin A, Kyne PM, Walker TI, McAuley RB (2010) An integrated risk assessment for climate change: analysing the vulnerability of sharks and rays on Australia's Great Barrier Reef. Glob Chang Biol 16:1936–1953

Collonello JH, Cortéz F, Belleggia M, Massa AM (2016) Reproduction and population parameters of spiny dogfish *Squalus acanthias* in the south-western Atlantic Ocean. J Fish Biol 88:1758–1775

Conrath CL, Musick JA (2002) Reproductive biology of the smooth dogfish, *Mustelus canis*, in the northwest Atlantic Ocean. Environ Biol Fishe 64:367–377

Driggers WB III, Hoffmayer ER (2009) Variability in the reproductive cycle of finetooth sharks, *Carcharhinus isodon*, in the northern Gulf of Mexico. Copeia 2009:390–393

Driggers WB III, Oakley DA, Ulrich G, Carlson JK, Cullum BJ, Dean JM (2004) Reproductive biology of *Carcharhinus acronotus* in coastal waters of South Carolina. J Fish Biol 64:1540–1551

Dulvy NK, Fowler SL, Musick JA, Cavanagh RD, Kyne PM, Harrison LR, Carlson JK, Davidson LNK, Fordham S, Francis MP, Pollock CM, Simpfendorfer CA, Burgess GH, Carpenter KE, Compagno LJV, Ebert DA, Gibson C, Heupel MR, Livingstone SR, Sanciangco JC, Stevens JD, Valenti S, White WT (2014) Extinction risk and conservation of the world's sharks and rays. eLife 3:e00590

Dulvy NK, Simpfendorfer CA, Davidson LNK, Fordham S, Bräutigam A, Sant G, Welch DJ (2017) Challenges and priorities in shark and ray conservations. Curr Biol 27:R565–R572

Ebert DA, Ebert TB (2005) Reproduction, diet and habitat use of leopard sharks, *Triakis semifasciata* (Girard), in Humboldt Bay, California, USA. Mar Freshw Res 56:1089–1098

Engel KB, Callard GV (2005) The testis and spermatogenesis. In: Hamlett WC (ed) Reproductive biology and phylogeny of chondrichthyes: sharks, batoids and chimaeras, vol 3. Science Publishers, Enfield, pp 171–200

Fahey DP, Spieler RE, Hamlett WC (2007) Preliminary observations of the reproductive cycle and uterine fecundity of the yellow stingray Urobatis jamaicensis (Elasmobranchii: Myliobatiformes: Urolophidae) in southeast Florida, U.S.A. Raffles Bull Zool Supplement Number 14:131–139

Farrell ED, Mariani S, Clark MW (2010) Reproductive biology of the starry smooth-hound shark *Mustelus asterias*: geographic variation and implications for sustainable exploitation. J Fish Biol 77:1505–1525

Finucci B, Dunn MR, Jones EG, Anderson J (2017) Reproductive biology of the two deep-sea chimaerids, longnose spookfish (*Harriotta raleighana*) and pacific spookfish (*Rhinochimaera pacifica*). Deep-Sea Res 1 Oceanogr Res Pap 120:76–87

Frick LH, Reina RD, Walker TI (2009) The physiological response of Port Jackson sharks and Australian swellsharks to sedation, gill-net capture, and repeated sampling in captivity. N Am J Fish Manag 29:127–139

Frick LH, Reina RD, Walker TI (2010a) Stress related physiological changes and post-release survival of Port Jackson sharks (*Heterodontus portusjacksoni*) and gummy sharks (*Mustelus antarcticus*) following gill-net and longline capture in captivity. J Exp Mar Biol Ecol 385:29–37

Frick LH, Walker TI, Reina RD (2010b) Trawl capture of Port Jackson sharks, *Heterodontus portusjacksoni*, and gummy sharks, *Mustelus antarcticus*, in a controlled setting: effects of tow duration, air exposure and crowding. Fish Res 106:344–350

Frick LH, Walker TI, Reina RD (2012) Immediate and delayed effects of gill-net capture on acid–base balance and intramuscular lactate concentration of gummy sharks, *Mustelus antarcticus*. Comp Biochem Physiol A 162:88–93

Gelsleichter J (2004) Hormonal regulation of elasmobranch physiology. In: Carrier JC, Musick JA, Heithaus MR (eds) Biology of sharks and their relatives. CRC Press, Boca Raton, pp 287–323

Gilmore RG, Putz O, Dodrill JW (2005) Oophagy, intrauterine cannibalism and reproductive strategy in lamnoid sharks. In: Hamlett WC (ed) Reproductive biology and phylogeny of chondrichthyes: sharks, batoids and chimaeras, vol 3. Science Publishers, Enfield, pp 435–462

Girard M, Du Buit M-H (1999) Reproductive biology of two deep-water sharks from the British Isles, *Centroscymnus coelolepis* and *Centrophorus squamosus* (Chondrichthyes: Squalidae). J Mar Biol Assoc U K 79:923–931

Guallart J, Vincent JJ (2001) Changes in composition during embryo development of the gulper shark, *Centrophorus granulosus* (Elasmobranchii, *Centrophoridae*): an assessment of maternal-embryonic nutritional relationships. Environ Biol Fish 61:135–150

Guida L, Walker TI, Reina RD (2016) The adenylate energy charge as a new and useful indicator of capture stress in chondrichthyans. J Comp Physiol B 186:193–204

Guida L, Auruck C, Walker TI, Reina RD (2017) Prenatal stress from trawl capture affects mothers and neonates: a case study using the southern fiddler ray (*Trygonorrhina dumerilii*). Sci Rep 7:46300. https://doi.org/10.1038/srep46300. 10 pp

Hamlett WC (1999) Male reproductive system. In: Hamlett WC (ed) Sharks, skates and rays the biology of elasmobranch fishes. The Johns Hopkins University Press, Baltimore, pp 444–470

Hamlett WC, Koob TJ (1999) Female reproductive system. In: Hamlett WC (ed) Sharks, skates and rays the biology of elasmobranch fishes. The Johns Hopkins University Press, Baltimore, pp 398–443

Hamlett WC, Musick JA, Hysell CK, Sever DM (2002a) Uterine epithelial-sperm interaction, endometrial cycle and sperm storage in the terminal zone of the oviducal gland in the placental smoothhound, *Mustelus canis*. J Exp Zool 292:129–144

Hamlett WC, Reardon M, Clark J, Walker TI (2002b) Ultrastructure of sperm storage and male genital ducts in a male holocephalan, the elephant fish, *Callorhynchus milii*. J Exp Zool 292:111–128

Hamlett WC, Jones CJP, Paulesu LR (2005a) Placentatrophy in sharks. In: Hamlett WC (ed) Reproductive biology and phylogeny of chondrichthyes: sharks, batoids and chimaeras, vol 3. Science Publishers, Enfield, pp 463–502

Hamlett WC, Knight DP, Pereira FTV, Steele J, Sever DM (2005b) Oviducal glands in chondrichthyans. In: Hamlett WC (ed) Reproductive biology and phylogeny of chondrichthyes: sharks, batoids and chimaeras, vol 3. Science Publishers, Enfield, pp 301–335

Hamlett WC, Kormanik G, Storrie MT, Stevens B, Walker TI (2005c) Chondrichthyan parity, lecithotrophy and matrotrophy. In: Hamlett WC (ed) Reproductive biology and phylogeny of chondrichthyes: sharks, batoids and chimaeras, vol 3. Science Publishers, Enfield, pp 395–434

Hanchet S (1988) Reproductive biology of *Squalus acanthias* from the east coast, South Island, New Zealand. N Z J Mar Freshw Res 22:537–549

Hazin F, Oliveira PG, Broadhurst MK (2002) Reproduction of the blacknose shark (*Carcharhinus acronotus*) in coastal waters off northeastern Brazil. Fish Bull 100(1):143–148

Heard M, Van Rijn JA, Reina RD, Huveneers C (2014) Impacts of crowding, trawl duration and air exposure on the physiology of stingarees (family: Urolophidae). Conser Physiol 2:cou040. https://doi.org/10.1093/conphy/cou040. 14 pp

Hobday AJ, Smith ADM, Stobutzki IC, Bulman C, Daley R, Dambacher J, Deng R, Dowdney JM, Fuller M, Furlani D, Griffiths SP, Johnson D, Kenyon RK, Knuckey IA, Ling SD, Pitcher R, Sainsbury KJ, Sporcic M, Smith T, Turnball C, Walker TI, Wayte SE, Webb H, Williams A, Wise BS, Zhou S (2011) Ecological risk assessment for the effects of fishing. Fish Res 108:372–384

Huveneers C, Walker TI, Otway NM, Harcourt RG (2007) Reproductive synchrony of three sympatric species of wobbegong shark (genus *Orectolobus*) in New South Wales, Australia: reproductive parameter estimates necessary for population modelling. Mar Freshw Res 58:765–777

Irvine SB, Daley RK, Graham KJ, Stevens JD (2012) Biological vulnerability of two exploited sharks of the genus *Deania* (Centrophoridae). J Fish Biol 80:1181–1206

Jones TS, Ugland KI (2001) Reproduction of female spiny dogfish, *Squalus acanthias*, in the Oslofjord. Fish Bull 99(4):685–690

Jones CJP, Walker TI, Bell JD, Reardon MB, Ambrosio CE, Almeida A, Hamlett WC (2005) Male genital ducts and copulatory appendages in chondrichthyans. In: Hamlett WC (ed) Reproductive biology and phylogeny of chondrichthyes: sharks, batoids and chimaeras, vol 3. Science Publishers, Enfield, pp 361–393

Joung S-J, Chen C-T, Clark E, Uchida S, Huang WYP (1996) The whale shark, *Rhincodon typus*, is a livebearer: 300 embryos found in one 'megamamma' supreme. Environ Biol Fish 46:219–223

Kyne PM, Simpfendorfer CA (2010) Deepwater chondrichthyans. In: Carrier JC, Musick JA, Heithaus MR (eds) Sharks and their relatives II. CRC Press LLC, Boca Raton, pp 37–114

Lacy ER (2005) Alkaline glands and clasper glands of batoids. In: Hamlett WC (ed) Reproductive biology and phylogeny of chondrichthyes: sharks, batoids and chimaeras, vol 3. Science Publishers, Enfield, pp 337–360

Lenanton RCJ, Heald DI, Platell M, Cliff M, Shaw J (1990) Aspects of the reproductive biology of the gummy shark, *Mustelus antarcticus* Günther, from waters off the south coast of Western Australia. Aust J Mar Freshwat Res 41:807–822

Lyon WS, Francis RICC, Francis MP (2011) Calculating incubation times and hatching dates for embryonic elephantfish (*Callorhinchus milii*). N Z J Mar Freshw Res 45:59–72

Marshall AD, Bennett MB (2010) Reproductive ecology of the reef manta ray *Manta alfredi* in southern Mozambique. J Fish Biol 77:169–190

Martins CL (2017) Stress, survival and movement following fishing gear capture in chondrichthyan species. Ph. D. Thesis, Monash University, Clayton, Victoria, Australia

Martins CL, Walker TI, Reina RD (2018) Stress-related physiological changes and post-release survival of elephant fish (*Callorhinchus milii*) after longlining, gillnetting, angling and handling in a controlled setting. Fish Res 204:116–124

Mejía-Falla PA, Navia AF, Cortéz E (2012) Reproductive variables of *Urotrygon rogersi* (Batoidea: Urotrygonidae): a species with a triannual reproductive cycle in the eastern tropical Pacific Ocean. J Fish Biol 80:1246–1266

Mellinger J (1965) Stades de la spermatogenese cher *Scyliorhinus caniculus* (L.): description, donnees histochimigues, variations normales et experimentales. Z Zellforsch 67:653–673

Musick JA, Ellis JK (2005) Reproductive evolution of chondrichthyans. In: Hamlett WC (ed) Reproductive biology and phylogeny of chondrichthyes: sharks, batoids and chimaeras, vol 3. Science Publishers, Enfield, pp 45–79

Olsen AM (1954) The biology, migration, and growth rate of the school shark, *Galeorhinus australis* (Macleay) (Carcharhinidae) in south-eastern Australian waters. Aust J Mar Freshwat Res 5:353–410

Parsons GR, Grier HJ (1992) Seasonal changes in shark testicular structure and spermatogenesis. J Exp Zool 261:173–184

Pratt HL (1988) Elasmobranch gonad structure: a description and survey. Copeia 3:719–729

Pribac F, Punt AE, Walker TI, Taylor BL (2005) Using length, age and tagging data in a stock assessment of a length selective fishery for gummy shark (*Mustelus antarcticus*). J Northwest Atl Fish Sci 35:267–290

Punt AE, Walker TI (1998) Stock assessment and risk analysis for the school shark (*Galeorhinus galeus*) off southern Australia. Mar Freshw Res 49:719–731

Punt AE, Pribac F, Walker TI, Taylor BL, Prince JD (2000) Stock assessment of school shark *Galeorhinus galeus* based on a spatially-explicit population dynamics model. Mar Freshw Res 51:205–220

Ramírez-Mosqueda E, Pérez-Jiménez JC, Mendoza-Carranza M (2012) Reproductive parameters of the southern stingray *Dasyatis americana* in southern Gulf of Mexico. Lat Am J Aquat Res 40:335–344

Reardon MB, Walker TI, Hamlett WC (2002) Microanatomy of spermatophore formation and male genital ducts in the holocephalan, *Callorhynchus milii*. Mar Freshw Res 53:591–600

Ricker WE (1958) Handbook of computations for biological statistics of fish populations. Bull Fish Res Board Can 119:1–191

Rochowski BEA, Graham KJ, Day RW, Walker TI (2015a) Reproductive biology of the greeneye spurdog (*Squalus chloroculus*) from the continental slope of southern Australia. J Fish Biol 86:734–754. https://doi.org/10.1111/jfb.12593

Rochowski BEA, Walker TI, Day RW (2015b) Geographical variability in life-history traits of a midslope dogfish: the brier shark (*Deania calcea*). J Fish Biol 87:728–747. https://doi.org/10.1111/jfb.12756

Romero-Caicedo AF, Carrera-Fernández M (2015) Reproduction of the whitesnout guitarfish *Rhinobatos leucorhynchus* in the Ecuadorian Pacific Ocean. J Fish Biol 87:1434–1448

Simpfendorfer CA (1992) Reproductive strategy of the Australian sharpnose shark, *Rhizoprionodon taylori* (Elasmobranchii: Carcharhinidae), from Cleveland Bay, Northern Queensland. Aust J Mar Freshwat Res 43:67–75

Skomal GB, Mandelman JW (2012) The physiological response to anthropogenic stressors in marine elasmobranch fishes: a review with a focus on secondary response. Comp Biochem Physiol A 162:146–155

Smith RM, Walker TI, Hamlett WC (2004) Microscopic organization of the oviducal gland of the holocephalan elephant fish, *Callorhynchus milii*. Mar Freshw Res 55:155–164

Snelson FF, Burgess GH, Roman BL (2008) The reproductive biology of pelagic elasmobranchs. In: Camhi MD, Pikitch EK, Babcock EA (eds) Sharks of the open ocean: biology, fisheries and conservation. Blackwell Publishing, Oxford, pp 369–392

Storrie MT (2004) Microscopic modifications of the reproductive tissues of the female gummy shark (*Mustelus antarcticus*) throughout maturation and gestation. Ph.D. Thesis, Deakin University, Warrnambool, Victoria, Australia

Storrie MT, Walker TI, Laurenson LJ, Hamlett WC (2008) Microscopic organisation of the sperm storage tubules in the oviducal gland of the female gummy shark (*Mustelus antarcticus*), with observations on sperm distribution and storage. J Morphol 269:1308–1324

Tovar-Ávila J, Walker TI, Day RW (2007) Reproduction of *Heterodontus portusjacksoni* in Victoria, Australia: evidence of two populations and reproductive parameters for the eastern population. Mar Freshw Res 58:956–965

Trinnie FI, Walker TI, Jones PL, Laurenson LJ (2009) Reproductive biology of the eastern shovelnose stingaree *Trygonoptera imitata* from South-Eastern Australia. Mar Freshw Res 60:845–860

Trinnie FI, Walker TI, Jones PL, Laurenson LJ (2012) Biennial reproductive cycle in an extensive matrotrophic viviparous batoid: the sandyback stingaree *Urolophus bucculentus* from South-Eastern Australia. J Fish Biol 80:1267–1291

Trinnie FI, Walker TI, Jones PL, Laurenson LJ (2014) Regional differences in the reproductive parameters of the sparsely-spotted stingaree, *Urolophus paucimaculatus*, from South-Eastern Australia. Mar Freshw Res 98:943–958

Trinnie FI, Walker TI, Jones PL, Laurenson LJ (2015) Asynchrony and regional differences in the reproductive cycle of the greenback stingaree *Urolophus viridis* from South-Eastern Australia. Environ Biol Fish 98:421–441

Trinnie FI, Walker TI, Jones PL, Laurenson LJ (2016) Reproductive cycle of *Urolophus cruciatus* in South-Eastern Australia: does the species exhibit obligate or facultative diapause? Mar Biol 163:00–00

Van Rijn JA, Reina RD (2010) Distribution of leukocytes as indicators of stress in the Australian swellshark, *Cephaloscyllium laticeps*. Fish Shellfish Immunol 29:234–238

Walker TI (2005a) Management measures. Management techniques for elasmobranch fisheries. (251 pp). FAO Fisheries Technical Paper 474, pp. 216–242

Walker TI (2005b) Reproduction in fisheries science. In: Hamlett WC (ed) Reproductive biology and phylogeny of chondrichthyes: sharks, batoids and chimaeras. Science Publishers, Inc., Enfield, pp 81–127

Walker TI (2007) Spatial and temporal variation in the reproductive biology of gummy shark *Mustelus antarcticus* (Chondrichthyes: Triakidae) harvested off southern Australia. Mar Freshw Res 58:67–97

Waltrick DS, Awruch CA, Simpfendorfer CA (2012) Embryonic diapause in elasmobranchs. Rev Fish Biol Fish 22:849–859

Waltrick DS, Jones SM, Simpfendorfer CA, Awruch CA (2014) Endocrine control of embryonic diapause in the Australian sharpnose shark *Rhizoprionodon taylori*. PLoS One 9(7):e101234. https://doi.org/10.1371/jounalpone.0101234

Waltrick DS, Simpfendorfer CA, Awruch CA (2017) A review on the morphology of ovarian follicles in elasmobranchs: a case study in *Rhizoprionodon taylori*. J Morphol 278:486–499

Wourms JP (1977) Reproduction and development in chondrichthyan fishes. Am Zool 17:379–410

Yamaguchi A, Taniuchi T, Shimizu M (2000) Geographic variations in reproductive parameters of the starspotted dogfish, *Mustelus manazo*, from five localities in Japan and Taiwan. Environ Biol Fish 57:221–233

Yano K, Tanaka S (1988) Size at maturity, reproductive cycle, fecundity and depth segregation of the deep sea squaloid sharks *Centroscymnus owstoni* and *C. coelolepis* in Suruga Bay, Japan. Nippon Suisan Gakkaishi 54:167–174

Chapter 11
Fertilization in Amphibians: The Cellular and Molecular Events from Sperm Approach to Egg Activation

Yasuhiro Iwao and Mami Watabe

Abstract The mechanism of fertilization is important for understanding the role of sexual reproduction in animals. Amphibians are one of the most suitable systems to investigate the cellular and molecular mechanisms of fertilization. Inseminated sperm approach the egg by initiation of motility and chemotactic guidance in response to substances secreted from oviducts on the egg surface. Sperm undergoing the acrosome reaction penetrate the vitelline envelope, and then fuse with the egg plasma membrane. The fertilizing sperm induces an increase in the intracellular Ca^{2+} concentration, causing egg activation, and initiation of embryonic development. The activated egg elicits several blocks to polyspermy, ensuring development from only one sperm. Recent advances in cellular and molecular mechanisms in amphibian fertilization are summarized and their important roles for establishment of amphibian fertilization are discussed.

Keywords Sperm motility · Acrosome reaction · Egg activation · Polyspermy block

11.1 Introduction

Most amphibians are adapted for life on land and live in moist environments, such as under trees in the forest. However, they reproduce in fresh water ponds and small streams. Amphibians can be classified into three orders: Anurans (frogs and toads), Urodela (newts and salamanders), and Caecilian (limbless amphibians) (Table 11.1,

---

Y. Iwao (✉) · M. Watabe
Laboratory of Reproductive Developmental Biology, Division of Earth Sciences, Biology, and Chemistry, Graduate School of Sciences and Technology for Innovation, Yamaguchi University, Yamaguchi, Yamaguchi, Japan
e-mail: iwao@yamaguchi-u.ac.jp

M. Yoshida, J. F. Asturiano (eds.), *Reproduction in Aquatic Animals*,
https://doi.org/10.1007/978-981-15-2290-1_12

Table 11.1 Mode of fertilization in amphibians

Order	Typical species	Mode of fertilization		Fast block by positive fertilization potential	Slow block by fertilization envelope	Very slow block in egg cytoplasm
Anura	*Ascaphus truei*	Internal	nd[a]	nd	nd	nd
	Discoglossus pictus	External	Polyspermy[b]	–[c]	+[d]	–
	Xenopus laevis	External	Monospermy	+	+	–
	Bufo japonicus	External	Monospermy	+	+	–
	Rana pipiens	External	Monospermy	+	+	–
Urodela	*Hynobius nebulosus*	External	Monospermy	+	–	–
	Andrias japonicus	External	Polyspermy	nd	–	+
	Cynops pyrrhogaster	Internal	Polyspermy	–	–	+
Caecilian	*Siphonops annulatus*	Internal[e]	nd	nd	nd	nd

[a]Not determined
[b]Occasionally polyspermy
[c]With a positive-going fertilization potential
[d]Animal dimple granules
[e]Viviparity in most species

Duellman 2003; Larson et al. 2003; Wilkinson and Nussbaum 2006). Anurans seem to have branched early from Urodela and Caecilian at the beginning of the Mesozoic period and the divergence between Urodela and Caecilian is estimated to have happened in the late Mesozoic period (Feller and Hedges 1998). As amphibians evolved to adapt to various environments on land, their mode of reproduction varied among species. Most anurans exhibit monospermy, but internal fertilization is observed in a few species, such as the tailed frog, *Ascaphus truei* (Sever et al. 2003). In urodeles, most of species exhibit internal fertilization, although a small number of primitive species, such as the clouded salamander, *Hynobius nebulosus*, and the Japanese giant salamander, *Andrias japonicus*, exhibit external fertilization (Iwao 2000a). Caecilians exhibit internal fertilization and are viviparous in most species (Gomes et al. 2012).

Full-grown oocytes, surrounded only with a thin vitelline envelope (VE) 5–10 μm in thickness, are ovulated from the ovary into the body cavity (coelom). They pass through the oviduct before oviposition from the cloaca. Mature oocytes are surrounded by several layers of oviductal sections on the VE (Fig. 11.1a). These extracellular coats have several important roles for establishment of fertilization, such as

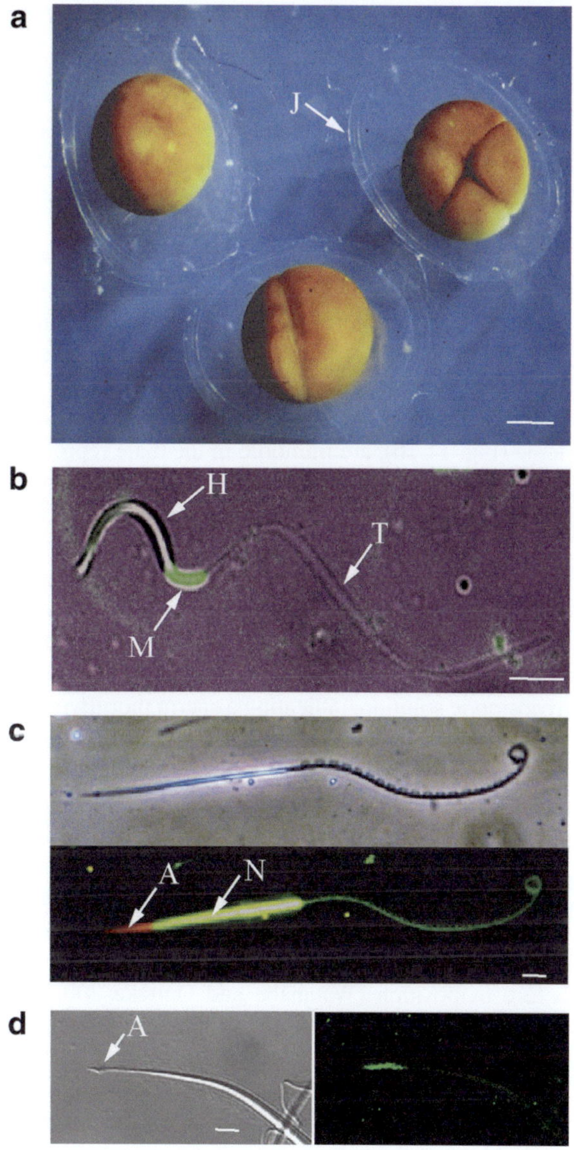

Fig. 11.1 (**a**) Fertilized *C. pyrrhogaster* eggs surrounding several jelly layers (J). (**b**) The *X. laevis* sperm consisting of head (H), midpiece (M), and tail (axoneme, T), showing mitochondria stained by MitoTracker Green (green). A merge between DIC and fluorescence images. (**c**) The *H. nebulosus* sperm stained with acridine orange, showing a nucleus (N, yellow) and an acrosome (A, red) at the top of sperm head. Upper panel, DIC image; Lower panel, fluorescence image. (**d**) The acrosome of *C. pyrrhogaster* sperm (A, left panel, DIC image), showing a protease activity against fluorescence-quenching casein (right panel, green). Bar, 1 mm (**a**), 10 μm (**b** and **c**), 5 μm (**d**)

activation of sperm motility, sperm chemotaxis, acrosome reaction, and polyspermy block. Ovulated oocytes undergo resumption of meiosis to reach metaphase of the second meiotic division before fertilization. Thus, oocytes (eggs) acquire fertility during passage through the oviduct.

11.2 Initiation of Sperm Motility and Sperm-Binding on the Vitelline Envelope

The sperm of amphibians are released outside of males and acquire fertility in response to the egg. The extracellular coat surrounding amphibian eggs includes an outer thick jelly layer and an inner thin VE (Fig. 11.1a). The jelly layers of the Japanese toad, *Bufo japonicus*, consist of a sugar (60–70%) and a protein (30–40%) portion (Katagiri 1973), and a major component of the high-molecular sugar moieties is mucin-like proteins with *o*-linked glycosylation (Guerardel et al. 2000). Oligosaccharide structures of the jelly layers of anuran eggs are well investigated (Plancke et al. 1995; Strecker et al. 1995; Tseng et al. 1997, 2001; Guerardel et al. 2000; Zhang et al. 2004; Li et al. 2011). The jelly layers contain several important factors for establishment of fertilization. The sperm, with mitochondria and axoneme (Fig. 11.1b), are immotile in the male reproductive organs, such as the testes and sperm ducts, or in the storage organs of sperm, such as spermathecae and the sperm storage tubes in females exhibiting internal fertilization (Sever et al. 2003).

In the South African clawed frog, *Xenopus laevis*, sperm motility is suppressed by seminal osmolality in the male body and initiated by a decrease in osmolality when the eggs are inseminated in hypotonic fresh water (Inoda and Morisawa 1987). Sperm of the Mediterranean painted frog, *Discoglossus pictus*, initiate motility immediately after immersion in egg jelly extract, as well as after immersion in hyposmotic solution, indicating a sperm motility-activating factor in the jelly layers (Takayama-Watanabe et al. 2012). In the South American toad, *Rhinella* (*Bufo*) *arenarum*, hypotonic shock triggers activation of sperm motility, but components of the egg jelly layers induce motility shifts to a progressive pattern (Krapf et al. 2014) concomitantly with a rise in the intracellular cAMP level and activation of protein kinase A (PKA) (O'Brien et al. 2011). Calcineurin and protein kinase C (PKC) seem to be involved in the acquisition of progressive motility (Krapf et al. 2014). In the red-bellied newt, *Cynops pyrrhogaster*, which exhibits internal fertilization, the motility of sperm stored in the spermathecae is initiated after insemination just before oviposition (Watanabe and Onitake 2003; Watanabe et al. 2003). Sperm motility increases in response to sperm motility-initiating substance (SMIS) on the surface of the egg jelly layers (Watanabe et al. 2010). SMIS, consisting of 150 amino acids (34 kDa), induces an increase in sperm motility independent of a hypotonic condition. An active site on SMIS localized to the C-terminal region of the second loop of the cysteine knot (CK) motif binds to the midpiece of sperm, initiating and enhancing the circular motion of sperm (Yokoe et al. 2016). SMIS is distributed in sequestered granules on the surface of the egg jelly layers and may be released by digestion with acrosomal enzymes (Watanabe et al. 2010; Yokoe et al. 2014). A T-type voltage-dependent Ca^{2+} channel is involved in the initiation of sperm motility by SMIS (Takahashi et al. 2013; Takayama-Watanabe et al. 2015). SMIS-like molecules may be involved in regulation of sperm motility in other anurans and urodeles (Ohta et al. 2010; Takayama-Watanabe et al. 2012; Yokoe et al. 2016; Sato et al. 2017). When sperm reach the egg, they penetrate the jelly layers and reach the VE, associating with a sperm chemotactic molecule. In *X. laevis* and the Western clawed frog, *Xenopus tropicalis*, allurin, a cysteine-rich secretory protein family (21 kDa), binds to sperm and modifies the sperm-orienting behavior (al-Anzi and Chandler 1998; Olson and Chandler 1999; Olson et al. 2001; Burnett et al. 2008, 2011a, b; Sugiyama et al. 2009). Allurin is secreted from the oviduct in the outermost jelly layer (Xiang et al. 2004, 2005). Thus, sperm of amphibians reach the VE with support from several molecules secreted from the oviduct in the egg jelly layers.

The VE surrounding *X. laevis* eggs consists of several glycoproteins (gp) named according to their size and their gene types, such as major components of gp37 (42%, ZPB, polypeptide: 36 kDa) and gp41 (47%, ZPC, 36 kDa); and minor ones of gp69/64 (3%, ZPA, 53 kDa), gp80 (<1%, ZPD, 32 kDa), and gp112–120 (7%, ZPX-Z, 75–83 kDa) (for a review, Hedrick 2008). Similar components are found in the VE of other anurans. Among them, gp37 (ZPB), gp69/64 (ZPA), and gp41 (ZPC) are homologous with zona pellucida proteins, ZP1, ZP2, and ZP3, in mammalian eggs, respectively. As sperm have to bind and pass through the VE before contacting the egg plasma membrane, the VE is a barrier for the sperm to fertilize the egg. Indeed, sperm are unable to bind and penetrate the VE of oocytes ovulated into the coelom (coelomic envelope, CE), but readily bind and penetrate the VE after passing through the oviduct. In *X. laevis*, ZPA and ZPC on the VE are proposed to function as ligands for sperm-surface receptors (Tian et al. 1997a, b; Vo and Hedrick 2000; Kubo et al. 2002; Vo et al. 2003). In *B. japonicus*, *R. arenarum*, and *D. pictus*, ZPC is probably a major ligand for sperm-binding on the VE (Omata and Katagiri 1996; Barisone et al. 2007; Caputo et al. 2001). ZPC (gp43) in the CE is converted to ZPC (gp41) by the oviductal enzyme, oviductin (Gerton and Hedrick 1986a, b; Kubo et al. 1999), indicating that partial proteolysis of ZPC unmasks existent ligand binding sites via conformational changes (Lindsay et al. 1999a; Hedrick 2008). Similar changes of CE into VE during passage though the oviduct are reported in other species (Yoshizaki and Katagiri 1981; Katagiri et al. 1999; Hiyoshi et al. 2002; Llanos et al. 2006; Barrera et al. 2012). A ZP protein-associated protein, dicalcin, is a 26 kDa Ca^{2+}-binding protein present in the VE of unfertilized *X. laevis* eggs (Miwa et al. 2007). Dicalcin suppresses the efficiency of fertilization through its interaction with gp41 (Miwa et al. 2010; Miwa 2015a, b). The interaction between dicalcin and gp41 alters the structural architecture of the VE meshwork by changing the binding affinity between ZP protein molecules, which regulates fertilization efficiency (Miwa 2015a, b; Miwa et al. 2015a, b). In addition, a trypsin-type protease, as well as an aminopeptidase B-like enzyme on the sperm plasma membrane may be involved in the binding of sperm and the VE (Kubo et al. 2008). Egg jelly components are necessary for binding of acrosome-reacted sperm to the VE in *C. pyrrhogaster* (Hiyoshi et al. 2007). Furthermore, a sperm-surface glycoprotein (SGP) is involved in binding of sperm to the VE in *X. laevis* (Kubo et al. 2010). SGP is also important for the interactions of sperm with the egg plasma membrane (Nagai et al. 2009) as described in Sect. 11.3, indicating that the SGP has a bifunctional role in fertilization.

11.3 Sperm Acrosome Reaction and Membrane Fusion Between the Sperm and Egg

Amphibian sperm has a flattened, acidic vesicle, called an acrosome, at the top of the sperm head (Fig. 11.1c and d). The acrosome contains several proteases which are discharged after acrosome reaction (Fig. 11.1d, Katagiri et al. 1982). The acrosome reaction results in the fusion of the sperm plasma membrane with an outer acrosomal membrane to expose an inner acrosome membrane which fuses with the egg plasma membrane. In *D. pictus*, the acrosome reaction occurs in the jelly coat layers (Campanella et al. 1997), whereas the sperm of most anurans undergo acrosome reaction after reaching the VE (Raisman et al. 1980; Katagiri et al. 1982; Martinez and Cabada 1996; Ueda et al. 2002, 2003). In *X. laevis*, an acrosome reaction-inducing substance in *Xenopus* (ARISX) is secreted onto the VE from the uppermost region of the oviduct, pars recta (Ueda et al. 2002, 2003). ARISX is a large molecular weight glycoprotein (300 kDa) and terminal α-D-GalNAc residues are essential to its activity (Ueda et al. 2007). The activity of ARISX is species specific. Most anuran sperm undergo acrosome reaction by ARISX, but some urodele sperm do not. The incidence of acrosome reaction by ARISX corresponds well with the occurrence of cross-fertilization (Ueda et al. 2007). Therefore, the species specificity of the acrosome reaction acts as a barrier for preventing cross-fertilization in amphibians. In urodeles, acrosome reaction is induced by substances of the high-molecular-weight components (500 kDa) in the outer jelly layers of *C. pyrrhogaster* eggs (Sasaki et al. 2002). Both 122 and 90 kDa proteins in the substance, recognized with wheat germ agglutinin (WGA), are the acrosome reaction-inducing substances in the egg jelly (Watanabe et al. 2009). N-linked carbohydrate moieties in these proteins may be responsible for the activity.

The acrosome of amphibian sperm contains proteases (Fig. 11.1d), called VE lysins, which are released after the acrosome reaction and may be responsible for sperm penetration into the VE (Elinson 1971; Penn and Gledhill 1972; Raisman and Cabada 1977; Iwao and Katagiri 1982). A tryptic protease (32 kDa) in *B. japonicus* sperm is involved in the envelope penetration process (Iwao and Katagiri 1982; Yamasaki et al. 1988). The tryptic protease is also found in other anurans (Cabada et al. 1978, 1989; Raisman et al. 1980). A gelatin-digesting protease activity in the acrosome of *D. pictus* sperm may be released after the acrosome reaction to act at the egg surface (Infante et al. 2001). In urodeles, *C. pyrrhogaster* sperm contains strong tryptic protease activity in the acrosome (Fig. 11.1d, Iwao et al. 1994; Mizote et al. 1999). Although its role in VE penetration remains unknown, it may be involved in Ca^{2+} transient necessary for egg activation during fertilization (Mizote et al. 1999; Harada et al. 2011).

SGP is distributed over nearly the entire surface of *X. laevis* sperm (Nagai et al. 2009). It mediates sperm-egg membrane binding and is responsible for establishment of fertilization. SGP is complex of molecular masses of 65–150 kDa and minor molecules with masses of 20–28 kDa (Nagai et al. 2009). More membrane vesicles containing SGP bind to the surface in the animal hemisphere compared with the vegetal hemisphere in unfertilized eggs, which confirms a finding that sperm entry is restricted to the animal hemisphere or equatorial region in *X. laevis* eggs (Grey et al. 1982; Kline 1988). In addition, SGP is closely associated with matrix metalloproteinase-2 (MMP-2) on the sperm membrane (Iwao et al. 2014). SGP seems to be responsible for maintenance of the secreted type of MMP-2 on the sperm membrane. MMP-2 associated with SGP may participate in sperm-binding not only on the VE, but also on the egg plasma membrane. The interaction between MMP-2 on the sperm membrane and ganglioside GM1 on the egg membrane regulates sperm-egg binding/fusion as well as signal transduction at egg activation, as described in Sect. 11.4. In other anurans, a lipovitellin/glycoproteins complex in *D. pictus* (Campanella et al. 2011) and integrins and HSP70 in *R. arenarum* (Coux and Cabada 2006) are reportedly involved in sperm-egg membrane binding and/or fusion. However, the molecule responsible for the membrane fusion process, such as fusogen, remains unknown.

11.4 Egg Activation by a Fertilizing Sperm

Unfertilized eggs, which are arrested at the second meiotic metaphase, undergo resumption of meiosis, followed by cleavage to develop embryos after fertilization. A fertilizing sperm must provide a signal to stimulate the initiation of development, called egg activation. A major trigger for egg activation is an increase in the intracellular concentration of Ca^{2+} ($[Ca^{2+}]_i$) during amphibian fertilization (for reviews, Iwao 2000a, b, 2012; Tokmakov et al. 2014; Iwao and Izaki 2018). The increase in $[Ca^{2+}]_i$ propagates from the sperm entrance site and spreads over the whole egg cytoplasm as a Ca^{2+} wave in the fertilized *X. laevis* egg (Fig. 11.2a and b). The concentration of Ca^{2+} in the egg cytoplasm increases to 0.5–2.2 μM, during fertilization of *X. laevis* (Fontanilla and Nuccitelli 1998), *D. pictus* (Nuccitelli et al. 1988), and the Spanish newt, *Pleurodeles waltl* (Grandin and Charbonneau 1992). The velocity of the Ca^{2+} waves propagating in the egg cytoplasm is 5.7–8.9 μm/s in *X. laevis* (Fontanilla and Nuccitelli 1998), 2.1 μm/s in *X. tropicalis* (Watabe et al. 2019), or 5.0–6.0 μm/s in *C. pyrrhogaster* (Yamamoto et al. 1999; Harada et al. 2011). In physiologically polyspermic *C. pyrrhogaster* eggs, several Ca^{2+} waves are propagated toward the egg cortex with an interval of several minutes during normal fertilization (Harada et al. 2011).

How does a fertilizing sperm induce the initial signal for the $[Ca^{2+}]_i$ increase during amphibian fertilization? It is known that extracellular divalent cations, such as Ca^{2+} and Mg^{2+}, are necessary for frog fertilization (Ishihara et al. 1984; Wozniak et al. 2017). The egg jelly layers store divalent cations at a millimolar level, even in fresh water (Ishihara et al. 1984; Medina et al. 2010), which are probably important for the acrosome reaction (Ueda et al. 2002) and sperm-egg binding and/or fusion. In addition, Ca^{2+} is indispensable for initiation of embryonic development and blocks to polyspermy (Wozniak et al. 2017). Amphibian eggs can be activated by an artificial increase in $[Ca^{2+}]_i$ by treatment with a Ca^{2+} ionophore (Steinhardt et al.

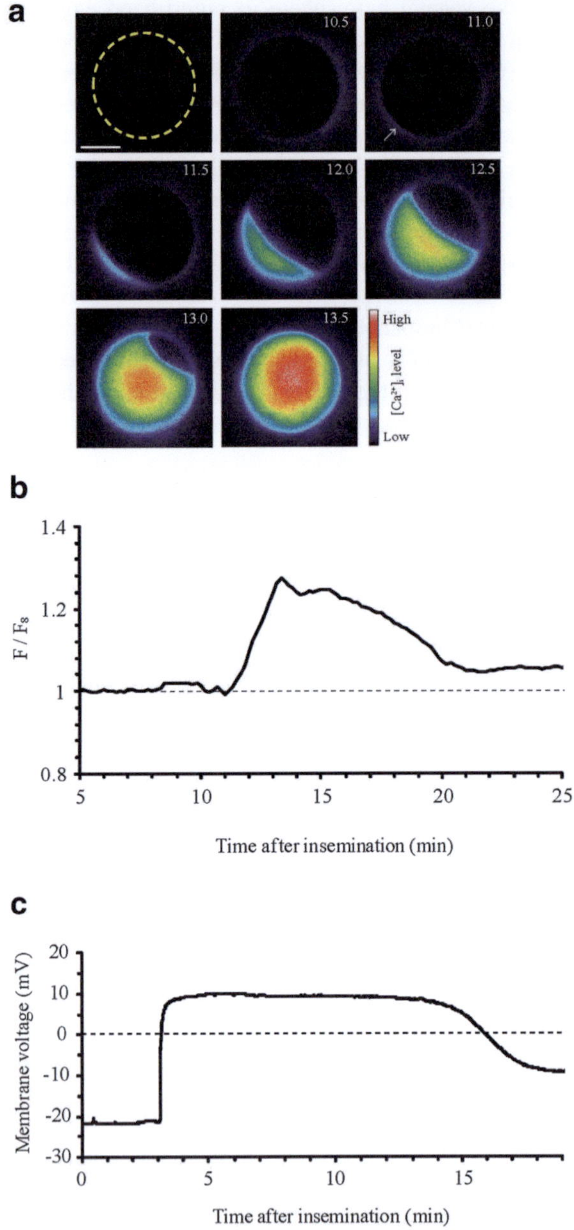

Fig. 11.2 (**a**) The Ca^{2+} wave during fertilization of *X. laevis* egg, showing the propagative increase in fluorescence of Ca^{2+}-sensitive dye in egg cytoplasm (arrow). Numbers on the right of figures show time (min) after insemination. Bar, 0.5 mm. (**b**) Changes in fluorescence intensity in the whole egg shown in **a**. (**c**) The positive-going fertilization potential during fertilization of *X. laevis* egg

1974; Iwao 1982; Charbonneau and Picheral 1983; Charbonneau et al. 1983; Andreuccetti et al. 1984; Iwao and Masui 1995). Anuran eggs are activated by pricking with a fine glass needle to cause a Ca^{2+} influx (Goldenberg and Elinson 1980; Iwao 1982; Talevi et al. 1985). However, most urodele eggs are resistant to the pricking stimulus (Fankhauser 1967; Iwao and Masui 1995), except for

H. nebulosus (Iwao 1989) and *P. waltl* (Aimar and Labrousse 1975). Egg activation is completely inhibited by the injection of Ca^{2+} chelators in *X. laevis* (Kline 1988) and *C. pyrrhogaster* (Yamamoto et al. 1999). Although external Ca^{2+} is necessary for amphibian fertilization (Ishihara et al. 1984; Wozniak et al. 2017), eggs can be activated by Ca^{2+} released from internal Ca^{2+} stores, such as from the endoplasmic reticulum (ER), in the absence of external Ca^{2+} (Steinhardt et al. 1974; Yamamoto et al. 1999). The Ca^{2+} wave occurs independent of external Ca^{2+} (Fontanilla and Nuccitelli 1998; Yamamoto et al. 1999; Harada et al. 2011). Thus, the increase in $[Ca^{2+}]_i$ from internal Ca^{2+} stores is sufficient for amphibian egg activation, but an initial Ca^{2+} influx from the external solution might occur around the sperm entry site.

The initiation of egg activation seems to be caused by two different mechanisms in monospermic eggs and physiologically polyspermic eggs, the membrane receptor model and the soluble factor model, respectively (for reviews, Iwao 2000b, 2012). In both cases, it is important that the fertilizing sperm stimulate phospholipase Cγ (PLCγ) to produce inositol-1,4,5-trisphosphate (IP3) to initiate the $[Ca^{2+}]_i$ increase in the egg cytoplasm (for reviews, Hasan et al. 2011; Sato 2018). The amount of IP3 in the egg cytoplasm increases 3- to 5-fold during *X. laevis* fertilization (Stith et al. 1993, 1994; Snow et al. 1996). IP3 binds an IP3-receptor on the ER and Ca^{2+} is released from the ER into the cytoplasm by opening Ca^{2+} channels (IP3 receptors) (Nuccitelli et al. 1993). A local increase in $[Ca^{2+}]_i$ may sensitize IP3 receptors in adjacent ER and/or stimulate PLCγ to induce further increases of IP3 to propagate an increase of $[Ca^{2+}]_i$ over the egg surface as a Ca^{2+} wave (Iwao 2012). In monospermic frogs, increasing evidence supports the notion that an agonist (ligand) on the sperm membrane binds to a receptor on the egg membrane to transmit a signal for IP3 production into the egg cytoplasm. Peptides containing a ligand for integrins, such as an arginyl-glycyl-aspartic acid (RGD) amino acid sequence, trigger the $[Ca^{2+}]_i$ increase (Iwao and Fujimura 1996). RGD-containing peptides cause egg activation in monospermic urodele *H. nebulosus* (Iwao 2014). Treatment with a peptide containing a CRMPKTE sequence derived from a metalloprotease/disintegrin/cysteine-rich (MDC) family (xMDC16) causes a $[Ca^{2+}]_i$ increase and egg activation (Shilling et al. 1998). It is also reported that a RGDS peptide induces protein tyrosine phosphorylation, but not egg activation in *R. arenarum* eggs (Mouguelar et al. 2011). Treatment of *X. laevis* eggs with a peptide corresponding to a partial amino acid sequence of MMP-2 hemopexin domain (HPX) induces a $[Ca^{2+}]_i$ increase and egg activation (Iwao et al. 2014). The positively charged MMP-2 HPX peptide (473–485, GMSQIRGETFFFK, pI: 10.2) can bind ganglioside GM1, which contains negatively charged sialic acid in the membrane microdomain (MD) of the egg plasma membrane. A signal for egg activation is then transmitted into the egg. Another candidate sperm ligand is a tryptic protease that efficiently hydrolyzes the C-terminus of double arginine residues to induce a Ca^{2+} wave in *X. laevis* eggs (Iwao et al. 1994; Mizote et al. 1999). Sato and co-workers demonstrated that uroplakin III (UPIII) in the MD plays an important role in the $[Ca^{2+}]_i$ increase during fertilization (Sakakibara et al. 2005; Mahbub Hasan et al. 2007, 2014; Sato et al. 2002). Cleavage of UPIII by the protease at the double arginine residues induces activation of Src kinase, followed by activation of PLCγ to produce IP3, and then a propagative Ca^{2+} release from the ER (Sato et al. 1999, 2003; Mahbub Hasan et al. 2005; Hasan et al. 2011). Both MMP-2 HPX and sperm tryptic protease are most likely responsible for the Ca^{2+} increase during *X. laevis* fertilization (Hasan et al. 2011; Sato 2014, 2018). Phosphatidic acid produced by phospholipase D1b may be involved in egg activation during *X. laevis* fertilization (Bates et al. 2015; Fees and Stith 2019).

The mechanism of egg activation in physiologically polyspermic urodeles is quite different from that in monospermic anurans (Iwao 2012, 2014). An initial Ca^{2+} rise at each site of sperm entry propagates a Ca^{2+} wave into the egg cytoplasm but does not reach the opposite side of the egg in *C. pyrrhogaster* (Harada et al. 2011). The Ca^{2+} wave induced by a single sperm only spreads to one-eighth to one-quarter of the egg surface (Harada et al. 2011). Multiple sperm entry causes an increase in $[Ca^{2+}]_i$ over the entire egg and then initiates egg activation. Although a transient $[Ca^{2+}]_i$ spike is induced by the tryptic protease activity in sperm, the Ca^{2+} wave is induced by an egg-activation factor found in the sperm cytoplasm which is introduced into the egg cytoplasm after fusion of sperm and egg (Yamamoto et al. 2001; Harada et al. 2007, 2011). The egg-activation factor is considered sperm-specific as well as extra mitochondrial citrate synthase (CS) (Harada et al. 2007, 2011; Ueno et al. 2014). The enzymatic activity of CS plays a central role in the Ca^{2+} increase because acetyl-CoA and oxaloacetate have sufficient activity to induce Ca^{2+} waves. Sperm-specific CS is highly phosphorylated and binds the components containing microtubules in the axial fiber in the middle piece of the sperm (Harada et al. 2007; Ueno et al. 2014). Propagation of the Ca^{2+} wave is inhibited by microtubule depolymerization in the egg cytoplasm (Ueno et al. 2014). CS accumulates at the sperm entry site and PLCγ in the egg cytoplasm accumulates around the midpiece region. Thus, sperm CS forms a complex of microtubules and ER with the IP3 receptor in addition to PLCγ to induce Ca^{2+} release from the ER. However, the detailed molecular mechanism for induction of the Ca^{2+} wave remains to be investigated.

11.5 Blocks to Polyspermy

Fertilization for amphibians occurs in fresh water in ponds or streams, and the conditions at time of insemination vary upon environmental conditions. To avoid failure of fertilization due to a decrease in sperm concentration, the male usually releases a large number of sperm toward the eggs. In vitro fertilization of *X. laevis* eggs requires insemination with a high concentration of sperm (more than 10^6 cells/mL), with more than 10^3 sperm reaching the vicinity of the egg. Thus, although the incidence of fertilization is ensured by a high collision rate between sperm and eggs, polyspermy seems to be probable in most eggs. To ensure monospermy with a diploid configuration, the egg has several mechanisms (polyspermy blocks) to reject extra sperm after the first sperm enters into the egg cytoplasm (Elinson 1986; Iwao 2000a, b; Iwao and Izaki 2018).

Egg investments such as jelly layers and a VE provide fertility to the sperm, but also reduce the number of sperm which can reach the egg surface (Reinhart et al. 1998; Miwa 2015a, b; Miwa et al. 2015a, b). It is estimated that 1–20 sperm reach the egg at an interval of 2–10 s in the leopard frog, *Rana pipiens* (Schlichter and Elinson 1981) and *B. japonicus* (Iwao 1987), or 10–40 s in *X. laevis* (Watabe et al. 2019). The egg needs time to prepare the block mechanism in order to prevent polyspermy, which is supplied by the jelly layers and the VE. The jelly layers absorb water in fresh water to swell, preventing additional sperm from penetrating the jelly layers 10–15 min after insemination (Katagiri 1962; Wolf and Hedrick 1971; McLaughlin and Humphries 1978; Matsuda and Onitake 1984). Thus, the jelly layers function as a very slow extracellular block to polyspermy.

Several sperm are, however, able to reach the egg surface during fertilization. In the physiologically polyspermic *C. pyrrhogaster* eggs, which lack any extracellular blocks except for the swelling of the jelly layers, more than ten sperm are able to enter an egg (Iwao 1985). Thus, additional extracellular blocks to polyspermy are necessary to ensure monospermy. Two different types of polyspermy blocks operate during the monospermic fertilization of amphibians, the fast electrical block at the egg plasma membrane, followed by the slow permanent block by the formation of the fertilization envelope. As the anuran eggs exposing plasma membrane exhibit monospermy (Elinson 1973; Katagiri 1974; Stewart-Savage and Grey 1984; Mizote et al. 1999), a polyspermy block at the sperm-egg binding and/or fusion is most important for ensuring monospermy (Table 11.1).

In monospermic eggs, the increase in $[Ca^{2+}]_i$ in the egg cytoplasm, induced by a fertilizing sperm, causes a positive shift of the egg membrane potential, called a positive-going fertilization potential (Iwao and Izaki 2018). In *X. laevis*, the membrane potential of an unfertilized egg (−17 mV) increases up to 7 mV during fertilization (Fig. 11.2c, Webb and Nuccitelli 1985; Glahn and Nuccitelli 2003; Iwao et al. 2014; Watabe et al. 2019). Similar positive-going fertilization potentials are observed during the fertilization of not only monospermic anurans (Iwao 2000a), but also the monospermic urodele, *H. nebulosus* (Iwao 1989). The positive-going fertilization potential is caused by the opening of Ca^{2+}-activated Cl^- channels, such as the transmembrane protein 16a (TMEM16A) channels in *X. laevis* eggs (Kline and Nuccitelli 1985; Wozniak et al. 2018a; Watabe et al. 2019). A decrease in fertilization potential by addition of halide ions or Cl^- channel inhibitors also results in polyspermy (Cross and Elinson 1980: Grey et al. 1982; Wozniak et al. 2018a; Watabe et al. 2019). Clamping the membrane potential of unfertilized eggs at a positive level inhibits fertilization, whereas at a negative level causes polyspermy (Cross and Elinson 1980; Iwao and Jaffe 1989; Glahn and Nuccitelli 2003: Iwao et al. 2014; Watabe et al. 2019). In addition, the inhibition of IP3-induced Ca^{2+} release from the ER inhibits fertilization-triggered depolarization and increases the incidence of polyspermy (Wozniak et al. 2018b). Thus, fertilization of monospermic amphibians is dependent on the voltage of the egg membrane, called voltage-dependent fertilization. As the membrane potential increases at the positive level within several seconds after fertilization of the first sperm (Fig. 11.2c), the positive-going fertilization potential can prevent extra sperm from entering the egg. The membrane potential, however, returns to negative 10–15 min after fertilization. The positive-going fertilization potential provides a fast, but temporal block to polyspermy at sperm-egg fusion. Although the molecular mechanism of voltage-dependent fertilization remains to be investigated, a positively charged hemopexin domain of metalloproteinase-2 (MMP-2) on the sperm membrane probably plays an important role in *X. laevis* (Iwao et al. 2014). The voltage-dependent phosphatase might also be involved in voltage-dependent fertilization (Ratzan et al. 2011).

The Ca^{2+} wave during fertilization causes the discharge of contents in cortical granules into the perivitelline space, resulting in formation of the fertilization envelope (FE) (Hedrick 2008). N-acetyl glucosaminidase is released from these cortical granules (Prody et al. 1985), rendering the eggs unfertilizable, probably due to the digestion of gp41 (ZPC) sugar chains on the VE. The most notable change, however, is the formation of the fertilization layer (F layer) on the FE by cortical granule lectin (CG lectin) within most inner jelly layer (J1) (Grey et al. 1974; Quill and Hedrick 1996). CG lectin consisting of 38 kDa and 48 kDa subunits binds terminal α-galactose of the jelly materials to form a precipitant through which sperm are unable to penetrate. It is also reported that CG lectin binds a pre-fertilization layer (PF layer) on the VE (Yoshizaki 1989a, b). In addition, the VE of anuran eggs is modified into fertilization envelope (FE) following fertilization, resulting in hardening of VE to prevent polyspermy (Lindsay and Hedrick 1989; Lindsay et al. 1988). Both trypsin-like (45 kDa) and chymotrypsin-like (30 kDa) proteases are released in perivitelline space after cortical granule exocytosis (Lindsay and Hedrick 1989; Lindsay et al. 1992) and the chymotrypsin-like protease (ovochymase) is probably involved in envelope hardening after activation by the trypsin-like protease (Lindsay and Hedrick 1989, 1995; Lindsay et al. 1999b). Furthermore, partial proteolysis of ZPA (gp69/64 into gp66/61) by a zinc metalloprotease from *X. laevis* eggs may also result in envelope hardening (Lindsay and Hedrick 2004). Taken together, the VE is modified following fertilization by its hardening and the loss of sperm receptors to prevent polyspermy. The transformation of VE into FE occurs several minutes after fertilization, followed by the formation of S-layer on the egg plasma membrane (Stewart-Savage and Grey 1987), which functions as the slow, but permanent block to polyspermy after the fast electrical block.

Physiologically polyspermic urodele eggs lack the fast electrical block on the egg plasma membrane, since no positive-going fertilization potential is elicited and fertilization is insensitive to positive membrane potential (Table 11.1, Charbonneau et al. 1983; Jaffe et al. 1983; Iwao 1985; Iwao and Jaffe 1989). The slow block at the FE also disappears because the eggs lack cortical granules. In *C. pyrrhogaster*, for example, 2–20 sperm enter an egg at normal fertilization (Iwao et al. 1985, 1993, 2002). Not only a sperm nucleus, but also centrioles, mitochondria, and axoneme are incorporated into the egg cytoplasm. After nuclear membrane breakdown, highly condensed sperm chromatins undergo decondensation to form a swelling sperm pronucleus (Fig. 11.3). Mitochondria and axoneme are destructed before cleavage, but γ-tubulin in the egg cytoplasm is accumulated in sperm centrioles and initiates polymerization of microtubules to develop a sperm aster (Iwao et al. 2002). In physiologically polyspermic eggs, all fertilized sperm form sperm pronuclei in the egg

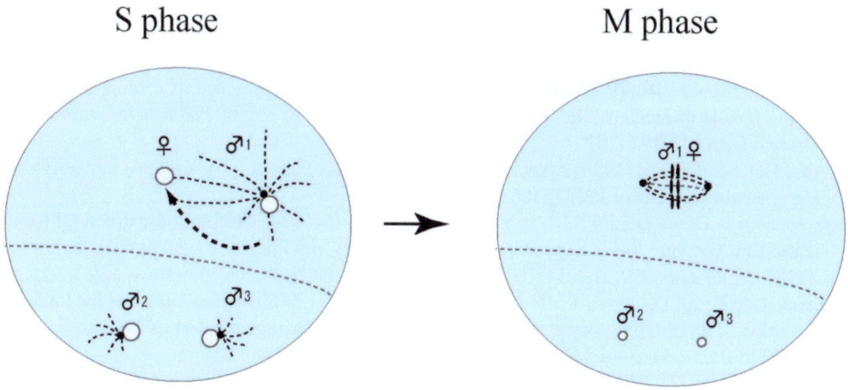

Fig. 11.3 Schematic illustration of behavior of sperm and egg nuclei in the physiologically polyspermic *C. pyrrhogaster* egg, showing the approach of a primary sperm pronucleus with a large aster to form a zygote nucleus with an egg nucleus during S phase, and the formation of a metaphase plate in the zygote nucleus for the first cleavage and degeneration of accessory sperm nuclei after M phase

cytoplasm and spread small sperm asters from each sperm centriole (Fankhauser 1948; Iwao et al. 1985, 1993, 1997, 2002). Only one sperm pronucleus forms a zygote nucleus with an egg pronucleus in the center of the animal hemisphere (Fig. 11.3). A sperm pronucleus with the largest sperm aster most likely makes contact with an egg pronucleus, but how the sperm nucleus is selected to participate in zygote nucleus formation is unknown. Both sperm and egg pronuclei enter the S phase, but progress in DNA synthesis is faster in the zygote nucleus than in the accessory sperm pronuclei (Wakimoto 1979; Iwao et al. 1993). When the zygote nucleus enters M phase to form a bipolar spindle for the first cleavage, the accessory sperm nuclei undergo degeneration (pycnosis) without the separation of the centrosome (Fig. 11.3, Iwao and Elinson 1990, Iwao et al. 1993, 2002). The nuclear degeneration seems to be caused by insufficient exposure of MPF, cdk1(cdc2) kinase activity, in the accessory sperm pronuclei in the periphery of the egg (Iwao and Elinson 1990; Iwao et al. 2002), but the molecular mechanisms remain unknown. Thus, diploid configuration development is ensured by a very slow, cytoplasmic block to polyspermy in physiologically polyspermic eggs.

Acknowledgements We thank Dr. Akihiko Watanabe, Yamagata University, for reviewing the manuscript. This work was supported by JSPS KAKENHI Grant Numbers, JP16K07373 and JP19K06690 to Y.I. and by The YU "Pump-Priming Program" for Fostering Research Activities.

References

Aimar C, Labrousse JP (1975) DNA synthesis and evolution, in presence of a somatic nucleus, of the female pronucleus after experimental activation of the egg of *Pleurodeles waltlii*. Dev Growth Differ 17:197–207

al-Anzi B, Chandler DE (1998) A sperm chemoattractant is released from *Xenopus* egg jelly during spawning. Dev Biol 198(2):366–375

Andreuccetti P, Denis-Donini S, Burrini AG, Campanella C (1984) Calcium ultrastructural localization in *Xenopus laevis* eggs following activation by pricking or by calcium ionophore A 23187. J Exp Zool 229(2):295–308

Barisone GA, Krapf D, Correa-Fiz F, Arranz SE, Cabada MO (2007) Glycoproteins of the vitelline envelope of amphibian oocyte: biological and molecular characterization of ZPC component (gp41) in *Bufo arenarum*. Mol Reprod Dev 74(5):629–640

Barrera D, Llanos RJ, Miceli DC (2012) Participation of the 39-kDa glycoprotein (gp39) of the vitelline envelope of *Bufo arenarum* eggs in sperm-egg interaction. Zygote 20(2):159–171

Bates RC, Fees CP, Holland WL, Winger CC, Batbayar K, Ancar R, Bergren T, Petcoff D, Stith BJ (2015) Activation of Src and release of intracellular calcium by phosphatidic acid during *Xenopus laevis* fertilization. Dev Biol 386(1):165–180

Burnett LA, Boyles S, Spencer C, Bieber AL, Chandler DE (2008) *Xenopus tropicalis* allurin: expression, purification, and characterization of a sperm chemoattractant that exhibits cross-species activity. Dev Biol 316(2):408–416

Burnett LA, Sugiyama H, Bieber AL, Chandler DE (2011a) Egg jelly proteins stimulate directed motility in *Xenopus laevis* sperm. Mol Reprod Dev 78(6):450–462

Burnett LA, Tholl N, Chandler DE (2011b) Two types of assays for detecting frog sperm chemoattraction. J Vis Exp 58:e3407

Cabada MO, Mariano MI, Raisman JS (1978) Effect of trypsin inhibitors and concanavalin A on the fertilization of *Bufo arenarum* coelomic oocytes. J Exp Zool 204(3):409–416

Cabada MO, Manes ME, Gomez MI (1989) Spermatolysins in *Bufo arenarum*: their activity on oocyte surface. J Exp Zool 249(2):229–234

Campanella C, Carotenuto R, Infante V, Maturi G, Atripaldi U (1997) Sperm-egg interaction in the painted frog (*Discoglossus pictus*): an ultrastructural study. Mol Reprod Dev 47(3):323–333

Campanella C, Caputo M, Vaccaro MC, De Marco N, Tretola L, Romano M, Prisco M, Camardella L, Flagiello A, Carotenuto R, Limatola E, Polzonetti-Magni A, Infante V (2011) Lipovitellin constitutes the protein backbone of glycoproteins involved in sperm-egg interaction in the amphibian *Discoglossus pictus*. Mol Reprod Dev 78(3):161–171

Caputo M, Infante V, Talevi R, Vaccaro MC, Carotenuto R, Campanella C (2001) Following passage through the oviduct, the coelomic envelope of *Discoglossus pictus* (amphibia) acquires fertilizability upon reorganization, conversion of gp 42 to gp 40, extensive glycosylation, and formation of a specific layer. Mol Reprod Dev 58(3):318–329

Charbonneau M, Picheral B (1983) Early events in anuran amphibian gertilization: an ultrastructural study of changes occurring in the course of monospermic fertilization and artificial activation. Dev Growth Differ 25(1):23–37

Charbonneau M, Moreau M, Picheral B, Vilain JP, Guerrier P (1983) Fertilization of amphibian eggs: a comparison of electrical responses between anurans and urodeles. Dev Biol 98(2):304–318

Coux G, Cabada MO (2006) Characterization of *Bufo arenarum* oocyte plasma membrane proteins that interact with sperm. Biochem Biophys Res Commun 343(1):326–333

Cross NL, Elinson RP (1980) A fast block to polyspermy in frogs mediated by changes in the membrane potential. Dev Biol 75(1):187–198

Duellman WE (2003) An overview of anuran phylogeny, classification and reproductive modes. In: Jamieson BGM (ed) Reproductive biology and phylogeny of anura. Reproductive biology and phylogeny, vol 2. Science Publishers, Plymouth, pp 1–18

Elinson EP (1971) Sperm lytic activity and its relation to fertilization in the frog Rana pipiens. J Exp Zool 177(2):207–217

Elinson RP (1973) Fertilization of frog body cavity eggs enhanced by treatment affecting the vitelline coat. J Exp Zool 183:291–302

Elinson RP (1986) Fertilization in amphibians: the ancestry of the block to polyspermy. Int Rev Cytol 101:59–100

Fankhauser G (1948) The organization of the amphibian egg during fertilization and cleavage. Ann N Y Acad Sci 49(Art 5):684–708

Fankhauser G (1967) Urodeles. In: Wilt FH, Wessels NK (eds) Methods in developmental biology. Thomas Crowell Publishers, New York, pp 85–99

Fees CP, Stith BJ (2019) Insemination or phosphatidic acid induces an outwardly spiraling disk of elevated Ca^{2+} to produce the Ca^{2+} wave during *Xenopus laevis* fertilization. Dev Biol 448(1):59–68

Feller AE, Hedges SB (1998) Molecular evidence for the early history of living amphibians. Mol Phylogenet Evol 9(3):509–516

Fontanilla RA, Nuccitelli R (1998) Characterization of the sperm-induced calcium wave in *Xenopus* eggs using confocal microscopy. Biophys J 75(4):2079–2087

Gerton GL, Hedrick JL (1986a) The coelomic envelope to vitelline envelope conversion in eggs of *Xenopus laevis*. J Cell Biochem 30(4):341–350

Gerton GL, Hedrick JL (1986b) The vitelline envelope to fertilization envelope conversion in eggs of *Xenopus laevis*. Dev Biol 116(1):1–7

Glahn D, Nuccitelli R (2003) Voltage-clamp study of the activation currents and fast block to polyspermy in the egg of *Xenopus laevis*. Dev Growth Differ 45(2):187–197

Goldenberg M, Elinson RP (1980) Animal/vegetal difference in cortical granules exocytosis during activation of the frog egg. Dev Growth Differ 22:345–356

Gomes AD, Moreira RG, Navas CA, Antoniazzi MM, Jared C (2012) Review of the reproductive biology of caecilians (Amphibia, Gymnophiona). South Am J Herpetol 7(3):191–202

Grandin N, Charbonneau M (1992) Intracellular free Ca^{2+} changes during physiological polyspermy in amphibian eggs. Development 114(3):617–624

Grey RD, Wolf DP, Hedrick JL (1974) Formation and structure of the fertilization envelope in *Xenopus laevis*. Dev Biol 36(1):44–61

Grey RD, Bastiani MJ, Webb DJ, Schertel ER (1982) An electrical block is required to prevent polyspermy in eggs fertilized by natural mating of *Xenopus laevis*. Dev Biol 89(2):475–484

Guerardel Y, Kol O, Maes E, Lefebvre T, Boilly B, Davril M, Strecker G (2000) O-glycan variability of egg-jelly mucins from *Xenopus laevis*: characterization of four phenotypes that differ by the terminal glycosylation of their mucins. Biochem J 352 Pt 2:449–463

Harada Y, Matsumoto T, Hirahara S, Nakashima A, Ueno S, Oda S, Miyazaki S, Iwao Y (2007) Characterization of a sperm factor for egg activation at fertilization of the newt *Cynops pyrrhogaster*. Dev Biol 306(2):797–808

Harada Y, Kawazoe M, Eto Y, Ueno S, Iwao Y (2011) The Ca^{2+} increase by the sperm factor in physiologically polyspermic newt fertilization: its signaling mechanism in egg cytoplasm and the species-specificity. Dev Biol 351(2):266–276

Hasan AK, Fukami Y, Sato K (2011) Gamete membrane microdomains and their associated molecules in fertilization signaling. Mol Reprod Dev 78(10–11):814–830

Hedrick JL (2008) Anuran and pig egg zona pellucida glycoproteins in fertilization and early development. Int J Dev Biol 52(5–6):683–701

Hiyoshi M, Takamune K, Mita K, Kubo H, Sugimoto Y, Katagiri C (2002) Oviductin, the oviductal protease that mediates gamete interaction by affecting the vitelline coat in *Bufo japonicus*: its molecular cloning and analyses of expression and posttranslational activation. Dev Biol 243(1):176–184

Hiyoshi W, Sasaki T, Takayama-Watanabe E, Takai H, Watanabe A, Onitake K (2007) Egg jelly of the newt, *Cynops pyrrhogaster* contains a factor essential for sperm binding to the vitelline envelope. J Exp Zool A Ecol Genet Physiol 307(6):301–311

Infante V, Amirante R, Vaccaro MC, Wilding M, Campanella C (2001) Enzyme activity in anuran spermatozoa upon induction of the acrosome reaction. Zygote 9(4):293–298

Inoda T, Morisawa M (1987) Effect of osmolality on the initiation of sperm motility in *Xenopus laevis*. Comp Biochem Physiol A Comp Physiol 88(3):539–542

Ishihara K, Hosono J, Kanatani H, Katagiri C (1984) Toad egg-jelly as a source of divalent cations essential for fertilization. Dev Biol 105(2):435–442

Iwao Y (1982) Differential emergence of cortical granule breakdown and electrophysiological responses during meiotic maturation of toad oocytes. Dev Growth Differ 24(5):467–477

Iwao Y (1985) The membrane potential changes of amphibian eggs during species- and cross-fertilization. Dev Biol 111(1):26–34

Iwao Y (1987) The spike component of the fertilization potential in the toad, *Bufo japonicus*: changes during meiotic maturation and absence during cross-fertilization. Dev Biol 123(2):559–565

Iwao Y (1989) An electrically mediated block to polyspermy in the primitive urodele *Hynobius nebulosus* and phylogenetic comparison with other amphibians. Dev Biol 134(2):438–445

Iwao Y (2000a) Fertilization in amphibians. In: Tarin JJ, Cano A (eds) Fertilization in protozoa and metazoan animal. Springer, Berlin, pp 147–191

Iwao Y (2000b) Mechanisms of egg activation and polyspermy block in amphibians and comparative aspects with fertilization in other vertebrates. Zool Sci 17(6):699–709

Iwao Y (2012) Egg activation in physiological polyspermy. Reproduction 144(1):11–22

Iwao Y (2014) Chapter 15: egg activation in polyspermy: its molecular mechanisms and evolution in vertebrates. In: Sawada H, Inoue N, Iwano M (eds) Sexual reproduction in animals and plants. Springer Open, pp 171–180

Iwao Y, Elinson RP (1990) Control of sperm nuclear behavior in physiologically polyspermic newt eggs: possible involvement of MPF. Dev Biol 142(2):301–312

Iwao Y, Fujimura T (1996) Activation of *Xenopus* eggs by RGD-containing peptides accompanied by intracellular Ca^{2+} release. Dev Biol 177(2):558–567

Iwao Y, Izaki K (2018) Universality and diversity of a fast, electrical block to polyspermy during fertilization in animals. In: Kobayashi K, Kitano T, Iwao Y, Kondo M (eds) Reproductive and developmental strategies. Diversity and commonality in animals. Springer, Tokyo

Iwao Y, Jaffe LA (1989) Evidence that the voltage-dependent component in the fertilization process is contributed by the sperm. Dev Biol 134(2):446–451

Iwao Y, Katagiri C (1982) Properties of the vitelline coat lysin from toad sperm. J Exp Zool 219(1):87–95

Iwao Y, Masui Y (1995) Activation of newt eggs in the absence of Ca^{2+} activity by treatment with cycloheximide or D_2O. Dev Growth Differ 37:641

Iwao Y, Yamasaki H, Katagiri C (1985) Experiments pertaining to the suppression of accessory sperm in fertilized newt eggs. Dev Growth Differ 27:323–331

Iwao Y, Sakamoto N, Takahara K, Yamashita M, Nagahama Y (1993) The egg nucleus regulates the behavior of sperm nuclei as well as cycling of MPF in physiologically polyspermic newt eggs. Dev Biol 160(1):15–27

Iwao Y, Miki A, Kobayashi M, Onitake K (1994) Activation of *Xenopus* eggs by an extract of *Cynops* sperm. Dev Growth Differ 36(5):469–479

Iwao Y, Yasumitsu K, Narihira M, Jiang J, Nagahama Y (1997) Changes in microtubule structures during the first cell cycle of physiologically polyspermic newt eggs. Mol Reprod Dev 47(2):210–221

Iwao Y, Murakawa T, Yamaguchi J, Yamashita M (2002) Localization of gamma-tubulin and cyclin B during early cleavage in physiologically polyspermic newt eggs. Dev Growth Differ 44(6):489–499

Iwao Y, Shiga K, Shiroshita A, Yoshikawa T, Sakiie M, Ueno T, Ueno S, Ijiri TW, Sato K (2014) The need of MMP-2 on the sperm surface for *Xenopus* fertilization: its role in a fast electrical block to polyspermy. Mech Dev 134:80–95

Jaffe LA, Cross NL, Picheral B (1983) Studies of the voltage-dependent polyspermy block using cross-species fertilization of amphibians. Dev Biol 98(2):319–326

Katagiri C (1962) On the fertilizability of the frog egg, II. Change of the jelly envelopes in water. Jap. J Zool 8:365–374

Katagiri C (1973) Chemical analysis of toad egg-jelly in relation to its 'sperm-capacitating' activity. Dev Growth Differ 15:81–92

Katagiri C (1974) A high frequency of fertilization in premature and mature coelomic toad eggs after enzymic removal removal removal of vitelline membrane. J Embryol Exp Morphol 31(3):573–587

Katagiri C, Iwao Y, Yoshizaki N (1982) Participation of oviducal pars recta secretions in inducing the acrosome reaction and release of vitelline coat lysin in fertilizing toad sperm. Dev Biol 94(1):1–10

Katagiri C, Yoshizaki N, Kotani M, Kubo H (1999) Analyses of oviductal pars recta-induced fertilizability of coelomic eggs in *Xenopus laevis*. Dev Biol 210(2):269–276

Kline D (1988) Calcium-dependent events at fertilization of the frog egg: injection of a calcium buffer blocks ion channel opening, exocytosis, and formation of pronuclei. Dev Biol 126(2):346–361

Kline D, Nuccitelli R (1985) The wave of activation current in the *Xenopus* egg. Dev Biol 111(2):471–487

Krapf D, O'Brien E, Maidagan PM, Morales ES, Visconti PE, Arranz SE (2014) Calcineurin regulates progressive motility activation of *Rhinella* (*Bufo*) *arenarum* sperm through dephosphorylation of PKC substrates. J Cell Physiol 229(10):1378–1386

Kubo H, Matsushita M, Kotani M, Kawasaki H, Saido TC, Kawashima S, Katagiri C, Suzuki A (1999) Molecular basis for oviductin-mediated processing from gp43 to gp41, the predominant glycoproteins of *Xenopus* egg envelopes. Dev Genet 25(2):123–129

Kubo H, Kotani M, Suzuki H, Yoshizaki N (2002) Immunohistochemical localisation of gp69/64 molecules in *Xenopus* egg envelopes in relation to their sperm binding activity. Zygote 10(2):131–140

Kubo H, Kotani M, Yamamoto Y, Hazato T (2008) Involvement of sperm proteases in the binding of sperm to the vitelline envelope in *Xenopus laevis*. Zool Sci 25(1):80–87

Kubo H, Shiga K, Harada Y, Iwao Y (2010) Analysis of a sperm surface molecule that binds to a vitelline envelope component of *Xenopus laevis* eggs. Mol Reprod Dev 77(8):728–735

Larson A, Weisrock DW, Hozak KH (2003) Phylogentice systematics of salamanders (Amphiba: Urodela), a review. In: Sever DM (ed) Reproductive biology and phylogeny of Urodela. Reproductive biology and phylogeny, vol 1. Science Publishers, Plymouth, pp 32–108

Li B, Russell SC, Zhang J, Hedrick JL, Lebrilla CB (2011) Structure determination by MALDI-IRMPD mass spectrometry and exoglycosidase digestions of O-linked oligosaccharides from *Xenopus borealis* egg jelly. Glycobiology 21(7):877–894

Lindsay LL, Hedrick JL (1989) Proteases released from *Xenopus laevis* eggs at activation and their role in envelope conversion. Dev Biol 135(1):202–211

Lindsay LL, Hedrick JL (1995) Isolation and characterization of ovochymase, a chymotrypsin-like protease released during *Xenopus laevis* egg activation. Dev Biol 167(2):513–516

Lindsay LL, Hedrick JL (2004) Proteolysis of *Xenopus laevis* egg envelope ZPA triggers envelope hardening. Biochem Biophys Res Commun 324(2):648–654

Lindsay LL, Yamasaki H, Hedrick JL, Katagiri C (1988) Egg envelope conversion following fertilization in *Bufo japonicus*. Dev Biol 130(1):37–44

Lindsay LL, Larabell CA, Hedrick JL (1992) Localization of a chymotrypsin-like protease to the perivitelline space of *Xenopus laevis* eggs. Dev Biol 154(2):433–436

Lindsay LL, Wieduwilt MJ, Hedrick JL (1999a) Oviductin, the *Xenopus laevis* oviductal protease that processes egg envelope glycoprotein gp43, increases sperm binding to envelopes, and is translated as part of an unusual mosaic protein composed of two protease and several CUB domains. Biol Reprod 60(4):989–995

Lindsay LL, Yang JC, Hedrick JL (1999b) Ovochymase, a *Xenopus laevis* egg extracellular protease, is translated as part of an unusual polyprotease. Proc Natl Acad Sci U S A 96(20):11253–11258
Llanos RJ, Barrera D, Valz-Gianinet JN, Miceli DC (2006) Oviductal protease and trypsin treatment enhance sperm-envelope interaction in *Bufo arenarum* coelomic eggs. J Exp Zool A Comp Exp Biol 305(10):872–882
Mahbub Hasan AK, Sato K, Sakakibara K, Ou Z, Iwasaki T, Ueda Y, Fukami Y (2005) Uroplakin III, a novel Src substrate in *Xenopus* egg rafts, is a target for sperm protease essential for fertilization. Dev Biol 286(2):483–492
Mahbub Hasan AK, Ou Z, Sakakibara K, Hirahara S, Iwasaki T, Sato K, Fukami Y (2007) Characterization of *Xenopus* egg membrane microdomains containing uroplakin Ib/III complex: roles of their molecular interactions for subcellular localization and signal transduction. Genes Cells 12(2):251–267
Mahbub Hasan AK, Hashimoto A, Maekawa Y, Matsumoto T, Kushima S, Ijiri TW, Fukami Y, Sato K (2014) The egg membrane microdomain-associated uroplakin III-Src system becomes functional during oocyte maturation and is required for bidirectional gamete signaling at fertilization in *Xenopus laevis*. Development 141(8):1705–1714
Martinez ML, Cabada MO (1996) Assessment of the acrosome reaction in *Bufo arenarum* spermatozoa by immunostaining: comparison with other methods. Zygote 4(3):181–190
Matsuda M, Onitake K (1984) Fertilization of the eggs of *Cynops pyrrhogaster* (japanese newt) after immersion in water. Wilehm Roux Arch Dev Biol 193(2):61–63
McLaughlin EW, Humphries AA Jr (1978) The jelly envelopes and fertilization of eggs of the newt, Notophthalmus viridescens. J Morphol 158(1):73–90
Medina MF, Crespo CA, Ramos I, Fernández SN (2010) Role of cations as components of jelly coats in *Bufo arenarum* fertilization. Zygote 18(1):69–80
Miwa N (2015a) Dicalcin, a zona pellucida protein that regulates fertilization competence of the egg coat in *Xenopus laevis*. J Physiol Sci 65(6):507–514
Miwa N (2015b) Protein-carbohydrate interaction between sperm and the egg-coating envelope and its regulation by Dicalcin, a *Xenopus laevis* zona pellucida protein-associated protein. Molecules 20(5):9468–9486
Miwa N, Shinmyo Y, Kawamura S (2007) Cloning and characterization of *Xenopus* dicalcin, a novel S100-like calcium-binding protein in *Xenopus* eggs. DNA Seq 18(5):400–404
Miwa N, Ogawa M, Shinmyo Y, Hiraoka Y, Takamatsu K, Kawamura S (2010) Dicalcin inhibits fertilization through its binding to a glycoprotein in the egg envelope in *Xenopus laevis*. J Biol Chem 285(20):15627–15636
Miwa N, Ogawa M, Hanaue M, Takamatsu K (2015a) Corrigendum: fertilization competence of the egg-coating envelope is regulated by direct interaction of dicalcin and gp41, the *Xenopus laevis* ZP3. Sci Rep 5:15303
Miwa N, Ogawa M, Hanaue M, Takamatsu K (2015b) Fertilization competence of the egg-coating envelope is regulated by direct interaction of dicalcin and gp41, the *Xenopus laevis* ZP3. Sci Rep 5:12672
Mizote A, Okamoto S, Iwao Y (1999) Activation of *Xenopus* eggs by proteases: possible involvement of a sperm protease in fertilization. Dev Biol 208(1):79–92
Mouguelar VS, Cabada MO, Coux G (2011) The integrin-binding motif RGDS induces protein tyrosine phosphorylation without activation in *Bufo arenarum* (Amphibia) oocytes. Reproduction 141(5):581–593
Nagai K, Ishida T, Hashimoto T, Harada Y, Ueno S, Ueda Y, Kubo H, Iwao Y (2009) The sperm-surface glycoprotein, SGP, is necessary for fertilization in the frog, *Xenopus laevis*. Dev Growth Differ 51(5):499–510
Nuccitelli R, Kline D, Busa WB, Talevi R, Campanella C (1988) A highly localized activation current yet widespread intracellular calcium increase in the egg of the frog, *Discoglossus pictus*. Dev Biol 130(1):120–132
Nuccitelli R, Yim DL, Smart T (1993) The sperm-induced Ca^{2+} wave following fertilization of the *Xenopus* egg requires the production of Ins(1, 4, 5)P_3. Dev Biol 158(1):200–212

O'Brien ED, Krapf D, Cabada MO, Visconti PE, Arranz SE (2011) Transmembrane adenylyl cyclase regulates amphibian sperm motility through protein kinase A activation. Dev Biol 350(1):80–88

Ohta M, Kubo H, Nakauchi Y, Watanabe A (2010) Sperm motility-initiating activity in the egg jelly of the externally-fertilizing urodele amphibian, *Hynobius lichenatus*. Zool Sci 27(11):875–879

Olson JH, Chandler DE (1999) Xenopus laevis egg jelly contains small proteins that are essential to fertilization. Dev Biol 210(2):401–410

Olson JH, Xiang X, Ziegert T, Kittelson A, Rawls A, Bieber AL, Chandler DE (2001) Allurin, a 21-kDa sperm chemoattractant from *Xenopus* egg jelly, is related to mammalian sperm-binding proteins. Proc Natl Acad Sci U S A 98(20):11205–11210

Omata S, Katagiri C (1996) Involvement of carbohydrate moieties of the toad egg vitelline coat in binding with fertilizing sperm. Dev Growth Differ 38:663–672

Penn A, Gledhill BL (1972) Acrosomal proteolytic activity of amphibian sperm. A direct demonstration. Exp Cell Res 74(1):285–288

Plancke Y, Wieruszeski JM, Alonso C, Boilly B, Strecker G (1995) Structure of four acidic oligosaccharides from the jelly coat surrounding the eggs of *Xenopus laevis*. Eur J Biochem 231(2):434–439

Prody GA, Greve LC, Hedrick JL (1985) Purification and characterization of an N-acetyl-beta-D-glucosaminidase from cortical granules of *Xenopus laevis* eggs. J Exp Zool 235(3):335–340

Quill TA, Hedrick JL (1996) The fertilization layer mediated block to polyspermy in *Xenopus laevis*: isolation of the cortical granule lectin ligand. Arch Biochem Biophys 333(2):326–332

Raisman JS, Cabada MO (1977) Acrosomic reaction and proteolytic activity in the spermatozoa of an anuran amphibian *Leptodactylus chaquensis* (CEI). Dev Growth Differ 19:227–232

Raisman JS, Cunio RW, Cabada MO, Del Pino EF, Mariano MI (1980) Acrosome breakdown in *Leptodactylus chaquensis* (amphibia anura) spermatozoa. Dev Growth Differ 22:289–297

Ratzan WJ, Evsikov AV, Okamura Y, Jaffe LA (2011) Voltage sensitive phosphoinositide phosphatases of *Xenopus*: their tissue distribution and voltage dependence. J Cell Physiol 226(11):2740–2746

Reinhart D, Ridgway J, Chandler DE (1998) Xenopus laevis fertilisation: analysis of sperm motility in egg jelly using video light microscopy. Zygote 6(2):173–182

Sakakibara K, Sato K, Yoshino K, Oshiro N, Hirahara S, Mahbub Hasan AK, Iwasaki T, Ueda Y, Iwao Y, Yonezawa K, Fukami Y (2005) Molecular identification and characterization of *Xenopus* egg uroplakin III, an egg raft-associated transmembrane protein that is tyrosine-phosphorylated upon fertilization. J Biol Chem 280(15):15029–15037

Sasaki T, Kamimura S, Takai H, Watanabe A, Onitake K (2002) The activity for the induction of the sperm acrosome reaction localises in the outer layers and exists in the high-molecular-weight components of the egg-jelly of the newt, *Cynops pyrrhogaster*. Zygote 10(1):1–9

Sato K (2014) Transmembrane signal transduction in oocyte maturation and fertilization: focusing on *Xenopus laevis* as a model animal. Int J Mol Sci 16(1):114–134

Sato K (2018) Fertilization and protein tyrosine kinase signaling: are they merging or emerging? In: Kobayashi K, Kitano T, Iwao Y, Kondo M (eds) Reproductive and developmental strategies. Diversity and commonality in animals. Springer, Tokyo

Sato K, Iwao Y, Fujimura T, Tamaki I, Ogawa K, Iwasaki T, Tokmakov AA, Hatano O, Fukami Y (1999) Evidence for the involvement of a Src-related tyrosine kinase in *Xenopus* egg activation. Dev Biol 209(2):308–320

Sato K, Iwasaki T, Ogawa K, Konishi M, Tokmakov AA, Fukami Y (2002) Low density detergent-insoluble membrane of *Xenopus* eggs: subcellular microdomain for tyrosine kinase signaling in fertilization. Development 129(4):885–896

Sato K, Tokmakov AA, He CL, Kurokawa M, Iwasaki T, Shirouzu M, Fissore RA, Yokoyama S, Fukami Y (2003) Reconstitution of Src-dependent phospholipase Cγ phosphorylation and transient calcium release by using membrane rafts and cell-free extracts from *Xenopus* eggs. J Biol Chem 278(40):38413–38420

Sato T, Yokoe M, Endo D, Morita M, Toyama F, Kawamura Y, Nakauchi Y, Takayama-Watanabe E, Watanabe A (2017) Sperm motility initiating substance may be insufficient to induce forward motility of *Cynops ensicauda* sperm. Mol Reprod Dev 84(8):686–692

Schlichter LC, Elinson RP (1981) Electrical responses of immature and mature *Rana pipiens* oocytes to sperm and other activating stimuli. Dev Biol 83(1):33–41

Sever DM, Hamlett C, Slabach, R, Stephenson B, Verrell PA (2003) Internal fertilization in the anura with special reference to mating and female sperm storage in Ascaphus. In: Jamieson BGM (ed) Reproductive biology and phylogeny of anura. Reproductive biology and phylogeny, vol 2. Science Publishers, Plymouth, pp 319–341

Shilling FM, Magie CR, Nuccitelli R (1998) Voltage-dependent activation of frog eggs by a sperm surface disintegrin peptide. Dev Biol 202(1):113–124

Snow P, Yim DL, Leibow JD, Saini S, Nuccitelli R (1996) Fertilization stimulates an increase in inositol trisphosphate and inositol lipid levels in *Xenopus* eggs. Dev Biol 180(1):108–118

Steinhardt RA, Epel D, Carroll EJ Jr, Yanagimachi R (1974) Is calcium ionophore a universal activator for unfertilised eggs? Nature 252(5478):41–43

Stewart-Savage J, Grey RD (1984) Fertilization of investment-free *Xenopus* eggs. Exp Cell Res 154(2):639–642

Stewart-Savage J, Grey RD (1987) Loss of functional sperm entry into *Xenopus* eggs after activation correlates with a reduction in surface adhesivity. Dev Biol 120(2):434–446

Stith BJ, Goalstone M, Silva S, Jaynes C (1993) Inositol 1,4,5-trisphosphate mass changes from fertilization through first cleavage in *Xenopus laevis*. Mol Biol Cell 4(4):435–443

Stith BJ, Espinoza R, Roberts D, Smart T (1994) Sperm increase inositol 1,4,5-trisphosphate mass in *Xenopus laevis* eggs preinjected with calcium buffers or heparin. Dev Biol 165(1):206–215

Strecker G, Wieruszeski JM, Plancke Y, Boilly B (1995) Primary structure of 12 neutral oligosaccharide-alditols released from the jelly coats of the anuran *Xenopus laevis* by reductive beta-elimination. Glycobiology 5(1):137–146

Sugiyama H, Burnett L, Xiang X, Olson J, Willis S, Miao A, Akema T, Bieber AL, Chandler DE (2009) Purification and multimer formation of allurin, a sperm chemoattractant from *Xenopus laevis* egg jelly. Mol Reprod Dev 76(6):527–536

Takahashi T, Kutsuzawa M, Shiba K, Takayama-Watanabe E, Inaba K, Watanabe A (2013) Distinct Ca^{2+} channels maintain a high motility state of the sperm that may be needed for penetration of egg jelly of the newt, *Cynops pyrrhogaster*. Dev Growth Differ 55(7):657–667

Takayama-Watanabe E, Campanella C, Kubo H, Watanabe A (2012) Sperm motility initiation by egg jelly of the anuran, *Discoglossus pictus* may be mediated by sperm motility-initiating substance of the internally-fertilizing newt, *Cynops pyrrhogaster*. Zygote 20(4):417–422

Takayama-Watanabe E, Ochiai H, Tanino S, Watanabe A (2015) Contribution of different Ca(2)(+) channels to the acrosome reaction-mediated initiation of sperm motility in the newt *Cynops pyrrhogaster*. Zygote 23(3):342–351

Talevi R, Dale B, Campanella C (1985) Fertilization and activation potentials in *Discoglossus pictus* (Anura) eggs: a delayed response to activation by pricking. Dev Biol 111(2):316–323

Tian J, Gong H, Thomsen GH, Lennarz WJ (1997a) Gamete interactions in *Xenopus laevis*: identification of sperm binding glycoproteins in the egg vitelline envelope. J Cell Biol 136(5):1099–1108

Tian J, Gong H, Thomsen GH, Lennarz WJ (1997b) Xenopus laevis sperm-egg adhesion is regulated by modifications in the sperm receptor and the egg vitelline envelope. Dev Biol 187(2):143–153

Tokmakov AA, Stefanov VE, Iwasaki T, Sato K, Fukami Y (2014) Calcium signaling and meiotic exit at fertilization in *Xenopus* egg. Int J Mol Sci 15(10):18659–18676

Tseng K, Lindsay LL, Penn S, Hedrick JL, Lebrilla CB (1997) Characterization of neutral oligosaccharide-alditols from *Xenopus laevis* egg jelly coats by matrix-assisted laser desorption Fourier transform mass spectrometry. Anal Biochem 250(1):18–28

Tseng K, Xie Y, Seeley J, Hedrick JL, Lebrilla CB (2001) Profiling with structural elucidation of the neutral and anionic O-linked oligosaccharides in the egg jelly coat of *Xenopus laevis* by Fourier transform mass spectrometry. Glycoconj J 18(4):309–320

Ueda Y, Yoshizaki N, Iwao Y (2002) Acrosome reaction in sperm of the frog, *Xenopus laevis*: its detection and induction by oviductal pars recta secretion. Dev Biol 243(1):55–64

Ueda Y, Kubo H, Iwao Y (2003) Characterization of the acrosome reaction-inducing substance in *Xenopus* (ARISX) secreted from the oviductal pars recta onto the vitelline envelope. Dev Biol 264(1):289–298

Ueda Y, Imaizumi C, Kubo H, Sato K, Fukami Y, Iwao Y (2007) Analysis of terminal sugar moieties and species-specificities of acrosome reaction-inducing substance in *Xenopus* (ARISX). Dev Growth Differ 49(7):591–601

Ueno T, Ohgami T, Harada Y, Ueno S, Iwao Y (2014) Egg activation in physiologically polyspermic newt eggs: involvement of IP(3) receptor, PLCgamma, and microtubules in calcium wave induction. Int J Dev Biol 58(5):315–323

Vo LH, Hedrick JL (2000) Independent and hetero-oligomeric-dependent sperm binding to egg envelope glycoprotein ZPC in *Xenopus laevis*. Biol Reprod 62(3):766–774

Vo LH, Yen TY, Macher BA, Hedrick JL (2003) Identification of the ZPC oligosaccharide ligand involved in sperm binding and the glycan structures of *Xenopus laevis* vitelline envelope glycoproteins. Biol Reprod 69(6):1822–1830

Wakimoto BT (1979) DNA synthesis after polyspermic fertilization in the axolotl. J Embryol Exp Morphol 52:39–48

Watabe M, Izaki K, Fujino S, Maruyama M, Kojima C, Hiraiwa A, Ueno S, Iwao Y (2019) The electrical block to polyspermy induced by an intracellular Ca^{2+} increase at fertilization of the clawed frogs, *Xenopus laevis* and *Xenopus tropicalis*. Mol Reprod Dev 86:387–403

Watanabe A, Onitake K (2003) Sperm activation. In: Sever DM (ed) Reproductive biology and phylogeny of Urodel. Reproductive biology and phylogeny, vol 1. Science Publishers, Plymouth, pp 425–445

Watanabe T, Itoh T, Watanabe A, Onitake K (2003) Characteristics of sperm motility induced on the egg-jelly in the internal fertilization of the newt, *Cynops pyrrhogaster*. Zool Sci 20(3):345–352

Watanabe A, Fukutomi K, Kubo H, Ohta M, Takayama-Watanabe E, Onitake K (2009) Identification of egg-jelly substances triggering sperm acrosome reaction in the newt, *Cynops pyrrhogaster*. Mol Reprod Dev 76(4):399–406

Watanabe T, Kubo H, Takeshima S, Nakagawa M, Ohta M, Kamimura S, Takayama-Watanabe E, Watanabe A, Onitake K (2010) Identification of the sperm motility-initiating substance in the newt, *Cynops pyrrhogaster*, and its possible relationship with the acrosome reaction during internal fertilization. Int J Dev Biol 54(4):591–597

Webb DJ, Nuccitelli R (1985) Fertilization potential and electrical properties of the *Xenopus laevis* egg. Dev Biol 107(2):395–406

Wilkinson M, Nussbaum RA (2006) Caecilian phylogeny and classification. In: Exbrayat JM (ed) Reproductive biology and phylogeny of Gymnophiona: (caecilians). Reproductive biology and phylogeny, vol 5. Science Publishers, Plymouth, pp 39–78

Wolf DP, Hedrick JL (1971) A molecular approach to fertilization. 3. Development of a bioassay for sperm capacitation. Dev Biol 25(3):360–376

Wozniak KL, Mayfield BL, Duray AM, Tembo M, Beleny DO, Napolitano MA, Sauer ML, Wisner BW, Carlson AE (2017) Extracellular Ca^{2+} is required for fertilization in the African clawed frog, *Xenopus laevis*. PLoS One 12(1):e0170405

Wozniak KL, Phelps WA, Tembo M, Lee MT, Carlson AE (2018a) The TMEM16A channel mediates the fast polyspermy block in *Xenopus laevis*. J Gen Physiol 150(9):1249–1259

Wozniak KL, Tembo M, Phelps WA, Lee MT, Carlson AE (2018b) PLC and IP3-evoked Ca^{2+} release initiate the fast block to polyspermy in *Xenopus laevis* eggs. J Gen Physiol 150(9):1239–1248

Xiang X, Burnett L, Rawls A, Bieber A, Chandler D (2004) The sperm chemoattractant "allurin" is expressed and secreted from the *Xenopus* oviduct in a hormone-regulated manner. Dev Biol 275(2):343–355

Xiang X, Kittelson A, Olson J, Bieber A, Chandler D (2005) Allurin, a 21 kD sperm chemoattractant, is rapidly released from the outermost jelly layer of the *Xenopus* egg by diffusion and medium convection. Mol Reprod Dev 70(3):344–360

Yamamoto S, Yamashita M, Iwao Y (1999) Rise of intracellular Ca^{2+} level causes the decrease of cyclin B1 and Mos in the newt eggs at fertilization. Mol Reprod Dev 53(3):341–349

Yamamoto S, Kubota HY, Yoshimoto Y, Iwao Y (2001) Injection of a sperm extract triggers egg activation in the newt *Cynops pyrrhogaster*. Dev Biol 230(1):89–99

Yamasaki H, Takamune K, Katagiri C (1988) Classification, inhibition, and specificity studies of the vitelline coat lysin from toad sperm. Gamete Res 20(3):287–300

Yokoe M, Sano M, Shibata H, Shibata D, Takayama-Watanabe E, Inaba K, Watanabe A (2014) Sperm proteases that may be involved in the initiation of sperm motility in the newt, *Cynops pyrrhogaster*. Int J Mol Sci 15(9):15210–15224

Yokoe M, Takayama-Watanabe E, Saito Y, Kutsuzawa M, Fujita K, Ochi H, Nakauchi Y, Watanabe A (2016) A novel cysteine knot protein for enhancing sperm motility that might facilitate the evolution of internal fertilization in amphibians. PLoS One 11(8):e0160445

Yoshizaki N (1989a) Comparison of two lectins isolated from *Xenopus* cortical granules. Zool Sci 6:507–514

Yoshizaki N (1989b) Immunoelectron microscopic demonstration of cortical granule lectins in coelomic, unfertilized and fertilized eggs of *Xenopus laevis*. Dev Growth Differ 31(4):325–330

Yoshizaki N, Katagiri C (1981) Oviducal contribution to alteration of the vitelline coat in the frog *Rana japonica*: an electron microscopic study. Dev Growth Differ 23:495–506

Zhang J, Xie Y, Hedrick JL, Lebrilla CB (2004) Profiling the morphological distribution of O-linked oligosaccharides. Anal Biochem 334(1):20–35

Part IV
Behavior, Ecology and Reproductive Strategies

Chapter 12
Motility and Guidance of Sea Urchin Sperm

Adán Guerrero, Hermes Gadêlha, Héctor Vicente Ramírez-Gómez, Roberto Ramírez, Carmen Beltrán, and Idan Tuval

Abstract The sperm tail movement is a direct result from contributions of fluid mechanics, elasticity, and molecular-motor activity. Within the flagellum, the axonemal engine yields overall mechanical response and, ultimately, motility. This chapter attempts to provide a comprehensive and integrative overview of the relationship between the mechanics, signaling of sperm propulsion, and the physiological function of these cells in 3D. Sperm swimming, with its intricate coupling between the regulations of the flagellar beating has to ultimately fulfill its evolutionary function honed in their natural environment, the open sea. The strategies that are being employed to unravel this fascinating and fundamental process are revisited, where the sliding of water bodies shape chemical landscapes sensed by sperms during their journey, affecting motility patterns and directly determining gamete encounter rates.

Keywords Sperm chemotaxis · Fertilization · Flagella · Turbulence

A. Guerrero (✉) · R. Ramírez
Laboratorio Nacional de Microscopía Avanzada, Instituto de Biotecnología, Universidad Nacional Autónoma de México (UNAM), Cuernavaca, Morelos, Mexico
e-mail: adanog@ibt.unam.mx

H. Gadêlha
Department of Engineering Mathematics, University of Bristol, Bristol, UK
e-mail: hermes.gadelha@bristol.ac.uk

H. V. Ramírez-Gómez · C. Beltrán
Departamento de Genética del Desarrollo y Fisiología Molecular, Instituto de Biotecnología, Universidad Nacional Autónoma de México (UNAM), Cuernavaca, Morelos, Mexico
e-mail: beltran@ibt.unam.mx

I. Tuval
Mediterranean Institute for Advanced Studies, IMEDEA (CSIC-UIB), Esporles, Balearic Islands, Spain
e-mail: ituval@imedea.uib-csic.es

M. Yoshida, J. F. Asturiano (eds.), *Reproduction in Aquatic Animals*,
https://doi.org/10.1007/978-981-15-2290-1_13

12.1 Introduction

Broadcast spawning organisms, such as marine invertebrates, release their gametes into open sea, where they are often subject to extensive dilution that reduces the probability of gamete encounter (Lotterhos et al. 2010). In many marine organisms, female gametes release diffusible molecules that attract homologous spermatozoa (Lillie 1913; Miller 1985; Suzuki 1995). Propelled by their beating flagella, spermatozoa detect and respond to chemoattractant concentration gradients by steering their swimming trajectory toward the gradient source: the egg. Though it was in bracken ferns where sperm chemotaxis was first identified (Pfeffer 1884), sea urchins are currently the best-characterized model system for studying sperm chemotaxis at a molecular level (Alvarez et al. 2012; Cook et al. 1994; Darszon et al. 2008; Strünker et al. 2015; Wood et al. 2015).

Decoding the female gamete's positional information is a process that spermatozoa undergo in a three-dimensional (3D) space. In this chapter, we present a compendium of literature showing the internal architecture of the sperm axoneme in marine invertebrates. The motility implication of having such architecture, considering the elastohydrodynamical properties of the flagella and its interaction with the surrounding milieu. The consequences of sea urchin sperm bearing an almost naked motor apparatus, which drives helical swimming behaviors, are discussed, with particular interest on consequences of the periodic sampling of chemoattractant plumes shaped by the sliding of water bodies.

12.2 The Low Reynolds Number World

The locomotion of microorganisms is dominated by the excessive viscous friction of the fluid, even when the viscosity of the fluid is as low as the viscosity of water. Due to their diminutive microscopic sizes, the fluid is "felt" by the microorganism as an incredibly thick environment. This is the so-called world of low Reynolds number. The Reynolds number measures the ratio between inertial and viscous effects in a fluid. When the Reynolds number is low, viscous friction dominates the movement. This means that none of our experiences based on our daily inertia-dominated regime may be used to understand this microscopic world with tremendous consequences on the motility of any microbe. To start with, any persistent net movement requires the continuous action of a force to drive it. If the total force vanishes at any instant the movement halts abruptly. This is very different from the inertia-dominated world we live. For example, while riding a bicycle, if a cyclist stops pedaling, the bicycle continues its movement all due to Newton's inertia.

Cilia and flagella are structurally closely-related organelles that act as motors to propel a wide range of swimming microorganisms. Their propulsion is generated by the repetitive propagation of bends along the length of the flagellum, which generates sufficient thrust to propel the microorganism forward (Gray 1955; Gray and

Hancock 1955). The locomotive properties of microorganisms have been analyzed under different perspectives (Lauga and Powers 2009; Purcell 1977; Reynolds 1883; Stokes 1851). In general, mathematical models study the swimming behavior of flagellated microorganisms by imposing a periodic deformation of their flagellum. Mathematically, this is achieved by balancing the equations that describe the total hydrodynamic forces and torques acting on the microorganism. Interesting, if the movement of the flagellum is *time-reversible*, no net propulsion can be generated, as viscous friction damps any movement arising via inertial effects in low Reynolds number regime. The latter is known as the "scallop theorem." Traveling waves, however, are not reversible in time: a video-recording of a swimming sperm played in reverse shows a very different nonsymmetric movement when compared with the normal direction of the video. This is not the case for an angular back-and-forth flapping motion of rigid rod in low Reynolds number, for example. In this case, the flapping movement always looks the same in respect of the direction of time in which the video recording is played. As a result, such simple articulated arm swimmer cannot move forward in inertialess environment. Traveling waves, however, are used across multiple scales in nature and are especially relevant for the microscopic world, as a spermatozoon would not go anywhere if the flagellum movement was reciprocal in time.

12.3 Two Historical Tales of the Sperm Tail

Sea urchin spermatozoa are pivotal for our understanding on how spermatozoa swim and flagellar bending waves emerge. Sea urchin sperm is not only conveniently available for empirical studies, but also its internal structure is simpler when compared with other species. The flagellum is composed of the axoneme surrounded by a plasma membrane, as detailed in Sect. 12.4. These are perhaps the reasons as to why sea urchin spermatozoa are a canonical model for flagella across many disciplines.

The fundamental discovery that underpins our current understanding of flagellar wave generation, the so-called sliding-filament mechanism, took place about 50 years ago. Gibbons demonstrated that ATP-induced inter-microtubule motion within the flagellar axoneme is converted into bending and was first observed in sea urchin spermatozoa (Gibbons and Rowe 1965). Gregg Gundersen described that during the annual ASCB meeting, the recordings of the ATP-induced microtubule sliding were received with "a standing ovation by the audience." Indeed, up to this point in science, flagellar waves were believed to be generated by angular actuation of the sperm head junction, rather than a distribution of forcing elements along the flagellum. The active motion of microtubules inside the axoneme observed by Gibbon's stands today as the "central dogma" behind the flagellar wave generation and self-organization. This opened the field to the fast-growing interdisciplinary research we see today, from cell biology, physiology, and microbiology to mathematics, physics, and chemistry.

Interestingly, almost one decade before Gibbons' discovery, the existence of "contractile" elements along the sperm flagellum was hypothesized by Machin (1958). Using a simple mathematical model coupling the hydrodynamic viscous forces and the elastic bending resistance experienced by the flexible tail, he demonstrated that this combined action of both viscous and elastic resistance rapidly dampens any wave generated by oscillating one end only of a passive elastic tail. By comparing his mathematical prediction with observations of sea urchin sperm, Machin theorized that the action of a *distribution* of contractile elements would be required in order to sustain the large amplitude waves observed in experiments (Gray and Hancock 1955). This early mathematical prediction demonstrated the incredible power of theoretical research. While these two developments unfolded in parallel, we can only wonder what would have happened with our understanding of flagella if Machin and Gibbons have had the chance to meet at early stages. The cornerstone of the modern interdisciplinary research pursued today embraces the two distinct, but complementary, perspectives of Gibbons and Machin, that is, observations and theory.

12.4 Flagellar Axonemes

Today we know that the axonemal flagellum is composed of an arrangement of nine pairs of microtubule doublets surrounding a central microtubule pair (Fig. 12.1) (Gibbons 1981; Lindemann and Lesich 2010). The outer doublet microtubules form a cylindrical structure interconnect via nexin links, with radial spokes linking the outer doublets to the central pair, and joint also through the active binding and unbinding of molecular motors, the dynein arms.

The axoneme is however far from being just a passive elastic structure. The shear forces are provided by dynein ATPases that power the sliding of the outer doublet microtubules relative to one another alongside the axoneme—the so-called *sliding filament mechanism* (Brokaw 1989; Gibbons and Rowe 1965; Porter and Sale 2000). In a visual perspective from the base to the tip, the dynein arms always extend in a clockwise direction, pointing toward the adjacent doublet (Gibbons 1961). This chiral distribution has direct implication on the shape of the flagellar beat cycle, as the activity of the dynein arms alternates cyclically, hence driving periodic bending of the axoneme in alternate directions.

Due to the presence of constraints such as the nexin links, radial spokes, and the basal anchor, this relative sliding motion is converted into bends of the axoneme (Brokaw 1971, 1972a, b). As a consequence, there is a cyclical bend generation and propagation of bending along the flagellum. The flagellar bends also regulate the activity of the dyneins therein (Hayashibe et al. 1997; Ishikawa and Shingyoji 2007; Lin and Nicastro 2018; Okuno and Hiramoto 1976), which concomitantly initiates self-regulatory feedbacks that alternates the microtubule sliding on opposites sides of the axoneme (Hayashi and Shingyoji 2008; Morita and Shingyoji 2004; Nakano et al. 2003).

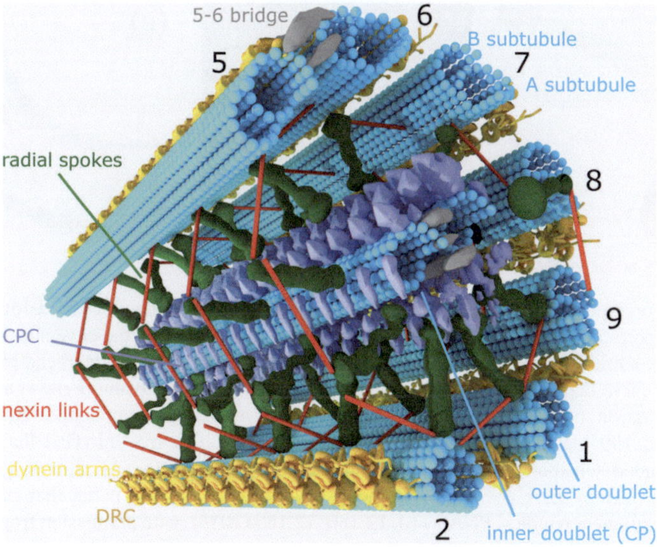

Fig. 12.1 The axoneme of sea urchin spermatozoa. Schematic 3D diagram of the axoneme of sea urchin, inspired by cryo-electron microscope studies of Nicastro et al. (2006, 2011). Lateral view of the axoneme for an observer located at the basal body. *DRC* dynein regulatory complex; *CPC* central pair complex

12.5 The Mechanical Role of Axonemal Structure

Despite the inherent complexity of flagellar axonemes described above (Afzelius 1959; Fawcett 1975; Warner and Satir 1974), and early observations of the actual microtubule sliding during wave propagation (Brokaw 1991), flagellar axonemes have been traditionally considered to behave as a simple elastic rod (Howard 2008); in other words, their bending resistance to deformation is proportional to the curvature of the bend—only true for simple elastic filaments. The striking similarity between the shape deformation of axonemes and an elastic rod under buckling, observed by Okuno (Fig. 12.2) (Okuno 1980) led to the assumption that such simplistic linear relation between bending resistance and curvature (Antman 2005) could capture the mechanical properties of flagellar axonemes. In this case, the axonemal bending stiffness was considered to be proportional to the number of constituent microtubules. It was not until the discovery of paradoxical *counterbend phenomenon* (Pelle et al. 2009) by Lindemann and co-workers the true mechanical complexity of axonemes was unveiled. The counterbend phenomenon instigates a bend in opposition to any imposed curvature in passive axonemes, i.e., axonemes in which molecular motors have been rendered passive (Fig. 12.2b). This unveiled the inadequacy of the over simplistic mechanical view of flagellar axonemes—almost 30 years after the first buckling experiments with axonemes by Okuno and Hiramoto (1979).

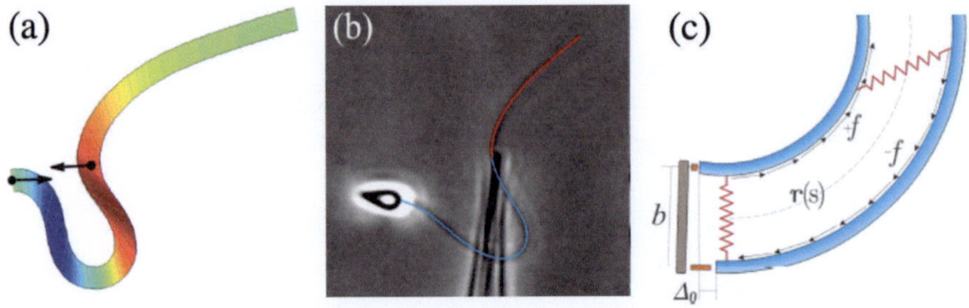

Fig. 12.2 Force-displacement experiments of flagellar axonemes showing the deformation of a sea urchin sperm axoneme rendered passive with the head attached to the coverslip while forced externally by a micro-probe. (**a**) Mathematical model prediction of the counterbend phenomenon, (**b**) overlaid curve depicts the model fit with the experiment by Lindemann (Pelle et al. 2009), (**c**) 2D representation of the sliding filament mechanism and basal compliance (Brokaw 1971; Everaers et al. 1995; Riedel-Kruse and Hilfinger 2007). We refer the reader to Gadelha et al. (2013) for an expanded schematic diagram depicting all model parameters. (All figures adapted from Gadêlha, H., Gaffney, E. A. & Goriely, A. The counterbend phenomenon in flagellar axonemes and cross-linked filament bundles. PNAS 110, 12,180–12,185 (2013), with permission from Copyright (2013) National Academy of Sciences)

The counterbend phenomenon depicted in Fig. 12.2 was later described mathematically by incorporating the interfilament sliding mechanism (Brokaw 1971; Gadelha et al. 2013) (model fitting is depicted by the red curve in Fig. 12.2b), demonstrating how the induction of curvature in one section of the axoneme instigates compensatory counter-curvature elsewhere. The flagellar counter-bending observed by Lindemann established the large-scale role of the nanometric interfilament elastic connectors within the axoneme. This established the generation of large-amplitude deformations via the non-local transmission of elastic bending moments to distant parts of the flagellum (Coy and Gadêlha 2017; Gadelha et al. 2013; Lindemann et al. 2005; Pelle et al. 2009). In particular, it was shown that counterbend deformations could be conveniently used to extract multiple axonemal material quantities, such as the bending stiffness, basal compliance, and the interfilament sliding resistance, which not be possible otherwise (Gadelha et al. 2013).

The sliding filament model is a mathematical abstraction of a rail-track, as depicted in Fig. 12.2c, d. Upon bending deformation, each filament travels distinct contour lengths (Brokaw 1971). Since points of equal contour length are connected by elastic linking proteins, a shearing force is generated along the structure proportional to the sliding displacement (Gadelha et al. 2013). At the basal end, the connection between the sperm head and the flagellum also contributes to the overall interfilament sliding. The constituent microtubules, however, are not rigidly cemented at the sperm head junction, but rather linked via connecting proteins, thus allowing some interfilament sliding compliance at base (Fawcett 1975). As a result, non-local moments generated by the sliding movement of filaments have a dramatic effect on the overall structure. This is observed in both experiments and model predictions (Fig. 12.2). The interplay between the sliding resistance and basal compliance induce a bimodal mechanical response of the axoneme, which depends on whether the filaments at the base are free to slide past each other or restricted (Coy and Gadêlha 2017; Gadelha et al. 2013). The counterbend phenomenon is maximized for axonemes that are rigidly attached at the sperm head. The amplitude of the couterbends are greatly reduced when the filaments are free to slide at base, for example, when the axoneme is detached from the sperm head (homogeneous axonemes). Interestingly, counterbends were shown to prevent violent jumps if the axoneme is abruptly forced to buckle (Gadelha 2018). These empirical and theoretical results highlight the critical importance of the nanometric structures necessary to hold the axonemal assembly together, which despite their diminutive scale, the linking proteins effectively contributes to a much larger length scale, thus dictating how the mechanical information is transmitted along the flagellum.

12.6 The Sperm Flagellum Elastohydrodynamics

The sperm flagellum undulation is a direct result of a triad of contributions coupling fluid mechanics, elasticity, and molecular motor activity (Camalet and Jülicher 2000; Gaffney et al. 2011; Hines and Blum 1972; Machin 1958; Moreau et al. 2018; Tornberg and Shelley 2004). For slender structures moving in a fluid, such as cilia and the sperm flagellum, the hydrodynamic forces experienced by a small cylindrical section of the flagellum is anisotropic and proportional to its velocity (Gray and Hancock 1955). A slender cylinder moving perpendicularly in a fluid experiences a drag force almost twice as large as if the movement is along the axis of the cylinder. This approximation is known as the resistive force theory, first demonstrated for sea urchin spermatozoa. Mathematically, the dynamics of a passive sperm flagellum in a fluid is governed by the balance of hydrodynamic and elastic moments arising from the axonemal structure, as we discussed above. The elastic and hydrodynamic combined interactions are customary termed elastohydrodynamics. The elastohydrodynamic formulation tends to be very complex, as it negotiates costs to shape changes and movement. The combined elastic and hydrodynamic effects make the tail feel a "hyperdiffusive" dissipation, that is, if the flagellum is initially deformed into a shape, the rate by which the shape unbends toward the straight configuration is much faster than diffusion (Coy and Gadêlha 2017; Machin 1958; Wiggins and Goldstein 1998).

Figure 12.3 depicts the canonical solution for the elastohydrodynamics of a passive filament angularly actuated at the left end (Coy and Gadêlha 2017). This was first derived by Machin (1958) for very small deformations. Figure 12.3 (left panel) shows how bending waves initially generated at the left end are rapidly damped along the tail, demonstrating that such simple oscillatory mechanism at the sperm head junction could not explain the large amplitude waves observed in swimming spermatozoa, thus requiring motors along the tail to sustain the waving propagation along the flagellum. Indeed, the hyperdiffusive dissipation of bending is characterized by a "bending penetration length" which measures the length by which a wave is able to permeate along the filament. Because the penetration length is a direct

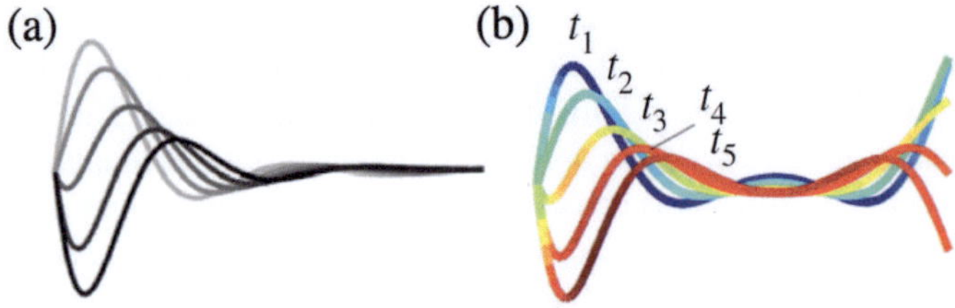

Fig. 12.3 Elastohydrodynamics of elastic filaments (**a**) and axonemes (**b**). (**a**) Purely hyperdiffusive elastohydrodynamics of a simple elastic tail as first obtained by Machin (1958) for driven angular oscillations at the left end of an elastic passive tail. Bending waves decay before reaching the end of the filament. (**b**) Counter-traveling wave phenomenon instigated by the non-local moments of sliding filament mechanism in flagellar axonemes. (Republished with permission of Royal Society, from [The counterbend dynamics of cross-linked filament bundles and flagella, Coy, R. & Gadêlha, H, 14, 20170065 (2017)]; permission conveyed through Copyright Clearance Center, Inc.)

result of both fluid and elastic properties, this may be exploited to characterize elastic properties from wide range of biological filaments (Wiggins and Goldstein 1998).

The axonemal structure however is modified by the sliding filament mechanics, as discussed above. The cross-linking moments introduce a "diffusion-like" effect into the elastohydrodynamics of an axoneme (Coy and Gadêlha 2017). When sliding movement is permitted at the base, the cross-linking axonemal diffusion only impacts the structure locally. This simply increases the overall bending stiffness of the axoneme. On the other hand, when the filament sliding is not permitted at the base, long-range, non-local curvature-reversal occurs, as shown in Fig. 12.3 (right panel). The flagellar counterbending is thus capable of generating waves that travel in opposition to the driven oscillations for passive flagella with no motor activity. This is in contrast with a simple elastic filament—compare left and right panels of Fig. 12.3. Shape deformations of passive axonemes in a viscous fluid relax a hundred times faster than simple elastic filaments. Furthermore, a relatively small inter-filament sliding displacement, up to only 30% of the axoneme diameter, was observed to induce large amplitude counterbends.

12.7 Molecular Motor Control Hypotheses and the Role of the Flagellum Length

The elastohydrodynamic response of flagellar axonemes are critically dependent on flagellar length. The sliding filament mechanism responds differently for flagella with different lengths. The counter-wave phenomena (Fig. 12.3) described above becomes increasingly more important for flagellar axonemes that are longer than 5 μm. Interestingly, the majority of eukaryotic flagella exceed this figure by a few orders of magnitude, from approximately 30 μm for *Chlamydomonas* and sea urchin sperm to almost 200 μm for quail sperm. Recently, the *local* curvature control hypothesis of molecular motor (Sartori et al. 2016) have regained empirical support when tested against short flagella experiments (*Chlamydomonas* flagella). In this length regime, the counterbend phenomenon, and thus the non-local effects introduced by the sliding filament mechanism are approximately negligible (Coy and Gadêlha 2017). This reemerging support for the curvature control hypothesis disagrees with the well-known negative support of curvature control models by Brokaw (1985), when tested against long flagella (Riedel-Kruse and Hilfinger 2007). This recurrent contradiction between interfilament sliding and curvature control hypothesis (Brokaw 1971; Brokaw and Rintala 1975; Hines and Blum 1972; Lindemann 2009; Sartori et al. 2016) may suggest the existence of distinct length-dependent self-organization regimes of axonemes.

A fundamental challenge, both experimentally and theoretically, is thus to understand how flagellar axonemes yields overall mechanical response and, ultimately, self-organization in active flagella (Lindemann 2009). This is aggravated by the increasing number of, repeatedly contradicting, control hypothesis for the flagellar wave coordination in the literature (Brokaw and Rintala 1975; Hines and Blum 1972; Lindemann 2009; Oriola and Gadêlha 2017; Sartori et al. 2016). Paradoxically, in order to induce bending waves, any flagellar control model rely on the implementation of axonemal deformations that are yet to be scrutinized in isolation, from curvature to interfilament sliding and axial distortions (Coy and Gadêlha 2017). This is aggravated by the strong coupling between the *unknown* molecular motor activity and the distinct passive components discussed above, across many length scales. The number of dissipative mechanisms involved may also lead to a non-identifiability of parameters when attempting to match the mathematical model with experiments (Plouraboué et al. 2017). Without the disentanglement between the passive and active elements, and without the rationalization of the resultant mechanical response of cross-linked axonemes, it is unclear, for example, which competing flagellar control hypothesis, if any, is able to provide a quantitative understanding of the flagellar regulation and, crucially, function of the internal components of the axoneme. This highlights the importance of future observational and theoretical studies focusing on the understanding of the underlying mechanics of this complex cell appendage, thus a fertile area for the combination of theoretical and empirical approaches (Moreau et al. 2018).

12.8 The Role of Viscosity and Potential Evolutionary Adaptations

Sea urchins are external fertilizers, and the sperm cells are released in open water for fertilization to occur. In contrast with human sperm, which is an internal fertilizer, sea urchin sperm is not required to swim in high-viscosity fluids. Nevertheless, their axoneme is a mechanotransducer, and the flagellar beat is modulated by changes in the external loading imposed by the increased fluid viscosity. The latter is a direct result of the force-velocity response of molecular motors (Oriola and Gadêlha 2017). In general, the higher the external load, the higher will be the force induced by a molecular motor, but the slower is the frequency of oscillation. Likewise, if the external load is low, less force is required, which in turn allows for high frequency of oscillation. This specific load-response of motors motivates further investigations on the impact of the fluid viscosity while modulating the waveform of the flagellum.

Woolley and Vernon (2001) observed that sea urchin sperm flagellum when swimming in a high viscosity fluid undergoes a wave-compression instability and asymmetric waveform formation (Fig. 12.4). They also observed extreme levels of wave confinement when the sperm head was attached to the coverslip Fig. 12.4e. This is in contrast with observations of human spermatozoa migrating in similar high viscosity fluids, as depicted in Fig. 12.4d, e. In this case, the flagellar waveform is characterized by a smooth meandering envelope with a gradual increase of the amplitude along its length. Sea urchin and human sperm flagella thus respond very differently to changes in fluid viscosity (Fig. 12.5). This difference between sea urchin and human sperm at high viscosities is attributed to the additional structural components that are found in mammalian sperm but not in sea urchin sperm. In

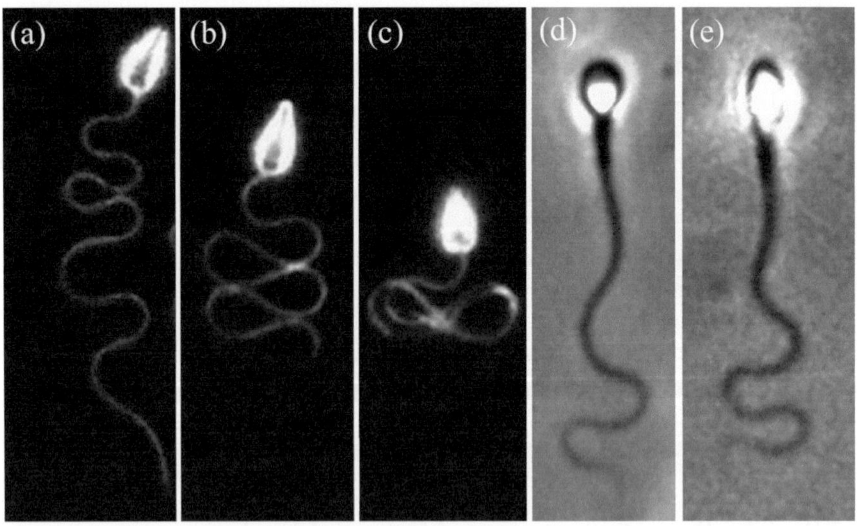

Fig. 12.4 Sea urchin sperm (**a**, **b**, **c**) and human spermatozoa (**d**, **e**) in a high viscosity fluid, containing 1% (**d**) and 2% (**a**, **b**, **c**, **e**) methylcellulose, with a nominal viscosity of 1.5 in (**a**), 2 in (**d**), and 4 in (**b**, **c**, **e**), according to the molecular mass of the methyl-cellulose. (**c**) Sea urchin sperm with its head adhered to the coverslip. ((**a–c**) Republished with permission of Company of Biologists, from [A study of helical and planar waves on sea urchin sperm flagella, with a theory of how they are generated Woolley, D. M. & Vernon, G. G., 204, 1333–1345 (2001)]; permission conveyed through Copyright Clearance Center, Inc. (**d–e**) Reprinted with permission from John Wiley and Sons from [Bend propagation in the flagella of migrating human sperm, and its modulation by viscosity, D. J. Smith, E. A. Gaffney, H. Gadêlha, et al., 66 4 17 (2009)]; license number 4603221112072)

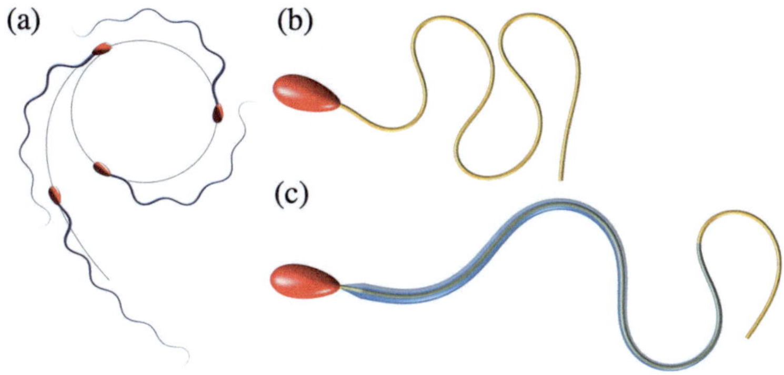

Fig. 12.5 Comparison between the virtual sea urchin-like flagellum (**a**, **b**) and human-like sperm flagellum (**c**) swimming in high-viscosity fluids. In (**c**) the flagellar axoneme is reinforced by outer-dense fibers present in internal fertilizers denoted by the blue coating. Sea urchin flagellum is only composed of the naked axoneme, thus is susceptible to wave compression instability and asymmetric waveforms capable of inducing circular swimming paths (**c**). (All figures are mathematical reconstructions from [Gadêlha, H. & Gaffney, E. A. Flagellar ultrastructure suppresses buckling instabilities and enables mammalian sperm navigation in high-viscosity media. Journal of The Royal Society Interface 16, 20180668 (2019)] and [Gadêlha, H., Gaffney, E. A., Smith, D. J. & Kirkman-Brown, J. C. Nonlinear Instability in Flagellar Dynamics: A Novel Modulation Mechanism in Sperm Migration? J. R. Soc. Interface 7, 1689–1697 (2010)] with permission from the author)

addition to the axonemal scaffolding, each of the nine concentric microtubule doublets in mammalian sperm is attached to an outer dense fiber that tapers along its length, forming a complex known as the 9 + 9 + 2 flagellum (Fawcett 1975). Little is known as to why sea urchin sperm flagella do not possess such additional structures, and likewise, what are the evolutionary pressures that lead to species-specific flagellar structural adaptation.

The elastohydrodynamic formulation described above provides a powerful tool to test different biological questions *in silico*. For example, why does the sea urchin sperm flagellum only has the canonical axoneme? Using a virtual sperm model, different features of the flagellum may be added or removed as to identify their impact on the resulting waveform. When the sea urchin-like sperm were made to swim through high-viscosity liquid, their tails quickly buckled under the excessive viscous friction (Fig. 12.5a, b)—which meant they were unable to propel themselves forward. In many cases, this forced the sperm to swim in circles (Fig. 12.5a). The sperm elastohydrodynamics predicts that under these conditions, asymmetric flagellar beating may emerge dynamically via compression-driven and buckling instabilities due to the large hydrodynamic drag experienced by the flagellar axoneme. The flagellar axoneme is thus unable to sustain the high internal compression, ultimately compromising the forward motion in a high viscosity (Fig. 12.5a, b). On the other hand, by adding the elastic contribution arising from the ultrastructural components present in the mammalian sperm flagellum, the flagellar elastic instability was suppressed (Fig. 12.5c). The latter reinforces and stabilizes the flagellum in regions where high compression may occur (Gadêlha et al. 2010), thus preventing buckling instabilities that would otherwise compromise spermatozoa swimming in high viscosity (Brokaw 2009; Kirkman-Brown and Smith 2011).

Under natural fertilizing conditions, sea urchin spermatozoa are not prone to flagellar to buckling instabilities, as they fertilize in low-viscosity seawater. This suggests that sea urchin sperm flagella could have evolved in a way to preserve the simplest form of the flagellar axoneme without changes, as sea urchins are not exposed to high viscosities during fertilization. On the other hand, quail spermatozoa also possess the simplest flagellar axoneme but, yet, is an internal fertilizer and swims in higher viscosities fluids than water. Nevertheless, its flagellar length is excessively long, about four times the length of a sea urchin sperm flagellum, for reasons that are still unknown. The latter emphasizes the challenging complexity while trying to establish the biological function of different flagellar components and its relationship with species-specific evolutionary adaptations.

12.9 The Asymmetry of the Flagellar Beat Shapes Swimming Behavior

The intrinsic asymmetry of the axoneme creates an imbalance of shear forces that results in unequal bending resistance promoting twisting and, as a consequence, the generation of local bending deflections (Chwang and Wu 1971; Cosson et al. 2003; Smith et al. 2009; Woolley 2003; Woolley and Vernon 2001). The bending generated within the axoneme is counterbalanced by the viscosity of the surrounding media, thus the overall flagellar beat pattern results from the dynamic interplay between these quantities (Brokaw 1989; Smith et al. 2009).

Since Gray (1955), it has been understood that if there is both asymmetry of the bending pattern and deflections of the swimming path, they will swim in a 3D helix, and close to boundaries will thus become "trapped" in their proximity (Gray 1955). Jennings (1901) provided us with the first report describing qualitatively that sperm display a helical trajectory in 3D. Thereafter, in 1978, Hiramoto and Baba studied sperm swimming behavior at two different depths (close to the microscope slide and 300 mm above) and observed that the speed and curvature of the trajectories were significantly different between cells swimming close to surface, and free in a volume without surface boundary effects (Hiramoto and Baba 1978). Later, the sperm swimming of both marine invertebrates (Corkidi et al. 2008; Crenshaw 1990; Jikeli et al. 2015) and mammalian spermatozoa (Su et al. 2012) was finally recorded and characterized. Hence, confirming that marine spermatozoa explore their environment via helical swimming, whereupon encountering boundaries these helices collapse to circular trajectories.

12.10 Sperm Guidance by Diffusible Cues

The intrinsic periodicity of either helical or circular swimming behavior commonly results in the periodic sampling of the cell chemical environment with direct implications for their ability to accurately determine the source of guidance molecules, which for sperm cells is the egg. This biased motility response uphill the source of chemical signals is known as sperm chemotaxis (Guerrero et al. 2010b).

Classical experiments performed by Miller (1985) showed sperm chemotaxis in several species in response to egg-jelly extracts, which were presented as diffusible compounds emanating from a glass pipette. In those elegant studies, sperm cells describing drifting circles approached toward the capillary tip (2D confinement), and in some cases, the sperm path trajectories resembled the projection of a helix pointing toward the tip of the pipette (3D swimming).

Sperm chemotaxis in confined sperm has been demonstrated for several echinoderm species (Miller 1985). From these studies, and from more recent studies, it is now understood that sperms can reorient themselves by adjusting their motion according to the intensity of an external stimulus (Böhmer et al. 2005; Brokaw 1958; Crenshaw 1989, 1990, 1993a, b; Crenshaw and Edelstein-Keshet 1993; Friedrich and Jülicher 2007, 2008, 2009; Guerrero et al. 2010a; Shiba et al. 2008). A spermatozoon can adjust its movement by changing either the direction of the principal axis of the helix (in 3D) or by altering the local curvature of its path trajectory (for confined 2D swimming). The cellular and molecular mechanisms governing sperm chemorientation and chemotaxis will be the focus of the following sections.

12.11 Sperm Chemotaxis Requires an Internal Ca^{2+} Pacemaker

The homeostasis of Ca^{2+} is fundamental for sperm chemotaxis. The blockade of Ca^{2+} exchange through the plasma membrane inhibits the Ca^{2+}-dependent motility alterations required for chemotaxis (Guerrero et al. 2010a; Kaupp et al. 2003; Wood et al. 2005, 2007; Yoshida et al. 2002). The extent of axoneme shearing affect the symmetry of the beat, higher concentrations of Ca^{2+} increase shearing, leading to an asymmetrical wave propagation during the flagellar beating (Brokaw 1979). High concentrations of this ion inhibit the activity of dyneins of doublets 3/4 on the axoneme, without affecting dynein activity on the opposite side of the axoneme (Nakano et al. 2003). Among the myriad of accessory proteins that have Ca^{2+} binding domains associated with dyneins (DiPetrillo and Smith 2009; King 2010), calaxin has been shown that has a role on regulating sperm chemotaxis (Mizuno et al. 2009). In the presence of Ca^{2+}, calaxin binds to the beta-dynein heavy chain of the outer arm dynein. Treatment of *Ciona* spermatozoa with repaglinide, a drug that inhibits Calaxin activity, impairs their chemotactic responses (Mizuno et al. 2012).

The understanding of the molecular and cellular links between Ca^{2+} signaling and sperm motility was established through the use of optochemical techniques which allow to expose swimming sea urchin spermatozoa to their conspecific attractant, in a well-controlled experimental regime (Böhmer et al. 2005; Wood et al. 2005). Currently, it is well established that chemoattractants trigger an intracellular Ca^{2+} pacemaker that drive a sequence of turns and straight swimming episodes (turn-and-run), where each turning event results from the increase in the $[Ca^{2+}]_i$ (Alvarez et al. 2012; Böhmer et al. 2005; Shiba et al. 2008; Wood et al. 2005).

12.12 A Key Element of Sperm Chemotaxis

When searching for the egg, sperm navigate following periodic paths in a chemoattractant concentration field, in which the sampling of a periodic signal, provided by the recruitment of chemoattractants, is translated by intracellular signaling into the periodic modulation of the swimming path (Friedrich and Jülicher 2007, 2008). As a result, the periodic swimming path drifts in a direction affected by the internal dynamics of the signaling system (i.e., the $[Ca^{2+}]_i$ signal) (Fig. 12.6). In this conceptual framework, the latency of the intracellular signaling is a crucial determinant of the directed looping of the swimming trajectory up the chemical concentration field. The latency can be understood as the time required for all molecular processes that occur in between the stimulus and the motility response.

This conceptual framework provides insights into the mechanism governing sperm chemotaxis; however, it does not explore the scenario whereby chemoattractants trigger an autonomous $[Ca^{2+}]_i$ pacemaker operating without the need of a periodic stimulus (Aguilera et al. 2012; Espinal et al. 2011; Priego-Espinosa et al. 2018; Wood et al. 2003). The existence of an internal Ca^{2+} pacemaker triggered by chemoattractants indicates that sperm chemotaxis operates in a dynamical space where, at least, two autonomous oscillators reach frequency entrainment.

To better understand how sperm chemotaxis operates, it is important to assess the nature of the two major driving forces governing this phenomenon (Fig. 12.7). The first is the stimulus function which results from the periodic sampling of a chemoattractant concentration field. The second oscillator is the internal Ca^{2+}-dependent pacemaker, which is triggered by the binding of chemoattractants to their receptor. This Ca^{2+} oscillator regulates the sperm swimming behavior, by altering the degree of axonemal shearing. An increase of $[Ca^{2+}]_i$ leads to an increase of asymmetrical wave propagation during the flagellar beating, hence to sharp turning events; the decrease of $[Ca^{2+}]_i$ coincides with a more symmetrical wave propagation, which drives straighter swimming responses. Both oscillators can exist without the presence of the other. The internal Ca^{2+} oscillator occurs even for immotile cells, for which the stimulus is presented as spatially uniform chemoattractant field (Wood et al. 2003; Ramírez-Gómez et al. 2018). The stimulus oscillator exists under two tested experimental negative controls: sperms swimming in high K^+ or low Ca^{2+}

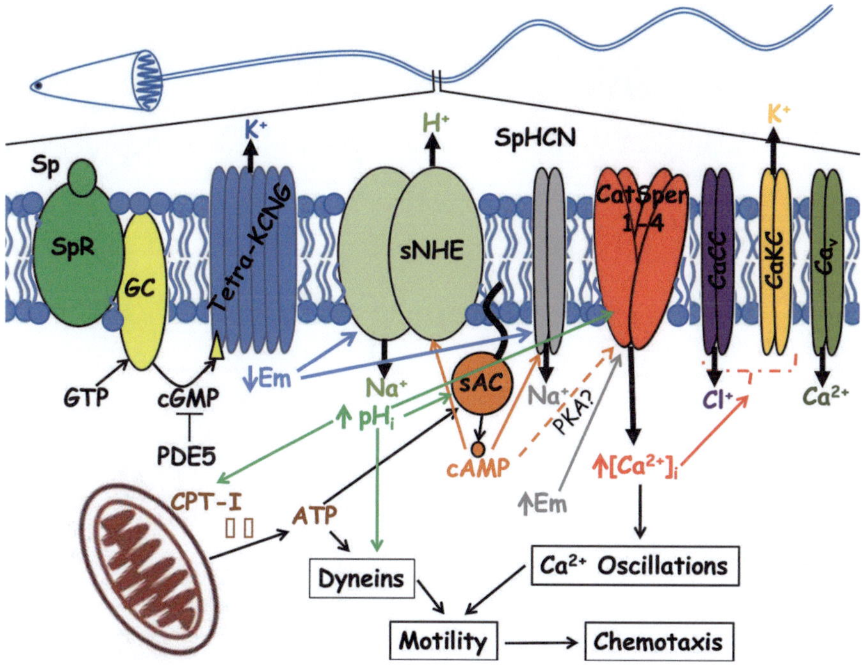

Fig. 12.6 The speract signaling cascade. The interaction of the motility regulator speract (Sp; from the outer layer of the egg), to its receptor (SpR) in the sperm flagella membrane, coupled to guanylyl cyclase (GC), elevates cGMP levels which activates (Rybalkin et al. 2003) and increases the activity of phosphodiesterase 5 (PDE5; Su and Vacquier 2006), and opens the K^+ ion channel (Tetra-KCNG), producing a K^+ efflux and a transient decrease (hyperpolarization) in the membrane potential (Em) which activates a sperm-specific Na^+/H^+ exchanger (sNHE), raising the intracellular pH (pH_i), which stimulates soluble adenylyl cyclase (sAC; Vacquier et al. 2014) increasing the cAMP levels, dynein ATPases, the cationic sperm-specific Ca^{2+} channel, CatSper (the most complex of known ion channels and whose absence causes infertility in mice) (Chung et al. 2017; Espinal-Enríquez et al. 2017; Ren et al. 2001; Seifert et al. 2015) and the carnitine palmitoyl transferase I (CPT-I; associated with the mitochondrial outer membrane) that regulates the mitochondrial metabolism to produce ATP (García-Rincón et al. 2016). The initial hyperpolarization triggered by the speract and the increase in the cAMP levels open the Na^+ channel, SpHCN, allowing the influx of Na^+ and a depolarization of the sperm. This depolarization, together with the rise in pH_i and cAMP, activates the CatSper channel that induces Ca^{2+} oscillations in the flagellum of the cell. Elevated intracellular Ca^{2+} ($[Ca^{2+}]_i$) enhances flagellar bending, leading the spermatozoon to turn. Possibly, the $[Ca^{2+}]_i$ increase also opens Ca^{2+}-regulated Cl channels (CaCC) and/or Ca^{2+}-regulated K channels (CaKC), which then contribute to hyperpolarize again the Em, removing inactivation from voltage-gated Ca^{2+} channels (Ca_v) and opening HCN channels. This mechanism is then cyclically repeated to generate a train of Ca^{2+} increases. The sequence continues until one or more of the molecular components in the pathway are downregulated. Besides, cAMP activates a poorly characterized Ca^{2+} influx pathway, which may contribute to a tonic $[Ca^{2+}]_i$ increase. Finally, the dyneins hydrolyze ATP in concert with the oscillations of $[Ca^{2+}]_i$ to regulate sperm swimming (based on Darszon et al. 2011; Espinal-Enríquez et al. 2017; González-Cota et al. 2015; Guerrero et al. 2010b; Nishigaki et al. 2014). The rise in pH_i triggered by the binding of speract to its receptor also dephosphorylates (inactivates) the GC (Ward et al. 1985)

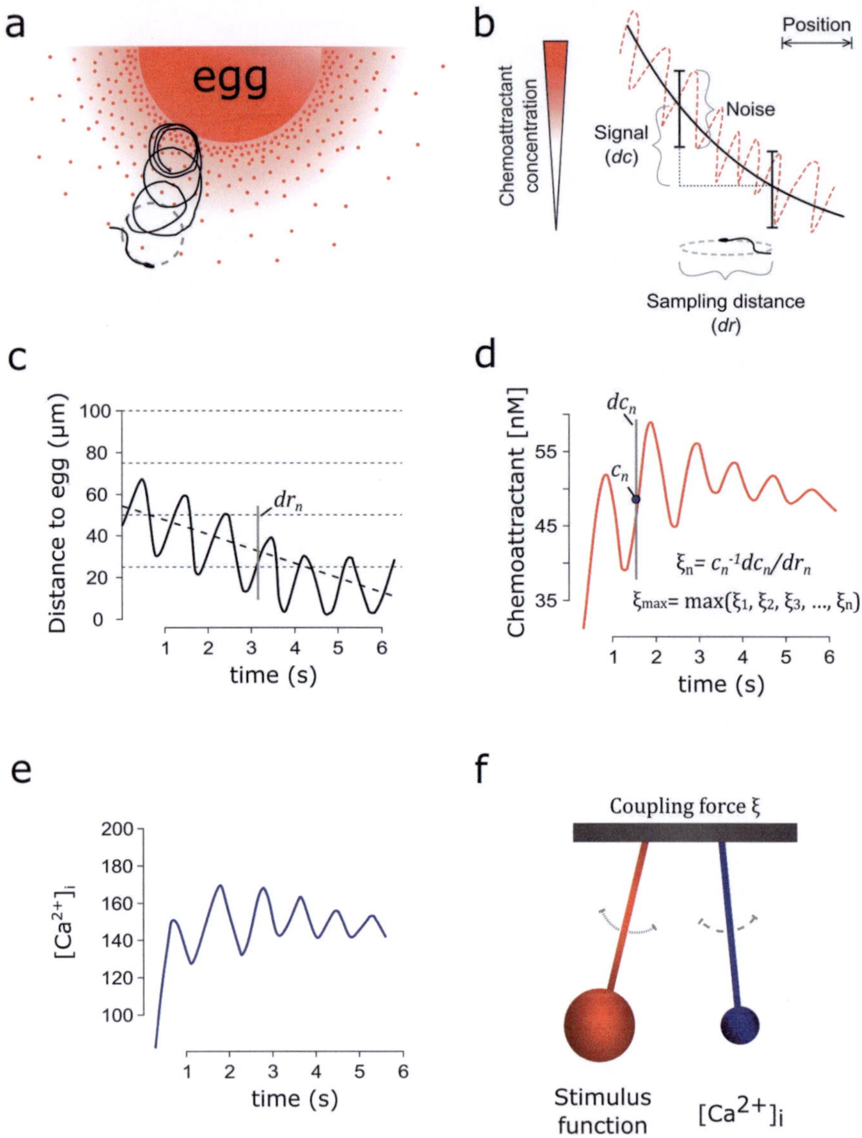

Fig. 12.7 Biophysics of chemodetection. (**a**) A sperm cell is swimming near to a chemoattractant source, in this case an egg, sensing the difference in the number of the chemoattractant molecules along a sampling distance. (**b**) Sea urchin sperm swimming in circles in a plane along a linear sampling distance, given by the diameter of the circle, is sensing a difference between the number of chemoattractant molecules that arise from the noise. (**c**) Distance of sperm to the egg while swimming to the source of the chemoattractant (see panel a, black line). (**d**) Stimulus function that a sperm swimming to the source of the chemoattractant experiments, where ξn represents each of the relative slopes of the gradient. Note that stimulus function itself is an oscillator. (**e**) $[Ca^{2+}]_i$ oscillations that a sperm experiments while swimming in the chemoattractant gradient. (**f**) Coupled oscillators model, where one oscillator is represented by the stimulus function (**d**) and the other by the changes in $[Ca^{2+}]_i$ (**e**) that the sperm experiments during its swimming toward the egg

artificial sea water, both of which inhibit the internal Ca^{2+} oscillator (Guerrero et al. 2010a; Ramírez-Gómez et al. 2018).

Once they become coupled, the two oscillators are somehow interconnected (Ramírez-Gómez et al. 2018). The stimulus oscillator implies an oscillatory recruitment of chemoattractant molecules, hence an oscillator input to the internal Ca^{2+} oscillator. Additionally, the internal Ca^{2+} oscillator drives a steering response, which by consequence leads the sperm cell to alter its way to sample the chemoattractant field.

In summary, for chemotaxis to occur, the timing of the Ca^{2+} transients triggered by the chemoattractants must also be kept in phase with the polarity of the chemoattractant concentration field (Böhmer et al. 2005; Friedrich and Jülicher 2008; Guerrero et al. 2010a; Kaupp et al. 2008). This requisite coupling ensures that the turning events start at the descending phase of the chemoattractant concentration field; otherwise, spermatozoa are driven away by Ca^{2+}-dependent motility adjustments. The periodic sampling of chemoattractants by the sperm flagellum continuously feeds back to the signaling pathway governing the intracellular Ca^{2+} oscillator, hence providing a potential coupling mechanism for sperm chemotaxis.

12.13 The Slope of the Chemical Gradient Is the Driving Force for Sperm Chemotaxis

Regardless of the swimming scenario, 2D confined vs. 3D free helical-like, sperm chemotactic responses require the coupling between the stimulus and response oscillators (Guerrero et al. 2011; Jikeli et al. 2015). The strength of coupling between both oscillators is provided by the slope of the chemoattractant concentration field (Ramírez-Gómez et al. 2018). A steep slope of the chemoattractant concentration gradient ensures the entrainment of the frequencies of the stimulus function and the internal Ca^{2+} oscillator. There is a minimum strength required for the two oscillators to couple, bellow which, the driving force for entrainment the motility response with the polarity of the gradient is lost. Within the synchronous region, the phase difference between the oscillators follows a simple relation to the strength of coupling, which is determined by the time required by the signaling events that occur in-between the binding of the chemoattractants and the motility response.

12.14 A Unified Mechanistic Signaling Model for Sperm Chemotaxis

In 1994, Cook et al. proposed that (1) decreasing chemoattractant concentration gradients increase $[Ca^{2+}]_i$ to generate chemotactic turns. (2) However, $[Ca^{2+}]_i$ is maintained low by swimming up sufficiently steep increasing chemoattractant gradients, allowing linear swimming trajectories (the "runs") until the egg is reached. Furthermore, it was proposed that (3) once the sperm enters a descending gradient, bound chemoattractants may dissociate from their receptors, allowing the resetting of the signaling pathway, hence providing free receptor for gathering fresh attractants resulting in a $[Ca^{2+}]_i$ rise (Cook et al. 1994; Miller 1985).

Even though the essence of the Ca^{2+}-dependent "turn"-and-"straight" swimming mechanism, which must be biased by the polarity of the gradient, was right, it has some conceptual inconsistencies. For example, it was subsequently demonstrated that the overall sperm navigation does not strictly follow $[Ca^{2+}]_i$ dynamics. Only the fast initial $[Ca^{2+}]_i$ rise correlates with the increase of curvature (the turning events); thereafter, the straighter swimming episodes occur even though $[Ca^{2+}]_i$ levels remain elevated, indicating that they are governed by other factors apart from those related to $[Ca^{2+}]_i$ dynamics (Darszon et al. 2008; Guerrero et al. 2010a; Wood et al. 2005). In addition, now it is known that speract binding is essentially irreversible (k_{off} 10^{-2} to 10^{-6} s^{-1}; k_{on} 10^7 M^{-1} s^{-1}); hence, receptor occupancy is practically unaltered while spermatozoa swim down the chemical gradient, thus making the proposal of signaling resetting through chemoattractant dissociation unlikely (Nishigaki et al. 2001; Nishigaki and Darszon 2000). The latter likely reflects the complex signaling network governing sperm chemotaxis.

We propose a unified model for sperm chemotaxis of marine invertebrates, the heart of which lies at the interconnection between the feedback loops that connect both the stimulus and Ca^{2+} oscillators (Fig. 12.8). In this model, the changes in membrane potential act as governor for the activity of either CatSper or voltage-gated Ca^{2+}-channels (Ca_Vs) that translates the state of the chemoattractant gradient through the rate of recruitment of chemoattractants during sperm voyage (Fig. 12.8a) (Guerrero et al. 2010a, b). In an ascending chemical gradient, the increase of receptor stimulation may lead to an extended hyperpolarization that suppresses calcium channel opening and thus prevents the $[Ca^{2+}]_i$ increase (Fig. 12.8b (i)). The hyperpolarization reverses once the sperm swims down the gradient, which—after a typical 150 ms delay (red line)—leads to the generation of a chemotactic turn that reorients it toward the gradient source (Fig. 12.8b (ii)). This repolarization results from several signaling contributions such as the inactivation of guanylyl cyclase, the reduction of cGMP levels by degradation, Na^+ influx through HCN channels, and results in the opening of CaV and/or CatSper channels (Granados-Gonzalez et al. 2005; Strünker et al. 2006; Wood et al. 2003, 2005) (Fig. 12.8b (ii)). Subsequently, at a certain time while swimming up in the ascending chemoattractant gradient, the hyperpolarized membrane potential is reestablished and maintained by a sustained cGMP synthesis due to sustained recruitment of chemoattractants (Fig. 12.8b (iii)),

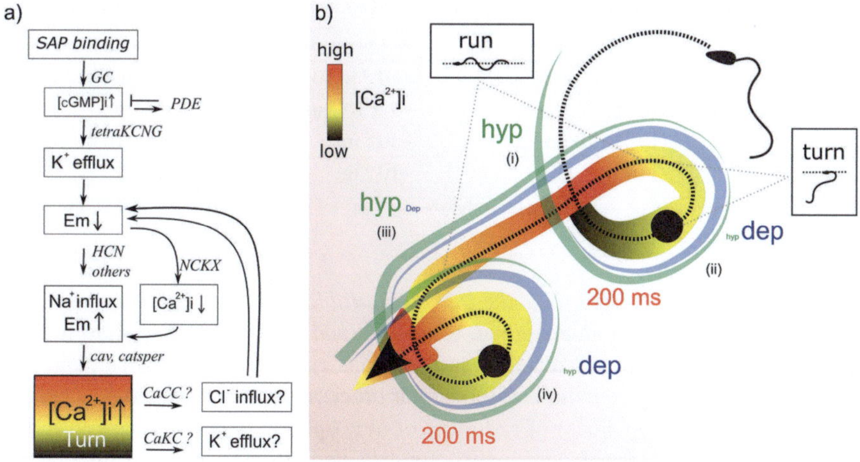

Fig. 12.8 Model of the molecular mechanisms that drive chemotaxis. (**a**) Simplified model of the major components which participate on driving the Ca^{2+} pacemaker triggered by chemoattractant such as speract or resact, both members of the family of sperm-activating peptides (SAPs) (see Fig. 12.6 for details). (**b**) A sperm swimming (depicted as a black arrow) in a chemoattractant gradient (background) undergoes cyclic changes in Em from resting to a hyperpolarized state (Hyp, green shadow) and then to a depolarized state (Dep, blue shadow) that control CaV activity. The pseudo-color bar represents the $[Ca^{2+}]_i$ changes experienced by the sperm; red and gray indicate low and high $[Ca^{2+}]_i$, respectively. Note that the straight swimming periods coincide with an interval of elevated $[Ca^{2+}]_i$ (see main text for details)

which later and once again reverts to depolarized membrane potentials as sperm starts the descending phase of the chemoattractant gradient. This sets up a sequence of chemotactic drifts, triggered by hyperpolarization/depolarization cycles translate the state of the chemoattractant gradient to $[Ca^{2+}]_i$ dynamics and hence to the Ca^{2+}-dependent regulation of axonemal components (Fig. 12.8).

Sperm swimming as described throughout this chapter, with its intricate coupling between the regulation of the flagellar beating, 3D kinematics, and axonemal elastohydrodynamics, has to ultimately fulfill its evolutionary function in the natural environment: the open sea. How do cells perceive the chemical signals released by the egg in agitated waters? To what extent oceanic turbulence affects the mechanisms described so far? Our current understanding of all these all fundamental questions is briefly dissected in Sects. 12.15 and 12.16.

12.15 Ocean Turbulence, Life in a Drop of Water

Turbulence is the prevailing physical mechanism in the transfer of energy and momentum, and in dispersing solutes, nutrients, microorganisms, and chemical signals, in the atmosphere and oceans of our planet. Although a complete definition of

Table 12.1 Typical turbulence parameters encountered by sperm cells in natural turbulent flows

Location	Energy dissipation rate, ε	Kolmogorov scale, lK	Batchelor scale, lB
Abyssal ocean	~10^{-10} to 10^{-6} m^2 s^{-3}	~10^{-2} to 10^{-3} m	~10^{-4} to 5×10^{-5} m
Continental shelf	~10^{-5} to 10^{-4} m^2 s^{-3}	~5×10^{-4} m	~10^{-5} m
Intertidal zone	~10^{-3} to 10^{-1} m^2 s^{-3}	~10^{-4} m	~5×10^{-6} m
Rocky shore	~1 m^2 s^{-3}	~10^{-5} m	~10^{-6} m

The kinematic viscosity of water is taken to be $\nu = 10^{-6}$ m^2 s^{-1} and the molecular diffusivity of the dispersed chemical $D = 10^{-9}$ m^2 s^{-1}

turbulence is often referred to as *the last open problem of classical mechanics*, we generally accept it to be an energetic, rotational, and eddying state of fluid motion that elicits the dispersion of material and the transfer of momentum, heat, and solutes at rates far higher than those of molecular processes alone. It disperses, stresses, and strains clusters (or flocs) of sediment or atmospheric dust particles and living organisms within the ocean, and it stirs, spreads, and dilutes the chemicals that are dissolved in the seawater or released into the ocean from natural and anthropogenic sources. Perhaps its most notable attribute, and one that is generally used to characterize it, is that by generating relatively large gradients of velocity at the smallest scales, turbulence promotes conditions in which viscous dissipation rapidly transfers the kinetic energy of fluid motion into heat, a process of energy transfer and dissipation. To characterize the intensity of this complex energetics, physicists use a single scalar quantity: the turbulent dissipation rate, ε, i.e., the rate at which the kinetic energy of the fluid is converted into heat.

The oceanic values of ε have a vast range, extending over ten orders of magnitude, from about 10^{-10} m^2 s^{-3} in the abyssal ocean, 10^{-1} m^2 s^{-3} in the surf zone or in fast tidal currents through straits, and up to 1 m^2 s^{-3} in the most actively turbulent regions such as for high waves breaking on the steeper slope of a typical rocky shore (see Table 12.1).

For many organisms living in the benthic zone such marine invertebrates that reproduce by external fertilization, i.e., by broadcasting gametes into the open ocean, turbulence affects fertilization success rates through a number of interconnected mechanisms: directly determining ***encounter rates***, dispersing ***chemical signals***, and affecting ***motility patterns***.

While turbulent fluctuating eddying motion spans a wide range of spatial scales, the size of sperm cells is smaller than the smallest scale at which turbulent velocity fluctuations occur: the so-called Kolmogorov scale, $l_K = (\nu^3/\varepsilon)^{1/4}$, where ν is the kinematic viscosity of water and ε the turbulent dissipation rate introduced above (Fig. 12.9). Below l_K, viscous dissipation dominates, flow is smooth, and shear is linear. Accordingly, sperm cells experience turbulence as smooth, slowly varying velocity gradients and respond solely to their local microenvironment, as ***it is relative motion that matters***. This idea pertains not only to the spread of attractants but also to the motility of cells. Discarding sperm motility as irrelevant on the basis that the swimming speed of spermatozoa is negligible compared with typical speeds of macroscopic marine flows is therefore inaccurate: with macroscopic flows the entire

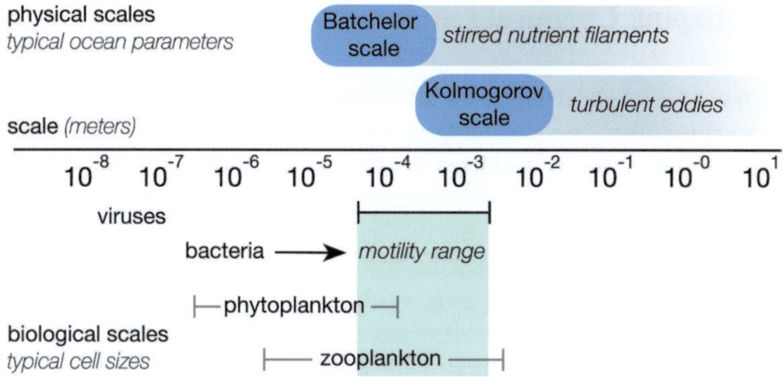

Fig. 12.9 Physical and biological length scales in the ocean. Turbulent stirring generates variance in the distribution of dissolved chemicals on scales as small as the Batchelor scale but does not directly affect the diffusive flux of chemoattractant on the scales of sperm cells. However, motile spermatozoa sample spatial scales considerably larger than their size: their "motility range" is the distance that they can travel during the lifetime of a typical chemoattractant patch, while moving up chemical gradients

system is traveling at the same speed. In other words, sperm in turbulence travel jointly with the attractants surrounding them; only relative motion, in the form of local velocity gradients or motility, can favor or hinder fertilization success. The accumulation of bacteria within the nutrient filaments produced by turbulence, recently demonstrated in numerical simulations, is an example of this process (Taylor and Stocker 2012). As eggs and sperm are released far from each other, compared to both their size and their motility range, the prime rate-limiting step in external fertilization is the encounter rate of male and female gametes. This is proportional to the instantaneous local concentration of both which, in turn, depends to a great extent on turbulent conditions. At low levels of turbulence, fertilization rates have been shown to increase with turbulent intensity (Denny et al. 2002; Mead and Denny 1995). This result contradicts elementary intuition by which dilution of gametes by mixing and, hence low fertilization, would have been predicted. One plausible explanation stems from a widely known aspect of fluid mixing in chaotic flows: it proceeds by repetitious stretching and folding of fluid parcels (Aref et al. 2017). In this fashion, distant gamete-containing blobs are also stretched into elongated filaments of high cell concentration before being folded into one another, enhancing long-range encounters between gametes. However, fertilization rates plummet above a certain level of intense turbulence (Denny et al. 2002; Mead and Denny 1995) as at excessively high shear rates (>3 s^{-1} for abalone (Riffell and Zimmer 2007; Zimmer and Riffell 2011)), sperm cells cannot progress longer steadily toward the egg but rather rotate incessantly in Jeffery orbits imposed by fluid torques (Marcos et al. 2012). Still, and as for today, little is known about how flow affects motility, especially during chemotaxis.

12.16 Shaping Chemical Landscapes

Flow further affects fertilization by deforming chemoattractant plumes. By the same incessant underlying stretching and folding stirring mechanism that brings gametes together, turbulence creates small-scale heterogeneity into our oceanic chemical landscapes. Turbulent whirls erratically weave chemoattractants into ever-finer sheets and filaments down to the Batchelor scale, the length scale below which molecular diffusion dissipates gradients, truly mixing solutes (Aref et al. 2017).

The Batchelor scale is defined as $l_B = (\nu D^2/\varepsilon)^{1/4}$. It increases with the kinematic viscosity of water, ν, and with the diffusivity of the solute, D, and decreases with the turbulent dissipation rate, ε, namely with the strength of turbulence (Aref et al. 2017; Batchelor 2000).

For typical marine conditions, the Batchelor scale ranges from 10 to 100 μm (see Table 12.1). This is a fundamental scale for a plethora of oceanographic processes that inevitably depend on the distribution of chemicals. Turbulence will ultimately stir any source of solutes into Batchelor-scale filaments. For microscopic solute patches smaller than l_B, turbulence amounts to a simple stretching. For macroscopic solute patches larger than l_B, such as the plumes of released chemoattractants, turbulence acts to fragment the patch into smaller, disconnected patches, increasing the heterogeneity of the chemical field. Beyond its effect in ultimately mixing solutes, turbulence can therefore contribute to produce a rich fabric of microscale gradients.

At the scale of gametes, fertilization can be enhanced by the ability of sperm to follow chemical signals, i.e., to perform chemotaxis (Hussain et al. 2016; Riffell and Zimmer 2007). Eggs release attractants that sperm sense by swimming in helical trajectories (Crenshaw 1996). The chemoattractant plumes released by female gametes are therefore distorted by shear, which extends the range of the chemoattractant but also dilutes its concentration. It is to these turbulence-induced Batchelor-scale microgradients that spermatozoa have to respond in their voyage toward fertilizing the egg.

The recently described differences in sensitivity to chemoattractant concentration gradients between spermatozoa of different sea urchin species (Ramírez-Gómez et al. 2018) point in this direction. Species primarily found in the low intertidal zone, amid strong wave action, relatively high levels of turbulence ($\varepsilon \sim 10^{-4}$ m^2 s^{-3}) and, hence, steep chemoattractant gradients, show relatively low sensitivity to speract. On the contrary, for species mostly found at the edge of or inside kelp beds, where the levels of turbulence are much more moderate ($\varepsilon \sim 10^{-6}$ m^2 s^{-3}) sensitivity is significantly enhanced. The observed relative sensitivity to chemoattractant fits considerably well, given the significant differences between their ecological reproduction niches and, hence, the gradients they might naturally encounter during their voyage searching for the egg in their particular habitat.

Furthermore, microscale gradients sensed by sperm cells in the oceans are time-dependent and short-lived: first, because relative motion limits the contact window between gradients and cells; second, because they eventually fade by molecular diffusion. This imposes a temporal constraint on chemotaxis, whereby the response must be sufficiently rapid to allow cells to seize the moment and move into patches of chemoattractants before these are dissipated. This contrasts with experimental conditions in most classical studies of chemotaxis, which have typically been performed for steady gradients or without explicit regard to the temporal component of the cellular response. More importantly, this temporal constraint has likely resulted in specific adaptations in marine organisms.

12.17 Conclusions

To conclude, and despite the manifest role of fluid motion in the form of turbulence in shaping ecological dynamics in the oceans, the effect of flow on chemotaxis in general—and its implication for the success of external fertilization in particular—has received too little attention to date. Thus, it represents an important avenue for future research with potentially great implications for the ecology of marine organisms.

Acknowledgments A.G. and C.B. acknowledge grants from the Programa de Apoyo a Proyectos de Investigación e Innovación Tecnológica UNAM (PAPIIT/DGAPA) IA202417 to A.G. and IN206016 and IN215519 to C.B. A.G. acknowledge the Consejo Nacional de Ciencia y Tecnologıa (CONACyT, 252213). We thank CONACYT and PAPIIT for fellowships to H.R. I.T. acknowledges the support from the Spanish Ministry of Economy and Competitiveness Grants No. FIS2016-77692-C2-1-P. We also would like to thank Shirley Ainsworth and all her group for library services and to Juan Manuel Hurtado, Roberto Rodríguez, Omar Arriaga, and Arturo Ocádiz for computer services in IBT-UNAM.

References

Afzelius B (1959) Electron microscopy of the sperm tail; results obtained with a new fixative. J Biophys Biochem Cytol 5:269–278

Aguilera LU, Galindo BE, Sánchez D, Santillán M (2012) What is the core oscillator in the speract-activated pathway of the Strongylocentrotus purpuratus sperm flagellum? Biophys J 102:2481–2488

Alvarez L, Dai L, Friedrich BM, Kashikar ND, Gregor I, Pascal R, Kaupp UB (2012) The rate of change in Ca^{2+} concentration controls sperm chemotaxis. J Cell Biol 196:653–663

Antman SS (2005) Nonlinear problems of elasticity. Applied mathematical sciences. Springer, New York

Aref H, Blake JR, Budišić M, Cardoso SSS, Cartwright JHE, Clercx HJH, El Omari K, Feudel U, Golestanian R, Gouillart E, van Heijst GF, Krasnopolskaya TS, Le Guer Y, MacKay RS, Meleshko VV, Metcalfe G, Mezić I, de Moura APS, Piro O, Speetjens MFM, Sturman R, Thiffeault J-L, Tuval I (2017) Frontiers of chaotic advection. Rev Mod Phys 89:025007

Batchelor GK (2000) An introduction to fluid dynamics. Cambridge University Press, Cambridge

Böhmer M, Van Q, Weyand I, Hagen V, Beyermann M, Matsumoto M, Hoshi M, Hildebrand E, Kaupp UB (2005) Ca^{2+} spikes in the flagellum control chemotactic behavior of sperm. EMBO J 24:2741–2752

Brokaw CJ (1958) Chemotaxis of bracken spermatozoids. The role of bimalate ions. J Exp Biol 35:192–196

Brokaw CJ (1971) Bend propagation by a sliding filament model for flagella. J Exp Biol 55:289–304
Brokaw CJ (1972a) Flagellar movement: a sliding filament model. Science (80-) 178:455–462
Brokaw CJ (1972b) Computer simulation of flagellar movement I Demonstration of stable bend propagation. Biophys J 12:564–586
Brokaw CJ (1979) Calcium-induced asymmetrical beating of triton-demembranated sea urchin sperm flagella. J Cell Biol 82:401–411
Brokaw CJ (1985) Computer simulation of flagellar movement. VI. Simple curvature-controlled models are incompletely specified. Biophys J 48:633–642
Brokaw CJ (1989) Direct measurements of sliding between outer doublet microtubules in swimming sperm flagella. Science (80-) 243:1593–1596
Brokaw CJ (1991) Microtubule sliding in swimming sperm flagella: direct and indirect measurements on sea urchin and tunicate spermatozoa [published erratum appears in *J Cell Biol* 1991 Nov;115(4):1204]. J Cell Biol 114:1201–1215
Brokaw CJ (2009) Thinking about flagellar oscillation. Cell Motil Cytoskeleton 66:425–436
Brokaw CJ, Rintala DR (1975) Computer simulation of flagellar movement. III. Models incorporating cross-bridge kinetics. J Mechanochem Cell Motil 3:77–86
Camalet S, Jülicher F (2000) Generic aspects of axonemal beating. New J Phys 2:24. https://doi.org/10.1088/1367-2630/2/1/324
Chung JJ, Miki K, Kim D, Shim SH, Shi HF, Hwang JY, Cai X, Iseri Y, Zhuang X, Clapham DE (2017) Catsperς regulates the structural continuity of sperm Ca^{2+} signaling domains and is required for normal fertility. eLife 6:1–25
Chwang AT, Wu TY (1971) A note on the helical movement of micro-organisms. Proc R Soc Lond Ser B Biol Sci 178:327–346
Cook SP, Brokaw CJ, Muller CH, Babcock DF (1994) Sperm chemotaxis: egg peptides control cytosolic calcium to regulate flagellar responses. Dev Biol 165:10–19
Corkidi G, Taboada B, Wood CD, Guerrero A, Darszon A (2008) Tracking sperm in three-dimensions. Biochem Biophys Res Commun 373:125–129
Cosson J, Huitorel P, Gagnon C (2003) How spermatozoa come to be confined to surfaces. Cell Motil Cytoskeleton 54:56–63
Coy R, Gadêlha H (2017) The counterbend dynamics of cross-linked filament bundles and flagella. J R Soc Interface 14:20170065
Crenshaw HC (1989) Kinematics of helical motion of microorganisms capable of motion with four degrees of freedom. Biophys J 56:1029–1035
Crenshaw HC (1990) Helical orientation—a novel mechanism for the orientation of microorganisms. Springer, Berlin, pp 361–386
Crenshaw HC (1993a) Orientation by helical motion—I. Kinematics of the helical motion of organisms with up to six degrees of freedom. Bull Math Biol 55:197–212
Crenshaw HC (1993b) Orientation by helical motion—III. Microorganisms can orient to stimuli by changing the direction of their rotational velocity. Bull Math Biol 55:231–255
Crenshaw HC (1996) A new look at locomotion in microorganisms: rotating and translating. Am Zool 36:608–618
Crenshaw HC, Edelstein-Keshet L (1993) Orientation by helical motion—II. Changing the direction of the axis of motion. Bull Math Biol 55:213–230
Darszon A, Guerrero A, Galindo BE, Nishigaki T, Wood CD (2008) Sperm-activating peptides in the regulation of ion fluxes, signal transduction and motility. Int J Dev Biol 52:595–606
Darszon A, Nishigaki T, Beltran C, Trevino CL, Treviño CL (2011) Calcium channels in the development, maturation, and function of spermatozoa. Physiol Rev 91:1305–1355
Denny MW, Nelson EK, Mead KS (2002) Revised estimates of the effects of turbulence on fertilization in the purple sea urchin, Strongylocentrotus purpuratus. Biol Bull 203:275–277
DiPetrillo C, Smith E (2009) Calcium regulation of ciliary motility analysis of axonemal calcium-binding proteins. Methods Cell Biol 92:163–180

Espinal J, Aldana M, Guerrero A, Wood C, Darszon A, Martínez-Mekler G (2011) Discrete dynamics model for the speract-activated Ca^{2+} signaling network relevant to sperm motility. PLoS One 6:e22619

Espinal-Enríquez J, Priego-Espinosa DA, Alberto D, Beltrán C, Martínez-Mekler G, Darszon A, Beltrán C, Martínez-Mekler G (2017) Network model predicts that CatSper is the main Ca^{2+} channel in the regulation of sea urchin sperm motility. Sci Rep 7:4236

Everaers R, Bundschuh R, Kremer K (1995) Fluctuations and stiffness of double-stranded polymers: railway-track model. EPL (Europhys. Lett.) 29:263–268

Fawcett DW (1975) The mammalian spermatozoon. Dev Biol 44:394–436

Friedrich BM, Jülicher F (2007) Chemotaxis of sperm cells. Proc Natl Acad Sci 104:13256–13261

Friedrich BM, Jülicher F (2008) The stochastic dance of circling sperm cells: sperm chemotaxis in the plane. New J Phys 10:123025

Friedrich BM, Jülicher F (2009) Steering chiral swimmers along noisy helical paths. Phys Rev Lett 103:068102

Gadelha HAB (2018) The filament-bundle elastica. IMA J Appl Math 83:634–654

Gadêlha H, Gaffney EA, Smith DJ, Kirkman-Brown JC (2010) Nonlinear instability in flagellar dynamics: a novel modulation mechanism in sperm migration? J R Soc Interface 7:1689–1697

Gadelha H, Gaffney EA, Goriely A (2013) The counterbend phenomenon in flagellar axonemes and cross-linked filament bundles. Proc Natl Acad Sci U S A 110:12180–12185

Gaffney EA, Gadêlha H, Smith DJ, Blake JR, Kirkman-Brown JC (2011) Mammalian sperm motility: observation and theory. Annu Rev Fluid Mech 43:501–528

García-Rincón J, Darszon A, Beltrán C (2016) Speract, a sea urchin egg peptide that regulates sperm motility, also stimulates sperm mitochondrial metabolism. Biochim Biophys Acta Bioenerg 1857:415–426

Gibbons IR (1961) Structural asymmetry in cilia and flagella. Nature 190:1128–1129

Gibbons IR (1981) Cilia and flagella of eukaryotes. J Cell Biol 91(3 Pt 2):107–124

Gibbons IR, Rowe AJ (1965) Dynein: a protein with adenosine triphosphatase activity from cilia. Science (80-) 149:424–426

González-Cota AL, Silva PÂ, Carneiro J, Darszon A (2015) Single cell imaging reveals that the motility regulator speract induces a flagellar alkalinization that precedes and is independent of Ca^{2+} influx in sea urchin spermatozoa. FEBS Lett 589:2146–2154

Granados-Gonzalez G, Mendoza-Lujambio I, Rodriguez E, Galindo BE, Beltrán C, Darszon A (2005) Identification of voltage-dependent Ca^{2+} channels in sea urchin sperm. FEBS Lett 579:6667–6672

Gray J (1955) The movement of sea-urchin spermatozoa. J Exp Biol 32:775–801

Gray J, Hancock GJ (1955) The propulsion of sea-urchin spermatozoa. J Exp Biol 32:802–814

Guerrero A, Nishigaki T, Carneiro J, Tatsu Y, Wood CD, Darszon A, Yoshiro T, Wood CD, Darszon A (2010a) Tuning sperm chemotaxis by calcium burst timing. Dev Biol 344:52–65

Guerrero A, Wood CD, Nishigaki T, Carneiro J, Darszon A (2010b) Tuning sperm chemotaxis. Biochem Soc Trans 38:1270–1274

Guerrero A, Carneiro J, Pimentel A, Wood CD, Corkidi G, Darszon A (2011) Strategies for locating the female gamete: the importance of measuring sperm trajectories in three spatial dimensions. Mol Hum Reprod 17:511–523

Hayashi S, Shingyoji C (2008) Mechanism of flagellar oscillation-bending-induced switching of dynein activity in elastase-treated axonemes of sea urchin sperm. J Cell Sci 121:2833–2843

Hayashibe K, Shingyoji C, Kamiya R (1997) Induction of temporary beating in paralyzed flagella of Chlamydomonas mutants by application of external force. Cell Motil Cytoskeleton 37:232–239

Hines M, Blum JJ (1972) Bend propagation in flagella. Biophys J 23:41–57

Hiramoto Y, Baba S (1978) A quantitative analysis of flagellar movement in echinoderm spermatozoa. J Exp Biol 76:85–104

Howard J (2008) Molecular mechanics of cells and tissues. Cell Mol Bioeng 1:24–32

Hussain YH, Guasto JS, Zimmer RK, Stocker R, Riffell JA (2016) Sperm chemotaxis promotes individual fertilization success in sea urchins. J Exp Biol 219:1458–1466

Ishikawa R, Shingyoji C (2007) Induction of beating by imposed bending or mechanical pulse in demembranated, motionless sea urchin sperm flagella at very low ATP concentrations. Cell Struct Funct 32:17–27

Jennings HS (1901) On the significance of the spiral swimming of organisms. Am Nat 35:369–378

Jikeli JF, Alvarez L, Friedrich BM, Wilson LG, Pascal R, Colin R, Pichlo M, Rennhack A, Brenker C, Kaupp UB (2015) Sperm navigation along helical paths in 3D chemoattractant landscapes. Nat Commun 6:7985

Kaupp UB, Solzin J, Hildebrand E, Brown JE, Helbig A, Hagen V, Beyermann M, Pampaloni F, Weyand I (2003) The signal flow and motor response controling chemotaxis of sea urchin sperm. Nat Cell Biol 5:109–117

Kaupp UB, Kashikar ND, Weyand I (2008) Mechanisms of sperm chemotaxis. Annu Rev Physiol 70:93–117

King SM (2010) Sensing the mechanical state of the axoneme and integration of Ca^{2+} signaling by outer arm dynein. Cytoskeleton 67(4):207–213

Kirkman-Brown JC, Smith DJ (2011) Sperm motility: is viscosity fundamental to progress? Mol Hum Reprod 17:539–544

Lauga E, Powers TR (2009) The hydrodynamics of swimming microorganisms. Rep Prog Phys 72:096601

Lillie FR (1913) The mechanism of fertilization. Science (80-) 38:524–528

Lin J, Nicastro D (2018) Asymmetric distribution and spatial switching of dynein activity generates ciliary motility. Science (80-) 360:eaar1968

Lindemann CB (2009) Heart of the beat (the flagellar beat, that is). Biophys J 97:2865–2866

Lindemann CB, Lesich KA (2010) Flagellar and ciliary beating: the proven and the possible. J Cell Sci 123:519–528

Lindemann CB, Macauley LJ, Lesich KA (2005) The counterbend phenomenon in dynein-disabled rat sperm flagella and what it reveals about the interdoublet elasticity. Biophys J 89:1165–1174

Lotterhos K, Levitan D, Traits G (2010) Gamete release and spawning behavior in broadcast spawning marine invertebrates. In: The evolution of primary sexual characters in animals. pp 99–120

Machin KE (1958) Wave propagation along flagella. J Exp Biol 35:796–806

Marcos FHC, Powers TR, Stocker R (2012) Bacterial rheotaxis. Proc Natl Acad Sci 109:4780–4785

Mead KS, Denny MW (1995) The effects of hydrodynamic shear stress on fertilization and early development of the purple sea urchin Strongylocentrotus purpuratus. Biol Bull 188:46–56

Miller RL (1985) Sperm chemo-orientation in the metazoa. In: Metz CB, Monroy A (eds) Biology of fertilization. Academic, New York, pp 275–337

Mizuno K, Padma P, Konno A, Satouh Y, Ogawa K, Inaba K (2009) A novel neuronal calcium sensor family protein, calaxin, is a potential Ca^{2+} –dependent regulator for the outer arm dynein of metazoan cilia and flagella. Biol Cell 101:91–103

Mizuno K, Shiba K, Okai M, Takahashi Y, Shitaka Y, Oiwa K, Tanokura M, Inaba K (2012) Calaxin drives sperm chemotaxis by Ca^{2+}–mediated direct modulation of a dynein motor. Proc Natl Acad Sci 109:20497–20502

Moreau C, Giraldi L, Gadêlha H (2018) The asymptotic coarse-graining formulation of slender-rods, bio-filaments and flagella. J R Soc Interface 15:20180235

Morita Y, Shingyoji C (2004) Effects of imposed bending on microtubule sliding in sperm flagella. Curr Biol 14:2113–2118

Nakano I, Kobayashi T, Yoshimura M, Shingyoji C (2003) Central-pair-linked regulation of microtubule sliding by calcium in flagellar axonemes. J Cell Sci 116:1627–1636

Nicastro D, Schwartz C, Pierson J, Gaudette R, Porter ME, Mcintosh R (2006) The molecular architecture of axonemes revealed by cryoelectron tomography. Science (80-) 313:944–948

Nicastro D, Fu X, Heuser T, Tso A, Porter ME, Linck RW (2011) Cryo-electron tomography reveals conserved features of doublet microtubules in flagella. Proc Natl Acad Sci 108:E845–E853

Nishigaki T, Darszon A (2000) Real-time measurements of the interactions between fluorescent speract and its sperm receptor. Dev Biol 223:17–26

Nishigaki T, Zamudio FZ, Possani LD, Darszon A (2001) Time-resolved sperm responses to an egg peptide measured by stopped-flow fluorometry. Biochem Biophys Res Commun 284:531–535

Nishigaki T, José O, González-Cota AL, Romero F, Treviño CL, Darszon A (2014) Intracellular pH in sperm physiology. Biochem Biophys Res Commun 450:1149–1158

Okuno M (1980) Inhibition and relaxation of sea urchin sperm flagella by vanadate. J Cell Biol 85:712–725

Okuno M, Hiramoto Y (1976) Mechanical stimulation of starfish sperm flagella. J Exp Biol 65:401–413

Okuno BYM, Hiramoto Y (1979) Direct measurements of the stiffness of echinoderm sperm flagella. J Exp Biol 79:235–244

Oriola D, Gadêlha H (2017) Nonlinear amplitude dynamics in flagellar beating. R Soc Open Sci 4:160698

Pelle DW, Brokaw CJ, Lesich KA, Lindemann CB (2009) Mechanical properties of the passive sea urchin sperm flagellum. Cell Motil Cytoskeleton 66:721–735

Pfeffer W (1884) Locomotorische richtungsbewegungen durch chemische reize. Untersuchungen aus dem Bot Inst zu Tübingen 1, H.3:363

Plouraboué F, Thiam EI, Delmotte B, Climent E (2017) Identification of internal properties of fibres and micro-swimmers. Proc R Soc A Math Phys Eng Sci 473:20160517

Porter ME, Sale WS (2000) The 9 + 2 axoneme anchors multiple inner arm dyneins and a network of kinases and phosphatases that control motility. J Cell Biol 151:37–42

Priego-Espinosa DA, Darszon A, Guerrero A, González-Cota AL, Nishigaki T, Martínez-Mekler G, Carneiro J (2018) Modular mathematical analysis of the control of flagellar Ca^{2+}-spike trains produced by CatSper and Ca_V channels in sea urchin sperm. bioRxiv. pp 1–43

Purcell EM (1977) Life at low Reynolds number. Am J Phys 45:3–11

Ramírez-Gómez HV, Jiménez-Sabinina V, Tuval I, Velázquez-Pérez M, Beltrán C, Carneiro J, Wood C, Darszon A, Guerrero A (2018) Sperm chemotaxis is driven by the slope of the chemoattractant concentration field bioRxiv. p 148650

Ren D, Navarro B, Perez G, Jackson AC, Hsu S, Shi Q, Tilly JL, Clapham DE (2001) A sperm ion channel required for sperm motility and male fertility. Nature 413:603–609

Reynolds O (1883) XXIX. An experimental investigation of the circumstances which determine whether the motion of water shall be direct or sinuous, and of the law of resistance in parallel channels. Philos Trans R Soc Lond 174:935–982

Riedel-Kruse IH, Hilfinger A (2007) How molecular motors shape the flagellar beat. HFSP J 1:192–208

Riffell JA, Zimmer RK (2007) Sex and flow: the consequences of fluid shear for sperm egg interactions. J Exp Biol 210:3644–3660

Rybalkin SD, Rybalkina IG, Shimizu-Albergine M, Tang XB, Beavo JA (2003) PDE5 is converted to an activated state upon cGMP binding to the GAF A domain. EMBO J 22:469–478

Sartori P, Geyer VF, Scholich A, Jülicher F, Howard J (2016) Dynamic curvature regulation accounts for the symmetric and asymmetric beats of Chlamydomonas flagella. eLife 5:1–26

Seifert R, Flick M, Bönigk W, Alvarez L, Trötschel C, Poetsch A, Müller A, Goodwin N, Pelzer P, Kashikar ND, Kremmer E, Jikeli J, Timmermann B, Kuhl H, Fridman D, Windler F, Kaupp UB, Strünker T (2015) The CatSper channel controls chemosensation in sea urchin sperm. EMBO J 34:379–392

Shiba K, Baba SA, Inoue T, Yoshida M (2008) Ca^{2+} bursts occur around a local minimal concentration of attractant and trigger sperm chemotactic response. Proc Natl Acad Sci U S A 105:19312–19317

Smith DJ, Gaffney EA, Gadelha H, Kapur N, Kirkman-Brown JC (2009) Bend propagation in the flagella of migrating human sperm, and its modulation by viscosity. Cell Motil Cytoskeleton 66:220–236

Stokes GG (1851) On the effect of the internal friction of fluids on the motion of pendulums. Mathematical and physical papers. Cambridge University Press, Cambridge, pp 1–10

Strünker T, Weyand I, Bönigk W, Van Q, Loogen A, Brown JE, Kashikar N, Hagen V, Krause E, Kaupp UB (2006) A K^+-selective cGMP-gated ion channel controls chemosensation of sperm. Nat Cell Biol 8:1149–1154

Strünker T, Alvarez L, Kaupp UB (2015) At the physical limit—chemosensation in sperm. Curr Opin Neurobiol 34:110–116

Su Y-H, Vacquier VD (2006) Cyclic GMP-specific phosphodiesterase-5 regulates motility of sea urchin Spermatozoa. Mol Biol Cell 17:114–121

Su TT-W, Xue L, Ozcan A (2012) High-throughput lensfree 3D tracking of human sperms reveals rare statistics of helical trajectories. Proc Natl Acad Sci 109:16018–16022

Suzuki N (1995) Structure, function and biosynthesis of sperm-activating peptides and fucose sulfate glycoconjugate in the extracellular coat of sea urchin eggs. Zool Sci 12:13–27

Taylor JR, Stocker R (2012) Trade-offs of chemotactic foraging in turbulent water. Science (80-) 338:675–679

Tornberg AK, Shelley MJ (2004) Simulating the dynamics and interactions of flexible fibers in Stokes flows. J Comput Phys 196:8–40

Vacquier VD, Loza-Huerta A, García-Rincón J, Darszon A, Beltrán C (2014) Soluble adenylyl cyclase of sea urchin spermatozoa. Biochim Biophys Acta Mol basis Dis 1842:2621–2628

Ward GE, Brokaw CJ, Garbers DL, Vacquier VD (1985) Chemotaxis of Arbacia punctulata spermatozoa to resact, a peptide from the egg jelly layer. J Cell Biol 101(6):2324–2329

Warner FD, Satir P (1974) The structural basis of ciliary bend formation. J Cell Biol 63:35–63

Wiggins CH, Goldstein RE (1998) Flexive and propulsive dynamics of elastica at low reynolds number. Phys Rev Lett 80:3879–3882

Wood CD, Darszon A, Whitaker M (2003) Speract induces calcium oscillations in the sperm tail. J Cell Biol 161:89–101

Wood CD, Nishigaki T, Furuta T, Baba SA, Darszon A (2005) Real-time analysis of the role of Ca^{2+} in flagellar movement and motility in single sea urchin sperm. J Cell Biol 169:725–731

Wood CD, Nishigaki T, Tatsu Y, Yumoto N, Baba SA, Whitaker M, Darszon A (2007) Altering the speract-induced ion permeability changes that generate flagellar Ca^{2+} spikes regulates their kinetics and sea urchin sperm motility. Dev Biol 306:525–537

Wood CD, Guerrero A, Priego-Espinosa DA, Martínez-Mekler G, Carneiro J, Darszon A (2015) Sea urchin sperm chemotaxis. In: Flagellar mechanics and sperm guidance. Bentham Science Publishers, Sharjah

Woolley DM (2003) Motility of spermatozoa at surfaces. Reproduction 126:259–270

Woolley DM, Vernon GG (2001) A study of helical and planar waves on sea urchin sperm flagella, with a theory of how they are generated. J Exp Biol 204:1333–1345

Yoshida M, Murata M, Inaba K, Morisawa M (2002) A chemoattractant for ascidian spermatozoa is a sulfated steroid. Proc Natl Acad Sci U S A 99:14831–14836

Zimmer RK, Riffell JA (2011) Sperm chemotaxis, fluid shear, and the evolution of sexual reproduction. Proc Natl Acad Sci 108:13200–13205

Chapter 13
Behavior and Fertilization of Squids

Yoko Iwata and Noritaka Hirohashi

Abstract Recent studies have highlighted the intriguing reproductive behaviors in some coastal squid species. Here we review briefly pre-copulative male–male and male–female interactions, alternative mating tactics and their associating sperm traits in squids. In one scenario, males compete with one another for mating opportunities by conventional means such as body pattern displays. After the male–male contests being settled, the winners are allowed to mate with the female more predominantly over others by transferring their spermatophores to the particular location of the female body. In another scenario, males pursue alternative reproductive tactics by which male individuals display either mate guarding or sneaking behavior, resulting in insemination dimorphism with different fertilization success, i.e., sperm transfer occurs near the oviduct or around the mouth adjacent to the seminal receptacle, in favor of leaching to eggs (immediate fertilization) or the seminal receptacle (storage), respectively. These characteristics of male mating behavior can lead to diverged evolution in sperm morphology, physiology, and behavior. Hence, the reproductive systems of squids would provide rare opportunities to analyze sperm evolution mechanisms.

Keywords Squid · Sperm competition · Alternative reproductive tactics · Sperm dimorphism · Chemotaxis

13.1 Introduction

To achieve reproductive success, males need to conquest several behavioral steps. First, to achieve successful mating, males need to search females, fight with other males for females (male–male competition), court females, be chosen by the females

Y. Iwata (✉)
Atmosphere and Ocean Research Institute, The University of Tokyo, Kashiwa, Chiba, Japan
e-mail: iwayou@aori.u-tokyo.ac.jp

N. Hirohashi
Oki Marine Biological Station, Shimane University, Oki, Shimane, Japan

M. Yoshida, J. F. Asturiano (eds.), *Reproduction in Aquatic Animals*,
https://doi.org/10.1007/978-981-15-2290-1_14

(mate choice) and release adequate amount of sperm. Even if males can mate with females successfully, especially in copulative species, transferred sperm still need to conquest several steps to achieve fertilization. Sperm need to migrate to sperm storage organ (Orr and Zuk 2012), survive therein until spawning, swim toward ova with competition with rival sperm (sperm competition; Parker 1970) and be chosen by females (cryptic female choice; Eberhard 1996). Each step of this process can be a target for sexual selection.

Squid (Cephalopoda, Coleoidea, Decapodiforms) are sexually reproducing organisms and terminal spawners (die after single spawning season). As squids are highly mobile and do not have reproductive territory, the information about reproductive biology in natural spawning ground has been limited. However, accumulation of knowledge by both underwater survey and captive experiments in aquarium during last two decades have revealed many unique characteristics in the reproduction of this group. In this chapter, we introduce reproductive biology, mainly about coastal squid in the family loliginidae having alternative reproductive tactics (ARTs) and sperm traits adapting to an each mating tactic.

13.2 Reproductive Behavior in Squids

13.2.1 Courtship, Male Competition, and Copulation

Males often conduct conventional behaviors, involving visual, auditory, olfactory, and vibratory displays (Riechert 1984). Such stereotyped signaling can be favored by male–male competition and mate choice to indicate their competitive ability to other males or their attractiveness to females (Andersson 1994). Cephalopods are regarded as the most intelligent animal group among invertebrates with a developed nerve system. Their visual communication is also highly developed, i.e., they can express various body patterns by combining chromatic, postural, and locomotor components (Hanlon and Messenger 2018). Chromatophores are principal elements of body patterning controlled directly by the brain. These animals use body patterning not only for camouflage but also for intra- and inter-specific communications, and coastal species tend to show more various, complex body patterns than oceanic species. The chromatophoric components are described in some coastal squids, *Loligo reynaudii* (Hanlon et al. 1994), *L. pealei* (Hanlon et al. 1999), *L. opalescens* (Hunt et al. 2000), and *Sepioteuthis lessoniana* (Lin and Chao 2017; Lin et al. 2017).

Although the definite meaning of each body pattern in cephalopods is largely unclear, some body patterns are sex specific (Hanlon et al. 1994), suggesting that the patterns would closely be associated with sexual behaviors such as male–male competition for mates and courtship. Figure 13.1 shows sex-specific body patterns observed in mature males and females in *Heterololigo bleekeri*. Lateral chromatophoric patterns are often used for reproductive behaviors in common with some species in this group. Male–male competition in loliginid squid has escalating structure from visual display to physical contact (Di Marco and Hanlon 1997). The outcome of the competition is influenced by the relative body size of competing males, and male–male competition escalates more when the body size difference between competing males is smaller. The behavioral characteristics, displaying lateral line (mantle margin stripe and arm stripe, Fig. 13.1), positioning parallel to rival male, and adjusting the tip of mantle and stretch tentacles would be useful to compare their body size that closely correlate with their competitive abilities. Agonistic display with lateral chromatophoric pattern often follows physical contact with fins ("fin-beating" behavior). Allometric analysis of various body parts in loliginid squid showed that the fins are the body parts having the strongest correlation with mantle

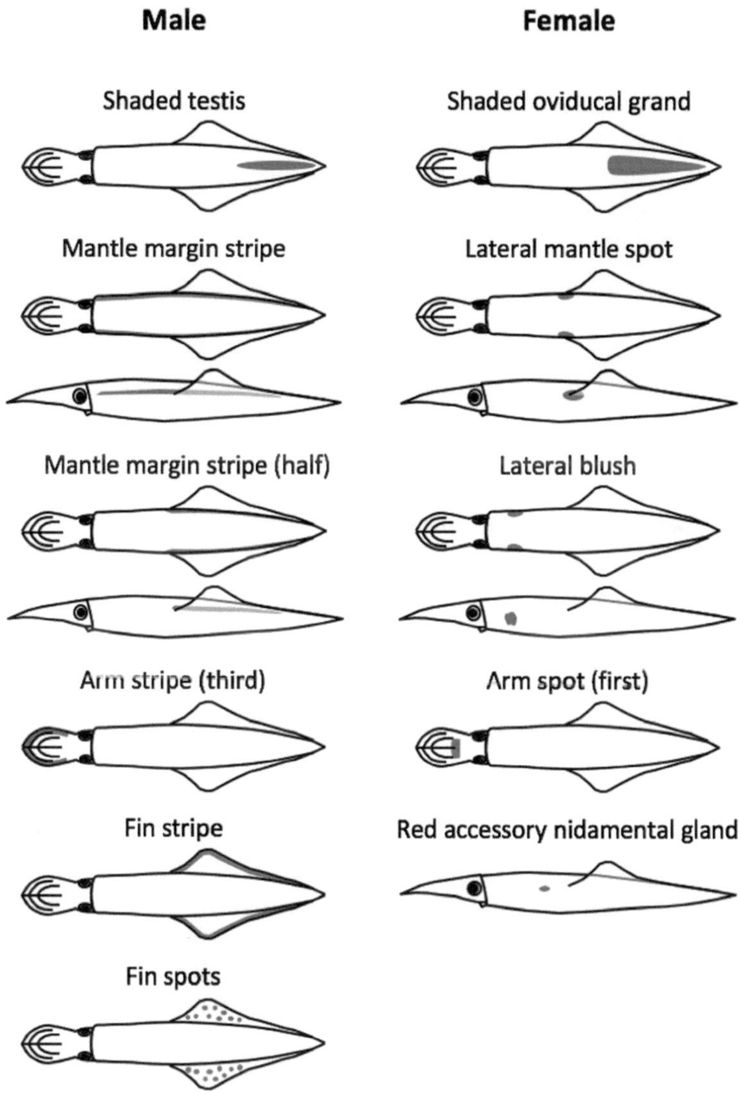

Fig. 13.1 Sex-specific body pattern in *Heterololigo bleekeri*

length (Iwata and Sakurai 2007), suggesting that the fin-beating would be an effective behavior to compare the competitive abilities.

Females of some loliginid species and other species such as *Todarodes pacificus* are known to show various sorts of spot, blush, or bar on the lateral mantle, probably showing the female choice toward approaching males (Hanlon and Messenger 2018). Schnell et al. (2015) showed that in cuttlefish *Sepia apama*, females that display white lateral stripe rejected 78% of mating, whereas those without displaying it accepted 80% of mating.

Both male and female have a body pattern which shadow their gonad by dark color chromatophores ("shaded testis" and "shaded oviducal gland" in Fig. 13.1). Because gonads are visible through translucent body muscle, it is possible to demonstrate one's reproductive status to others as an honest signal. By contrast, one might shadow gonad if showing the honest signal is somewhat inconvenient for one's mating opportunities. Shadowing the gonad color might be a negative signal for males to avoid unnecessary or hopeless fights for mating with other males, and for females to avoid unpleasant mating approach.

In addition to the intriguing pre-copulative male–male and male–female interactions, one of the characteristics in cephalopod reproduction is to transfer sperm from male to female. Males produce the spermatophores, which are cylindrical capsules containing sperm (Drew 1911). During mating, the spermatophores undergo the "spermatophoric reaction," a complex process of the spermatophoric tunics and membranes leading to extrusion of the sperm mass (Drew 1911; Mann 1984; Marian 2012). The spermatangium, the everted spermatophore containing the sperm mass, attaches to the female body by the mechanical anchorage provided by the ejaculatory apparatus and chemical adhesion by the cement body on the head of spermatangium (Hoving et al. 2008, 2009; Marian 2012, 2015).

13.2.2 Mating Systems

Squids are as a rule polyandrous (a single female mates with multiple males) (Franklin and Stuart-Fox 2017; Iwata et al. 2011; Naud et al. 2016; Sato et al. 2017) and semelparous (death after reproduction), which may bring about a situation where a male–male competition toward female becomes so intense that alternative male mating behavior could be introduced. This phenomenon is collectively called alternative reproductive tactics (ARTs) by which extra-pair (or sneak) copulations by small males could have some sort of reproductive fitness.

Several species in the family loliginidae such as *L. pealeii* (Hanlon 1996), *L. reynaudii* (Hanlon et al. 2002), *H. bleekeri* (Iwata et al. 2005), *S. lessoniana* (Wada et al. 2005; Lin et al. 2017), *S. australis* (Jantzen and Havenhand 2003), and perhaps *Uroteuthis edulis* (Hirohashi et al. 2016a) are known to adopt this strategy that gives rise to insemination dimorphism, i.e., large consort and small sneaker males inseminate at the proximal (on/near oviduct in mantle cavity) and distal site (seminal receptacle near mouth) for fertilization, respectively, resulting in difference in fertilization success. Paternity rates are 90% by consorts and 10% by sneakers in *H. bleekeri* (Iwata et al. 2005).

ARTs are known not only in those coastal squid species but also in an octopus *Abdopus aculeatus* (Huffard et al. 2008) and a cuttlefish *S. apama* (Hall and Hanlon 2002; Naud et al. 2004; Hanlon et al. 2005). However, it remains unknown what condition could have driven ARTs in these species because there are large number of highly polyandrous species without ARTs.

Squid ARTs are unique among others (i.e., fish, insects, even cuttlefish and octopus) in that females have developed the seminal receptacle, sperm storage organ specific to extra-pair copulations. It is worthwhile to mention that females of some non-ARTs species also have the seminal receptacle and males transfer sperm around there, and even in ARTs species, the seminal receptacle is used to store sperm by small males even in the absence of rival consort males. This suggests that sneak copulations by extra- or small males are unlikely to be an alternative, derivative method against an authentic mating tactic; rather it is a prototypical mating form succeeded in the particular cephalopod lineages.

13.3 Characteristic Postcopulatory Sexual Selection Leads to Dimorphic Sperm Strategies in Squids

13.3.1 Sperm Competition and Sperm Allocation Theory

Male competitions continue after mating—multiple mating by female generates the competition between ejaculates to fertilize a set of ova (Parker 1970; Birkhead and Møller 1998). Energetic cost to produce gamete cells is greatly different between males and females, and the cost to produce sperm is greatly cheaper than that of ova. However, sperm production generates nontrivial costs (Dewsbury 1982; Nakatsuru and Kramer 1982); males should use minimum number of sperm to ensure fertilization if there is no sperm competition. Parker (1998) established a theoretical framework on sperm allocation strategy. He used game models to seek evolutionarily stable strategy on sperm expenditures, and predicted that individual males should change the number of sperm ejaculated per copulation depending on sperm competition risk at the certain copulation (Parker 1990a). Differences in sperm competition risk among alternative mating behaviors would affect internal morphology and physiology. Against consort males that have strong competitive ability with full-developed external secondary sexual character, sneaker males expend reproductive effort on sperm production to compensate disadvantage in fertilization with constantly high sperm competition risk. Parker (1990b) also predicted that sneaker males should expend more sperm than consort males do, as sneakers always face sperm competition but consort males sometimes can monopolize females and might not always face sperm competition with sneakers.

Many studies in various species support the prediction that sperm competition risks affect internal secondary sexual character such as testis size (mammals, Dixson and Anderson 2004; birds Møller 1991; fish, Stockley et al. 1997) and ejaculate characters (externally fertilizing fishes: Leach and Montgomerie 2000; internally fertilizing fishes: Pilastro and Bisazza 1999, Evans et al. 2003; birds: Hunter et al. 2000, Nicholls et al. 2001). In these species, male individuals are different from each other in sperm release timing and/or mating order; however, they share the same sperm transfer site on female, thus fertilization environment should not differ among males.

13.3.2 Sperm Allocation in Squids

In loliginid squid, however, sperm transfer sites associating with ARTs are absolutely independent. It is highly possible that males specialize ejaculation strategy in each sperm transfer site. The lifespan of loliginid squid is 1 year, and males produce and store a few hundreds of the spermatophores in the storage organ (Needham's sac) until use. Therefore, ejaculate characters in male individuals could be determined when they mature, and the sperm allocation strategy is measurable with the size of the spermatophores. Therefore, loliginid species is ideally suited to study ejaculate strategies in relation to individual competitive ability, mating behavior, and sperm competition risk.

The relationship between mantle length and spermatophore length is clearly discontinuous, suggesting that it has dimorphism in sperm allocation strategy in *H. bleekeri* (Iwata and Sakurai 2007). There was a morphological switch point where the relationship between mantle length and spermatophore length changed, and males reached at the switch point produced relatively longer spermatophores, but males without reaching at the switch point produced relatively shorter spermatophores.

In *H. bleekeri*, not only spermatophore size but also spermatangium shows dimorphism (Iwata et al. 2015). The spermatangia produced by small males are drop-like, and the spermatangia produced by large males are rope-like. Furthermore, spermatangia produced by small males are more likely to have a spine-shaped oral extremity on the head (Iwata et al. 2015).

The spermatangium dimorphism is closely associated with the alternative sperm transfer sites, which potentially present two different environments between consort and sneaker males. The drop-like shape of the sneaker male's spermatangium could render the surface area to be small relative to its volume, which might be an adaptive trait to reduce the risk of detachment by water flow. Although the strength of water flow in the arm crown and in the mantle cavity of living females is difficult to be measured, there must be some, but less, water flow at both sperm-attachment sites compared with the other parts of the female's body surface. Naud and Havenhand (2006) proposed a term "confined external fertilization" to describe the fertilization situation in cuttlefish having a single sperm transfer site in the arm crown—the fertilization is external but contained within a confined volume formed by the animal. Because consort males of *H. bleekeri* transfer spermatophores to the inside wall of the oviduct, their spermatangia must experience far less turbulence than they would elsewhere in the mantle cavity where water is continuously circulated. However, it is unclear which sperm-attachment site has higher risk of spermatangium loss and/or sperm dilution. A similar dimorphism in spermatophore and spermatangia has been reported in some loliginid species (*Doryteuthis plei*; Apostólico and Marian 2017, *L. reynaudii*; Iwata et al. 2018), suggesting that the dimorphisms associating with the ARTs are the common phenomena in this family.

13.3.3 Sperm Dimorphism

It has been recently suggested that males under sperm competition invest energy not only in sperm number but also in sperm quality, such as size, velocity, and mobility (e.g., Leach and Montgomerie 2000; Vladić and Järvi 2001; Uglem et al. 2001). Males should invest minimum number of sperm required for fertilization. However, when available resources for reproduction are limited, trade-off between sperm number and sperm quality could happen (Snook 2005).

In *H. bleekeri*, the length of the spermatophore is longer in consorts than in sneakers and sperm number in a spermatophore is ~5-fold greater in consorts than in sneakers (Iwata et al. 2011). This is contrary to the general prediction that sneaker males having strong sperm competition risk inseminate more sperm per mate than consort males do (Parker 1990b). Immature female loliginid squid already stores sperm in the seminal receptacle (Drew 1911), suggesting that the seminal receptacle is not only for sneaking copulation. Even if consort males can prevent mating by other consort and sneaker males from females before spawning, females already have some sperm in the seminal receptacle. In squid, consort males always face sperm competition same as sneakers. In addition, sperm competition between sneakers and consorts would occur at different stage other than insemination, therefore ejaculation volume would not simply correlate with fertilization success. Because of the difference of sperm transfer sites, sperm of consort males can meet egg earlier, therefore sperm of sneaker males are under higher sperm competition risk. Physical (water flow, etc.) and biological (chemical component, etc.) environments at the two sperm storage sites would be different. In addition, sneaker sperm must travel from the spermatangium to the seminal receptacle where sperm become quiescently stored until time of spawning. Considering such postcopulatory circumstances, allocating more energy to increase sperm quality would be adaptive for sneaker males to achieve fertilization success.

In *H. bleekeri*, sperm of sneaker males is about 50% longer than that of consort males (Iwata et al. 2011). Sperm dimorphism, i.e., two different forms of sperm within a species, has been in some species, but in all cases, each individual produces both types of sperm simultaneously, one is eusperm (true fertile sperm), and the other is parasperm (fake sperm) that will help sibling eusperm or interfere with sperm from rival males. Dimorphic sperm observed in squid are both fertile and one individual has only one of the two types of sperm, therefore it is a novel discovery of dimorphic eusperm.

A positive relationship between sperm size and swimming velocity is reported in many species (e.g., Fitzpatrick et al. 2009). However, the difference in flagellar length does not result in the difference in the swimming velocity in *H. bleekeri*. Large sperm have an advantage to occupy the limited space within the seminal receptacle to outcompete rival sperm (LaMunyon and Ward 1998). However, it is unlikely that sperm with longer flagellum are preferentially stored within the seminal receptacle because the size distribution of sperm population within the seminal receptacle did not differ from that within the sperm mass. Therefore, sperm competition does not simply explain why sneaker males produce longer sperm. Considering the previous studies with other taxa that species with ARTs did not show clear difference between sneaker and consort sperm, the characteristic having different fertilization environments at the two sperm storage sites would be important for the evolution of sperm dimorphism in the squid. Nonetheless, why sneaker sperm have a longer flagellum still remains to be determined.

13.3.4 Sperm Behavior and Chemotaxis

Despite unsolved problems in ARTs ontogeny in squids, there are predominant sperm traits that differ between two tactics in *H. bleekeri, L. reynaudii,* and *U. edulis*, i.e., sperm from sneaker males are capable of clustering when released from the spermatangium, whereas sperm from consort males do not show such a clustering behavior (Hirohashi et al. 2013). Serendipitously, we found that clustering is caused by sperm chemotaxis in response to carbon dioxide emitted as a result of aerobic respiration. Experimentally, sperm were capable of swarming at vicinity of a bubble of CO_2 or a flush spot where aqueous CO_2 emerged instantaneously by uncaging the caged-carbonate (Hirohashi et al. 2013).

CO_2 is a rare but ubiquitous gas in the biosphere of the current Earth planet. Atmospheric CO_2 concentrations over past 0.4 M years were fluctuated in the range between 180 and 280 ppt; however, currently it continues to increase and exceeded 400 ppm, causing global warming and ocean acidification (the Intergovernmental Panel on Climate Change; http://www.ipcc.ch/). Massive absorption of atmospheric CO_2 by ocean resulted in carbonation (carbonic acid) in water and hence lowering pH. Environmental CO_2 serves as a chemical cue for various animal behaviors such as host seeking in mosquitos (McMeniman et al. 2014), food searching in fruit flies (Jones 2013), and avoidance behavior in rodents (Mietz 1997) and *Caenorhabditis elegans* (Hallem and Sternberg 2008). In aquatic (marine) environment, catfish could detect a minute change of $\leq$0.1 pH to find cryptic respiring prey (Caprio et al. 2014).

Neural cells are the primary sensors that can sense CO_2, hydrogen carbonate or proton due to rapid chemical equilibrium: $CO_2 + H_2O \leftrightarrows HCO_3^- + H^+$. Carbonic anhydrases (CAs) accelerate CO_2 hydration at rates as high as a million times a second. Because CO_2 gas can permeate cell membranes, CO_2 hydration can occur either at the plasma membranes or inside the cell. For example, in mouse olfactory neurons, CO_2 is hydrated by a cytoplasmic CAII and resulting bicarbonate activates a membrane guanylyl cyclase-D that elicits downstream signaling (Mietz 1997). In mouse amygdala, extracellular protons generated from CO_2 hydration by a membrane-anchored CA are taken up into the cell through an ASIC1a channel, causing intracellular acidification and therefore evoking fear sensation (Yang et al. 2013; Coryell et al. 2008).

In case of squid sperm, a membrane-bound CA acidifies the microenvironments of the preformed sperm cluster, making a proton gradient developed radially from center of sperm mass (Iida et al. 2017). Sperm swimming along a descendant proton gradient makes a quick turn at a pH threshold of ~5.5 to reorient to the source of acidity, thus maintaining self-clustering (Iida et al. 2017). Cell signaling that underlies pH/CO_2-taxis is unknown; however, modulating ion permeability for proton and Ca^{2+} is key to regulate flagellar movements as indicated by sperm chemotaxis in sea urchins and other marine invertebrates (Hirohashi et al. 2013).

Besides ionic control, there are number of similarities between egg-guided sperm chemotaxis and squid's sperm pH/CO_2-taxis: (1) a chemical gradient guides swimming direction and (2) reorientation involves switching between two modes of flagellar motions known for "straight" and "turn," (3) which is enabled by flagellar beat symmetry and its break (Iida et al. 2017). In contrast, prominent differences are the presence/absence of the target (an egg) and receptor for a chemoattractant. In common, chemotactic cells can modulate sensitivity over several orders of magnitude, facilitating long-distance directional movement. By contrast, squid sperm can only detect a threshold pH at around 5.5 and rate of pH change to decide whether or not they make a turn, thus turning points are strictly limited (Iida et al. 2017).

Why sperm from only sneaker male squid form the cluster? This perhaps relates to insemination environment, postcopulatory sperm storage or their transporting process. Before meeting the egg, sperm must travel from the spermatangium to the female seminal receptacle where sperm await ovulation/spawning of the eggs. In species without ARTs such as *Todarodes pacificus* and *Idiosepius paradoxus*, modes of insemination and sperm storage resemble with that of sneaker males of *H. bleekeri*, and sperm of these species display a clustering trait, suggesting that post-mating circumstances rather than ARTs must be engaged in this trait. In rodents, sperm ejaculated in the female reproductive tract are capable of clustering as an evolutionary consequence of sperm competition and kin selection (Higginson and Pitnick 2011). Notably, it is hypothesized that group swimming is better in moving ascent though the oviduct than sole sperm swimming to reach the site of fertilization. Regardless of mechanistic differences, sperm from diverse taxa (mammals, birds, and insects) show sperm bundling after ejaculation and it occurs mostly inside the female. Under in vitro conditions, the sperm cluster of *H. bleekeri* was capable of moving *en masse*, supporting the hypothesis that collective sperm migration may operate *en route* to the site of fertilization, per se.

13.3.5 Sperm Longevity and Metabolism

Another striking difference in two types of sperm (sneaker and consort) is their longevity. Upon dilution in seawater, sperm become motile and sustain their motility for a certain period of time. As a role, the capacity of sperm to sustain their motility is endowed differently by species. Sperm motility lifetime of cichlid fish *Lamprologus callipterus* is no longer than 300 s (Taborsky et al. 2018) and in marine mussel, *Mytilus sp.*, sperm remain swimming for 3–4 days (Lillie 1919). In *H. bleekeri*, sperm from sneaker males sustain substantial motility as long as 30 min, whereas consort sperm lose their motility within 10 min (Hirohashi et al. 2016b). This difference was accounted for abundance of stored glycogen within sperm, and a tight correlation was found between motility loss and glycogen shortage. Unexpectedly, this result suggests that glycolysis fuels primary energy for flagellar movement as reported in octopus and mouse. Furthermore, the sperm of squid, as well as octopus, are able to use extracellular glucose to drive glycolysis, and both sneaker and consort sperm were able to maintain motility at high level for at least 3 days in the presence of 10 mM D-glucose. Utilization of extracellular nutrition by sperm is a hallmark of internal fertilization system. Mammalian sperm can uptake fructose from the seminal plasma and octopus female reproductive tract may contain similar nutrients that sperm can use. Similarly, squid sperm may also be nourished from either the spermatophores or the females after copulation.

Curiously, sperm longevity is affected by sperm density, especially for sneaker sperm. Experimentally, when sperm were subject to a series of dilutions, then motility reduced as dilution factor increased. Owning to an intrinsic trait, sneaker sperm are capable of autonomously condensing in the water column even after dilution. If this clustering effect was abolished, the sperm longevity became shortened. This would be another reason why sneaker sperm form a cluster. Although its mechanism remains to be elucidated, cell density may determine switching of the respiratory mode, i.e., aerobic to anaerobic respiration in response to oxygen availability. In a physiological aspect, sperm of sneaker males must be survived long after copulation. In squids, post-mating sperm behaviors are complex and elaborate but each should be fitted in order to maximize fertilization success. Future studies should address precisely what environmental factor could affect sperm behavior and metabolism.

13.4 Conclusion

As we described above, squid reproduction has many unique features both in precopulatory and postcopulatory processes (Fig. 13.2). These highly mobile Mollusca with developed brain use elaborate body patterning for intra-specific visual communications. Males use mating tactics properly depending on their own status in male–male competition and female mate choice, and adjust not only their behaviors

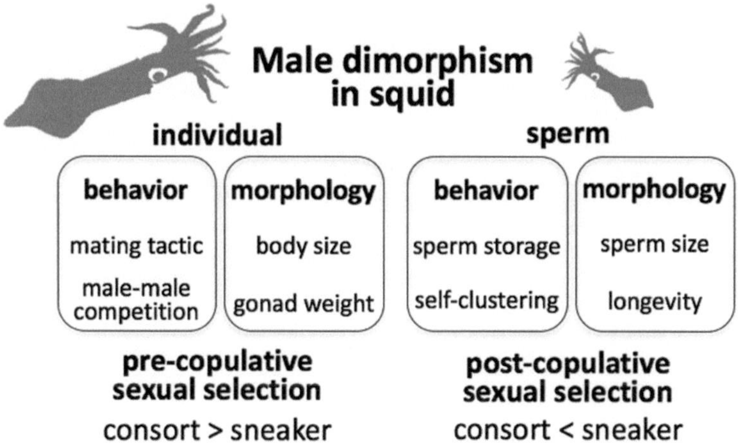

Fig. 13.2 Male dimorphism under pre- and post-copulative sexual selections in squid mating system

but also the morphology and physiology to a mating tactic that males have adopted. Especially in loliginid species, the presence of alternative sperm storage sites leads to diverged evolution in various aspects of sperm traits. It provides a novel opportunity to analyze proximate and ultimate factors in sperm evolution by comparison between consort and sneaker sperm, having different sperm competition risk, sperm storage, and fertilization conditions.

References

Andersson (1994) Sexual selection. Princeton University Press, Princeton

Apostólico LH, Marian JEAR (2017) Dimorphic male squid show different gonadal and ejaculate expenditure. J Morphol 278:1490–1505

Birkhead TRM, Møller AP (1998) Sperm competition, sexual selection and different routes to fitness. In: Birkhead TRM, Møller AP (eds) Sperm competition and sexual selection. Academic, New York, pp 757–781

Caprio J, Shimohara M, Marui T, Harada S, Kiyohara S (2014) Marine teleost locates live prey through pH sensing. Science 344:1154–1156

Coryell MW, Wunsch AM, Haenfler JM, Allen JE, McBride JL, Davidson BL, Wemmie JA (2008) Restoring acid-sensing ion channel-1a in the amygdala of knock-out mice rescues fear memory but not unconditioned fear responses. J Neurosci 28:13738–13741

Dewsbury DA (1982) Ejaculate cost and male choice. Am Nat 119:601–610

Di Marco FP, Hanlon RT (1997) Agonistic behavior in the squid *Loligo plei* (Loliginidae, Teuthoidea): fighting tactics and the effect of size and resource value. Ethology 103:89–108

Dixson AF, Anderson MJ (2004) Sexual behavior, reproductive physiology and sperm competition in male mammals. Physiol Behav 83:361–371

Drew GA (1911) Sexual activities of the squid, *Loligo pealii* (Les.). J Morphol 22:327–359

Eberhard WG (1996) Female control: sexual selection by cryptic female choice. Princeton University Press, Princeton

Evans JP, Pierotti M, Pilastro A (2003) Male mating behavior and ejaculate expenditure under sperm competition risk in the eastern mosquitofish. Behav Ecol 14:268–273

Fitzpatrick JL, Montgomerie R, Desjardins JK, Stiver KA, Kolm N, Balshine S (2009) Female promiscuity promotes the evolution of faster sperm in cichlid fishes. Proc Natl Acad Sci U S A 103:15113–15117

Franklin AM, Stuart-Fox D (2017) Single and multiple mating reduces longevity of female dumpling squid (*Euprymna tasmanica*). J Evol Biol 30:977–984

Hall KC, Hanlon RT (2002) Principal features of the mating system of a large spawning aggregation of the giant Australian cuttlefish *Sepia apama* (Mollusca: Cephalopoda). Mar Biol 140:533–545

Hallem EA, Sternberg PW (2008) Acute carbon dioxide avoidance in *Caenorhabditis elegans*. Proc Natl Acad Sci U S A 105:8038–8043

Hanlon RT (1996) Evolutionary games that squids play: fighting, courting, sneaking, and mating behaviors used for sexual selection in Loligo pealei. Biol Bull 191(2):309–310

Hanlon RT, Messenger JB (2018) Cephalopod behavior, 2nd edn. Cambridge University Press, Cambridge

Hanlon RT, Smale MJ, Sauer WHH (1994) An ethogram of body patterning behavior in the squid *Loligo vulgaris reynaudii* on spawning grounds in South Africa. Biol Bull 187:363–372

Hanlon RT, Maxwell MR, Shashar N, Loew ER, Boyle KL (1999) An ethogram of body patterning behavior in the biomedically and commercially valuable squid *Loligo pealei* off Cape Cod, Massachusetts. Biol Bull 197:49–62

Hanlon RT, Smale MJ, Sauer WHH (2002) The mating system of the squid Loligo vulgaris reynaudii (Cephalopoda, Mollusca) off South Africa: fighting, guarding, sneaking, mating and egg laying behavior. Bull Mar Sci 71:331–345

Hanlon RT, Naud MJ, Shaw PW, Havenhand JN (2005) Behavioural ecology: transient sexual mimicry leads to fertilization. Nature 430:212

Higginson DM, Pitnick S (2011) Evolution of intra-ejaculate sperm interactions: do sperm cooperate? Biol Rev Camb Philos Soc 86:249–270

Hirohashi N, Alvarez L, Shiba K, Fujiwara E, Iwata Y, Mohri T, Inaba K, Chiba K, Ochi H, Supuran CT, Kotzur N, Kakiuchi Y, Kaupp UB, Baba SA (2013) Sperm from sneaker male squids exhibit chemotactic swarming to CO_2. Curr Biol 23:775–781

Hirohashi N, Iida T, Sato N, Sauer WHH, Iwata Y (2016a) Complex adaptive traits between mating behavior and post-copulatory sperm behavior in squids. Rev Fish Biol Fish 26:601–607

Hirohashi N, Tamura-Nakano M, Nakaya F, Iida T, Iwata Y (2016b) Sneaker male squid produce long-lived spermatozoa by modulating their energy metabolism. J Cell Biol 291:19324–19334

Hoving HJT, Lipinski MR, Videler JJ, Bolstad KS (2008) Reproductive system and the spermatophoric reaction of the mesopelagic squid *Octopoteuthis sicula* (Rüppell, 1844) (Cephalopoda: Octopoteuthidae) from southern African waters. Afr J Mar Sci 30:603–612

Hoving HJT, Nauwelaerts S, Van Genne B, Stamhuis EJ, Zumholz K (2009) Spermatophore implantation in *Rossia moelleri* Steenstrup, 1856 (Sepiolidae: Cephalopoda). J Exp Mar Biol Ecol 372:75–81

Huffard CL, Caldwell RL, Boneka F (2008) Mating behavior of *Abdopus aculeatus* (d'Orbigny 1834) (Cephalopods: Octopodidae) in the wild. Mar Biol 154:353–362

Hunt JC, Zeidberg LD, Hamner WM, Robison BH (2000) The behaviour of *Loligo Opalescens* (Mollusca: Cephalopoda) as observed by a remotely operated vehicle (ROV). J Mar Biol Assoc UK 80:873–883

Hunter FM, Harcourt R, Wright M, Davis LS (2000) Strategic allocation of ejaculates by male Adelie penguins. Proc R Soc Lond B 267:1541–1545

Iida T, Iwata Y, Mori T, Baba SA, Hirohashi N (2017) A coordinated sequence of distinct flagellar waveforms enables a sharp flagellar turn mediated by squid sperm pH-taxis. Sci Rep 7:12938

Iwata Y, Sakurai Y (2007) Threshold dimorphism in ejaculate characteristics in the squid *Loligo bleekeri*. Mar Ecol Prog Ser 345:141–146

Iwata Y, Munehara H, Sakurai Y (2005) Dependence of paternity rates on alternative reproductive behaviors in the squid *Loligo bleekeri*. Mar Ecol Prog Ser 298:219–228

Iwata Y, Shaw P, Fujiwara E, Shiba K, Kakiuchi Y, Hirohashi N (2011) Why small males have big sperm: dimorphic squid sperm linked to alternative mating behaviours. BMC Evol Biol 11:236

Iwata Y, Sakurai Y, Shaw PW (2015) Dimorphic sperm-transfer strategies and alternative mating tactics in loliginid squid. J Molluscan Stud 81:147–151

Iwata Y, Sauer WHH, Sato N, Shaw PW (2018) Spermatophore dimorphism in the chokka squid *Loligo reynaudii* associated with alternative mating tactics. J Molluscan Stud 84:157–162

Jantzen TM, Havenhand JN (2003) Reproductive behavior in the squid *Sepioteuthis australis* from South Australia: interactions on the spawning grounds. Biol Bull 204:305–317

Jones W (2013) Olfactory carbon dioxide detection by insects and other animals. Mol Cells 35:87–92

LaMunyon CW, Ward S (1998) Larger sperm outcompete smaller sperm in the nematode *Caenorhabditis elegans*. Proc R Soc Lond B 265:1997–2002

Leach B, Montgomerie R (2000) Sperm characteristics associated with different male reproductive tactics in bluegills (*Lepomis macrochirus*). Behav Ecol Sociobiol 49:31–37

Lillie FR (1919) Problems of fertilization, vol 38. University of Chicago Press, Chicago, p 246

Lin CY, Chiao CC (2017) Female choice leads to a switch in oval squid male mating tactics. Biol Bull 233:219–226

Lin CY, Tsai YC, Chiao CC (2017) Quantitative analysis of dynamic body patterning reveals the grammar of visual signals during the reproductive behavior of the oval squid *Sepioteuthis lessoniana*. Front Ecol Evol 5:30

Mann T (1984) Chapter 3 mollusca. In: Spermatophores: development, structure, biochemical attributes and role in the transfer of spermatozoa. Springer, Berlin, pp 18–71

Marian JEAR (2012) Spermatophoric reaction reappraised: novel insights into the functioning of the loliginid spermatophore based on *Doryteuthis plei* (Mollusca: Cephalopoda). J Morphol 273:248–278

Marian JEAR (2015) Evolution of spermatophore transfer mechanisms in cephalopods. J Nat Hist 49:1423–1455

McMeniman CJ, Corfas RA, Matthews BJ, Ritchie SA, Vosshall LB (2014) Multimodal integration of carbon dioxide and other sensory cues drives mosquito attraction to humans. Cell 156:1060–1071

Mietz H (1997) Filtering glaucoma operations. Can growth factor blockers replace antimetabolites? Ophthalmologe 94:242–247

Møller AP (1991) Sperm competition, sperm depletion, paternal care, and relative testis size in birds. Am Nat 137:882–906

Nakatsuru K, Kramer DL (1982) Is sperm cheap? Limited male fertility and female choice in the lemon tetra (Pisces, Characidae). Science 216:753–755

Naud MJ, Havenhand JN (2006) Sperm motility and longevity in the giant cuttlefish, *Sepia apama* (Mollusca: Cephalopoda). Mar Biol 148:559–566

Naud MJ, Hanlon RT, Hall KC, Shaw PW, Havenhand JN (2004) Behavioral and genetic assessment of mating success in a natural spawning aggregation of the giant cuttlefish (*Sepia apama*) in southern Australia. Anim Behav 67:1043–1050

Naud MJ, Sauer WH, McKeown NJ, Shaw PW (2016) Multiple mating, paternity and complex fertilisation patterns in the Chokka Squid *Loligo reynaudii*. PLoS One 11:e0146995

Nicholls EH, Burke T, Birkhead TR (2001) Ejaculate allocation by male sand martins, *Riparia riparia*. Proc R Soc Lond B 268:1265–1270

Orr TJ, Zuk M (2012) Sperm storage. Curr Biol 22:R8–R10

Parker GA (1970) Sperm competition and its evolutionary consequences in the insects. Biol Rev 45:525–567

Parker GA (1990a) Sperm competition games: raffles and roles. Proc R Soc Lond B 242:120–126

Parker GA (1990b) Sperm competition games: sneaks and extra-pair cop- ulations. Proc R Soc Lond B 242:127–133

Parker GA (1998) Sperm competition and the evolution of ejaculates: towards a theory base. In: Birkhead TR, Møller AP (eds) Sperm competition and sexual selection. Academic, London, pp 3–54

Pilastro A, Bisazza A (1999) Insemination efficiency of two alternative male mating tactics in the guppy (*Poecilia reticulata*). Proc R Soc Lond B 266:1887–1891

Riechert SE (1984) Games spiders play. III: cues underlying context-associated changes in agonistic behaviour. Anim Behav 32:1–15

Sato N, Yoshida MA, Kasugai T (2017) Impact of cryptic female choice on insemination success: larger sized and longer copulating male squid ejaculate more, but females influence insemination success by removing spermatangia. Evolution 71:111–120

Schnell AK, Smith CL, Hanlon RT, Harcourt RT (2015) Female receptivity, mating history, and familiarity influence the mating behavior of cuttlefish. Behav Ecol Sociobiol 69:283–292

Snook RR (2005) Sperm in competition: not playing by the numbers. Trends Ecol Evol 20:46–53

Stockley P, Gage MJG, Parker GA, Møller AP (1997) Sperm competition in fishes: the evolution of testis size and ejaculate characteristics. Am Nat 149:933–954

Taborsky M, Schütz D, Goffinet O, van Doorn GS (2018) Alternative male morphs solve sperm performance/longevity trade-off in opposite directions. Sci Adv 4:eaap8563

Uglem I, Galloway TF, Rosenqvist G, Folstad I (2001) Male dimorphism, sperm traits and immunology in the corkwing wrasse (*Symphodus melops* L.). Behav Ecol Sociobiol 50:511–518

Vladić TV, Järvi T (2001) Sperm quality in the alternative reproductive tactics of Atlantic salmon: the importance of the loaded raffle mechanism. Proc R Soc Lond B 268:2375–2381

Wada T, Takeshige T, Mori T, Natsukari Y (2005) Alternative male mating behaviors dependent on relative body size in captive oval squid *Sepiotheuthis lessoniana* (Cephalopoda, Loliginidae). Zool Sci 22:645–651

Yang MT, Chien WL, Lu DH, Liou HC, Fu WM (2013) Acetazolamide impairs fear memory consolidation in rodents. Neuropharmacology 67:412–418

Part V
Biotechnology in Aquatic Species

Chapter 14
Improvements on the Reproductive Control of the European Eel

Juan F. Asturiano

Abstract A combination of several factors have produced an intense decline of the European eel populations until becoming an endangered species, but the European eel is a demanded species with high market values, and wild glass eels are still the base of the eel aquaculture. Nowadays, reproduction in captivity seems the only realistic alternative to reduce the pressure on natural populations and supply glass eels to eel farms. Altogether, trying to get its reproduction in captivity is still an exciting scientific challenge for physiologists, endocrinologists, nutritionists, embryologists, etc. This chapter tries to give an overview of the recent research developed in some of the key areas related with the reproduction of this species, and the main improvements, conclusions and hypothesis emerging from that research, making an especial emphasis in those papers published from 2000.

Keywords Anguilla · Reproduction · Aquaculture · Physiology

14.1 Introduction

The European eel (*Anguilla anguilla* L., 1758) is a catadromous fish species having a very complex life cycle, including two transoceanic migrations of several thousands of kilometers separated by a growth period in continental waters, and several metamorphoses, making its reproductive process an interesting model for the research on the regulatory mechanisms of reproductive physiology. Moreover, from an evolutionary point of view, this species belongs to the Elopomorpha superorder, a diverse group of predominantly marine fishes branching at the base of the teleosts, the largest group of vertebrates. Due to its phylogenetic position, the studies made in the eel may provide insights into ancestral regulatory functions in teleosts, providing information on ancestral vertebrate physiological regulations.

J. F. Asturiano (✉)
Aquaculture and Biodiversity Research Group, Institute for Animal Science and Technology, Universitat Politècnica de València, Valencia, Spain
e-mail: jfastu@dca.upv.es

M. Yoshida, J. F. Asturiano (eds.), *Reproduction in Aquatic Animals*,
https://doi.org/10.1007/978-981-15-2290-1_15

The European eel is a demanded species with high market values, which is intensively fished at all its life stages, including glass, yellow, and silver eels. Moreover, wild glass eels are still the base of the eel aquaculture, which supplies most of the eels consumed. But this species reproduces only once during its lifetime and the habitat where it lives has suffered a high deterioration due especially to pollution and dam construction (that makes many freshwater habitats inaccessible). These factors together with others such as overfishing, parasitism, changes in oceanographic conditions, etc. have produced a hard decline of the eel's stock. In fact, the International Council for the Exploration of the Sea (ICES) advised in 2009 that the level of the eel stock for all stages including glass eel, yellow eel, and silver eel was at a historical minimum. The European eel stock has decreased by 95–99%, compared to its levels in 1960–1980 (ICES 2013) leading to the list of the species as "Critically Endangered" on the Red List of Threatened Species, by the International Union for Conservation of Nature (IUCN). The annual recruitment of glass eel to European waters in 2017 remained low, at 1.6% of the 1960–1979 level in the "North Sea" series and 8.7% in the "Elsewhere Europe" series (ICES 2017).

The European eel has also received attention from the European Union, which categorized it as critically endangered and published in 2007 a regulation (Regulation 1100/2007, 18th September 2007) establishing measures for the recovery of the stock, which every state has to follow as a guide in order to elaborate a proper management plan by member states in 2009. The final objective was to allow the escapement to the sea of 40% of the biomass of silver eels.

Moreover, this species was listed by Convention on International Trade in Endangered Species of Wild Fauna and Flora (CITES) in 2007 as species "not necessarily threatened with extinction, but in which trade must be controlled to avoid utilization incompatible with their survival" under Appendix II, prohibiting international trade of European eel into and out of the EU, including the exportation of glass eels to Asia.

Nowadays, reproduction in captivity seems the only realistic alternative to reduce the pressure on natural populations and supply glass eels to eel farms. However, despite the evident interest in this species, and even after many years of research, different aspects related with the control of its reproduction are still unexplored, have received a small attention, or have rendered confusing results (Fig. 14.1). Research on this field got some results, but has proved difficult due to the complex reproductive physiology of the eels. Thus, improvements are crucial on the use of environmental factors, recirculation systems, hormonal treatments, broodstock management, specific diets, gamete evaluation and handling, breeding and hatchery methods, and specific culture techniques for a self-sustained eel aquaculture. The development of biotechnology techniques, such as production of recombinant hormones, germ cells xenotransplantation, or spermatogonia and sperm cryopreservation, could be another alternatives, or at least complementary tools.

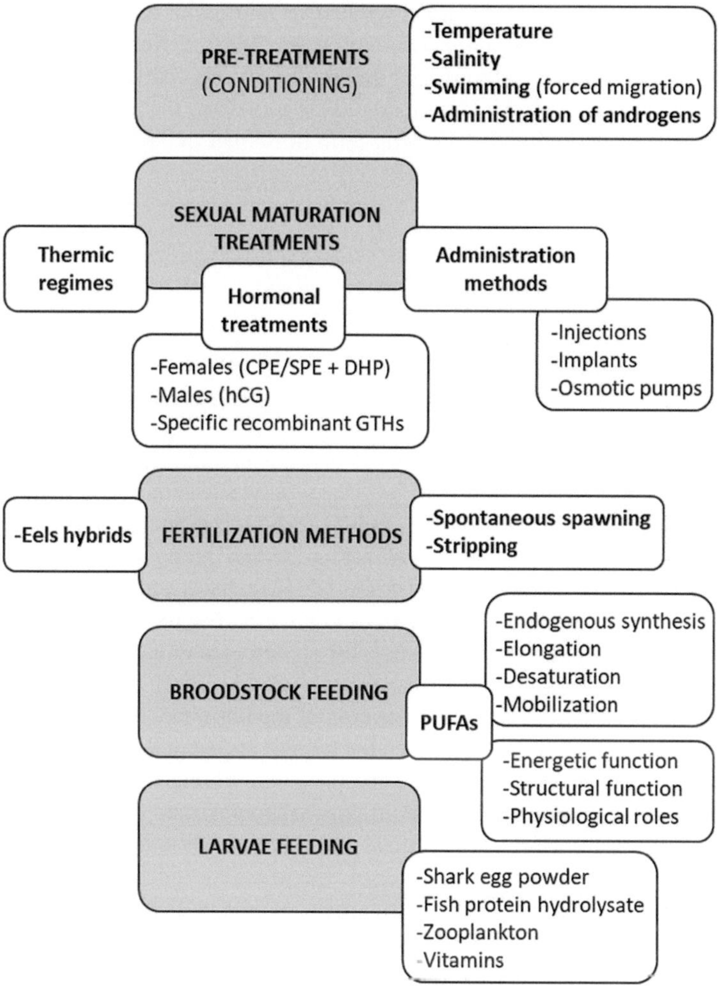

Fig. 14.1 Aspects related with the control of the European eel reproduction and reviewed in this chapter

14.2 Treatments for the Induction of Sexual Maturation

The European eel migration has an estimated duration of 5–7 months, although a recent study suggests that part of them start the migration one year before the spawning period (Righton et al. 2016). When they reach the Sargasso Sea, where they make the spawning, testes will be 55 times bigger and ovaries approximately 23 times their initial size (estimated from data obtained after sexual maturation in captivity). Considering the length of their transatlantic travel, it is probable that at least the beginning of the sexual maturation happen during their migration. However, when the European eels start their reproductive migration, they still are sexually immature (Tesch 2003; van Ginneken and Maes 2005). Silvering metamorphosis process is the beginning of the sexual maturation and reproductive phase (Dufour et al. 2003; Aroua et al. 2005; Durif et al. 2005). However, the process never occurs spontaneously in captivity. In fact, long-term hormonal treatments are required to induce the sexual maturation of the eels.

14.2.1 Treatments for Eel Males

Immature males of the *Anguilla* genus (both Japanese and European eels) have testes containing only type A and early-type B spermatogonia (Miura et al. 2002; Peñaranda et al. 2016; Morini et al. 2017) and to develop spermatogenesis, spermiogenesis, and spermiation, it is necessary to administer a hormonal treatment during several weeks. Hormonal treatments are based on weekly intraperitoneal administration of hormones, normally after a gradual acclimatization to seawater, maintaining the water temperature at 20 °C and fasting the fish throughout the treatment.

Human chorionic gonadotropin (hCG; 1.5 IU/g fish), purified from pregnant women urine, has been traditionally used with eel males, inducing good results in terms of sperm quantity, density ($3–6 \times 10^9$ spermatozoa/mL), and quality (Pérez et al. 2000; Müller et al. 2004; Asturiano et al. 2005, 2006), collecting the sperm samples always 24 h after the administration of the hormone, as Pérez et al. (2000) demonstrated that this is the time when the highest sperm quality is obtained.

However, some hCG supply problems happened during the last years, and the arrival of human recombinant gonadotropins to the pharmacy market (for humans treatment) caused the assessment of these recombinant hormones, with good results of sperm volume, density, motility and spermatozoa kinetic parameters were obtained throughout most weeks of treatment (Gallego et al. 2012). In a recent work, Herranz-Jusdado et al. (2019) compared the effects on the reproductive performance of European eel males of a human chorionic gonadotropin, purified from female urine, with a recombinant α-choriogonadotropin (testing different doses, 0.25–1.5 IU/g fish, of the last one). Results indicated that the type of hormone used significantly affected the progression of spermiation and that the recombinant hormone produced better results in terms of sperm quantity and quality in most of the weeks of the treatment, thus remaining an effective treatment to induce spermiation in this species. All the doses of rhCG were able to induce the whole spermiation process, but a weekly dose of 1.5 IU/g fish was necessary to provide a notable amount (volume and density) of high-quality (motility and velocity) samples throughout the treatment. Moreover, rhCG has been found as the best cost-effective treatment, considering both the hormone costs and the volume of high-quality sperm produced (Gallego et al. 2012).

14.2.2 Treatments for Eel Females

The trials done with purified gonadotropins had no positive results in the case of the eel females. In fact, the method traditionally used to induce eel females maturation consists basically on the weekly administration of carp or salmon pituitary extracts (~20 mg/kg body weight) around 20 weeks (or increasing from 0 to 40 mg/kg BW in the case of Mordenti et al. 2012). A final maturation-inducing treatment, administered once the oocytes reach the nuclear migration stage, consists in a dose of the maturation-inducing steroid (MIS; 17α,20ß-dihydroxy-4-pregnen-3-one or DHP; 2 mg per/kg body weight) administered intraperitoneally ~24 h after receiving the priming dose. Approximately 12 h after the MIS dosis, eggs can be stripped and fertilization done (Asturiano et al. 2002; Pedersen 2003, 2004; Palstra et al. 2005; Schmitz et al. 2005; Pérez et al. 2008, 2011; Butts et al. 2014; Vílchez et al. 2014).

The hypophysation can be considered as a primitive treatment because, although pituitary contains mainly LH and less FSH, it is not possible to know the precise dosage of gonadotropins used and, for example, trying to standardize a protocol. The treatment of females with pituitary extracts can induce an abnormal hormonal profile that jeopardizes the normal development of the oogenesis, including a reduction of the endogenous expression of FSHβ (Suetake et al. 2002; Ozaki et al. 2007; Jeng et al. 2007) and an overexpression of LHβ. Moreover, the length of the required treatments for eel females can provoke the induction of immunologic responses in some fish, having potential negative effects on their maturation process. Altogether, these factors, linked to the variation on the eel females individual response to the treatments, can explain the presence of a percentage of non-responders between the treated animals, or why the percentage of spawning females, or the egg quality and the rates of fertilization and hatching larvae are still low in this species (Palstra and van den Thillart 2009; Butts et al. 2014; Vílchez et al. 2014).

The economic aspect must be also considered due the high price of these treatments, especially when a high percentage of non-responders or low-quality eggs are the final results.

14.2.3 Treatments Based on the Combined Application of Thermic Regimes and Hormonal Treatments

Several groups have studied the influence of the rearing water temperature during artificial induction of ovulation and the spawning performance of artificially matured Japanese eels (Dou et al. 2008; Unuma et al. 2011, 2012). In this regard, our group considered the migratory biology of the European species, to carry out several experiments with females and males trying to determine the optimum thermic profile for the maturation, being pioneers on these studies in Europe. First, we compared a variable thermal regime that increased from 10 to 14 and 17 °C with a constant 20 °C regime during 12 weeks. We found that females maintained at low temperatures (10 °C) during the beginning of their induced maturation showed a higher expression of pituitary fshβ and lhβ expression, plasma 17βestradiol (E2) and vitellogenin (VTG), estrogen receptor 1 (esr1), and vitellogenin 2 (vtg2) expression in liver. The results strongly suggest that T10 is inducing a higher endogenous FSH level which increases the E2 circulating level during vitellogenesis (Pérez et al. 2011). The variable thermal regime induced an *fshβ* expression and E2 profile in vitellogenic hormonally matured eel females that were more similar to the profile observed in other naturally maturing fish (Pérez et al. 2011).

When 3 thermal regimes (2 variable between 10 and 15 or 15 and 18 °C, and one constant at 18 °C) were tested during the CPE females maturation, we found that low temperatures (10–15 °C) ease the onset of the ovarian development, inducing higher plasma levels of E2 and 11-ketotestosterone (11KT) and a higher expression of aromatase, while the high temperatures induced the acceleration of the maturation and a downregulation of ovarian cyp19a1 expression (Mazzeo et al. 2014).

Altogether, these results demonstrate that temperature can modulate eel ovarian development, and this knowledge could be used to manipulate the timing of vitellogenesis progression under laboratory conditions. In fact, more recently, we assessed the females maturation with CPE in combination with two treatments at low temperature (15 °C or changing weekly between 10 and 15 °C) and a final increase to 18 °C until reaching the spawnings, which happened after 18–23 weeks of treatment. The 10–15 °C changing treatment induced the best results, making us to obtain in 2012 the first larvae of European eel born in Spain (Vílchez et al. 2013, 2014), which survived until 3 days without exogenous feeding. Other European groups of research are using constant temperatures during the female maturation, obtaining good results. Butts et al. (2014) reported the use of 19–20 °C during a 10–20 weeks CPE treatment for females, while Mordenti et al. (2013) used 15 °C and obtained ovulations after 19–30 weeks of treatment with a larvae survival of 22–24 days after hatching.

In the case of males, the maturation-inducing treatments has been traditionally carried out at 20 °C, considering the ocean temperature in the putative spawning area of the European eel. However, after first evidences of thermic regimes experienced by the males during the transatlantic migration, Peñaranda et al. (2016) evaluated the effects of temperature on rhCG-induced males subjected to three thermal regimes: T10: 10 °C (first 4 weeks), 15 °C (next 3 weeks), and 20 °C (last 6 weeks); T15: 15 °C (first 4 weeks) and 20 °C (last 9 weeks); and T20: constant 20 °C for the duration of the experiment. Histological changes, steroidogenic enzyme gene expression, and steroid plasma levels were determined. This study demonstrated that temperature modulates the steroidogenesis, the gonad maturation, and the spermiation process in eels. They found that the androgen synthesis is not dependent on temperature because the synthesis of androgens (T and 11KT) happened even at low temperatures, linked to the gene expression of some steroidogenic enzymes (aacyp11a1, aacyp17-I, and aa11βHSD; Fig. 14.1) and to the stimulation of the proliferation of spermatogonia. Then, the testis maturation process seems to be stopped and maturation arrested at SPG1 until reaching higher temperatures (over a threshold of 15–20 °C) before inducing a change in the steroidogenesis from androgen synthesis to estrogen and progestin synthesis, with *aacyp19a1* and *aacyp21* gene expression increasing at 15 °C until completing spermatogenesis. This was proposed as the endocrine system regulating the testis maturation through the eel reproductive migration to the spawning areas in the West Atlantic.

14.2.4 Specific Recombinant Gonadotropins

From 2005, several studies have been carried out on the production and assay of specific (homologous) recombinant gonadotropins (rGtHs: rFSH and rLH) in different fish species (reviewed by Mylonas et al. 2017). In males of Japanese eel, the rGtHs stimulated the androgenesis in vitro, inducing a full spermatogenesis, with presence of spermatozoa in the testis (Kazeto et al. 2008). In Japanese eel females, rGtHs induce in vitro the production of steroids in the oocytes, but only when they are at least in mid vitellogenesis, and they are able to induce the germinal vesicle breakdown (GVBD) in mature oocytes (Kobayashi et al. 2010).

In vivo, the rGtHs have shown much more limited effects, both in Japanese eel females and males, in which only induced the full spermatogenesis, without reaching the spermiation (Kazeto et al. 2008). Only Ohta et al. (2017) in a recent paper reported the production of viable sperm using rLH.

Recently, Peñaranda et al. (2018) used new specific European eel recombinant gonadotropins (aarFSH and aarLH) produced in the ovarian cells of Chinese hamsters (CHO) to induce maturation of male eels. They demonstrated that these specific rGtHs (especially using a combination of both aarGths) are able to induce the full spermatogenesis and spermiation in immature fish, and confirmed that their half-life is long enough to induce these in vivo effects. However, they reported a variable sperm quality and suggested further experiments combining these recombinant hormones in order to improve the treatment. Moreover, the high price of production of these hormones restrict their use in aquaculture for the moment.

14.2.5 Long-Term Sustained Hormone Release Systems

Previous studies evidenced the role of androgens in the development of vitellogenesis in the Japanese eel (Lin et al. 1991, 1998). Later, studies with the short-finned eel (*Anguilla australis*) reported an increased development of silvering, lipid accumulation, and expression of the follicle-stimulating hormone receptor (Lokman et al. 2003, 2007; Thomson-Laing et al. 2018). These studies have been carried out using implants for the slow release of hormones in that eel species (Lokman et al. 2015; Damsteegt et al. 2016), although the same group concluded that 11KT implants can only partly mimic the effects of hCG in male eels in vivo (Lokman et al. 2016). Very recently, Lokman's group collaborated in an experiment using implants of cholesterol-cellulose to administer 17-methyltestosterone (17MT) to

European eel females (Di Biase et al. 2017). These implants were used as a pre-treatment, followed by the traditional CPE treatment for females (as done by Lokman et al. 2015), and results are mentioned down (see Sect. 14.3.4).

Osmotic pumps have been used as a long-term sustained hormone release system in Japanese eels. Spermatogenesis and spermiation were stimulated in immature males treated with hCG (Kagawa et al. 2009). In the case of females, the use of osmotic pumps releasing SPE induced vitellogenesis and increased the GSI and, although more eels attained full maturity when they were SPE-injected, after a final treatment, the number of eels that ovulated was similar for both treatments. Moreover, more eggs were spawned in the group treated with SPE-releasing osmotic pumps than in the SPE-injected group, and egg quality was similar for both groups (Kagawa et al. 2013a). These results made osmotic pumps a reliable method for inducing sexual maturation in the Japanese eels. However, no assays have been published in the case of the European species, probably due to the high cost of this system.

14.3 Pre-treatments to the Induction of Sexual Maturation

As has been mentioned before, the hormonal treatments to induce eel maturation are expensive, jeopardizing their application at a commercial scale. It is necessary to look for alternative or complementary treatments allowing the reduction of the use of hormones. Moreover, it is accepted that the release of the dopaminergic inhibition of puberty (Vidal et al. 2004) in the migrating silver eels must be induced by changes in the environmental conditions found by eels (Dufour et al. 2003; Sébert et al. 2008). We call pre-treatments (often called conditioning) to the use of some environmental or hormonal factors trying to induce the first phases of the maturation process and the gonadal development or, at least, increasing the sensitivity of the eels to the hormonal treatments applied consecutively. These pre-treatments can be especially important in the case of maturing farmed eels, which have not faced changes in the environmental conditions (Palstra and van den Thillart 2009).

Broodstock feeding, treated here separately (see Sect. 14.4), has been often included as part of these pre-treatments.

14.3.1 Temperature

The European eel life cycle depends intensively on the environmental conditions. During their reproductive migration to the Sargasso Sea, the eels travel at a high depth, in cold waters (7–12 °C), but the spawning happens in warm waters, at 18–20 °C. Moreover, through their Atlantic migration they experience daily changes of temperature because they swim during the day in colder and dark waters (7–10 °C, 560 m), while at night swim up to warmer waters (12 °C, 280 m; Aarestrup et al.

2009). It has been proposed that eel gonad maturation does not progress during their oceanic journey until they reach the Sargasso Sea where temperatures are around 10 °C higher at the same depth (Palstra and van den Thillart 2009).

During the last years, our group have carried out several studies to evaluate the effect of the temperature during the sexual maturation of European eel females (Pérez et al. 2011; Mazzeo et al. 2014) and males (Gallego et al. 2012, 2015; Peñaranda et al. 2016). A thermic pre-treatment of females at 10, 15, or 18 °C during 1 week (without hormonal treatment) induced differences in their levels of steroids and in the expression of FSHβ, LHβ, and the receptor 2 of GnRH (GnRHr2). The lower temperatures caused a bigger expression of GnRHr2 in brain and pituitary, suggesting an increase of the sensitivity of these organs to the GnRH (Pérez et al. 2013). The biggest expression of FSHβ happened at 15 °C, and that of LHβ at 10 °C (Pérez et al. 2013). Moreover, the plasma levels of E2 and 11KT and the expression of ovarian aromatase (cyp191a) were reduced at the highest temperature, suggesting that a pre-treatment at 10–15 °C would be the most appropriate for the females maturation (Mazzeo et al. 2014).

Considering these previous results, we recently carried out an experiment comparing the effects of pre-treatments of 2 weeks at 10 or 20 °C on the male transcriptome in the brain-pituitary-gonad axis and the plasma levels of androgens. Males maintained at 10 °C showed a significantly different transcriptome profile with 377, 350, and 76 genes differentially expressed, relative to day zero control, in brain, pituitary, and testis, respectively (Rozenfeld et al. 2017). GO-terms significantly enriched for processes and functions related to circadian rhythm, histone modification, the production of steroids, or the activity of FSH receptors were found among these genes. Furthermore, the plasma levels of T and 11KT were higher in the group maintained at 10 °C. These results indicate that a pre-treatment with low temperatures activates the brain-pituitary-gonad axis, and especially the synthesis of androgens, which could ease later maturation using standard hormonal methods.

14.3.2 Salinity

Kagawa et al. (1998) placed farmed female Japanese eels in seawater during only one week or during 3 months, before starting a SPE treatment. They reported no differences between the groups in fertility and hatching rates, but females kept in seawater for 3 months required a lower number of SPE doses to finish the vitellogenesis and reach the ovulation, suggesting an increase of the sensitivity to the hormones. These were the first results evidencing that long-term rearing in seawater induces vitellogenesis without any hormone treatments in cultured immature eels, resulting in an ovarian stage similar to that of wild migrating silver eels.

Years later, our group of research found that a period of 2 weeks in seawater was enough to induce a reduction of the digestive tract and an increase of the eye size, both considered as typical changes during the eel silvering process, being the beginning of the eel puberty (Mazzeo et al. 2014). In other words, the rearing period in seawater was able to induce the beginning of the silvering process, which could ease the later maturation of the females treated with hormones. Moreover, when the increase of water salinity was linked to a reduction of the temperature, an increase of the responsiveness to GnRH was found in brain and pituitary, evidenced by a higher expression of the receptor GnRHR2, considered the active one in the brain-pituitary-gonad axis on the eel (Pérez et al. 2013).

As mentioned before, when male European eels were maintained during one week in seawater there was an increase of the plasma levels of estrogens, as well as an increase of the testicular expression of enzymes involved in the synthesis of androgens (aacyp11a1; aacyp17-I; 11βHSD) and estrogens (aacyp19a1; Peñaranda et al. 2016; Fig. 14.2), suggesting that the increase of salinity was able to induce the onset of the spermatogenesis in this species.

New experiments are required to corroborate the combined effects of these salinity-temperature pre-treatments and their real potential to induce an increase of the gonads' sensitivity to the treatments with exogenous hormones, allowing the reduction of the intensity and/or the length of the hormonal treatments.

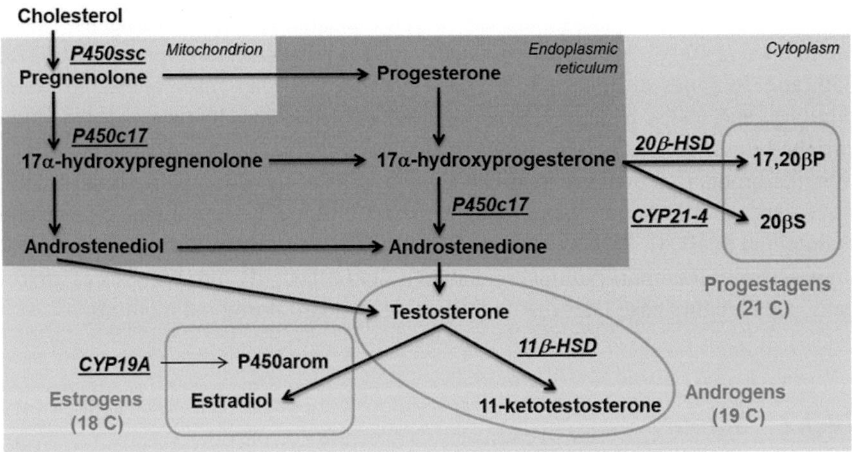

Fig. 14.2 Simplified schedule of the main steroid biosynthetic pathways of androgens, estrogens, and progestins in the fish gonad. Enzymes (underlined in the figure) are: P450scc (cytochrome P450 side-chain cleavage); P450c17 (17α-hydroxylase/C17-20lyase); CYP19A (gonadal aromatase promoter); P450arom (cytochrome P450 aromatase); 11β-HSD (11β-Hydroxysteroid dehydrogenase); 20b-HSD (20β-Hydroxysteroid dehydrogenase); CYP21-4 (21-hydroxylase). The level of expression of the underlined testis steroidogenic enzymes was evaluated by Peñaranda et al. (2016) to determine the combined effect of a hCG treatment and three thermal regimes

14.3.3 Swimming

As mentioned before, European eels have to swim long distances to reach their spawning area, and exercise has been proposed as a stimulating factor for eel maturation (reviewed by Rousseau et al. 2013). Using swim tunnels, eels have been shown as very efficient swimmers in terms of energy costs (van Ginneken and van den Thillart 2000; van den Thillart et al. 2004, 2009; Palstra et al. 2008a; 2010; Palstra and van den Thillart 2010). Moreover, short swimming (until 6 weeks) have induced the increase of eye diameter and testis and ovary histological aspect, including oocyte diameter, number of oil droplets, or gonadosomatic index, especially in older females (Palstra et al. 2007, 2008a, b).

On his hand, long swimming periods (1.5–3 months) caused stimulatory effects on maturation, as increases of pituitary levels of LH and plasma levels of 11KT and E2 (van Ginneken et al. 2007). Females showed an apparent suppression of hepatic vitellogenesis, with reduced expression of estrogen receptor α (ERα) vitellogenin 1 and 2, and a suppression of VTG deposition (Palstra et al. 2010). However, long-term swimming had a stimulatory effect on maturation because swimming stimulated oocyte growth and lipid deposition in the oocytes oil droplets, which has been proposed as the first steps of sexual maturation (Palstra et al. 2007). The induction of lipid deposition in the oocytes and the inhibition of vitellogenesis by swimming has been suggested as a natural mechanism to induce a sequence of events in females maturation process, with vitellogenesis and final maturation undergoing near or at the spawning grounds (Palstra and van den Thillart 2010; Palstra et al. 2010). Oppositely, simultaneous incorporation of lipids and VTG induced by artificial maturation protocols has been described as an unnatural process in Japanese (Adachi et al. 2003) and European eels (Palstra et al. 2009). Moreover, the same authors hypothesized that non-exercised European silver eels are probably still too premature to react to the traditional hormonal stimulation (Palstra et al. 2010).

Mes et al. (2016) carried out an experiment combining forced migration, in freshwater and seawater, of groups of farmed mixed-sex European eels under photothermal conditions mimicking the vertical migrations during their marine migration. They found that simulation of migration under mimicked photothermal conditions has significant stimulating effects (higher T plasma levels or changes in the GSI and other maturation parameters) on early maturation, although vitellogenesis was not induced. Thus, this method could be used to condition farmed eels for the use as broodstock eels for further hormonal stimulation with gonadotropins in maturation protocols.

14.3.4 Administration of Androgens

Years ago, Lokman et al. (2007) reported that 11KT (but not E2 or hCG) administered with implants induces, in a dose-dependent manner, an increase of 10–20% in the diameter of Australian eel (*A. australis*) previtellogenic oocytes and their nucleus. Later, Lokman et al. (2015) combined an androgen pre-treatment of female silver Australian eels with traditional CPE hypophysation to evaluate its utility as part of the maturation protocols. In this case, they used implants with different doses of 17MT and obtained increased oocyte diameters prior to commencing CPE treatment. Moreover, once the hypophysation started, less time, money, and handling were required to reach the pre-ovulatory stage, while retaining comparable amounts of lipid in ovulated eggs.

In a recent work, in collaboration with Italian colleagues, Lokman's group used cholesterol-cellulose implants to administer 17MT to European eel females (Di Biase et al. 2017) before starting the traditional CPE treatment until getting spontaneous spawnings from females. Using this pre-treatment they increased the female fecundity, the quality of the oocytes, the hatching rate, and the larval survival 15 days after hatching. Moreover, they reported a reduction of the time required to reach the spawning, reducing the treatment cost.

Very recently, Palstra et al. (2018) tested the separate and combined effects of swimming-simulated migration and hormonal treatments on sexual maturation in European eels with different background (farmed, feminized, and wild silver eels). Simulated migration enhanced early maturation (higher EI and GSI), but the eels that received 17MT implants without forced swimming had stronger effects and increased GSI up to values of 4, indicating an enhancement of early maturation in eels and reducing the subsequent hypophysation period.

In conclusion, androgen pre-treatments are demonstrating their effectiveness as part of the eels maturation protocols, showing a good cost-effectivity and reducing the required time for the artificial induction of maturation.

14.4 Broodstock Feeding and Fatty Acids Mobilization During the Gonad Maturation

14.4.1 Effects of Starvation

Dietary composition conditions the animal reproductive performance and, in the case of ovoviviparous species as most of fish, dietary constituents are transported into the egg yolk, jeopardizing the egg quality, the embryo development, and even the offspring survival. In many fish species, lipids and fatty acids have been identified as key nutrients for the successful gametogenesis.

Eels cease feeding during migration and that is why their energy reserves such as lipids play a crucial role to obtain good quality gametes (eggs and sperm) as the first step for their reproduction (in nature or in captivity). In fact, Damsteegt et al. (2015) described the spawning eels migration as "a textbook example of negative energy balance, forcing these fish to rely on stored fats (triacylglycerides, TAGs) to provide their muscles with energy for swimming and their growing oocytes with the nutrients needed to develop and support healthy offspring." They explained that when eels cease feeding, they must switch from using exogenous TAGs to body-stored ones, and as part of that change, TAGs need to be mobilized from peripheral stores, transported to the liver and packaged into lipoprotein complexes, and finally transported to gonads (for gametogenesis) or muscles (for migration).

Diet provides fish with the essential fatty acids for their metabolism. However, in species such as eel, in which reproduction is associated with a period of starvation, a change in the body fatty acid profile occurs by deposition, elongation, desaturation, and oxidation. During fasting, fatty acid elongation, endogenous synthesis, and desaturation are vital to meet the fish needs (Turchini et al. 2006).

14.4.2 Effects of Diets for Broodstock

Broodstock feeding and their effects on the gamete (eggs and sperm) quality are crucial to get the reproduction of the eels in captivity. Especially, when farmed eels are used as broodstock, a conditioning period could be very important for the success of the reproductive process (Palstra and van den Thillart 2009).

The importance of lipid composition and fatty acids in eggs of the Japanese eel was evidenced by Furuita et al. (2003, 2006, 2007), who found that high-quality eggs presented a lower content of polar lipids, a lower content of arachidonic acid (ARA), and a higher level of docosahexaenoic acid (DHA). Moreover, eels fattened in captivity presented a higher content of highly unsaturated fatty acids (HUFA) in the liver and ovary, while wild eels presented a higher content of n6-HUFAs in muscle, liver, and ovary (Ozaki et al. 2008). The few studies conducted with male and female European eels are mentioned in the following subsections (Sects. 14.4.2.1 and 14.4.2.2).

14.4.2.1 Males

Fatty acids are important in male reproduction and it has been demonstrated that they can influence several aspects such as spermiation periods, sperm quality parameters, sperm peroxidation, and sperm viability in different fish species, including the European eel (Asturiano et al. 2001; Pérez et al. 2000). Fatty acids with C20, ARA and eicosapentaenoic acid (EPA), are precursors of eicosanoids whose metabolites, prostaglandins, play an important role in male reproduction. ARA generates 2-series prostaglandins, and EPA competes with ARA in eicosanoid production and is itself converted into 3-series prostaglandins which are less biologically active than those produced by ARA (Sargent et al. 2002). Eicosanoids modulate steroid synthesis and spermiation during sexual maturation (Asturiano et al. 2000; Norambuena et al. 2013). Furthermore, polyunsaturated fatty acids (PUFA) are the main components of the spermatozoa membrane (Whates et al. 2007), and constant membrane synthesis is required for spermatozoa production.

In 2010, Mazzeo et al. investigated the variations on fatty acids composition in different tissues of European eel males during 13 weeks of hCG-induced sexual maturation, with the main aim of determining which fatty acids could play a central role in sexual maturation. They sacrificed 10 eels per week and performed histological analysis to evaluate the spermatogenic stage of each male. In muscle, no variations or preferential utilization of fatty acids were detected. In liver, MUFAs (monounsaturated fatty acids) were the most abundant fatty acids in the first weeks, while in the last week PUFAs showed the highest concentrations. In testes, PUFAs content increased due to DHA (docosahexaenoic acid), EPA, and ARA accumulation, while MUFAs diminished especially due to oleic and palmitoleic acid decrease. Linolenic (EPA and DHA precursor) and linoleic (ARA precursor) acids also decreased. As evidenced by histology, main changes in testes matched the appearance of spermatids and first spermatozoa, at the end of meiotic divisions. In milt, no variations were appreciated and PUFAs were the most abundant fatty acids, close to 50%.

Later, Baeza (2015) reviewed in her PhD Thesis the roles of lipids and fatty acids through the spermatogenesis of European eel males. Baeza et al. (2014) studied as well the changes in the fat content, as well as fatty acids, in different tissues of European eel males hormonally induced to sexual maturation (with weekly hCG injections) under different thermal regimes (two variables and one constant). Males finished spermatogenesis earlier with a constant temperature of 20 °C, suggesting that eel spermatogenesis is tightly regulated by temperature. The fat content did not change significantly in the muscle (although there was a reduction of muscle weight), but it was increased in the liver and decreased in testes during testicular development. A specific use of fatty acids depending on the tissue and on the stage of testis development was demonstrated.

With regard to fatty acids, during sperm maturation, the liver was highlighted as the main site of synthesis. EPA was synthetized de novo in the liver, to be sent to the testis, which seems to be a crucial step for sperm maturation. Finally, EPA, ARA, and DHA remained constant in the testis through their development, while the level of the rest of fatty acids decreased significantly (used as energy source). This PUFA maintenance in the testis during the eel spermiation period was related to their physiological and structural functions such as the development of spermatozoa membranes (where DHA was the fatty acid found in greater proportion, indicating a structural role in sperm membranes).

Secondly, they used samples from the same experiment to determine the correlation between the main steroid hormones and fatty acids at different stages of testicular development (Baeza et al. 2015a). Similarly as occurs in mammals, EPA and DHA were highlighted as possible modulators of androgen synthesis. The set of the results obtained suggests new perspectives concerning the functions and interactions between fatty acids and steroids in fish spermatogenesis and spermiation.

Moreover, with samples from that experiment, a study was carried out seeking to establish the possible correlations between different fatty acids and several sperm quality parameters (Baeza et al. 2015b). Correlations between several highly unsaturated fatty acids with parameters such as volume, the percentage of motile sperm, and sperm velocity were found, especially between spermatozoa velocity and ARA present in the testis, and concluded that ARA could be used as energy source to increase sperm velocity.

Finally, a second experiment was carried out by designing feeds with different fatty acid percentages in order to evaluate the influence of the diets on European eel sperm quality. It was shown that after 10 weeks of hormonal treatment to induce sexual maturity, animals previously fed with different diets reached a specific level of PUFA in the sperm, highlighting the important role of fatty acids in sperm viability (Baeza 2015; Butts et al. 2015). After the hormonal treatment, the sperm showed more than 60% of motile spermatozoa, with PUFAs being the fatty acid class with the highest concentration in sperm, and DHA having the highest concentration among PUFAs. The results showed that high levels of EPA in the diets induce higher sperm motility, while DHA-rich diets induce higher sperm volumes. Moreover, diets with n3/n6 ratio close to or >2 improve European eel sperm quality parameters. This study allowed obtaining results applicable to the design of optimum broodstock diets for this species.

14.4.2.2 Females

In 2013, Heinsbroek et al. reviewed the available data on the eel broodstock nutrition until that moment. In example, they reported that eel eggs contain a large amount of total lipids, with a higher amount of ARA but lower EPA and DHA levels than in other fish. On the other hand, farmed and wild eels revealed differences in the egg lipid composition and fatty acid profile. Eggs from wild eels mostly contain more ARA and less EPA than those of farmed ones.

Nutritional intervention through the fatty acid profile of the broodstock diet using formulated feeds to influence egg composition has been the target of several studies carried out from then, but the high variability in reproductive success makes difficult to find connections between feeding and egg/larvae quality.

Støttrup et al. (2013) confirmed a slow but steady incorporation of EFA after feeding farmed eels with a formulated feed resembling the profile of essential fatty acids (EFA) found in wild fish. They determined that at least 14 weeks of feeding is required to change lipid EFA in broodstock eel and to allow the incorporation of EFA in the first cohort of matured oocytes. In a second experiment, and after 24 weeks of feeding different formulated diets, Støttrup et al. (2016) found the highest percentage of stripped females, producing viable eggs and larvae, in the group fed the highest dietary ARA levels.

On their hand, da Silva et al. (2016) fed farmed female eels with two experimental diets with similar proximate composition, similar levels of ARA, but different n-3 PUFA levels (EPA and DHA were respectively 4.5 and 2.6 times higher in one of the diets). After the feeding period, each diet group was divided into two, receiving a constant dose of salmon pituitary extract (SPE) for 13 weeks, or an increasing hormone dosage. Females fed with higher levels of n-3 PUFAs and stimulated with the constant hormonal treatment reached higher GSIs than those receiving the variable hormonal treatment. Moreover, plasma levels of sex steroids (E2, T, and 11KT) did not differ between diets and hormonal treatments, and independent of hormonal treatment, the diet with higher levels of n-3 PUFAs led to the most advanced stages of oocyte development, such as germinal vesicle migration.

14.5 Fertilization, Embryo Development, and Larval Culture

14.5.1 In Vivo Trials

In 1974, Yamamoto and Yamauchi reported for the first time the induction of sexual maturation and hatching of Japanese eel larvae in captivity. Then, although the research continued in Japan, no publications appeared for more than 20 years. Then, Tanaka et al. (2001) published the first protocol to get leptocephali of Japanese eel in captivity, and the same research group reported the first production of glass eels in this species (Tanaka et al. 2003; Kagawa et al. 2005) after improvements on the mechanisms of oocyte maturation and ovulation in female, and of spermatozoa maturation in male (i.e., Yamauchi 1990; Ohta et al. 1996, 1997; Kagawa et al. 2003). From then, several generations of eels have been obtained without depending on any supply from wild resources (Okamura et al. 2009, 2013; Masuda et al. 2011, 2012; Kagawa et al. 2013b). The progression of these methods was reviewed by Tanaka in 2015, who explained that currently, research continues trying to develop a stable technology for the mass production of glass eels.

Although previous research with other eel species was carried out from the middle of twentieth century (reviewed by Palstra and van den Thillart 2009), the protocol established by the Japanese scientists has been the model for further methods applied to European eel (Pérez et al. 2000; Pedersen 2003, 2004; Müller et al. 2004, 2018; Asturiano et al. 2005, 2006; Palstra et al. 2005; Peñaranda et al. 2010; Gallego et al. 2012; Mordenti et al. 2013; Vílchez et al. 2013; Butts et al. 2014), New Zealand eels (*A. dieffenbachii* and *A. australis*; Lokman and Young 2000), and American eel (*A. rostrata*; Oliveira and Hable 2010).

From 2006–2007, Tomkiewicz et al. (Technical University of Denmark) started to produce European eel pre-feeding larvae. Later, they used combinations of protocols supplied from different participants of the PRO-EEL project (2010–2014) to improve the method and going deeper in several aspects as the evaluation of sperm density (Sørensen et al. 2013), the standardization of fertilization protocols (Butts et al. 2014), the use of environmental factors during embryonic development, as the ligth (Politis et al. 2014; Butts et al. 2016), the salinity (Sørensen et al. 2016a; Politis et al. 2018a), the first production of larvae using cryopreserved sperm (Asturiano et al. 2016), the musculoskeletal anatomy and feeding performance of larvae (Bouilliart et al. 2015), and even bacteria and parasites (Sørensen et al. 2014a, b) related with eel egg production.

Once the hormonal treatment procedures, the hatchery techniques, and the environmental parameters were fixed, and they got a regular production of eggs and larvae, new studies have been possible, including a detailed study on the early life stages (eggs, embryos, and larvae) of this species in captivity (Sørensen et al. 2016b), the effects of temperature on the larvae gene expression and morphological development (Politis et al. 2017, 2018b), and the effect of parental origin (wild vs. farmed) of European eel on early life history traits (i.e., fertilization success, embryonic survival, hatch success, and larval deformities), demonstrating that short-term domestication of male eels does not negatively affect offspring quality and enables the consideration of cultured male broodstock in future breeding programs (Benini et al. 2018). Moreover, they concluded that eggs of one female can develop more successfully when crossed with a compatible male, highlighting the importance of mate choice for future breeding programs for this species.

Maternal gene products govern early embryonic development in fish. The abundance of β-tubulin, insulin-like growth factor 2, nucleoplasmin, prohibitin 2, phosphatidylinositol glycan biosynthesis class F protein 5, and carnitine *O*-palmitoyltransferase liver isoform-like 1 has been related with embryonic developmental competence in other teleosts. Rozenfeld et al. (2016) found an upregulated transcription of these genes in European eel embryos after the mid-blastula transition (30 h post-fertilization) and suggested that it may be needed to sustain embryonic development and hatching success.

14.5.2 Fertilization Methods

There are two methods to carry out the fertilization in eels: the stripping and the spontaneous spawning. The spontaneous spawning method, developed in the Japanese eel (Horie et al. 2008), has been used in the European species by Di Biase et al. (2016), getting better results than using the stripping. In the spontaneous spawning method, when the females receive the final injection of DHP (2 mg/kg) to induce the ovulation, they are transferred into a different tank (spawning tank), at 20 °C, with a proportion of 4 males:1 female. The spawning happens spontaneously around 12–16 h later, and the broodstock are removed from the tank.

In the stripping method, after the final injection of DHP, females are not allowed to deliver the eggs in the tank, evaluating their stage of development at 1 h intervals. When they start to ovulate, the eggs are extracted by abdominal massage and are fertilized with diluted sperm and seawater. Our group of research have experience with this last method, which is more useful to develop experiments with sperm in different conditions. With the stripping method we obtained 12 spawnings with a mean of 84% floating eggs (Vílchez et al. 2013).

14.5.3 Larvae Feeding

Despite the big interest and all the developed investments, the reproduction of Japanese eel is not fully resolved, although good advances has been carried out. In 2001, Tanaka et al. (National Research Institute of Aquaculture, Nansei, Mie, Japan) were able to produce leptocephali of Japanese eel and, shortly later, reported the methods to produce glass eels around 250 days after hatching (Tanaka et al. 2003; Kagawa et al. 2005). Tanaka et al. (2001) used a mix of shark (spiny dogfish *Squalus acanthias*) egg powder, soya peptide, minerals, vitamins, and krill paste to feed the larvae. From then, many experiments have been carried out trying to improve the feed quality and obtain the best reproduction results, although most of this research has never been published through the standard scientific ways.

Furuita et al. (2009) reported that the content of different vitamins (C and E) in the eggs could modulate the larvae survival, and proposed the nutritional enrichment of the diets with these vitamins. More recently, Masuda et al. (2013, 2016) have assayed the use of diets without shark egg, including fish protein hydrolysate pre-digested with integral enzymes from frozen krill. Their results are promising but still far from application.

The use of live preys has been studied as an alternative/complementary way to feed the larvae. Wullur et al. (2013) reported the ingestion of several zooplankton species by Japanese eel larvae and suggested the rotifer *Proales similis* as a potential initial food source for Japanese eel larvae. On their hand, Butts et al. (2016) evidenced the first-feeding of European eel larvae reared in captivity. They found that up to 50% of larvae ingested a diet composed of concentrated rotifer paste, with or without natural feeding stimulants. Obviously, new research is required in the feeding of broodstock and larvae to guarantee a bigger eel species reproductive success.

14.5.4 Eel Hybrids

Hybrids of Atlantic eels (*A. anguilla*, *A. rostrata*) has been found in the nature (Albert et al. 2006) and several techniques have been used for its detection (Frankowski and Bastrop 2010). But some attempts have been carried out to get unusual hybrids by in vitro fertilization. Burgerhout et al. (2011) got the hybrids of the eel species *A. australis* and *A. anguilla*, while Müller et al. (2018) used cryopreserved European eel sperm for hybridization of *A. japonica* × *A. anguilla*.

14.6 Conclusion

In the last years, many recent advances have happened in the European eel reproduction, coinciding with the funding of European projects (PRO-EEL, IMPRESS) or the COST Action AQUAGAMETE (see the acknowledgements section). New techniques are being incorporated in this area of research, allowing good advances in our knowledge. However, many improvements are still required and reaching a commercial production of European eels, based on the production of glass eels in captivity, seems still quite far if we consider the rhythm of results obtained with the Japanese and other eel species, or the funding limitations that European scientists are facing today.

Acknowledgements During the last years, the research of our team has been mainly funded by the EELGONIA project (Spanish Ministry of Science, Innovation and Universities, MICIU; RTI2018-096413-B-I00), REPRO-TEMP project (Spanish Ministry of Economy, Industry and Competitiveness, MINECO; AGL2013-41646-R), European Union's 7th Framework Program (PRO-EEL, grant agreement n° 245257), European Union's Horizon 2020 research and innovation program under the Marie Skłodowska-Curie (IMPRESS, grant agreement n° 642893), and by COST Office (COST Action FA1205: AQUAGAMETE).

References

Aarestrup K, Okland F, Hansen MM, Righton D, Gargan P, Castonguay M, Bernatchez L, Howey P, Sparholt H, Pedersen MI, McKinley RS (2009) Oceanic spawning migration of the European eel (*Anguilla anguilla*). Science 325(5948):1660. https://doi.org/10.1126/science.1178120

Adachi S, Ijiri S, Kazeto Y, Yamauchi K (2003) Oogenesis in the Japanese eel, *Anguilla japonica*. In: Aida K, Tsukamoto K, Yamauchi K (eds) Eel biology. Springer, Tokyo, pp 502–518

Albert V, Jónsson B, Bernatchez L (2006) Natural hybrids in Atlantic eels (*A. anguilla*, *A. rostrata*): evidence for successful reproduction and fluctuating abundance in space and time. Mol Ecol 15:1903–1916. https://doi.org/10.1111/j.1365-294X.2006.02917.x

Aroua S, Schmitz M, Baloche S, Vidal B, Rousseau K, Dufour S (2005) Endocrine evidence that silvering, a secondary metamorphosis in the eel, is a pubertal rather than a metamorphic event. Neuroendocrinology 82:221–232. https://doi.org/10.1159/000092642

Asturiano JF, Sorbera LA, Zanuy S, Carrillo M (2000) Effects of polyunsaturated fatty acids and gonadotropin on prostaglandin series E production in a primary testis cell culture system for the European sea bass. J Fish Biol 57:1563–1574. https://doi.org/10.1111/j.1095-8649.2000.tb02232.x

Asturiano JF, Sorbera LA, Carrillo M, Zanuy S, Ramos J, Navarro JC, Bromage N (2001) Reproductive performance in male European sea bass (*Dicentrarchus labrax*, L.) fed two PUFA-enriched experimental diets: a comparison with males fed a wet diet. Aquaculture 194:173–190. https://doi.org/10.1016/S0044-8486(00)00515-9

Asturiano JF, Pérez L, Tomás A, Zegrari S, Espinós FJ, Jover M (2002) Inducción hormonal de la maduración gonadal y puesta en hembras de anguila europea *Anguilla anguilla* L. 1758: cambios morfológicos y desarrollo oocitario. Boletín IEO 18:127–137. In Spanish with abstract in English

Asturiano JF, Pérez L, Garzón DL, Peñaranda DS, Marco-Jiménez F, Martínez-Llorens S, Tomás A, Jover M (2005) Effect of different methods for the induction of spermiation on semen quality in European eel. Aquac Res 36:1480–1487. https://doi.org/10.1111/j.1365-2109.2005.01366.x

Asturiano JF, Marco-Jiménez F, Pérez L, Balasch S, Garzón DL, Peñaranda DS, Vicente JS, Viudes de Castro MP, Jover M (2006) Effects of hCG as spermiation inducer on European eel semen quality. Theriogenology 66:1012–1020. https://doi.org/10.1016/j.theriogenology.2006.02.041

Asturiano JF, Sørensen SR, Pérez L, Lauesen P, Tomkiewicz J (2016) First production of larvae using cryopreserved sperm. Effects of preservation temperature and cryopreservation on European eel sperm fertilization capacity. Reprod Domest Anim 51:485–491. https://doi.org/10.1111/rda.12706

Baeza R (2015) Roles of lipids and fatty acids through the spermatogenesis of European eel (*Anguilla anguilla*). PhD Thesis, Universitat Politècnica de València (Spain), 197p. http://hdl.handle.net/10251/49988

Baeza R, Mazzeo I, Vílchez MC, Gallego V, Peñaranda DS, Pérez L, Asturiano JF (2014) Effect of thermal regime on fatty acid dynamics in male European eels (*Anguilla anguilla*) during hormonally-induced spermatogenesis. Aquaculture 430:86–97. https://doi.org/10.1016/j.aquaculture.2014.03.045

Baeza R, Peñaranda DS, Vílchez MC, Tveiten H, Pérez L, Asturiano JF (2015a) Exploring correlations between sex steroids and fatty acids and their potential roles in induced maturation of male European eel. Aquaculture 435:328–335. https://doi.org/10.1016/j.aquaculture.2014.10.016

Baeza R, Mazzeo I, Vílchez MC, Gallego V, Peñaranda DS, Pérez L, Asturiano JF (2015b) Relationship between sperm quality parameters and fatty acid composition of the muscle, liver and testis of European eel. Comp Biochem Physiol A Mol Integr Physiol 181:79–86. https://doi.org/10.1016/j.cbpa.2014.11.022

Benini E, Politis SN, Kottmann JS, Butts IAE, Sørensen SR, Tomkiewicz J (2018) Effect of parental origin on early life history traits of European eel. Reprod Domest Anim 53(5):1149–1158. https://doi.org/10.1111/rda.13219

Bouilliart M, Tomkiewicz J, Lauesen P, De Kegel B, Adriaens D (2015) Musculoskeletal anatomy and feeding performance of pre-feeding engyodontic larvae of the European eel (*Anguilla anguilla*). J Anat 227(3):325–340. https://doi.org/10.1111/joa.12335Document

Burgerhout E, Brittijn SA, Kurwie T, Decker P, Dirks RP, Palstra AP, Spaink HP, van den Thillart GEEJM (2011) First artificial hybrid of the eel species *Anguilla australis* and *Anguilla anguilla*. BMC Dev Biol 11:16. https://doi.org/10.1186/1471-213X-11-16

Butts IAE, Sørensen SR, Politis SN, Pitcher TE, Tomkiewicz J (2014) Standardization of fertilization protocols for the European eel, *Anguilla anguilla*. Aquaculture 426-427:9–13. https://doi.org/10.1016/j.aquaculture.2014.01.020

Butts IAE, Baeza R, Støttrup JG, Krüger-Johnsen M, Jacobsen C, Pérez L, Asturiano JF, Tomkiewicz J (2015) Impact of dietary fatty acids on muscle composition, liver lipids, milt composition and sperm performance in European eel. Comp Biochem Physiol A Mol Integr Physiol 183:87–96. https://doi.org/10.1016/j.cbpa.2015.01.015

Butts IAE, Sørensen SR, Politis SN, Tomkiewicz J (2016) First-feeding by European eel larvae: a step towards closing the life cycle in captivity. Aquaculture 464:451–458. https://doi.org/10.1016/j.aquaculture.2016.07.028

da Silva FFG, Støttrup JG, Kjørsvik E, Tveiten H, Tomkiewicz J (2016) Interactive effects of dietary composition and hormonal treatment on reproductive development of cultured female European eel, *Anguilla anguilla*. Anim Reprod Sci 171:17–26. https://doi.org/10.1016/j.anireprosci.2016.05.007

Damsteegt EL, Falahatimarvast A, McCormick SPA, Lokman PM (2015) Triacylglyceride physiology in the short-finned eel, *Anguilla australis*—changes throughout early oogenesis. Am J Physiol Regul Integr Comp Physiol 308:R935–R944. https://doi.org/10.1152/ajpregu.00436.2014

Damsteegt EL, Ozaki Y, McCormick SP, Lokman PM (2016) Triacylglyceride physiology in the shortfinned eel, *Anguilla australis*—the effects of androgen. Am J Physiol Regul Integr Comp Physiol 310(5):R422–R431. https://doi.org/10.1152/ajpregu.00149.2015

Di Biase A, Casalini A, Emmanuele P, Mandelli M, Lokman PM, Mordenti O (2016) Controlled reproduction in *Anguilla anguilla* (L.): comparison between spontaneous spawning and stripping-insemination approaches. Aquac Res 47:3052–3060. https://doi.org/10.1111/are.12755

Di Biase A, Lokman PM, Govoni N, Casalini A, Emmanuele P, Parmeggiani A, Mordenti O (2017) Co-treatment with androgens during artificial induction of maturation in female eel, *Anguilla anguilla*: effects on egg production and early development. Aquaculture 479:508–515. https://doi.org/10.1016/j.aquaculture.2017.06.030

Dou SZ, Yamada Y, Okamura A, Shinoda A, Tanaka S, Tsukamoto K (2008) Temperature influence on the spawning performance of artificially-matured Japanese eel, *Anguilla japonica*, in captivity. Environ Biol Fish 82:151–164. https://doi.org/10.1007/s10641-007-9268-8

Dufour S, Burzawa-Gerard E, Le Belle N, Sbaihi M, Vidal B (2003) Reproductive endocrinology of the European eel, *Anguilla anguilla*. In: Aida K, Tsukamoto K, Yamauchi K (eds) Eel biology. Springer, Tokyo, pp 373–383

Durif C, Dufour S, Elie P (2005) The silvering process of the eel: a new classification from the yellow resident stage to the silver migrating stage. J Fish Biol 66:1–19. https://doi.org/10.1111/j.0022-1112.2005.00662.x

Frankowski J, Bastrop R (2010) Identification of *Anguilla anguilla* (L.) and *Anguilla rostrata* (Le Sueur) and their hybrids based on a diagnostic single nucleotide polymorphism in nuclear 18S rDNA. Mol Ecol Resour 10:173–176. https://doi.org/10.1111/j.1755-0998.2009.02698.x

Furuita H, Ohta H, Unuma T, Tanaka H, Kagawa H, Suzuki N, Yamamoto T (2003) Biochemical composition of eggs in relation to egg quality in the Japanese eel, *Anguilla japonica*. Fish Physiol Biochem 29:37–46. https://doi.org/10.1023/B:FISH.0000035897.58924.9d

Furuita H, Unuma T, Nomura K, Tanaka H, Okuzawa K, Sugita T, Yamamoto Y (2006) Lipid and fatty acid composition of eggs producing larvae with high survival rate in the Japanese eel. J Fish Biol 69:1178–1189. https://doi.org/10.1111/j.1095-8649.2006.01196.x

Furuita H, Hori K, Suzuki N, Sugita T, Yamamoto T (2007) Effect of n-3 and n-6 fatty acids in broodstock diet on reproduction and fatty acid composition of broodstock and eggs in the Japanese eel *Anguilla japonica*. Aquaculture 267:55–61. https://doi.org/10.1016/j.aquaculture.2007.01.039

Furuita H, Unuma T, Nomura K, Tanaka H, Sugita T, Yamamoto T (2009) Vitamin contents of eggs that produce larvae showing a high survival rate in the Japanese eel *Anguilla japonica*. Aquac Res 40:1270–1278. https://doi.org/10.1111/j.1365-2109.2009.02225.x

Gallego V, Mazzeo I, Vílchez MC, Peñaranda DS, Carneiro PCF, Pérez L, Asturiano JF (2012) Study of the effects of thermal regime and alternative hormonal treatments on the reproductive performance of European eel males (*Anguilla anguilla*) during induced sexual maturation. Aquaculture 354-355:7–16. https://doi.org/10.1016/j.aquaculture.2012.04.041

Gallego V, Vílchez MC, Peñaranda DS, Pérez L, Herráez MP, Asturiano JF, Martínez-Pastor F (2015) Subpopulation pattern of eel spermatozoa is affected by post-activation time, hormonal treatment and the thermal regimen. Reprod Fertil Dev 27:529–543. https://doi.org/10.1071/RD13198

Heinsbroek LTN, Støttrup JG, Jacobsen C, Corraze G, Kraiem MM, Holst LK, Tomkiewicz J, Kaushik SJ (2013) A review on broodstock nutrition of marine pelagic spawners: the curious case of the freshwater eels (*Anguilla* spp.). Aquac Nutr 19(1):1–24. https://doi.org/10.1111/anu.12091

Herranz-Jusdado JG, Rozenfeld C, Morini M, Pérez L, Asturiano JF, Gallego V (2019) Recombinant vs purified mammal gonadotropins as maturation hormonal treatments of European eel males. Aquaculture 501:527–536. https://doi.org/10.1016/j.aquaculture.2018.12.015

Horie N, Utoh T, Nimawa N, Yamada Y, Okamura A, Tanaka S, Tsukamoto K (2008) Influence of artificial fertilization methods of the hormone-treated Japanese eel *Anguilla japonica* upon the quality of eggs and larvae (comparison between stripping-insemination and spontaneous spawning methods). Nippon Suisan Gakkaishi 74:26–35. https://doi.org/10.2331/suisan.74.26. (in Japanese with English abstract)

ICES (2013) Working group on eels. Report of the ICES advisory committee on fisheries management 18. 253 p

ICES (2017) European eel (*Anguilla anguilla*) throughout its natural range. In: ICES advice on fishing opportunities, catch, and effort Ecoregions in the Northeast Atlantic. https://doi.org/10.17895/ices.pub.3440

Jeng SR, Yueh WS, Chen GR, Lee YH, Dufour S, Chang CF (2007) Differential expression and regulation of gonadotropins and their receptors in the Japanese eel, *Anguilla japonica*. Gen Comp Endocrinol 154:161–173. https://doi.org/10.1016/j.ygcen.2007.05.026

Kagawa H, Iinuma N, Tanaka H, Ohta H, Okuzawa K (1998) Effects of rearing period in seawater on induced maturation in female Japanese eel *Anguilla japonica*. Fish Sci 64(1):77–82. https://doi.org/10.2331/fishsci.64.77

Kagawa H, Tanaka H, Unuma T, Ohta H, Gen K, Okuzawa K (2003) Role of prostaglandin in the control of ovulation in the Japanese eel *Anguilla japonica*. Fish Sci 69:234–241. https://doi.org/10.1046/j.1444-2906.2003.00613.xl. Article234241BEES

Kagawa H, Tanaka H, Ohta H, Unuma T, Nomura K (2005) The first success of glass eel production in the world: basic biology on fish reproduction advances new applied technology in aquaculture. Fish Physiol Biochem 31:193–199. https://doi.org/10.1007/s10695-006-0024-3

Kagawa H, Kasuga Y, Adachi J, Nishi A, Hashimoto H, Imaizumi H, Kaji S (2009) Effects of continuous administration of human chorionic gonadotropin, salmon pituitary extract, and gonadotropin-releasing hormone using osmotic pumps on induction of sexual maturation in male Japanese eel, *Anguilla japonica*. Aquaculture 296:117–122. https://doi.org/10.1016/j.aquaculture.2009.07.023

Kagawa H, Fujie N, Imaizumi H, Masuda Y, Oda K, Adachi J, Nishi A, Hashimoto H, Teruya K, Kaji S (2013a) Using osmotic pumps to deliver hormones to induce sexual maturation of female Japanese eels, *Anguilla japonica*. Aquaculture 388:30–34. https://doi.org/10.1016/j.aquaculture.2013.01.025

Kagawa H, Sakurai Y, Horiuchi R, Kazeto Y, Gen K, Imaizumi H, Masuda Y (2013b) Mechanism of oocyte maturation and ovulation and its application to seed production in the Japanese eel. Fish Physiol Biochem 39:13–17. https://doi.org/10.1007/s10695-012-9607-3

Kazeto Y, Kohara M, Miura T, Miura C, Yamaguchi S, Trant JM, Adachi S, Yamauchi K (2008) Japanese eel follicle-stimulating hormone (Fsh) and luteinizing hormone (Lh): production of biologically active recombinant Fsh and Lh by *Drosophila* S2 cells and their differential actions on the reproductive biology. Biol Reprod 79:938–946. https://doi.org/10.1095/biolreprod.108.070052

Kobayashi M, Hayakawa Y, Park W, Banba A, Yoshizaki G, Kumamaru K, Kagawa H, Kaki H, Nagaya H, Sohn YC (2010) Production of recombinant Japanese eel gonadotropins by baculovirus in silkworm larvae. Gen Comp Endocrinol 167:379–386. https://doi.org/10.1016/j.ygcen.2010.01.003

Lin HR, Zhang ML, Zhang SM, Van Der Kraak G, Peter RE (1991) Stimulation of pituitary gonadotropin and ovarian development by chronic administration of testosterone in female Japanese silver eel, *Anguilla japonica*. Aquaculture 96(1):87–95. https://doi.org/10.1016/0044-8486(91)90141-S

Lin HR, Zhang LH, Wang XD, Chen LX (1998) Artificial induction of gonadal maturation and ovulation in the Japanese eel (*Anguilla japonica* T. et S.). Bull Fr Pêche Piscic 349:163–176. https://doi.org/10.1051/kmae:1998041

Lokman PM, Young G (2000) Induced spawning and early ontogeny of New Zealand freshwater eels (*Anguilla dieffenbachii* and *A. australis*). N Z J Mar Freshw Res 34(1):135–145. https://doi.org/10.1080/00288330.2000.9516921

Lokman PM, Rohr DH, Davie PS, Young G (2003) The physiology of silvering in anguillid eels—androgens and control of metamorphosis from the yellow to the silver stage. In: Aida K, Tsukamoto K, Yamauchi K (eds) Advances in eel biology. Springer, Tokyo, pp 331–349

Lokman PM, George KAN, Divers SL, Algie M, Young G (2007) 11-Ketotestosterone and IGF-I increase the size of previtellogenic oocytes from shortfinned eel, *Anguilla australis*, in vitro. Reproduction 133:955–967. https://doi.org/10.1530/REP-06-0229

Lokman PM, Wylie MJ, Downes M, Di Biase A, Damsteegt EL (2015) Artificial induction of maturation in female silver eels, *Anguilla australis*: the benefits of androgen pre-treatment. Aquaculture 437:111–119. https://doi.org/10.1016/j.aquaculture.2014.11.026

Lokman PM, Damsteegt EL, Wallace J, Downes M, Goodwin SL, Facoory LJ, Wylie MJ (2016) Dose-responses of male silver eels, *Anguilla australis*, to human chorionic gonadotropin and 11-ketotestosterone in vivo. Aquaculture 463:97–105. https://doi.org/10.1016/j.aquaculture.2016.05.009

Masuda Y, Imaizumi H, Oda K, Hashimoto H, Teruya K, Usuki H (2011) Japanese eel *Anguilla japonica* larvae can metamorphose into glass eel within 131 days after hatching in captivity. Nippon Suisan Gakkaishi 77:416–418. https://doi.org/10.2331/suisan.77.416. (in Japanese)

Masuda Y, Imaizumi H, Oda K, Hashimoto H, Usuki H, Teruya K (2012) Artificial completion of the Japanese eel, *Anguilla japonica*, life cycle: challenge to mass production. Bull Fish Res Agency 35:111–117

Masuda Y, Jinbo T, Imaizumi H, Furuita H, Matsunari H, Murashita K, Fujimoto H, Nagao J, Kawakami Y (2013) A step forward in development of fish protein hydrolysate-based diets for larvae of Japanese eel *Anguilla japonica*. Fish Sci 79(4):681–688. https://doi.org/10.1007/s12562-013-0637-2

Masuda Y, Yatabe T, Matsunari H, Furuita H, Kamoshida M, Shima Y, Kuwada H (2016) Rearing of larvae of Japanese eel *Anguilla japonica* to metamorphosis into glass eel by feeding with fish protein hydrolysate-based diets. Nippon Suisan Gakkaishi 82(2):131–133. https://doi.org/10.2331/suisan.15-00060

Mazzeo I, Gallego V, Pérez L, Peñaranda DS, Jover M, Asturiano JF (2010) Variations on fatty acids composition in different tissues of European eel (*Anguilla anguilla*) males during induced sexual maturation. J Appl Ichthyol 26:763–774. https://doi.org/10.1111/j.1439-0426.2010.01546.x

Mazzeo I, Peñaranda DS, Gallego V, Baloche S, Nourizadeh-Lillabadi R, Tveiten H, Dufour S, Asturiano JF, Weltzien FA, Pérez L (2014) Temperature modulates the vitellogenesis progression in European eel. Aquaculture 434:38–47. https://doi.org/10.1016/j.aquaculture.2014.07.020

Mes D, Dirks RP, Palstra AP (2016) Simulated migration under mimicked photothermal conditions enhances sexual maturation of farmed European eel (*Anguilla anguilla*). Aquaculture 452:367–372. https://doi.org/10.1016/j.aquaculture.2015.11.020

Miura T, Ando N, Miura C, Yamauchi K (2002) Comparative studies between in vivo and in vitro spermatogenesis of Japanese eel (*Anguilla japonica*). Zool Sci 19:321–329. https://doi.org/10.2108/zsj.19.321

Mordenti O, Di Biase A, Sirri R, Modugno S, Tasselli A (2012) Induction of sexual maturation in wild female European eels (*Anguilla anguilla*) in darkness and light. Isr J Aquacult Bamidgeh 64:9p. https://doi.org/10524/23179

Mordenti O, Di Biase A, Bastone G, Sirri R, Zaccaroni A, Parmeggiani A (2013) Controlled reproduction in the wild European eel (*Anguilla anguilla*): two populations compared. Aquac Int 21:1045–1063. https://doi.org/10.1007/s10499-012-9611-8

Morini M, Peñaranda DS, Vílchez MC, Nourizadeh-Lillabadi R, Lafont AG, Dufour S, Asturiano JF, Weltzien FA, Pérez L (2017) Nuclear and membrane progestin receptors in the European eel: characterization and expression in vivo through spermatogenesis. Comp Biochem Physiol A Mol Integr Physiol 207:79–92. https://doi.org/10.1016/j.cbpa.2017.02.009

Müller T, Urbányi B, Váradi B, Binder T, Horn P, Bercsényi M, Horváth Á (2004) Cryopreservation of sperm of farmed European eel *Anguilla anguilla*. J World Aquacult Soc 35:225–231. https://doi.org/10.1111/j.1749-7345.2004.tb01078.x

Müller T, Matsubara H, Kubara Y, Horváth Á, Kolics B, Taller J, Stéger V, Kovács B, Horváth L, Asturiano JF, Peñaranda DS, Kucharczyk D, Urbányi B (2018) Testing cryopreserved European eel sperm for hybridization (*A. japonica* × *A. anguilla*). Theriogenology 113:153–158. https://doi.org/10.1016/j.theriogenology.2018.02.021

Mylonas CC, Duncan NJ, Asturiano JF (2017) Hormonal manipulations for the enhancement of sperm production in cultured fish and evaluation of sperm quality. Aquaculture 472:21–44. https://doi.org/10.1016/j.aquaculture.2016.04.021

Norambuena F, Estévez A, Mañanós E, Bell JG, Carazo I, Duncan N (2013) Effects of graded levels of arachidonic acid on the reproductive physiology of Senegalese sole (*Solea senegalensis*): fatty acid composition, prostaglandins and steroid levels in the blood of broodstock bred in captivity. Gen Comp Endocrinol 191:92–101. https://doi.org/10.1016/j.ygcen.2013.06.006

Ohta H, Kagawa H, Tanaka H, Okuzawa K, Hirose K (1996) Changes in fertilization and hatching rate with time after ovulation induced by 17,20β-dihydroxy-4-pregnen-3-one in the Japanese eel, *Anguilla japonica*. Aquaculture 139:291–301. https://doi.org/10.1016/0044-8486(95)01167-6

Ohta H, Kagawa H, Tanaka H, Okuzawa K, Linuma N, Hirose K (1997) Artificial induction of maturation and fertilization in the Japanese eel, *Anguilla japonica*. Fish Physiol Biochem 17:163–169. https://doi.org/10.1023/A:1007720600588

Ohta H, Sato Y, Imaizumi H, Kazeto Y (2017) Changes in milt volume and sperm quality with time after an injection of recombinant Japanese eel luteinizing hormone in Japanese eel males. Aquaculture 479:150–154. https://doi.org/10.1016/j.aquaculture.2017.05.044

Okamura A, Yamada Y, Horita T, Horie N, Mikawa N, Utoh T, Tanaka S, Tsukamoto K (2009) Rearing eel leptocephali (*Anguilla japonica*) in a planktonkreisel. Aquac Res 40:509–512. https://doi.org/10.1111/j.1365-2109.2008.02127.x

Okamura A, Horie N, Mikawa N, Yamada Y, Tsukamoto K (2013) Recent advances in artificial production of glass eels for conservation of anguillid eel populations. Ecol Freshw Fish 23:95–110. https://doi.org/10.1111/eff.12086

Oliveira K, Hable WE (2010) Artificial maturation, fertilization, and early development of the American eel (*Anguilla rostrata*). Can J Zool 88(1):1121–1128. https://doi.org/10.1139/Z10-081

Ozaki Y, Ishida K, Saito K, Ura K, Adachi S, Yamauchi K (2007) Immunohistochemical changes in production of pituitary hormones during artificial maturation of female Japanese eel *Anguilla japonica*. Fish Sci 73(3):574–584. https://doi.org/10.1111/j.1444-2906.2007.01370.x

Ozaki Y, Koga H, Takahashi T, Adachi S, Yamauchi K (2008) Lipid content and fatty acid composition of muscle, liver, ovary and eggs of captive reared and wild silver Japanese eel *Anguilla japonica* during artificial maturation. Fish Sci 74:362–371. https://doi.org/10.1111/j.1444-2906.2008.01525.x

Palstra A, van den Thillart GE (2009) Artificial maturation and reproduction of the European eel. In: van den Thillart G, Dufour S, Rankin JC (eds) Spawning migration of the European eel. Springer Science + Business Media B.V, Dordrecht, pp 309–331

Palstra AP, van den Thillart GEEJM (2010) Swimming physiology of European silver eels (*Anguilla anguilla* L.): energetic costs and effects on sexual maturation and reproduction. Fish Physiol Biochem 36:297–322. https://doi.org/10.1007/s10695-010-9397-4

Palstra AP, Cohen EGH, Niemantsverdriet PRW, van Ginneken VJT, van den Thillart GEEJM (2005) Artificial maturation and reproduction of European silver eel: development of oocytes during final maturation. Aquaculture 249:533–547. https://doi.org/10.1016/j.aquaculture.2005.04.031

Palstra A, Curiel D, Fekkes M, De Bakker M, Székely C, Van Ginneken V, van den Thillart G (2007) Swimming stimulates oocyte development in European eel (*Anguilla anguilla* L.). Aquaculture 270:321–332. https://doi.org/10.1016/j.aquaculture.2007.04.015

Palstra A, van Ginneken V, van den Thillart G (2008a) Cost of transport and optimal swimming speed in farmed and wild European silver eels (*Anguilla anguilla*). Comp Biochem Physiol A Mol Integr Physiol 151:37–44. https://doi.org/10.1016/j.cbpa.2008.05.011

Palstra AP, Schnabel D, Nieveen MC, Spaink HP, van den Thillart GEEJM (2008b) Male silver eels mature by swimming. BMC Physiol 8:14. https://doi.org/10.1186/1472-6793-8-14

Palstra A, van Ginneken V, van den Thillart G (2009) Effects of swimming on silvering and maturation of the European eel, *Anguilla anguilla* L. In: van den Thillart G, Dufour S, Rankin JC (eds) Spawning migration of the European eel. Springer Science + Business Media B.V, Dordrecht, pp 229–250

Palstra AP, Schnabel D, Nieveen MC, Spaink HP, van den Thillart G (2010) Swimming suppresses hepatic vitellogenesis in European female silver eels as shown by expression of the estrogen receptor 1, vitellogenin1 and vitellogenin2 in the liver. Reprod Biol Endocrinol 8:27. https://doi.org/10.1186/1477-7827-8-27

Palstra AP, Jéhannet P, Kruijt L, Swinkels W, Heinsbroek LTN (2018) Separate and combined treatment effects of simulated reproductive migration and hormonal stimulation on sexual maturation in European eels. 11th international symposium on the reproductive physiology of fish, Manaus (Brazil), Book of abstracts, p 149

Pedersen BH (2003) Induced sexual maturation of the eel *Anguilla anguilla* and fertilization of the eggs. Aquaculture 224:323–328. https://doi.org/10.1016/S0044-8486(03)00242-4

Pedersen BH (2004) Fertilisation of eggs, rate of embryonic development and hatching following induced maturation of the European eel *Anguilla anguilla*. Aquaculture 237:461–473. https://doi.org/10.1016/j.aquaculture.2004.04.019

Peñaranda DS, Pérez L, Gallego V, Jover M, Baloche S, Dufour S, Asturiano JF (2010) Molecular and physiological study of the artificial maturation process in the European eel males: from brain to testis. Gen Comp Endocrinol 166:160–171. https://doi.org/10.1016/j.ygcen.2009.08.006

Peñaranda DS, Morini M, Tveiten H, Vílchez MC, Gallego V, Dirks RP, van den Thillart GEEJM, Pérez L, Asturiano JF (2016) Temperature modulates testis steroidogenesis in European eel. Comp Biochem Physiol A Mol Integr Physiol 197:58–67. https://doi.org/10.1016/j.cbpa.2016.03.012

Peñaranda DS, Gallego V, Rozenfeld C, Herranz-Jusdado JG, Pérez L, Gómez A, Giménez I, Asturiano JF (2018) Using specific recombinant gonadotropins to induce spermatogenesis and spermiation in the European eel (*Anguilla anguilla*). Theriogenology 107:6–20. https://doi.org/10.1016/j.theriogenology.2017.11.002

Pérez L, Asturiano JF, Tomás A, Zegrari S, Barrera R, Espinós FJ, Navarro JC, Jover M (2000) Induction of maturation and spermiation in the male European eel (*Anguilla anguilla*). Assessment of sperm quality throughout treatment. J Fish Biol 57:1488–1504. https://doi.org/10.1111/j.1095-8649.2000.tb02227.x

Pérez L, Peñaranda DS, Jover M, Asturiano JF (2008) Results of maturation and ovulation in European eel females. Cybium 32:320

Pérez L, Peñaranda DS, Dufour S, Baloche S, Palstra AP, van den Thillart GEEJM, Asturiano JF (2011) Influence of temperature regime on endocrine parameters and vitellogenesis during experimental maturation of European eel (*Anguilla anguilla*) females. Gen Comp Endocrinol 174:51–59. https://doi.org/10.1016/ygcen.2011.08.009

Pérez L., Mazzeo I, Peñaranda DS, Vílchez MC, Gallego V, Dufour S, Weltzien FA, Asturiano JF (2013) Modulatory effect of temperature on vitellogenesis and ovulation in the European eel. 17th international congress of comparative endocrinology (ICCE) 2013, Barcelona (Spain)

Politis SN, Butts IAE, Tomkiewicz J (2014) Light impacts embryonic and early larval development of the European eel, *Anguilla anguilla*. J Exp Mar Biol Ecol 461:407–415. https://doi.org/10.1016/j.jembe.2014.09.014

Politis SN, Mazurais D, Servili A, Zambonino-Infante JL, Miest JJ, Sørensen SR, Tomkiewicz J, Butts IAE (2017) Temperature effects on gene expression and morphological development of European eel, *Anguilla anguilla* larvae. PLoS One 12(8):e0182726. https://doi.org/10.1371/journal.pone.0182726

Politis SN, Mazurais D, Servili A, Zambonino-Infante JL, Miest J, Tomkiewicz J, Butts IAE (2018a) Salinity reduction benefits European eel larvae: insights at the morphological and molecular level. PLoS One 13(6):e0198294. https://doi.org/10.1371/journal.pone.0198294

Politis SN, Servili A, Mazurais D, Zambonino-Infante JL, Miest JJ, Tomkiewicz J, Butts IAE (2018b) Temperature induced variation in gene expression of thyroid hormone receptors and deiodinases of European eel (*Anguilla anguilla*) larvae. Gen Comp Endocrinol 259:54–65. https://doi.org/10.1016/j.ygcen.2017.11.003

Righton D, Westerberg H, Feunteun E, Økland F, Gargan P, Amilhat E, Metcalfe J, Lobon-Cervia J, Sjöberg N, Simon J, Acou A, Vedor M, Walker A, Trancart T, Brämick U, Aarestrup K (2016) Empirical observations of the spawning migration of European eels: the long and dangerous road to the Sargasso Sea. Sci Adv 2(10):e1501694. https://doi.org/10.1126/sciadv.1501694

Rousseau K, Lafont A-G, Pasquier J, Maugars G, Jolly C, Sébert M-E, Aroua S, Pasqualini C, Dufour S (2013) Advances in eel reproductive physiology and endocrinology. In: Trischitta F, Takei Y, Sébert P (eds) Eel physiology. CRC Press, Taylor & Francis Group, Boca Raton, pp 1–43

Rozenfeld C, Butts I, Tomkiewicz J, Zambonino-Infante JL, Mazurais D (2016) Abundance of specific mRNA transcripts impacts hatching success in European eel, *Anguilla anguilla* L. Comp Biochem Physiol A Mol Integr Physiol 191:59–65. https://doi.org/10.1016/j.cbpa.2015.09.011

Rozenfeld C, Pérez L, Gallego V, García-Carpintero V, Herranz-Jusdado JG, Cañizares J, Asturiano JF, Peñaranda DS (2017) Cold water rearing induces key reproductive mechanisms along BPG axis of European eel (*Anguilla anguilla*) males. Aquaculture Europe 2017. Book of abstracts, Dubrovnik (Croatia), pp 1012–1014

Sargent JR, Bell JG, Tocher DR (2002) The lipids. In: Halver JE, Hardy RW (eds) Fish nutrition. Academic, San Diego, pp 181–257

Schmitz M, Aroua S, Vidal B, Le Belle N, Elie P, Dufour S (2005) Differential regulation of luteinizing hormone and follicle-stimulating hormone expression during ovarian development and under sexual steroid feedback in the European eel. Neuroendocrinology 81:107–119. https://doi.org/10.1159/000086404

Sébert M-E, Weltzien F-A, Moisan C, Pasqualini C, Dufour S (2008) Dopaminergic systems in the European eel: characterization, brain distribution and potential role in migration and reproduction. Hydrobiologia 602:27–46. https://doi.org/10.1007/s10750-008-9288-1

Sørensen SR, Gallego V, Pérez L, Butts IAE, Tomkiewicz J, Asturiano JF (2013) Evaluation of methods to determine sperm density for the European eel, *Anguilla anguilla*. Reprod Domest Anim 48:936–944. https://doi.org/10.1111/rda.12189

Sørensen SR, Skov PV, Lauesen P, Tomkiewicz J, De Schryver P (2014a) Microbial interference and potential control in culture of European eel (*Anguilla anguilla*) embryos and larvae. Aquaculture 426-427:1–8. https://doi.org/10.1016/j.aquaculture.2014.01.011

Sørensen SR, Tomkiewicz J, Skovgaard A (2014b) *Ichthyodinium* identified in the eggs of European eel (*Anguilla anguilla*) spawned in captivity. Aquaculture 426-427:197–203. https://doi.org/10.1016/j.aquaculture.2014.02.002

Sørensen SR, Butts IA, Munk P, Tomkiewicz J (2016a) Effects of salinity and sea salt type on egg activation, fertilization, buoyancy and early embryology of European eel, *Anguilla anguilla*. Zygote 24:21–138. https://doi.org/10.1017/S0967199414000811

Sørensen SR, Tomkiewicz J, Munk P, Butts IAE, Nielsen A, Lauesen P, Graver C (2016b) Ontogeny and growth of early life stages of captive-bred European eel. Aquaculture 456:50–61. https://doi.org/10.1016/j.aquaculture.2016.01.015

Støttrup JG, Jacobsen C, Tomkiewicz J, Jarlbæk H (2013) Modification of essential fatty acid composition in broodstock of cultured European eel *Anguilla anguilla* L. Aquac Nutr 19(2):172–185. https://doi.org/10.1111/j.1365-2095.2012.00967.x

Støttrup JG, Tomkiewicz J, Jacobsen C, Butts I, Holst LK, Krüger-Johnsen M, Graver C, Lauesen P, Fontagné-Dicharry S, Heinsbroek LTN, Corraze G, Kaushik S (2016) Development of a broodstock diet to improve developmental competence of embryos in European eel, *Anguilla anguilla*. Aquac Nutr 22(4):725–737. https://doi.org/10.1111/anu.12299

Suetake H, Okubo K, Sato N, Yoshiura Y, Suzuki Y, Aida K (2002) Differential expression of two gonadotropin (GTH) b subunit genes during ovarian maturation induced by repeated injection of salmon GTH in the Japanese eel *Anguilla japonica*. Fish Sci 68:290–298. https://doi.org/10.1046/j.1444-2906.2002.00424.x

Tanaka H (2015) Progression in artificial seedling production of Japanese eel *Anguilla japonica*. Fish Sci 81:11–19. https://doi.org/10.1007/s12562-014-0821-z

Tanaka H, Kagawa H, Ohta H (2001) Production of leptocephali of Japanese eel (*Anguilla japonica*) in captivity. Aquaculture 201:51–60. https://doi.org/10.1016/S0044-8486(01)00553-1

Tanaka H, Kagawa H, Ohta H, Unuma T, Nomura K (2003) The first production of glass eel in captivity: fish reproductive physiology facilitates great progress in aquaculture. Fish Physiol Biochem 28:493–497. https://doi.org/10.1023/B:FISH.0000030638.56031.ed

Tesch F (2003) The eel. Blackwell, Oxford, pp 1–408

Thomson-Laing G, Jasoni CL, Lokman PM (2018) The effects of migratory stage and 11-ketotestosterone on the expression of rod opsin genes in the shortfinned eel (*Anguilla australis*). Gen Comp Endocrinol 257:211–219. https://doi.org/10.1016/j.ygcen.2017.06.025

Turchini G, Francis D, De Silva S (2006) Fatty acid metabolism in the freshwater fish Murray cod (*Maccullochella peelii peelii*) deduced by the whole-body fatty acid balance method. Comp Biochem Physiol B Biochem Mol Biol 144:110–118. https://doi.org/10.1016/j.cbpb.2006.01.013

Unuma T, Hasegawa N, Sawaguchi S, Tanaka T, Matsubara T, Nomura K, Tanaka H (2011) Fusion of lipid droplets in Japanese eel oocytes: stage classification and its use as a biomarker for induction of final oocyte maturation and ovulation. Aquaculture 322-323:142–148. https://doi.org/10.1016/j.aquaculture.2011.10.001

Unuma T, Sawaguchi S, Hasegawa N, Tsuda N, Tanaka T, Nomura K, Tanaka H (2012) Optimum temperature of rearing water during artificial induction of ovulation in Japanese eel. Aquaculture 358-359:216–223. https://doi.org/10.1016/j.aquaculture.2012.07.004

van den Thillart G, van Ginneken V, Körner F, Heijmans R, van der Linden R, Gluvers A (2004) Endurance swimming of the European eel. J Fish Biol 65:1–7. https://doi.org/10.1111/j.0022-1112.2004.00447.x

van den Thillart G, Palstra A, van Ginneken V (2009) Energy requirements of European eel for trans Atlantic spawning migration. In: van den Thillart G, Dufour S, Rankin JC (eds) Spawning migration of the European eel. Springer Science+Business Media B.V, Dordrecht, pp 179–198

van Ginneken VGT, Maes G (2005) The European eel (*Anguilla anguilla* Linnaeus), its lifecycle, evolution and reproduction: a literature review. Rev Fish Biol Fish 15:367–398. https://doi.org/10.1007/s11160-006-0005-8

van Ginneken VJT, van den Thillart GEEJM (2000) Eel fat stores are enough to reach the Sargasso. Nature 403:156–157. https://doi.org/10.1038/35003110

van Ginneken VJT, Dufour S, Sbaihi M, Balm P, Noorlander K, de Bakker M, Doornbos J, Palstra A, Antonissen E, Mayer I, van den Thillart G (2007) Does a 5500-km swim trial stimulate early sexual maturation in the European eel (*Anguilla anguilla* L.)? Comp Biochem Physiol 147:1095–1103. https://doi.org/10.1016/j.cbpa.2007.03.021

Vidal B, Pasqualini C, Le Belle N, Holland MCH, Sbaihi M, Vernier P, Zohar Y, Dufour S (2004) Dopamine inhibits luteinizing hormone synthesis and release in the juvenile European eel: a neuroendocrine lock for the onset of puberty. Biol Reprod 71:1491–1500. https://doi.org/10.1095/biolreprod.104.030627

Vílchez MC, Mazzeo I, Baeza R, Gallego V, Peñaranda DS, Asturiano JF, Pérez L (2013) Effect of thermal regime on the quality of eggs and larval development of European eel. 4th international workshop on biology of fish gametes. Book of abstracts, pp 228–229

Vílchez MC, Mazzeo I, Peñaranda DS, Gallego V, Dufour S, Weltzien FA, Asturiano JF, Pérez L (2014) Effect of thermal regime on vitellogenesis, ovulation and larval development of European eel. 10th international symposium on reproductive physiology of fish. Olhão (Portugal). Book of abstracts, 25–30 May 2014, p 190

Whates DC, Abayasekara DRE, Aitken RJ (2007) Polyunsaturated fatty acids in male and female reproduction. Biol Reprod 77:190–201. https://doi.org/10.1095/biolreprod.107.060558

Wullur S, Yoshimatsu T, Tanaka H, Ohtani M, Sakakura Y, Kim HJ, Hagiwara A (2013) Ingestion by Japanese eel *Anguilla japonica* larvae on various minute zooplanktons. Aquacult Sci 61(4):341–347. https://doi.org/10.11233/aquaculturesci.61.341

Yamamoto K, Yamauchi K (1974) Sexual maturation of Japanese eel and production of eel larvae in the aquarium. Nature 251:220–222. https://doi.org/10.1038/263412a0

Yamauchi K (1990) Studies of gonadal steroids involved in final gonadal maturation in the Japanese eel, *Anguilla japonica*, a review. Int Rev Gesamten Hydrobiol 75:859–860. https://doi.org/10.1002/iroh.19900750630

Chapter 15
Sperm Cryopreservation of Aquatic Species

Ákos Horváth and Béla Urbányi

Abstract Cryopreservation of fish sperm is feasible, and methods have been developed for several hundred species. The procedure includes the collection of sperm and assessment of its quality, dilution in suitable extenders, addition of cryoprotectants, loading into freezing devices, cooling, storage, thawing, and finally post-thaw use of sperm. An alternative to traditional freezing methods is the vitrification of sperm which is promising primarily in smaller model fish species. A wide variety of protocols are available in the literature; however, in spite of the significant progress made by the scientific community, commercial application of fish sperm cryopreservation is still very limited.

Keywords Sperm · Quality · Freezing · Cryopreservation · Vitrification

15.1 Introduction

Cryobiology is a branch of biology that focuses on the survival of living matter at ultra-low temperatures. Sperm was among the first cell types to be cryopreserved with the earliest reports dating back to the late 1940s. Thus, the survival of sperm from various mammalian and avian species exposed to dry ice (−79 °C) and the beneficial effect of glycerol on post-thaw survival was reported in 1949 (Polge et al. 1949). This was soon followed by the birth of the first calf born of insemination with cryopreserved sperm (Stewart 1951). The first study in fish has reported the use of cryopreservation for the hybridization of fall- and spring-spawning varieties of the herring (*Clupea harengus*) which would have obviously been impossible in natural conditions (Blaxter 1953). Since then, cryopreservation has become a subject of extensive scientific studies as well as a routine procedure in assisted reproduction.

Á. Horváth (✉) · B. Urbányi
Department of Aquaculture, Szent István University, Gödöllő, Hungary
e-mail: Horvath.Akos@mkk.szie.hu

M. Yoshida, J. F. Asturiano (eds.), *Reproduction in Aquatic Animals*,
https://doi.org/10.1007/978-981-15-2290-1_16

In aquatic species—primarily finfish—sperm cryopreservation faces specific challenges. These include the typically short motility period of spermatozoa, activation of spermatozoa upon release into the aquatic environment, as well as increased volumes of sperm and eggs related to the body weight in comparison to terrestrial mammals. These need to be taken into account when developing cryopreservation protocols for fish.

15.2 Principles of Cryopreservation

The main goal of cryopreservation is the arrest of biological time which is possible at extreme low temperatures. Below −130 °C water exists only in a crystalline or amorphous solid form where the movement of molecules is limited; thus, its effects can only be measured in geological time. Consequently, at the boiling point of liquid nitrogen (the most commonly used coolant and storage environment), −196 °C, no physical or chemical processes requiring heat energy exist (Mazur 1984).

The principles of cryopreservation of live matter are closely related to the principles of freezing of water or aqueous solutions. The freezing point of water is 0 °C; however, water rarely freezes at that temperature. Water and its solutions have the tendency to supercool beyond the freezing point and remain in the liquid state until ice formation begins along so called ice nuclei. Ice nucleation can start spontaneously or more commonly as a result of ice nucleating agents (Mazur 1990). Energy (called latent heat) is released during transition from the liquid state to the solid, which causes the temperature of solution to increase to the freezing point. During freezing, ice crystals are primarily built of water molecules (and not those of the solutes); thus, the concentration of the solution increases with the process of freezing. This is continued to the eutectic point when the entire solution solidifies (Denniston et al. 2011).

During cryopreservation of live cell and tissues, the cell membrane plays a key role in osmotic regulation that allow cells to survive the process. As there are no efficient ice nucleating agents within the cells, ice formation typically starts in the extracellular space (Mazur 1970) which leads to an increased concentration of solutes outside the cells. Cells try to compensate for the difference in osmotic pressure by releasing water. According to the two-factor hypothesis of freezing injury, cell death can occur either if cells are cooled too slow or too fast (Mazur et al. 1972). Too slow cooling results in prolonged exposure of cells to an increasing concentration of solutes (to the point of cytotoxicity) while too fast cooling does not allow the cells to release sufficient volumes of water which freezes inside the cells in the form of lethal ice crystals. Thus, an optimal cooling rate must be found to maximize the cryosurvival of cells.

Extreme fast cooling results in the formation of an amorphous glassy solid in the process of vitrification. In this case, the solid state is attained by a sudden increase of viscosity instead of crystallization. The absence of ice crystals—and the damages associated with them—would be ideal for cells; however, vitrification has several shortcomings. Successful vitrification requires high concentrations of toxic cryoprotectants (Fahy et al. 1984). Also, vitrified matter is metastable, it can only be sustained at very low temperatures (below −100 to −130 °C) as higher temperatures lead to recrystallization of ice. Nevertheless, even classical cell freezing protocols lead to partial vitrification, primarily in the intracellular space (Leibo et al. 1978; Hirsh et al. 1985).

The rate of thawing of cryopreserved cells or tissues is at least as important as that of their cooling. During slow thawing, small ice crystals present in cells can start growing and vitrified matter can recrystallize causing mechanical damage to the cells. When cells are thawed fast (at a rate of 1000 °C/min or faster), they quickly rehydrate and their survival improves (Denniston et al. 2011). For vitrified cells and tissues, it is commonly accepted that from the point of view of survival, the thawing rate is actually more important than the cooling rate (Mazur and Seki 2011; Seki et al. 2014).

15.3 Cryopreservation of Fish Sperm

Cryopreservation of fish sperm follows a set of important steps such as collection of sperm, determination of sperm quality, dilution with extenders and cryoprotectants, packaging (loading into straws or other devices), freezing or vitrification of sperm, storage, thawing, determination of post-thaw sperm quality, and finally fertilization with thawed sperm (Fig. 15.1). These steps are crucial for the development and application of a successful cryopreservation protocol.

15.3.1 Collection, Storage, and Determination of Sperm Quality

During the spawning season, the majority of fish species produces a relatively large volume of sperm that can be collected by hand-stripping. There are exceptions, such as many catfish species (members of the order Siluriformes) that are either oligospermic or stripping is impossible due to anatomical reasons. In these species, the testis or a part of it is surgically removed and sperm is squeezed through a sterile gauze or other mesh fabric (Legendre et al. 1996). A general rule during sperm collection is that contamination of sperm with water, urine, or feces should be avoided as these can either activate sperm or carry bacteria and enzymes that can damage spermatozoa. Collected sperm should be kept at a low temperature (0–4 °C) that slows the rate of chemical processes in it.

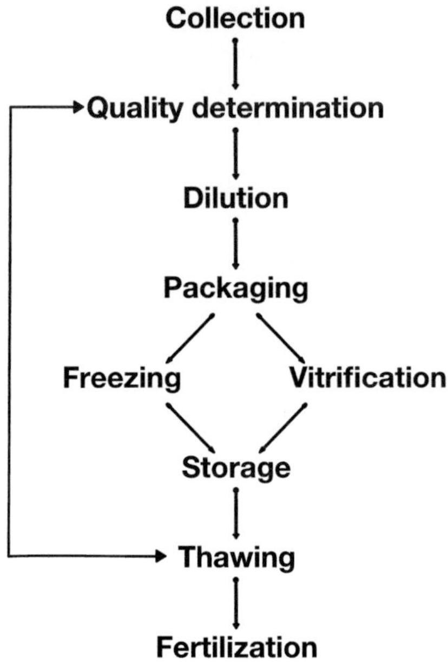

Fig. 15.1 Steps in the process of fish sperm cryopreservation

Following collection, the quality of sperm should be determined. In fish, sperm quality prior to cryopreservation is characterized almost exclusively by its motility determined either by visual estimation or by computer-assisted sperm analysis—CASA (Gallego and Asturiano 2018a,b). Visual estimation is often criticized for its subjectivity, especially in scientific circles. Sperm is typically activated on a microscope slide with an activating solution and observed under a light microscope at 10–25× magnification; finally the percentage of motile cells is estimated (Fauvel et al. 2010). The accuracy of visual estimation can be improved by pre-dilution of sperm in isotonic solutions prior to the investigations (Billard and Cosson 1992). CASA systems consisting of a phase contrast microscope, a camera, and a software analyzing the video sequences offer more objective results and more detailed information on the movement of cells such as motility or various velocity parameters (WHO 2010). On the other hand, they are more expensive, and their accuracy also depends on the analysis settings and dilution ratio of sperm; thus, their use also requires a training period. During the last decade, CASA became the standard procedure to determine sperm quality in fish sperm cryopreservation research (Beirão et al. 2011; Bernáth et al. 2016; Judycka et al. 2016; Riesco et al. 2017). This was further facilitated by the availability of free sperm analysis software packages such as the CASA plugin of the open-source image analysis software ImageJ (Wilson-Leedy and Ingermann 2007).

Sperm quality can be determined by other parameters, as well. These include sperm concentration expressed as the number of spermatozoa per milliliter of sperm that can be determined by the use of various hemocytometers (Bürker-Türk, Thoma, Neubauer, Neubauer Improved, etc.). Other devices whose primary function is not the determination of sperm concentration can also be successfully used for this task such as spectrophotometers (Ciereszko and Dabrowski 1993; Fauvel et al. 1999) or cell counters such as flow cytometers or Coulter counters (Fauvel et al. 2010). Sperm quality can also be determined by the intact morphology of cells; however, in fish these are restricted to species with a distinct sperm head shape such as sturgeons or eels (Asturiano et al. 2007). Membrane integrity of sperm cells (often referred to as viability) is also used to characterize sperm quality. Differential staining methods are used to distinguish cells with intact and damaged membranes. The eosin/nigrosin staining developed originally for mammalian sperm (Blom 1950) is rarely used in fish, although nigrosin alone was used to evaluate membrane integrity of sea bass (*Dicentrarchus labrax*) sperm (Zilli et al. 2004). Currently, differential fluorescent staining procedures combined with flow cytometry are commonly practiced. One example is the use of SYBR-14 to stain membrane-intact cells in fluorescent green and propidium iodide (PI) to counterstain membrane-damaged cells in fluorescent red (Segovia et al. 2000; Flajšhans et al. 2004; Cabrita et al. 2011). Other fluorescent dyes have also been tested on fish sperm such as Hoechst 33258 (Asturiano et al. 2007) to investigate membrane integrity or Rh123 (Ogier De Baulny et al. 1997) to test mitochondrial function. Genotoxic damage to cells can be tested using the Comet assay (Zilli et al. 2003; Cabrita et al. 2005), whereas the energy content of spermatozoa can be described by the concentration of ATP in the cells (Perchec et al. 1995; Boryshpolets et al. 2009).

15.3.2 Dilution of Sperm

Prior to freezing, sperm needs to be diluted in a cryomedium. The cryomedium consists of two principle components: the extender and the cryoprotectants (note that the expressions cryomedium and extender are used interchangeably in the literature). Extenders are solutions of salts, sugars (or the combination of the two), and other chemicals, and their main function is the reversible immobilization of sperm (Glenn III et al. 2011). They exert their function by providing isotonic conditions for the cells or by maintaining an adequate K^+ concentration in species requiring that (salmonids, sturgeons). Their buffer systems prevent fluctuations of pH during freezing. Extenders allow dilution of sperm to the required ratio.

Cryoprotectants—as shown by their name—protect cells from the damaging effects of freezing and thawing. They decrease the freezing point of the solution, they bind to water and thus prevent its incorporation into ice crystals and probably stabilize cell membranes. Cryoprotectants are classified as external or non-penetrating and internal or penetrating depending on whether they exert their action outside or inside cells (Denniston et al. 2011). External cryoprotectants can be sugars or polymers that can be part of the extender (contributing to the problems of the nomenclature); thus, they also have multiple roles: they maintain the osmolality of the solution around the cells, protect cells from osmotic shock (osmoprotectants), decrease the freezing point of the solution and provide nutrients for spermatozoa

(Lahnsteiner et al. 1993). They are, however, almost exclusively used in combination with internal cryoprotectants.

Internal cryoprotectants are low molecular weight chemicals that penetrate the cell membrane and exert their action inside the cells. Glycerol was the first internal cryoprotectant to be discovered and used both in livestock and in fish (Polge et al. 1949; Blaxter 1953); however, since then a multitude of chemicals were found to have cryoprotective function. Nevertheless, in fish only a handful of cryoprotectants are used regularly. The most common ones include dimethyl-sulfoxide (DMSO or Me_2SO) (Mounib 1978; Legendre and Billard 1980; Kurokura et al. 1984), methanol (MeOH) (Harvey et al. 1982; Lahnsteiner et al. 1996; Horváth et al. 2003; Kása et al. 2017), *N,N*-dimethyl-acetamide (DMA) (McNiven et al. 1993; Horváth and Urbányi 2000; Morris et al. 2003; Warnecke and Pluta 2003), ethylene glycol (EG) (Jähnichen et al. 1999), propylene glycol (PG) (Conget et al. 1996), and methyl glycol (also known as 2-methoxyethanol) (Viveiros et al. 2012, 2014, Gallego and Asturiano 2018b).

External cryoprotectants are typically characterized by low toxicity while internal cryoprotectants have a varying degree of toxicity. Thus, their concentration also has to be chosen according to different principles. While in case of external cryoprotectants, the most important goal is the maintenance of osmotic balance and prevention of hypo- or hyperosmotic shock, in the case of internal cryoprotectants toxicity also needs to be taken into account.

Sperm dilution is followed by equilibration of osmotic pressure between the extracellular and intracellular spaces. The importance of equilibration time is controversial. The cryoprotective effect of glycerol was evident after equilibration for 30–60 min (Suquet et al. 2000), and a positive effect was observed following a 10-min equilibration of rainbow trout sperm exposed to DMA (Babiak et al. 2001). In case of DMSO, equilibration seemed to be more harmful rather than beneficial (Stoss and Holtz 1983; Gwo 1994), sometimes causing premature activation of spermatozoa in marine species (Peñaranda et al. 2009). Equilibration time does not play a major effect on post-thaw sperm quality when methanol is used as a cryoprotectant (Lahnsteiner et al. 1997).

15.3.3 *Freezing of Sperm*

Liquid nitrogen and its vapors are currently used almost exclusively for the cryopreservation of fish sperm. Earlier, dry ice was also used as a coolant (Mounib et al. 1968; Chao et al. 1987; Billard et al. 1993); however, it is seldom employed for fish sperm now. Dry ice cannot be used for the storage of cells as cells gradually die at its sublimation temperature due to the possible presence of unfrozen solution to which cells are exposed (Mazur 1984). Liquid nitrogen, on the other hand, ensures adequate cooling rates and storage conditions at its boiling temperature (−196 °C).

Sperm can be packaged into various containers before freezing. These can be cryotubes, ampoules, or French straws developed for the freezing of dairy bull sperm. Straws were patented by Robert Cassou in France in 1949 and have seen worldwide use for dairy bull sperm. Straws are plastic tubes produced in various sizes (0.25, 0.5, and 1.2 mL) that contain a tripartite plug at one end consisting of a powder between two fibrous pads that gelifies upon contact with aqueous solutions and acts as a stopper. Macrotubes produced in volumes of 4–5 mL are of different construction, they are offered with metal or plastic balls to be inserted into one or both ends to act as stoppers.

Sperm filled into straws can be frozen in the vapor of liquid nitrogen in uncontrolled or controlled conditions. Uncontrolled cooling is typically done in polystyrene (Styrofoam) boxes filled with liquid nitrogen. Straws are laid at a certain distance from the surface of liquid nitrogen (e.g., on a polystyrene raft) and allowed to cool for several minutes before being plunged into the liquid. A standardized variant of these boxes was developed for the cryopreservation of salmonid sperm (Lahnsteiner 2011). Cooling rate can be regulated by the height and the duration of holding straws above the level of liquid nitrogen, and this can have a significant effect on post-thaw motility (Boryshpolets et al. 2017). An alternative to the cooling box is freezing in a dry shipper (a container used for the shipment of cryopreserved samples) that offers even less control over the cooling process; yet, it is used successfully for the freezing of fish sperm. The method was first used by Brazilian scientists (Viveiros and Godinho 2009) and was found to be more reliable than the cooling box when freezing varying numbers of straws (Horokhovatskyi et al. 2017).

A controlled-rate freezer allows a precise and replicable cooling of samples which is difficult to attain with the methods described above. These freezers consist of a freezing chamber and a computer. The software installed in the computer regulates the cooling rate according to a program set up by the user. In most system, individual probes measure the temperature of the chamber and the sample which gives the software feedback on the progress of cooling. Most controlled-rate freezers also use the vapor of liquid nitrogen; however, coolant-free devices such as an electric ultrafreezer (Diogo et al. 2018) also exist.

Following the required period of storage, samples need to be thawed before further use. Sperm samples need to be thawed quickly in order to avoid devitrification and recrystallization of partially vitrified solids that result in similar damages as the formation of large ice crystals during freezing. Typically, samples are thawed in a water bath at 25–40 °C for a varying period of time (Sarvi et al. 2006; Boryshpolets et al. 2017). For straws, thawing time can reach up to 30 s, while for macrotubes and cryovials up to 150 s (Cabrita et al. 2001; Riesco et al. 2017).

15.3.4 Post-thaw Use of Sperm

Following thawing, sperm quality is determined as described above (Sect. 15.3.1) and then sperm is used for fertilization. Post-thaw sperm motility is typically lower than that of fresh sperm; however, contrary results have also been reported (Viveiros et al. 2010). The maximum attainable post-thaw motility was earlier considered to be taxon-specific; however, due to methodical refinement, the threshold values keep improving.

The principles of fertilization with cryopreserved sperm are identical to those using fresh sperm. Sperm–egg ratio is regularly taken into account when maximizing fertilization success. As fertilizing capacity of frozen-thawed sperm is typically lower than that of fresh sperm, in many cases higher sperm–egg ratios are recommended and are crucial to optimize the use of cryopreserved sperm (Ciereszko et al. 2014; Nynca et al. 2017; Judycka et al. 2018). Fertilization success is usually measured as the hatch rate of larvae although sometimes larval deformities are also assessed (Horváth and Urbányi 2000; Young et al. 2009).

Unlike the use of cryopreserved sperm in dairy bull farming, fish sperm cryopreservation is not anticipated to be employed commercially for direct aquaculture production in the near future. This is mainly due to the fact that domestication and genetic improvement in aquatic species is still in its infancy, and higher yields can be achieved by other means (e.g., nutrition) (Asturiano et al. 2017). Also, the technologies developed for dairy bull sperm (intended to fertilize a single egg) cannot be adapted directly to fish species that produce much larger volumes of both sperm and eggs; thus, alternative methods are sought such as the use of 4–5 mL straws or macrotubes (Horváth et al. 2007, 2010). In spite of the recent efforts for the commercial use of cryopreserved sperm in the genetic improvement in the Atlantic salmon (*Salmo salar*, www.cryogenetics.com), the general consensus is that cryopreservation of sperm in fish species is expected to become part of aquaculture practice when the industry shows a legitimate demand for it. Until then, it will be applied on individual basis and mostly using public funding (Asturiano et al. 2017). Efforts to standardize fish sperm cryopreservation technologies using high-throughput methods contribute to the improvement of fertilization with cryopreserved sperm which is a further step towards commercial use (Tiersch et al. 2011; Hu et al. 2013; Matthews et al. 2018).

15.4 Vitrification of Fish Sperm

Vitrification of fish sperm has recently gained attention as an alternative to conventional freezing (Cuevas-Uribe et al. 2011a, b, 2015; Merino et al. 2012; Figueroa et al. 2013). What makes it interesting is applicability to the sperm of model fish species characterized by small individual body size such as the zebrafish (*Danio rerio*) or the Mexican swordtail (*Xiphophorus hellerii*). In these species, the volume of sperm that can be collected is very low, e.g., in zebrafish, it is in the range of 1–1.5 μL which makes most available freezing methods impractical, although recent systematic optimization of protocols resulted in stable and high fertilization rates (Matthews et al. 2018). Vitrification is efficient for the cooling of low liquid volumes such as 20–30 μL (Schuster et al. 2003) or according to our own experiences 2.5–5 μL (Kása et al. 2017). Successful vitrification requires the use of high concentrations (up to 40% v/v) of cryoprotectants which can be highly toxic to the cells;

thus, the combination of two or three cryoprotectants is recommended. For vitrification the use of devices in which the liquid forms a thin layer or film is recommended. These include devices specifically developed for vitrification such as Cryotops or even the inoculating loops used in cell cultures. Unfortunately, post-thaw motility and fertilization rates using vitrified fish sperm are typically very low; thus, the currently available vitrification methods are not suitable alternatives to conventional freezing in this field.

15.5 Conclusion

Fish sperm cryopreservation is an efficient tool in the preservation of genetic diversity of fish species. Protocols have been described in many species that result in viable spermatozoa following thawing; however, commercial application of sperm cryopreservation in aquatic species is still very limited. Future research is anticipated to concentrate on the standardization and development of high-throughput methods in order to improve fertilization results with cryopreserved sperm.

Acknowledgments This research was supported by the Higher Education Institutional Excellence Program (1783-3/2018/FEKUTSRAT) awarded by the Ministry of Human Capacities within the framework of water related researches of Szent István University as well as the EFOP-3.6.3-VEKOP-16-2017-00008 project co-financed by the European Union and the European Social Fund. Parts of this chapter are translations from the Hungarian language of the dissertation by Á. Horváth submitted to the Hungarian Academy of Sciences for the title of Doctor of the Hungarisn Academy of Sciences.

References

Asturiano JF, Marco-Jiménez F, Peñaranda DS, Garzón DL, Pérez L, Vicente JS, Jover M (2007) Effect of sperm cryopreservation on the European eel sperm viability and spermatozoa morphology. Reprod Domest Anim 42:162–166. https://doi.org/10.1111/j.1439-0531.2006.00746.x

Asturiano JF, Cabrita E, Horváth Á (2017) Progress, challenges and perspectives on fish gamete cryopreservation: a mini-review. Gen Comp Endocrinol 245:69–76. https://doi.org/10.1016/j.ygcen.2016.06.019

Babiak I, Glogowski J, Goryczko K, Dobosz S, Kuzminski H, Strzezek J, Demianowicz W (2001) Effect of extender composition and equilibration time on fertilization ability and enzymatic activity of rainbow trout cryopreserved spermatozoa. Theriogenology 56:177–192. https://doi.org/10.1016/S0093-691X(01)00553-2

Beirão J, Cabrita E, Pérez-Cerezales S, Martínez-Páramo S, Herráez MP (2011) Effect of cryopreservation on fish sperm subpopulations. Cryobiology 62:22–31. https://doi.org/10.1016/j.cryobiol.2010.11.005

Bernáth G, Bokor Z, Żarski D, Várkonyi L, Hegyi Á, Urbányi B, Radóczi Ifj J, Horváth Á (2016) Commercial-scale out-of-season cryopreservation of Eurasian perch (*Perca fluviatilis*) sperm and its application for fertilization. Anim Reprod Sci 170:170–177. https://doi.org/10.1016/j.anireprosci.2016.05.005

Billard R, Cosson MP (1992) Some problems related to the assessment of sperm motility in freshwater fish. J Exp Zool 261:122–131. https://doi.org/10.1002/jez.1402610203

Billard R, Cosson J, Crim LW (1993) Motility of fresh and aged halibut sperm. Aquat Living Resour 6:67–75. https://doi.org/10.1051/alr:1993008

Blaxter JHS (1953) Sperm storage and cross-fertilization of spring and autumn spawning herring. Nature 172:1189–1190

Blom E (1950) A one-minute live-dead sperm stain by means of eosin-nigrosin. Fertil Steril 1:176–177. https://doi.org/10.1016/0002-9378(51)91450-0

Boryshpolets S, Dzyuba B, Stejskal V, Linhart O (2009) Dynamics of ATP and movement of Eurasian perch (*Perca fluviatilis* L.) sperm in conditions of decreasing osmolality. Theriogenology 72:851–859

Boryshpolets S, Sochorová D, Rodina M, Linhart O, Dzyuba B (2017) Cryopreservation of Carp (*Cyprinus carpio* L.) sperm: impact of seeding and freezing rates on post–thaw outputs. Biopreserv Biobank 15:234–240. https://doi.org/10.1089/bio.2016.0065

Cabrita E, Robles V, Alvarez R, Herráez MP (2001) Cryopreservation of rainbow trout sperm in large volume straws: application to large scale fertilization. Aquaculture 201:301–314

Cabrita E, Robles V, Rebordinos L, Sarasquete C, Herráez MP (2005) Evaluation of DNA damage in rainbow trout (*Oncorhynchus mykiss*) and gilthead sea bream (*Sparus aurata*) cryopreserved sperm. Cryobiology 50:144–153. https://doi.org/10.1016/j.cryobiol.2004.12.003

Cabrita E, Ma S, Diogo P, Martínez-Páramo S, Sarasquete C, Dinis MT (2011) The influence of certain aminoacids and vitamins on post-thaw fish sperm motility, viability and DNA fragmentation. Anim Reprod Sci 125:189–195. https://doi.org/10.1016/j.anireprosci.2011.03.003

Chao N-H, Chao W-C, Liu K-C, Liao I-C (1987) The properties of tilapia sperm and its cryopreservation. J Fish Biol 30:107–118. https://doi.org/10.1111/j.1095-8649.1987.tb05737.x

Ciereszko A, Dabrowski K (1993) Estimation of sperm concentration of rainbow trout, whitefish and yellow perch using a spectrophotometric technique. Aquaculture 109:367–373. https://doi.org/10.1016/0044-8486(93)90175-X

Ciereszko A, Dietrich GJ, Nynca J, Dobosz S, Zalewski T (2014) Cryopreservation of rainbow trout semen using a glucose-methanol extender. Aquaculture 420–421:275–281. https://doi.org/10.1016/j.aquaculture.2013.11.014

Conget P, Fernández M, Herrera G, Minguell JJ (1996) Cryopreservation of rainbow trout (*Oncorhynchus mykiss*) spermatozoa using programmable freezing. Aquaculture 143:319–329

Cuevas-Uribe R, Leibo SP, Daly J, Tiersch TR (2011a) Production of channel catfish with sperm cryopreserved by rapid non-equilibrium cooling. Cryobiology 63:186–197. https://doi.org/10.1016/j.cryobiol.2011.06.004

Cuevas-Uribe R, Yang H, Daly J, Savage MG, Walter RB, Tiersch TR (2011b) Production of F_1 offspring with vitrified sperm from a live-bearing fish, the green swordtail *Xiphophorus hellerii*. Zebrafish 8:167–179. https://doi.org/10.1089/zeb.2011.0704

Cuevas-Uribe R, Chesney EJ, Daly J, Tiersch TR (2015) Vitrification of sperm from marine fish: effect on motility and membrane integrity. Aquacult Res 46:1770–1784. https://doi.org/10.1111/are.12337

Denniston RS, Michelet S, Bondioli KR, Godke RA (2011) Principles of embryo cryopreservation. In: Tiersch TR, Green CC (eds) Cryopreservation in aquatic species, 2nd edn. World Aquaculture Society, Baton Rouge, pp 274–290

Diogo P, Martins G, Quinzico I, Nogueira R, Gavaia PJ, Cabrita E (2018) Electric ultrafreezer (- 150 °C) as an alternative for zebrafish sperm cryopreservation and storage. Fish Physiol Biochem 44(6):1443–1455. https://doi.org/10.1007/s10695-018-0500-6

Fahy GM, MacFarlane DR, Angell CA, Meryman HT (1984) Vitrification as an approach to cryopreservation. Cryobiology 21:126–407

Fauvel C, Savoye O, Dreanno C, Cosson J, Suquet M (1999) Characteristics of sperm of captive seabass in relation to its fertilization potential. J Fish Biol 54:356–369. https://doi.org/10.1111/j.1095-8649.1999.tb00835.x

Fauvel C, Suquet M, Cosson J (2010) Evaluation of fish sperm quality. J Appl Ichthyol 26:636–643

Figueroa E, Risopatrón J, Sánchez R, Isachenko E, Merino O, Isachenko V, Valdebenito I (2013) Spermatozoa vitrification of sex-reversed rainbow trout (*Oncorhynchus mykiss*): effect of seminal plasma on physiological parameters. Aquaculture 372–375:119–126. https://doi.org/10.1016/j.aquaculture.2012.10.019

Flajšhans M, Cosson J, Rodina M, Linhart O (2004) The application of image cytometry to viability assessment in dual fluorescence-stained fish spermatozoa. Cell Biol Int 28:955–959. https://doi.org/10.1016/j.cellbi.2004.07.014

Gallego V, Asturiano JF (2018a) Sperm motility in fish: technical applications and perspectives through CASA-Mot systems. Reprod Fertil Dev 30:820–832. https://doi.org/10.1071/RD17460

Gallego V, Asturiano JF (2018b) Fish sperm motility assessment as a tool for aquaculture research: a historical approach. Rev Aquacult 11(3):697–724. https://doi.org/10.1111/raq.12253

Glenn DW III, Lang RP, Tiersch TR (2011) Evaluation of extenders for refrigerated storage of koi carp and goldfish sperm. In: Tiersch TR, Green CC (eds) Cryopreservation in aquatic species, 2nd edn. World Aquaculture Society, Baton Rouge, pp 107–124

Gwo J-C (1994) Cryopreservation of yellowfin seabream (*Acanthopagrus latus*) spermatozoa (Teleost, Perciformes, Sparidae). Theriogenology 41:989–1004. https://doi.org/10.1016/S0093-691X(05)80022-6

Harvey B, Kelley RN, Ashwood-Smith MJ (1982) Cryopreservation of zebra fish spermatozoa using methanol. Can J Zool 60:1867–1870

Hirsh AG, Williams RJ, Meryman HT (1985) A novel method of natural cryoprotection. Plant Physiol 79:41–56

Horokhovatskyi Y, Rodina M, Asyabar HD, Boryshpolets S (2017) Consequences of uncontrolled cooling during sterlet (*Acipenser ruthenus*) sperm cryopreservation on post-thaw motility and fertilizing ability. Theriogenology 95:89–95. https://doi.org/10.1016/j.theriogenology.2017.03.007

Horváth Á, Urbányi B (2000) The effect of cryoprotectants on the motility and fertilizing capacity of cryopreserved African catfish *Clarias gariepinus* (Burchell 1822) sperm. Aquacult Res 31:317–324. https://doi.org/10.1046/j.1365-2109.2000.00444.x

Horváth Á, Miskolczi E, Urbányi B (2003) Cryopreservation of common carp sperm. Aquat Living Resour 16:457–460. https://doi.org/10.1016/S0990-7440(03)00084-6

Horváth Á, Miskolczi E, Mihálffy S, Ősz K, Szabó K, Urbányi B (2007) Cryopreservation of common carp (*Cyprinus carpio*) sperm in 1.2 and 5 ml straws and occurrence of haploids among larvae produced with cryopreserved sperm. Cryobiology 54:251–257. https://doi.org/10.1016/j.cryobiol.2007.02.003

Horváth Á, Urbányi B, Wang C, Onders RJ, Mims SD (2010) Cryopreservation of paddlefish sperm in 5-mL straws. J Appl Icthyology 26:715–719. https://doi.org/10.1111/j.1439-0426.2010.01551.x

Hu E, Liao TW, Tiersch TR (2013) A quality assurance initiative for commercial-scale production in high-throughput cryopreservation of blue catfish sperm. Cryobiology 67:214–224. https://doi.org/10.1016/j.cryobiol.2013.07.001

Jähnichen H, Warnecke D, Trölsch E, Kohlman K, Bergler H, Pluta HJ (1999) Motility and fertilizing capability of cryopreserved *Acipenser ruthenus* L. sperm. J Appl Ichthyol 15:204–206. https://doi.org/10.1111/j.1439-0426.1999.tb00235.x

Judycka S, Cejko BI, Dryl K, Dobosz S, Grudniewska J, Kowalski RK (2016) The effect of supplementation of a trehalose-based extender with KCl on rainbow trout (*Oncorhynchus mykiss*) sperm freezability and post-thaw motility. Aquaculture 465:303–310. https://doi.org/10.1016/j.aquaculture.2016.09.029

Judycka S, Nynca J, Liszewska E, Dobosz S, Grudniewska J, Ciereszko A (2018) Optimal sperm concentration in straws and final glucose concentration in extender are crucial for improving the cryopreservation protocol of salmonid spermatozoa. Aquaculture 486:90–97. https://doi.org/10.1016/j.aquaculture.2017.12.019

Kása E, Bernáth G, Kollár T, Żarski D, Lujić J, Marinović Z, Bokor Z, Hegyi Á, Vílchez MC, Morini M, Peñaranda DS, Pérez L, Asturiano JF, Horváth Á (2017) Development of sperm vitrification protocols for freshwater fish (Eurasian perch, *Perca fluviatilis*) and marine fish (European eel, *Anguilla anguilla*). Gen Comp Endocrinol 245:102–107. https://doi.org/10.1016/j.ygcen.2016.05.010

Kurokura H, Hirano R, Tomita M, Iwahashi M (1984) Cryopreservation of carp sperm. Aquaculture 37:267–273. https://doi.org/10.1016/0044-8486(84)90159-5

Lahnsteiner F (2011) Cryopreservation protocols for sperm of salmonid fishes. In: Tiersch TR, Green CC (eds) Cryopreservation in aquatic species, 2nd edn. World Aquaculture Society, Baton Rouge, pp 409–420

Lahnsteiner F, Patzner RA, Weismann T (1993) Energy resources of spermatozoa of the rainbow trout *Oncorhynchus mykiss* (Pisces, Teleostei). Reprod Nutr Dev 33:349–360

Lahnsteiner F, Berger B, Weismann T, Patzner R (1996) The influence of various cryoprotectants on semen quality of the rainbow trout (*Oncorhynchus mykiss*) before and after cryopreservation. J Appl Ichthyol 12:99–106. https://doi.org/10.1111/j.1439-0426.1996.tb00070.x

Lahnsteiner F, Weismann T, Patzner RA (1997) Methanol as cryoprotectant and the suitability of 1.2 ml and 5 ml straws for cryopreservation of semen from salmonid fishes. Aquacult Res 28:471–479. https://doi.org/10.1046/j.1365-2109.1997.00886.x

Legendre M, Billard R (1980) Cryopreservation of rainbow trout sperm by deep-freezing. Reprod Nutr Dev 20:1859–1868. https://doi.org/10.2331/suisan.46.1493

Legendre M, Linhart O, Billard R (1996) Spawning and management of gametes, fertilized eggs and embryos in Siluroidei. Aquat Living Resour 9:59–80. https://doi.org/10.1051/alr:1996042

Leibo SP, McGrath JJ, Cravalho EG (1978) Microscopic observation of intracellular ice formation in unfertilized mouse ova as a function of cooling rate. Cryobiology 15:257–271. https://doi.org/10.1016/0011-2240(78)90036-6

Matthews JL, Murphy JM, Carmichael C, Yang H, Tiersch TR, Westerfield M, Varga ZM (2018) Changes to extender, cryoprotective medium, and in vitro fertilization improve zebrafish sperm cryopreservation. Zebrafish 15:279–290. https://doi.org/10.1089/zeb.2017.1521

Mazur P (1970) Cryobiology: the freezing of biological systems. Science (80-) 168:939–949

Mazur P (1984) Freezing of living cells: mechanisms and implications. Am J Physiol 247:C125–C142

Mazur P (1990) Equilibrium, quasi-equilibrium, and nonequilibrium freezing of mammalian embryos. Cell Biophys 17:53–92. https://doi.org/10.1007/BF02989804

Mazur P, Seki S (2011) Survival of mouse oocytes after being cooled in a vitrification solution to -196°C at 95° to 70,000°C/min and warmed at 610° to 118,000°C/min: a new paradigm for cryopreservation by vitrification. Cryobiology 62:1–7. https://doi.org/10.1016/j.cryobiol.2010.10.159

Mazur P, Leibo SP, Chu EHY (1972) A two-factor hypothesis of freezing injury. Evidence from Chinese hamster tissue-culture cells. Exp Cell Res 71:345–355. https://doi.org/10.1016/0014-4827(72)90303-5

McNiven MA, Gallant RK, Richardson GF (1993) Dimethyl-acetamide as a cryoprotectant for rainbow trout spermatozoa. Theriogenology 40:943–948

Merino O, Sánchez R, Risopatrón J, Isachenko E, Katkov II, Figueroa E, Valdebenito I, Mallmann P, Isachenko V (2012) Cryoprotectant-free vitrification of fish (*Oncorhynchus mykiss*) spermatozoa: first report. Andrologia 44(Suppl 1):390–395. https://doi.org/10.1111/j.1439-0272.2011.01196.x

Morris JP, Berghmans S, Zahrieh D, Neuberg DS, Kanki JP, Look AT (2003) Zebrafish sperm cryopreservation with N,N-dimethylacetamide. Biotechniques 35:956–968

Mounib MS (1978) Cryogenic preservation of fish and mammalian spermatozoa. J Reprod Fertil 53:13–18. https://doi.org/10.1530/jrf.0.0530013

Mounib MS, Hwang PC, Idler DR (1968) Cryogenic preservation of Atlantic cod (*Gadus morhua*) sperm. J Fish Res Board Can 25:2623–2632. https://doi.org/10.1139/f68-232

Nynca J, Judycka S, Liszewska E, Dobosz S, Ciereszko A (2017) Standardization of spermatozoa concentration for cryopreservation of rainbow trout semen using a glucose-methanol extender. Aquaculture 477:23–27. https://doi.org/10.1016/j.aquaculture.2017.04.036

Ogier De Baulny B, Le Vern Y, Kerboeuf D, Maisse G (1997) Flow cytometric evaluation of mitochondrial activity and membrane integrity in fresh and cryopreserved rainbow trout (*Oncorhynchus mykiss*) spermatozoa. Cryobiology 34:141–149

Peñaranda DS, Pérez L, Gallego V, Jover M, Asturiano JF (2009) Improvement of European eel sperm cryopreservation method by preventing spermatozoa movement activation caused by cryoprotectants. Cryobiology 59:119–126

Perchec G, Jeulin C, Cosson J, André F, Billard R (1995) Relationship between sperm ATP content and motility of carp spermatozoa. J Cell Sci 108:747–753

Polge C, Smith AU, Parkes AS (1949) Revival of spermatozoa after vitrification and dehydration at low temperatures. Nature 164:666

Riesco MF, Raposo C, Engrola S, Martínez-Páramo S, Mira S, Cunha ME, Cabrita E (2017) Improvement of the cryopreservation protocols for the dusky grouper, *Epinephelus marginatus*. Aquaculture 470:207–213. https://doi.org/10.1016/j.aquaculture.2016.12.027

Sarvi K, Niksirat H, Amiri BM, Mirtorabi SM, Rafiee G, Bakhtiyari M (2006) Cryopreservation of semen from the endangered Caspian brown trout (*Salmo trutta caspius*). Aquaculture 256:564–569

Schuster TG, Keller LM, Dunn RL, Ohl DA, Smith GD (2003) Ultra-rapid freezing of very low numbers of sperm using cryoloops. Hum Reprod 18:788–795. https://doi.org/10.1093/humrep/deg162

Segovia M, Jenkins JA, Paniagua-Chavez C, Tiersch TR (2000) Flow cytometric evaluation of antibiotic effects on viability and mitochondrial function of refrigerated spermatozoa of nile tilapia. Theriogenology 53:1489–1499. https://doi.org/10.1016/S0093-691X(00)00291-0

Seki S, Jin B, Mazur P (2014) Extreme rapid warming yields high functional survivals of vitrified 8-cell mouse embryos even when suspended in a half-strength vitrification solution and cooled at moderate rates to -196°C. Cryobiology 68:71–78. https://doi.org/10.1016/j.cryobiol.2013.12.001

Stewart DL (1951) Storage of bull spermatozoa at low temperatures. Vet Rec 63:65–66

Stoss J, Holtz W (1983) Cryopreservation of rainbow trout (*Salmo gairdneri*) sperm: IV. The effect of DMSO concentration and equilibration time on sperm survival, sucrose and KCl as extender components and the osmolality of the thawing solution. Aquaculture 32:321–330. https://doi.org/10.1016/0044-8486(83)90229-6

Suquet M, Dreanno C, Fauvel C, Cosson J, Billard R (2000) Cryopreservation of sperm in marine fish. Aquacult Res 31:231–243. https://doi.org/10.1046/j.1365-2109.2000.00445.x

Tiersch TR, Yang H, Hu E (2011) Outlook for development of high-throughput cryopreservation for small-bodied biomedical model fishes. Comp Biochem Physiol C Toxicol Pharmacol 154:76–81. https://doi.org/10.1016/j.cbpc.2011.03.004

Viveiros ATM, Godinho HP (2009) Sperm quality and cryopreservation of Brazilian freshwater fish species: a review. Fish Physiol Biochem 35:137–150. https://doi.org/10.1007/s10695-008-9240-3

Viveiros ATM, Nascimento AF, Orfão LH, Isaú ZA (2010) Motility and fertility of the subtropical freshwater fish streaked prochilod (*Prochilodus lineatus*) sperm cryopreserved in powdered coconut water. Theriogenology 74:551–556. https://doi.org/10.1016/j.theriogenology.2010.03.018

Viveiros ATM, Orfão LH, Nascimento AF, Corrêa FM, Caneppele D (2012) Effects of extenders, cryoprotectants and freezing methods on sperm quality of the threatened Brazilian freshwater fish pirapitinga-do-sul *Brycon opalinus* (Characiformes). Theriogenology 78:361–368. https://doi.org/10.1016/j.theriogenology.2012.02.015

Viveiros ATM, Nascimento AF, Leal MC, Gonçalves ACS, Orfão LH, Cosson J (2014) Methyl glycol, methanol and DMSO effects on post-thaw motility, velocities, membrane integrity and mitochondrial function of *Brycon orbignyanus* and *Prochilodus lineatus* (Characiformes) sperm. Fish Physiol Biochem 41:193–201. https://doi.org/10.1007/s10695-014-0016-7

Warnecke D, Pluta HJ (2003) Motility and fertilizing capacity of frozen/thawed common carp (*Cyprinus carpio* L.) sperm using dimethyl-acetamide as the main cryoprotectant. Aquaculture 215:167–185. https://doi.org/10.1016/S0044-8486(02)00371-X

WHO (2010) WHO laboratory manual for the examination and processing of human semen, 5th edn. World Health Organization, Geneva

Wilson-Leedy JG, Ingermann RL (2007) Development of a novel CASA system based on open source software for characterization of zebrafish sperm motility parameters. Theriogenology 67:661–672

Young WP, Frenyea K, Wheeler PA, Thorgaard GH (2009) No increase in developmental deformities or fluctuating asymmetry in rainbow trout (*Oncorhynchus mykiss*) produced with cryopreserved sperm. Aquaculture 289:13–18

Zilli L, Schiavone R, Zonno V, Storelli C, Vilella S (2003) Evaluation of DNA damage in *Dicentrarchus labrax* sperm following cryopreservation. Cryobiology 47:227–235

Zilli L, Schiavone R, Zonno V, Storelli C, Vilella S (2004) Adenosine triphosphate concentration and β-d-glucuronidase activity as indicators of sea bass semen quality. Biol Reprod 70:1679–1684. https://doi.org/10.1095/biolreprod.103.027177

Chapter 16
Specificity of Germ Cell Technologies in Sturgeons

Martin Pšenička and Taiju Saito

Abstract Sturgeons are one of the oldest, biggest, most valuable and today also most endangered group of fish species. Germ stem cells (GSCs), such us embryonic primordial germ cells (PGCs) or spermatogonial/oogonial stem cells, can be a key for an effective conservation and possible restoration of these unique and astonishing fishes. In this chapter, labeling, development, isolation, and transplantation of GSCs were studied in sturgeons. It was shown that the maternally supplied germ plasm, which determines the PGC origin, is localized in vegetal pole of ovulated egg and remains there throughout the cleavage period, and therefore, the PGC specification pattern is similar to that of anuran amphibians rather than teleostean fishes. This knowledge enabled to develop an original PGC labeling method using common cell tracer dye injection into the vegetal pole of two- to eight-cell stage embryo. Next inhibition of maternally supplied *dead end* RNA resulted in PGC mismigration and general sterilization of individuals. This method enables preparation of recipients for germ cell transplantation. Isolation and transplantation of spermatogonia and oogonia were developed as well. It was tested that one sturgeon juvenile (Siberian sturgeon) can provide approximately one million germ cells suitable for transplantation. Moreover, it was shown that these cells are capable of propagation via an in vitro culture system and of cryopreservation. After freezing/thawing of sturgeon gonadal tissue followed by enzymatic dissociation, above 90% of viable cells were obtained and used for transplantation. The technique of surrogate production can be applied for conservation and possibly restoration of critically endangered sturgeon species with a long term of maturation and a big body size (e.g.,

The chapter is written based on Pšenička's inaugural dissertation.

M. Pšenička (✉)
University of South Bohemia in České Budějovice, Faculty of Fisheries and Protection of Waters, Bohemian Research Center of Aquaculture and Biodiversity of Hydrocenoses, Vodňany, Czech Republic
e-mail: psenicka@frov.jcu.cz

T. Saito
Nishiura Station, South Ehime Fisheries Research Center, Ehime University, Uchidomari, Ainan, Ehime, Japan

M. Yoshida, J. F. Asturiano (eds.), *Reproduction in Aquatic Animals*,
https://doi.org/10.1007/978-981-15-2290-1_17

beluga), whereas a more common species with shorter term of maturation and smaller body size (e.g., sterlet) can be used as a recipient (surrogate parent).

Keywords Xenotransplantation · Germ-line chimera · Sturgeon · Surrogate production · Germ cells

16.1 Introduction

16.1.1 Sturgeons as the Most Critically Endangered Group of Species

The sturgeons, belonging to family Acipenseridae, are evolutionary very successful. They appear in the fossil record around 245 million years ago near the end of the Triassic period, but recently their populations have been drastically decreasing due to human activities such as overfishing and poaching for caviar, construction of dams that prevent their reproductive migration, damage of their natural habitats, and environmental pollutions. Recently 25 out of 27 sturgeon species are included in the red list of the International Union for Conservation of Nature (IUCN), and 16 are classified as critically endangered (the last degree of threat before extinction). Therefore, they are considered as the most endangered group of species in the world (IUCN, http://www.iucnredlist.org/news/sturgeons-highly-threatened). Nowadays, the existence of most sturgeon species is dependent on the artificial reproduction, which is complicated due to their specific characteristics such as (a) big body size—some species can grow more than 7 m, which means that huge space and labor are needed for keeping the fish; (b) late maturation—first maturation occurs after more than 20 years in some species, and long-term fish cultivation is needed to obtain gametes; (c) spawn once a few years—sturgeons do not repeat reproduction every year, and systematic collection of gametes needs an extra effort. Moreover, it is not possible to cryopreserve sturgeon eggs or embryos yet and a loss of maternal genetic information happens, if only sperm is cryopreserved. Due to these reasons, efficient techniques of sturgeon conservation and restoration are urgently required.

16.1.2 Germ Stem Cells

Germ stem cells (GSCs) such as embryonic primordial germ cells (PGCs) or subsequently spermatogonial/oogonial stem cells are precursors of gametes. They ensure gametogenesis throughout the life and have the ability to self-renew and differentiate into both types of gametes. In fish, PGCs migrate during embryonic development from the area where they are formed toward the genital ridge, future testes, or ovary. The undifferentiated gonial cells differentiate into spermatogonia and oogonia and undergo mitotic proliferation until meiotic phase of gametogenesis. For general view on fish gametogenesis see study written by Leal et al. (2009). Some years ago, the absence of PGC markers limited their identification to simple morphological characteristics such as the presence of a structure called germ plasm (Olsen et al. 1997; Yoon et al. 1997), which appears as an electrodense cytoplasmic inclusion, usually found in association with mitochondria (Eddy 1976). This electron-dense maternally supplied structure stores RNA and proteins needed for future PGC specification and development (Seydoux and Dunn 1997; Seydoux et al. 1996; Williamson and Lehmann 1996). One of the main components of germ plasm is the vasa transcript and protein, which nowadays serve as the main markers for the identification of the germ-line of many animals. Nowadays there are also several other genes relevant to PGC viability and migration. For example, knockdown of genes such as *nanos3* or *dead end* by antisense morpholino oligonucleotide caused PGC mismigration during embryonic development (Koprunner et al. 2001; Weidinger et al. 2003). It also has been demonstrated that Buc protein is crucial for germ plasm formation, and its overexpression induces ectopic germ cell formation (Bontems et al. 2009).

During the embryo development, at the blastula stage, PGCs undergo a transition from a non-motile to a motile mode (Blaser et al. 2005). The PGCs begin to respond to guidance cues provided by chemokine signals and start migrating toward the genital ridge. The chemokine SDF1a and its receptor CXCR4b are key players in PGC migration (Doitsidou et al. 2002; Kurokawa et al. 2006). Transplantation experiments suggest that molecular mechanisms controlling PGC migration are conserved among most fish species. Interestingly, PGCs isolated from one species and transplanted into another species with a completely different developmental and PGC migratory pattern (e.g., from European eel (*Anguilla anguilla*) to zebrafish (*Danio rerio*)) migrated toward the genital ridge of the host species (Saito et al. 2011). After migration, PGCs settle in the position where the gonads are formed. Although the mechanism how PGCs terminate their migration is still unknown, it seems that the duration of migratory activity of PGCs is controlled autonomously. When PGCs isolated from different stages were individually transplanted into blastula stage hosts, migration rates of PGCs toward the genital ridge were significantly lower at the later stages than earlier stages. It is likely that there is a difference between migration ability of germ cells before and after the somitogenesis period (Saito et al. 2010).

16.1.3 Surrogate Production Technology

In fish, the surrogate production means that a surrogate host parent produces gametes of a donor. The host individual carrying germ-line of a donor is called germ-line chimera. To obtain germ-line chimera, GSCs have to be isolated from a donor and transplanted into a host. The host become germ-line chimera if the transplanted GSCs succeed to become correctly localized to the genital ridge. Donor genotypes can then be restored at the next generation, if the transplanted germ cells differentiate into functional gametes. This technique has been attracting attention in the recent past, because it can lead to the production of gametes of endangered species using more common species as the surrogate parent (Okutsu et al. 2006; Yamaha et al. 2007).

The main key steps for the production of germ-line chimeras are (1) preparation of live isolates of donor GSCs; (2) sterilization of host; and (3) transplantation into host. The GSC transplantation can be generally distinguished into PGC transplantation and spermatogonia/oogonia transplantation.

16.1.3.1 Generation of a Germ-Line Chimera by Transplantation of PGCs

Production of germ-line chimeras in fish by transplantation of PGCs was first reported by Lin et al. (1992) in zebrafish. In this experiment, blastomeres at the blastula stage of the donor embryo were randomly aspirated into a needle and transplanted into host embryos of developmental stage identical to the donors. Thus, they transplanted PGCs together with somatic cells. Yoon et al. (1997) reported that PGCs are localized at the marginal region of the blastodisc at the blastula stage in zebrafish. Therefore, Yamaha et al. (2001) transplanted only the PGC-containing lower part of blastodisc into a host blastula embryo between diploid goldfish (*Carassius auratus*) and triploid crucian carp (*Carassius carassius*); this "sandwiching" technique was considered to likely improve the efficiency of germ-line chimera production.

The transplantation of PGCs together with other blastomeres between different fish species has a limitation: the somatic blastomeres disturb PGC migration, and PGCs are unable to arrive properly into genital ridge. To solve this problem, Saito et al. (2008) isolated and transplanted single PGCs. They used pearl danio (*Danio albolineatus*) as donor and sterilized zebrafish at blastula stage as host. The PGCs were visualized by the injection of *gfp-nos3* 3′UTR mRNA during one- to four-cell stages according to Koprunner et al. (2001) and Saito et al. (2006), prior to isolation. Ninety-four percent male and 66% female chimera were confirmed as germ-line chimera that carried and produced only donor-derived gametes. Later, Saito et al. (2010) showed a strong evidence that migratory mechanism of PGCs is well conserved among fish species. They transplanted single PGCs from zebrafish, pearl danio, goldfish, or loach (*Misgurnus anguillicaudatus*) into sterilized blastula-stage zebrafish embryos. The transplanted PGCs migrated together with intrinsic host PGCs toward the genital ridge, and the germ-line chimeras were produced between distantly related species, not only between different genera, but also between different families. Interestingly, a single PGC was sufficient for donor-derived gamete production. The migration success rate as ratio of transplanted PGC presenting migration activity to the host's genital ridge was estimated as more than 45% even in case of xenogeneic chimeras. In addition, this technique was applied to generate germ-line chimera between more distant species belonging to different orders such as Japanese eel (*Anguilla japonica*) to zebrafish (Saito et al. 2011). It was observed that donor PGCs could migrate into the genital ridges of hosts, beyond the "family barrier." However, the transplanted PGCs disappear after migration, and the chimera did not produce any donor-derived gametes probably due to incompatibility with host's somatic tissue in genital ridge.

Although PGCs possess the ability to induce germ-line chimera with such high efficiency, a main limiting factor is the number of PGCs available per embryo. It has been estimated that, in the case of fish, there are only a few tens of PGCs present in each embryo (Saito et al. 2006).

16.1.3.2 Induction of a Germ-Line Chimera by Transplantation of Spermatogonia or Oogonia

Alternatively, another technique to induce germ-line chimera has been reported in fish. The technique is based on transplantation of gonadal stem cells, such us undifferentiated spermatogonia or oogonia (SG/OG), so as to avoid the complexity of blastomere transplantation. Takeuchi et al. (2003, 2004) isolated undifferentiated gonocytes from the genital ridge of newly hatched rainbow trout (*Oncorhynchus mykiss*) and transplanted these cells into the peritoneal cavities of host masu salmon (*Oncorhynchus masou*) at the same stage as donor. With this approach, they obtained germ-line chimera that produced donor-derived gametes. Later this research group proved that SG collected from sexually immature trout could be used as donor cells too. Okutsu et al. (2006) performed intraspecific spermatogonia transplantation from transgenic rainbow trout juveniles carrying the green fluorescent protein (GFP) driven by the *vasa* promoter (expressing GFP in germ-line) into larvae of non-transgenic fish of the same species and later interspecific transplantation from rainbow trout to masu salmon (Okutsu et al. 2007). The transplanted SG colonized in a part of the gonad and differentiated into functional gametes. The germ-line chimeras were differentiated into both males and females and produced donor-derived offspring. In rainbow trout and zebrafish, it has been shown that the PGCs and early SG have the remaining sexual plasticity. The cells can be differentiated into both functional eggs and functional sperm in dependence on host sex (Nóbrega et al. 2010; Okutsu et al. 2006; Wong et al. 2011; Yoshizaki et al. 2011). A different procedure was applied for the transplantation of spermatogonia into adult Nile tilapia (*Oreochromis niloticus*) after a chemical treatment by busulfan through a genital pore (Lacerda et al. 2006).

Presently, SG/OG transplantations represent the major strategies of surrogate production via germ-line chimera among different fish species. The main advantage of this method is that a large amount of SG/OG can be obtained from juvenile or adult testes or ovaries. On the other hand, the disadvantage of this method is lack of suitable vital biomarkers for gonial stem cells.

Several techniques have been developed to sort the testicular cells in order to enrich early stages of undifferentiated SG/OG. The most common techniques of sorting are based on sedimentation or centrifugation in a density gradient (e.g., Percoll, Ficoll, BSA) (Lacerda et al. 2006; Pšenička et al. 2015) or centrifugal elutriation (Bellaiche et al. 2014). In the testicular tissue, undifferentiated spermatogonia are the largest cells, with the lowest relative weight and therefore remain at the upper layers of the density gradient. Nóbrega et al. (2010) established a technique to transplant SG from transgenic zebrafish (*vasa::egfp*) to wild-type one. They purified these cells by FACS (fluorescence-activated cell sorting) based on the GFP fluorescence and transplanted them into male and female adult zebrafish through the genital pore. Nevertheless, it should be mentioned that *vasa* is not a stem cell marker. Another possibility is offered by specific surface proteins. Ly75 protein in rainbow trout (Nagasawa et al. 2010) and Gfra1 receptor in Nile tilapia (Lacerda et al. 2013) were identified as markers of mitotic germ cells and were proposed as cell surface markers for spermatogonia sorting. Von Schönfeldt et al. (1999) described the enrichment of SG by magnetic-activated cell sorting using surface markers of spermatogonial stem cells in several mammalian species. Panda et al. (2011) introduced this technique in a cyprinid fish *Labeo rohita.*

16.1.4 Host Sterilization

Effective sterilization of the recipient is the prerequisite of effective use of germline chimera to produce donor-derived gametes only. There are several approaches to achieve effective sterilization in fish such as surgery, irradiation, chemical or hormonal treatments, and genome, gene, and gene expression manipulations. A non-transgenic genetic modification—gene knockdown strategy—can inhibit the function of genes involved in gonadal development. In many invertebrate and vertebrate species, *dead end* gene encodes an RNA-binding protein crucial for migration and survival of PGCs. It is an evolutionary conserved and maternally inherited protein present in male and female gonads. Therefore, it seems to be a good candidate not only for the identification of GSCs but also as a target gene to achieve sterilization using knockdown strategy (described in zebrafish by Weidinger et al. 2003). Nowadays several potential target genes were characterized in commercially valuable species such as Atlantic salmon (*Salmo salar*) (Nagasawa et al. 2013). Lacerda et al. (2006) induce sterilization for surrogate production purpose in Nile tilapia by busulfan treatment. Busulfan is a cell cycle non-specific alkylating agent used in cancer treatment that attaches an alkyl group (C_nH_{2n+1}) to DNA. Since cancer as well as spermatogonial cells proliferate faster and therefore with less error-correcting than the other cells in general, these cells are more sensitive to DNA damage, such as being alkylated (Iwamoto et al. 2004).

16.1.5 Germ Cell Cryopreservation

Probably the main potential of utilization of GSC technologies is its expansion into cryopreservation platform for conservation of genetic resources. Cryopreservation of fish eggs and embryos is not possible so far due to large size of egg, high content of yolk, and low membrane permeability. Therefore, most of the cryopreservation protocols have been developed for sperm, thus limiting the conservation to the paternal genome only. Although a new generation could be theoretically regenerated from cryopreserved sperm itself via androgenesis using genetically inactivated/irradiated eggs of similar species, the success rate of androgenesis is extremely low, and the resulting fish are homozygous nuclear-cytoplasmic hybrids with a loss of specific maternal genetic information. This altogether makes the androgenetic regeneration impractical. On the other hand, PGCs or SG/OG are diploid cells having the ability to preserve both paternally and maternally inherited genetic information. Cryopreservation of embryos or embryoids including PGCs and subsequent transplantation and germ-line chimera generation have been successfully achieved for example in zebrafish (Kawakami et al. 2010) or loach (Yasui et al. 2011). Nevertheless, as mentioned above, the number of PGCs in a single embryo or larva is very low. The cryopreservation of GSCs from juvenile or adult guarantees a higher number of cells that maintain their potential to contribute to the germ-line. Kobayashi et al. (2007) surgically isolated genital ridges containing gonial cells of transgenic rainbow trout (*vasa::egfp*) fry and cryopreserved the tissue. The thawed gonial cells were transplanted into the body cavity of host fry, and they found that the chimera produced donor-derived gametes. Cryopreserved rainbow trout spermatogonia were transplanted into peritoneal cavity of masu salmon by Yoshizaki et al. (2011), which gave rise to both functional sperm and eggs derived from the donor. Recently it was found that a successful regeneration of spermatogonia is possible to achieve also from frozen whole testis (Lee et al. 2013) and frozen whole fish in rainbow trout (Lee et al. 2015).

16.1.6 Sturgeon Germ Cells

The aim of this chapter is to summarize the recent findings about GSC visualization, isolation, cryopreservation, and transplantation in sturgeons in order to prepare a platform for their surrogate production, where the critically endangered sturgeon species with long term of maturation and a big body size (e.g., beluga) can be used as a donor of GSCs and a more common species, with shorter term of maturation and smaller body size (e.g., sterlet) as a recipient (surrogate parent) (see Fig. 16.1).

16.2 The Origin and Migration of Primordial Germ Cells in Sturgeons (Saito et al. 2014)

Although sturgeons as Chondrostei belong to the class Actinopterygii, their early developmental pattern differs from Teleostei. While eggs of teleostean fishes cleave meroblastically, the sturgeon eggs undergo holoblastic cleavage pattern, which is similar to that of amphibians. The mode of PGC formation is also altered according to the transition of cleavage from holoblastic to meroblastic. In amphibians, there are both known patterns of PGC specification, by inductive signals (epigenesis) or by maternally inherited determinants (preformation) (Extavour and Akam 2003). In urodeles, PGCs are likely formed in the animal cap, and they are induced rather epigenetically by interaction between animal and vegetal blastomeres (Sutasurja

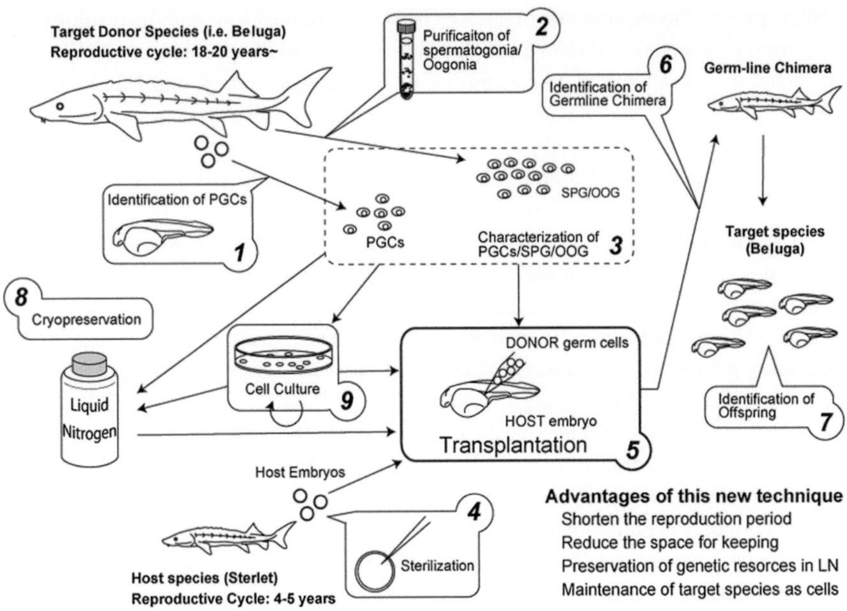

Fig. 16.1 An overview of strategy for the development of surrogate production technique in sturgeon. (Published in Robles et al. 2016)

and Nieuwkoop 1974). However, in anurans, PGCs are specified in the presumptive endoderm that is derived from blastomeres located at the vegetal pole (Ikenishi et al. 1974).

Saito et al. (2014) conducted several experiments to reveal whether PGCs in sturgeons are formed in animal or vegetal pole and whether PGCs are specified by inducing signals from somatic cells or predetermined by specific germ plasm (germ cell determinant). As the first, *gfp-nos3* 3′UTR mRNA was injected into vegetal and animal pole of sturgeon embryo in two- to eight-cell stage in order to visualize PGCs. Localization of orthologs of the *nanos* to germ cells have been found in nearly all taxa studied (Ewen-Campen et al. 2010). It has been previously shown that the three prime UTR of *nanos3* is subjected to degradation in somatic cells, but is stabilized in PGCs by interaction with the microRNA (miR-430), Dnd, and Dazl (Mishima 2012; Mishima et al. 2006). As a result, PGCs were observed only in embryos injected into vegetal pole, which was the first indication that PGCs origin in vegetal pole. In addition, this result strongly suggests that the mechanism of conditional regulation of *nanos3* mRNA should be conserved in sturgeons as in other teleosts (Saito et al. 2006). In the next experiment, the presence of germ plasm in sturgeon eggs was confirmed. In zebrafish, Buc protein has an important role in assembling the germ plasm in an oocyte and early stage embryo, and it is possible to visualize developing germ plasm at animal pole of a developing embryo along cleavage furrows by injecting chimeric RNA composed of GFP and *buc* open reading frame (Bontems et al. 2009). In sturgeons, germ plasm was observed along cleavage furrows (as seen in zebrafish) after injection of *gfp-buc* mRNA into vegetal pole, but not in animal pole. In addition, the same pattern of germ plasm accumulation was observed after examination of sturgeon embryos using transmission electron microscopy. This finding strongly suggests that the function of Buc protein is conserved in sturgeon, and their PGCs are specified by the inheritance of germ plasm, which is maternally supplied and deposited at the vegetal pole region of the egg. Pocherniaieva et al. (2018) confirmed the localization of PGC-specific RNAs in vegetal pole of sturgeon oocyte by qPCR analysis of the oocyte cryo-sections.

Sturgeon PGC migratory route from their formation site to the gonadal ridge differs from that of anurans, although these cells were specified in a similar manner. In *Xenopus*, PGCs are specified at the vegetal pole and migrate toward the gonadal ridge within the endoderm (Whitington and Dixon 1975). On the other hand, sturgeon PGCs migrate on the yolk ball and the yolk extension toward the gonadal ridge via the mesenchyme, which is similar to that observed in many teleosts (Saito et al. 2006). To prove whether the migration pattern is conserved among teleostean and chondrostean, sturgeon PGCs were isolated and transplanted into goldfish embryo at blastula stage. As a result, transplanted sturgeon PGCs migrated toward the gonadal ridge of the goldfish along with the endogenous goldfish PGCs (Fig. 16.2). It demonstrates that the migration mechanisms of PGCs are widely conserved in fish in spite of the huge differences such as their size, cleavage and developmental pattern, and localization of PGC specification. However, the transplanted PGCs that successfully arrived at the gonadal ridge disappeared as the embryo developed further.

16.3 Novel Technique for Visualizing Primordial Germ Cells in Sturgeons (Saito et al. 2014)

Generally, PGC labeling had been possible only using injection of aforementioned *gfp-nos3* 3′UTR mRNA. However, preparation of the artificial RNA is demanding, time consuming, and costly, and moreover, the RNA is unstable for storage, and the labeling remains relatively short time. In zebrafish, the expression of *nanos3* (*nos3*) mRNA declines to an undetectable level by 5 days post fertilization (Koprunner et al. 2001). In sturgeon, GFP expression in PGCs after injection of *gfp-nos3* 3′UTR

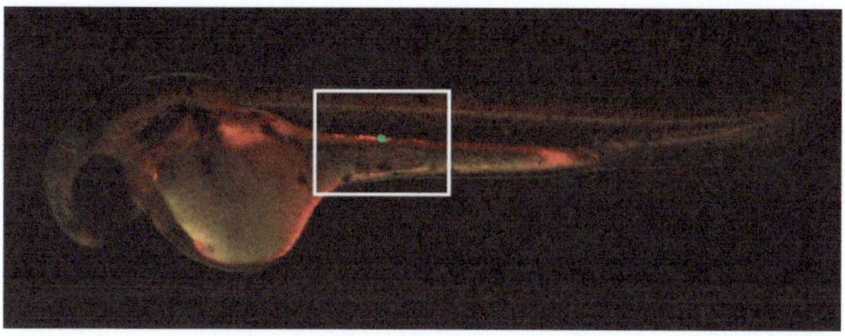

Fig. 16.2 Sturgeon PGC in goldfish embryo 3 days post transplantation. Goldfish and sturgeon PGCs are labeled by RFP and GFP-nos3 3′UTR (red and green), respectively (Saito et al. 2014)

mRNA did not last more than 2 months (Saito et al. 2014), probably because the injected mRNA was gradually degraded during PGC development. Second possibility for germ cell visualization is generation of a transgenic strain that possesses a reporter gene in germ cells such as the green fluorescent protein (GFP) driven by the *vasa* promoter (expressing GFP in germ-line). This technique was used, for example, for transplantation of spermatogonia in rainbow trout (Okutsu et al. 2006). However, currently, no transgenic strain is available in sturgeons, mainly because of the extremely long reproductive cycle differing among species (5–25 years). Therefore, it was needed to develop a new approach to enable visualization of sturgeon PGCs in vivo both to study the dynamics of their development and to enable isolation of viable PGCs for other downstream applications such as cryopreservation, cell culture, transplantation, and so on.

In a previous study, it was clearly showed that in sturgeons, PGCs are formed exclusively in vegetal pole of egg. We took advantage of this finding and developed an original, simple, and long-term non-transgenic technique of sturgeon PGC labeling. A general cell tracer dye with high molecular weight (MW) was injected into vegetal pole of sturgeon embryo, localization of PGC precursors, at two- to eight-cell stage. The high MW was necessary for keeping the tracer dye only in a strict area of vegetal pole. For the experiments, we used fluorescein isothiocyanate conjugated to high-molecular-weight dextran (FITC-dextran). We tested MW of FITC-dextran from 10,000 to 2,000,000, and the efficiency of PGC visualization was highest using FITC-dextran with MW 500,000. FITC-dextran with a lower MW was spread in whole embryo without any specific labeling. On the other hand, FITC-dextran with a higher MW than 500,000 labeled lower number of PGCs. This suggests that the germ plasm likely accumulates in a wider area of the vegetal hemisphere than that labeled by FITC-dextran with a high MW. To support this conclusion, it was reported that formation of GFP islands after injection of *gfp-buc* mRNA is observed distant to the vegetal pole (Saito et al. 2014). After injection of sturgeon embryos by FITC-dextran with MW 500,000 from two- to eight-cell stage, PGC-like cells appeared at fourth day post fertilization at 15 °C and migrate forward and form two germinal ridges similar to PGCs labeled by *gfp-nos3* 3′UTR mRNA. A strong signal was observed at least for 3 months post fertilization, when the cells started proliferation. The competence of labeled cells was proved by immunocytochemistry and immunohistochemistry against Vasa (germ-line-specific protein), whereby all FITC-labeled cells were positive to Vasa protein and were localized only in genital ridge 1 month post fertilization. It was clearly shown that FITC-labeled cells are PGCs (Fig. 16.3). This method gave us a powerful tool for visualization of PGCs, which could be used for the investigation of PGC migration, development of early stages of germ cells in larvae and young juveniles, and for other applications such as isolations and transplantations.

16.4 Sterilization of Sterlet (*Acipenser ruthenus*) by Using Knockdown Agent, Antisense Morpholino Oligonucleotide, Against Dead End Gene (Linhartová et al. 2015)

A prerequisite for the generation of germ-line chimera, producing the donor-derived gametes only, is sterilization of the host. Okutsu et al. (2006) transplanted spermatogonia in rainbow trout into normally fertile recipients. Fifty percent (13 from 26) of transplanted males produced only 5.46 ± 3.34% donor-derived progeny and 40% (16 from 40) females produced 2.14 ± 0.70% donor-derived progeny, while after a sterilization of recipient, in the case of successful transplantation, they produced only donor-derived gametes (Okutsu et al. 2007). Therefore, it is obvious that sterilization of recipient is crucial for the development of an efficient surrogate production technology.

The easiest and most common sterilization method in fish is triploidization or hybridization. It was reported in many fish species that retention of second polar body causes sterility of the individuals due to incapacity of three sets of homologous chromosome segregation. In the case that triploid fish produces gametes, those are usually aneuploidy (Ihssen et al. 1990). However, several exceptions from this rule have been reviewed by Maxime (2008) and Piferrer et al. (2009). Sturgeons are one of the exception, where the fertility of triploids as well as interspecific hybrids has been often reported, probably due to their polyploid nature and often hybridization events (Havelka et al. 2014; Piferrer et al. 2009). Therefore, an alternative sterilization method had to be developed for sturgeons. We applied a knockdown technology, namely temporary inhibition of *dnd1* translation using antisense morpholino oligonucleotide (MO). The *dnd1* is maternally supplied RNA crucial for the

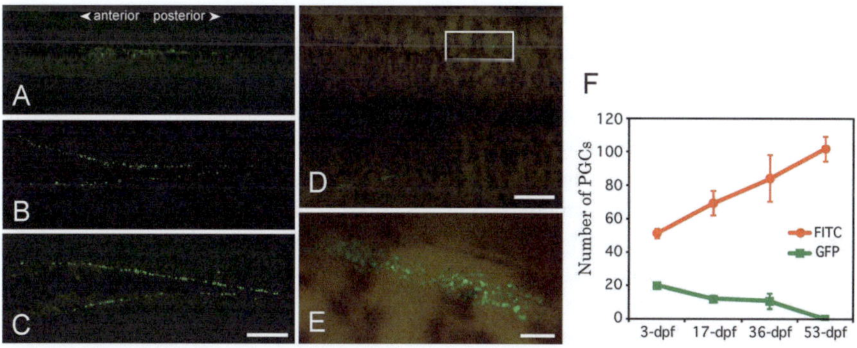

Fig. 16.3 PGC visualization during genital ridge formation in *Acipenser ruthenus*. (**a–d**) Ventral views of FITC-positive PGCs in the upper parts of the abdominal cavity after removal of the intestines in 17-dpf (**a**), 36-dpf (**b**), 53-dpf (**c**), and 90-dpf (**d**) embryos. (**e**) Magnified image of the boxed area in (**d**). The mingled PGCs in the upper part of the body cavity of the 17-dpf embryo separate into two lines as the fish develop. (**f**) Comparison of the average number of PGCs visualized by FD-500 or *gfp-nos3* 3'UTR at each developmental stage. Vertical bars indicate standard deviations. Bar = 500 μm (**c** and **d**) and 100 μm (**e**) (Saito et al. 2014)

migration and development of PGCs in zebrafish (Weidinger et al. 2003), and recently, it was also described in Chinese sturgeon (Yang et al. 2015). We injected different concentrations of *dnd1* MO together with FITC-dextran (for PGCs labeling) into vegetal pole of sturgeon embryo from two- to eight-cell stage according to the previous study by Saito and Pšenička (2015). *dnd1* MO was designed for 25 bp downstream from *dnd1* start codon. The embryos injected with 250 μM MO occasionally contained PGCs (0.1 ± 0.45) at fourth day after fertilization (at active migration stage) and completely disappeared at 21st day after fertilization. It was thought that PGCs, which survived until 4 days were mismigrated at later stages and did not reach the germinal ridge. The sterility was proved in 210 old treated fish vs. control fish by macroscopic observation, histology, and RT-PCR as well as in situ hybridization for germ-line-specific *dnd1* and *vasa*. As results, no germ cells were detected in *dnd1* MO-treated individuals. The embryos had significantly lower survival rate after injection with MO with concentration above 250 μM. It is known that a higher concentration of MO is toxic to organisms (Fig. 16.4).

In conclusion, we developed 100% successful method of sturgeon sterilization, which can be used for the preparation of recipient for surrogate production technology.

16.5 Isolation and Transplantation of Sturgeon Early-Stage Germ Cells (Pšenička et al. 2015)

The previous studies focused on sturgeon PGC migration, development, labeling, and also transplantation. A possibility of PGC utilization can be in single-cell-based applications, such as generation of isogenic lines originated form a single PGC; however, surrogate production technologies based on PGC transfer turned out to be relatively complicated due to high fragility and a low number of PGCs in an embryo. Therefore, nowadays, generation of germ-line chimera for surrogate production using gonial cells (spermatogonia and oogonia) is the preferred method.

The technique of spermatogenesis regeneration using spermatogonia transplantation was first introduced by Brinster and Zimmermann (1994) in a mouse model. They conducted dissociation of testicular cells using 0.1% collagenase and 0.25% trypsin in two steps. Spermatogonia were enriched using a sedimentation-based separation method in 2–4% bovine serum albumin (BSA) gradient developed by Bellve et al. (1977). Okutsu et al. (2006) applied a modified method to fish (rainbow trout *Oncorhynchus mykiss*). They dissociated trout testes in a medium containing 0.5% trypsin and transplanted them into the peritoneal cavity; testicular germ cells could colonize gonads of larvae and produce functional sperm and eggs. Psenicka et al. (2015) adopted the approaches and developed method for isolation of SG and OG from immature testis and ovary and transplantation into body cavity of larvae. As first, different enzymatic media were tested for the dissociation of Siberian sturgeon (*Acipenser baerii*) gonadal cells. The composition of best working dissociation

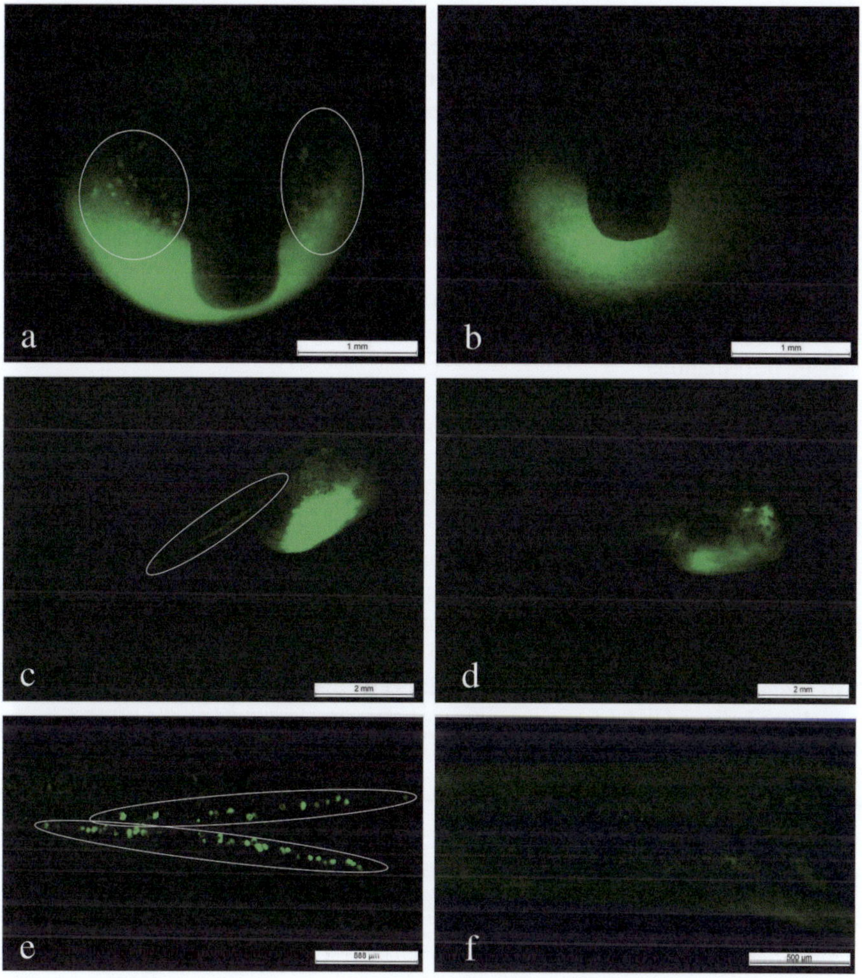

Fig. 16.4 Ablation of primordial germ cells (PGCs) in sterlet embryos and larvae treated with dnd-MO(250 mM) at 4, 7, and 21 days post fertilization (dpf) compared with controls. (**a**, **b**) Tail-bud stage, 4 dpf; (**c**, **d**) hatching period, 7 dpf; (**e**, **f**) dissection of body cavity, 21 dpf. (**a**, **c**, **e**) Control FITC–injected group with PGCs (white circles); (**b**, **d**, **f**) 250-mM dnd-MO–injected group. Scale bars are (**a**, **b**, **e**) 1 mm; (**c**, **d**) 2 mm, and (**e**) 500 mm. MO, morpholino olignucleotide (Linhartová et al. 2015)

media was phosphate-bufferred saline (PBS) containing 0.3% trypsin, adjusted to 238 mOsm kg^{-1} with pH 8. The dissociation lasted 2 h at 23 °C and was terminated by adding 1% BSA and 40 µg mL^{-1} DNAse. The suspension was filtrated through 40-µm filter, and the dissociated cells were sorted using density gradient centrifugation in Percoll solution. The sorted cells were tested by Vasa-antibody (DDX4) immuno-labeling. The highest amount of gonial cells was collected from 10 to 30% Percoll solution. The cells were washed by PBS and labeled with a general cell tracer PKH26 (according to Lacerda et al. 2006 in Nile tilapia) and injected into body cavity of sterlet (*Acipenser ruthenus*) larvae. The development of transplanted cells was recorded for 3 months until proliferation (Fig. 16.5). Three years post transplantation, the donor-derived germ cells were identified in the host using species-specific RT-PCR on *vasa*.

The isolation and transplantation of oogonia has been described in trout by Yoshizaki et al. (2010) and in zebrafish by Wong et al. (2011). The manipulation with oogonia should be considered especially in case of sturgeons, because they are suggested to have a female heterogamety (ZW female/ZZ male) sex-determination system (Flynn et al. 2006; Omoto et al. 2005). Due to this, conservation based only on male germ cells may not be effective because they may not include complete genetic information.

The sorting of gonadal cells is necessary especially in the case when mature individuals are used for the isolation and the gonadal cell suspension contains a high number of differentiated germ cells such as spermatocytes, spermatids, or spermatozoa. There are two general procedures to sort the germ cells: (1) density gradient centrifugation (as it was used in the present study) and (2) using a surface protein-based immuno-method. A promising approach of germ cell sorting was described by Von Schönfeldt et al. (1999). They conducted the enrichment of spermatogonial stem cells by magnetic-activated cell sorting (MACS) using surface markers in several mammalian species. Panda et al. (2011) introduced this technique in a cyprinid fish (*Labeo rohita*) and sorted the testicular cells using antibodies against proteins, which are believed to be conserved in eukaryotes and localized on the plasma membrane of spermatogonial stem cells, Thy1 (CD90) and Gfrα1. However, it is currently

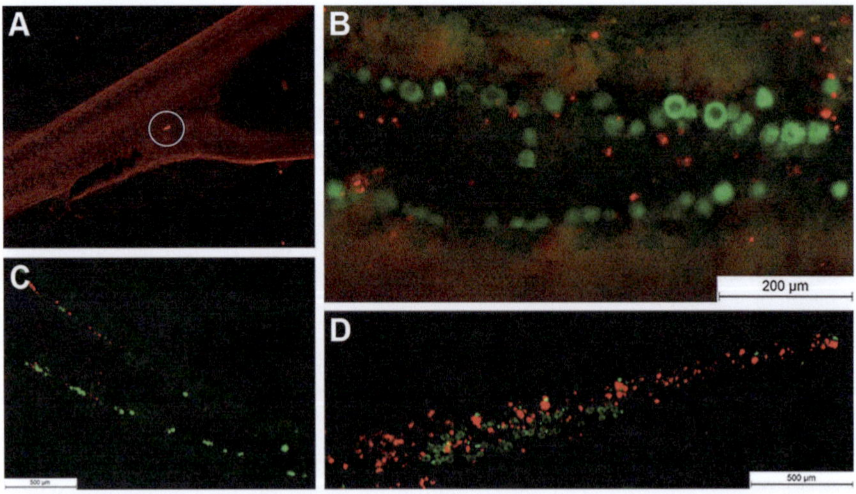

Fig. 16.5 Larvae of sterlet with red-labeled transplanted spermatogonia of Siberian sturgeon (**a**) 6 dpt in a living larva. Bar = 1 mm; (**b**) 30 dpt and (**c**) 50 dpt in a dissected larva with FITC-labeled (green) endogenous PGCs. Bar = 200 and 500 µm, respectively (Pšenička et al. 2015)

not possible to identify germ stem cells in sturgeons because of lack of the functional germ stem cell–specific marker.

It can be concluded that the technic for generation of germ-line chimera in sturgeon has been developed, when a juvenile is used as the donor. The effective method of GSC isolation from adult sturgeon individuals will be issue of future investigation. The gonadal tissue suitable for transplantation can be isolated from a critically endangered sturgeon species also by a surgery or an operation or when the individual is just dying from any reason. One 4-year-old Siberian sturgeon can provide about one million Vasa-positive germ cells, which is sufficient for the transplantation of approximately 5000 larvae.

16.6 Cryopreservation of Early Stages of Germ Cells in Sturgeon, Comparison Between Whole Tissue and Dissociated Cell Freezing (Pšenička et al. 2016)

We believe that GSC cryopreservation has a high potential in conservation biology. So far, it is not possible to successfully cryopreserve sturgeon eggs or embryos, and their cryobanking is based only on sperm cryopreservation. However, this strategy does not involve conservation of specific maternal genetic material such as mtDNA or a female-specific sex chromosome in the case of species with female heterogametic sex determination system. GSC-based cryobanking could involve both maternal and paternal genetic materials. In addition, spermatozoa cryopreservation can provide only one possible combination of genes from an individual male, whereas a single GSC will replicate and undergo recombination of genetic information in gametogenesis during the whole life of the fish, providing much wider genetic variability.

Yoshizaki et al. (2011) developed cryopreservation and transplantation of dissociated rainbow trout spermatogonia into peritoneal cavity of masu salmon and obtained both functional sperm and eggs derived from the donor. Lee et al. (2013, 2015) used frozen/thawed whole testis and frozen/thawed whole body of rainbow trout for regeneration of spermatogonia. This study compared different approaches such as cryopreservation of whole gonadal tissue (before dissociation) vs. gonadal cell suspension (after dissociation) and different cryoprotectants for Siberian sturgeon testes and ovary. The viability and number of cells were evaluated after thawing, and, in addition, the germ cells cryopreserved by the selected method were transplanted into peritoneal cavity of sterlet larvae to prove their ability to develop and proliferate in the recipient. Based on the results, the most suitable cryopreservation procedure for sturgeon early germ cells is surprisingly freezing of whole tissue from 10 to −80 °C at a cooling rate of 1 °C per minute followed by plunging into liquid nitrogen with extender based on PBS containing 0.5% BSA and 50 mM glucose with 1.5 M ethylene glycol as a cryoprotectant. The viability of cells after cryopreservation was about 60%. Moreover, the dead cells in the frozen/thawed tissue were digested by trypsin during the dissociation process, and the final suspension contained mostly only intact cells (over 90%). After intraperitoneal injection of the cryopreserved spermatogonia and oogonia, they started proliferation in 60% of sterlet larvae, which corresponds with results obtained after transplantation of fresh gonial cells.

This chapter provides a powerful tool for the conservation of sturgeons and can help with synchronization of donor and host stage.

If the technology based on cryopreservation and transplantation of GSCs should be applied in practice, it is necessary to introduce it to sturgeon farms, which possess the valuable sturgeon species. However, farmers usually do not use a programmable freezer neither liquid nitrogen. Also a transfer of frozen tissue between a farm and a laboratory or between two laboratories in liquid nitrogen is complicated. Therefore, it was necessary to develop a GSC cryopreservation and storage method in a more available condition. As it was described in a previous study (Pšenička et al. 2016), GSCs can be cryopreserved as whole tissue, which is advantageous for the work in field conditions. Then the cell dissociation, isolation, and transplantation can be performed in the laboratory conditions after thawing of frozen tissue. The tissue should be chilled from 10 to −80 °C at a cooling rate 1 °C per minute and then transferred and stored in liquid nitrogen. In the field conditions, the tissue can be frozen in a deep freezer or in a thermo-isolated box with dried ice, which is nearly −80 °C. The cooling rate can be achieved by a simple and available cell freezing container (e.g., CoolCell®). The present chapter describes if and for how long it is possible to keep the frozen gonadal tissue at −80 °C, without an adverse effect. It was found that 2 days storage at −80 °C has no significant effect on cell gain and viability from frozen/thawed gonadal tissue, while two- to threefold decline was observed after 7 days storage at −80 °C. It was concluded that the tissue can be chilled and kept for 2 days at −80 °C without any adverse effect and after that either used (thaw, dissociate and transplant) or transferred and stored in liquid nitrogen (approximately −196 °C) for a long time.

16.7 Conclusions

The development, labeling, isolation, transplantation, and cryopreservation of sturgeon early germ cells, namely PGC, SG, and OG, were introduced in the present chapter.

The origin and specification of sturgeon PGCs are rather similar to anuran amphibians than teleostean fishes. The PGCs are determined by maternally inherited germ plasm, which is specifically localized in the vegetal pole of oocyte. This finding could help us to develop original PGC labeling method by the injection of general cell tracer into the specific area of sturgeon oocyte. It was also confirmed that PGC migration mechanisms are conserved between chondrostean and teleostean, since PGC isolated from sturgeon and transplanted into goldfish embryo at blastula stage was capable to migrate into genital ridge.

The second part of the present work was dedicated to manipulation with gonial cells from juvenile sturgeons. Dissociation, purification, and transplantation of spermatogonia and oogonia were elaborated. This study enabled generation of germ-line chimera in sturgeons.

Next, sterilization of sturgeon recipient (sterlet) using *dead end* knockdown was developed, which provided us the last piece of puzzle to complete the surrogate production technology in sturgeons.

Cryopreservation is probably the most useful application in GSC technology, since it is the only possibility for the conservation of specific maternal genetic information in vitro not only for sturgeons, but for all bony fishes. Therefore, the cryopreservation of sturgeon early germ cells was elaborated. Moreover, it was shown that the cryopreserved germ cells were capable of transplantation after thawing, and the transplanted Siberian sturgeon germ cells could proliferate in a different species, sterlet.

In the last part of the present study, it was shown how to do cryopreservation of sturgeon GSCs practically in field conditions by using temperature around −80 °C, which can be obtained by either dried ice or a deep freezer. The gonadal tissue, either testis or ovary, can be frozen and transferred in the particular condition up to 2 days without any significant loss of cell viability. Afterwards the tissue can be placed in liquid nitrogen for long-term storage or thaw for a direct usage, usually transplantation.

As the final words, it can be concluded that the technique for the generation of germ-line chimera in sturgeons was developed. *dnd1* MO-treated sterlet with reproduction cycle about 4 years and body weight about 2 kg can be used as a surrogate parent for critically endangered sturgeon species with a long reproductive cycle (about 20 years) and a high body weight (above 100 kg) such as beluga or European sturgeon (*Acipenser sturio*).

16.8 Future Directions

It is evident from the work summarized above that the manipulation with germ cells in order to produce germ-line chimera is relatively complicated and require practice. Future work will be focused on simplification and transfer of the technology to practice. Sterilization of recipients for germ cells in sturgeon is one of the limiting steps. Linhartová et al. (2015) developed sterilization method using injection of antisense morpholino oligonucleotide knocking down *dead end* gene. However, this technique required injection into embryos from two- to eight-cell stage, which last only a few hour depending on temperature. Therefore, it would be advantageous to develop a sterilization method of sturgeon recipients without the demanding manipulation. One possibility can be using so-called vivo morpholino. Wong and Zohar (2015) successfully produced sterile zebrafish by a bath-immersion *dead end* knockdown approach. This method will be tested on sturgeons in next experiments. In addition, different sturgeon hybrids and triploids will be tested for their fertility/sterility, tolerance, and precocity, although the testing requires long time. Alternatively, induction of sterility of adult sterlet male using busulfan will be tested. This technique could provide a tool for fast production of sperm from SG isolated even from very young juvenile. For example, SG could be isolated from 2-year-old Russian sturgeon (*Acipenser gueldenstaedtii*) and transplanted into busulfan-treated adult sterlet males during summer. After resorption of endogenous sterlet germ cells, the Russian sturgeon SG can be injected directly into testis using a syringe. Theoretically the recipient should produce donor derived sperm already next spring after a completion of spermatogenic cycle. For comparison, Russian sturgeon matures about from 11 to 16 years (Chebanov and Galich 2011), but we could obtain mature sperm in 3 years. The sperm can be easily recognized and possibly sorted according to their differences in ploidy level (sterlet $2n$, Russian sturgeon $4n$).

Another problem is the efficiency of GSCs isolation techniques in fish, which is still low and deserve a special attention. Therefore, development of new fish GSC biomarkers and isolation methods, such us cell sorting methods based on cell density or specific cell surface proteins, will also be the issue of the further research.

In vitro culture of fish GSCs is still a challenging topic although it offers several promising advantages. Fan et al. (2008) isolated and cultured PGCs of a transgenic zebrafish line. The cells proliferated in culture with various growth factors. Culture conditions for the maintenance of the survival and mitotic activity of rainbow trout type A spermatogonia were established as well (Shikina et al. 2008). After transplantation, the cultured spermatogonia were colonized and proliferated in recipient gonads. The application of in vitro sturgeon GSC culture together with surrogate production technology can be a powerful tool in conservation genetics of these unique and vanishing species.

Another future direction will be development of somatic cell nuclear transfer in sturgeons. Our research team already managed artificial reproduction and manipulation with sturgeon gametes and germ cells. Using the technique of nuclear transfer, it could be possible to restore an individual genetic information collected noninvasively from a small piece of fin tissue (Fatira et al. 2018), which could be a huge advantage especially in the case of critically endangered species.

Acknowledgment This study has been financially supported by the Ministry of Education, Youth and Sports of the Czech Republic-projects CENAKVA (LM2018099), project Biodiversity (CZ.02.1.01/0.0/0.0/16_025/0007370), and the Czech Science Foundation (grant number 16-02407Y).

References

Bellaiche J, Lareyre J-J, Cauty C, Yano A, Allemand I, Le Gac F (2014) Spermatogonial stem cell quest: nanos2, marker of a subpopulation of undifferentiated a spermatogonia in trout testis. Biol Reprod 90:79. https://doi.org/10.1095/biolreprod.113.116392

Bellve AR, Cavicchia JC, Millette CF, O'Brien DA, Bhatnagar YM, Dym M (1977) Spermatogenic cells of the prepubertal mouse. Isolation and morphological characterization. J Cell Biol 74:68–85. https://doi.org/10.1083/jcb.74.1.68

Blaser H, Eisenbeiss S, Neumann M, Reichman-Fried M, Thisse B, Thisse C, Raz E (2005) Transition from non-motile behaviour to directed migration during early PGC development in zebrafish. J Cell Sci 118:4027–4038. https://doi.org/10.1242/jcs.02522

Bontems F, Stein A, Marlow F, Lyautey J, Gupta T, Mullins MC, Dosch R (2009) Bucky ball organizes germ plasm assembly in zebrafish. Curr Biol 19:414–422. https://doi.org/10.1016/j.cub.2009.01.038

Brinster RL, Zimmermann JW (1994) Spermatogenesis following male germ-cell transplantation. Proc Natl Acad Sci U S A 91:11298–11302. https://doi.org/10.1073/pnas.91.24.11298

Chebanov MS, Galich EV (2011) Sturgeon hatchery manual, FAO fisheries and aquaculture technical paper 558. Food and Agriculture Organisation of the United Nations, Ankara

Doitsidou M, Reichman-Fried M, Stebler J, Koprunner M, Dorries J, Meyer D, Esguerra CV, Leung T, Raz E (2002) Guidance of primordial germ cell migration by the chemokine SDF-1. Cell 111:647–659. https://doi.org/10.1016/S0092-8674(02)01135-2

Eddy EM (1976) Germ plasm and the differentiation of the germ cell line. Int Rev Cytol 43:229–280. https://doi.org/10.1016/S0074-7696(08)60070-4

Ewen-Campen B, Schwager EE, Extavour CGM (2010) The molecular machinery of germ line specification. Mol Reprod Dev 77:3–18. https://doi.org/10.1002/mrd.21091

Extavour CG, Akam M (2003) Mechanisms of germ cell specification across the metazoans: epigenesis and preformation. Development 130:5869–5884. https://doi.org/10.1242/dev.00804

Fan L, Moon J, Wong T-T, Crodian J, Collodi P (2008) Zebrafish primordial germ cell cultures derived from vasa: RFP transgenic embryos. Stem Cells Dev 17:585–597. https://doi.org/10.1089/scd.2007.0178

Fatira E, Havelka M, Labbé C, Depincé A, Iegorova V, Pšenička M, Saito T (2018) Application of interspecific somatic cell nuclear transfer (iSCNT) in sturgeons and an unexpectedly produced gynogenetic sterlet with homozygous quadruple haploid. Sci Rep 8(1):5997

Flynn SR, Matsuoka M, Reith M, Martin-Robichaud DJ, Benfey TJ (2006) Gynogenesis and sex determination in shortnose sturgeon, *Acipenser brevirostrum* Lesuere. Aquaculture 253:721–727. https://doi.org/10.1016/j.aquaculture.2005.09.016

Havelka M, Hulák M, Ráb P, Rábová M, Lieckfeldt D, Ludwig A, Rodina M, Gela D, Pšenička M, Bytyutskyy D, Flajšhans M (2014) Fertility of a spontaneous hexaploid male Siberian sturgeon, *Acipenser baerii*. BMC Genet 15:5. https://doi.org/10.1186/1471-2156-15-5

Ihssen PE, McKay LR, McMillan I, Phillips RB (1990) Ploidy manipulation and gynogenesis in fishes: cytogenetic and fisheries applications. Trans Am Fish Soc 119:698–717. https://doi.org/10.1577/1548-8659(1990)119<0698:PMAGIF>2.3.CO;2

Ikenishi K, Kotani M, Tanabe K (1974) Ultrastructural changes associated with UV irradiation in the "germinal plasm" of Xenopus laevis. Dev Biol 36:155–168. https://doi.org/10.1016/0012-1606(74)90198-5

Iwamoto T, Hiraku Y, Oikawa S, Mizutani H, Kojima M, Kawanishi S (2004) DNA intrastrand cross-link at the 5′-GA-3′ sequence formed by busulfan and its role in the cytotoxic effect. Cancer Sci 95:454–458. https://doi.org/10.1111/j.1349-7006.2004.tb03231.x

Kawakami Y, Goto-Kazeto R, Saito T, Fujimoto T, Higaki S, Takahashi Y, Arai K, Yamaha E (2010) Generation of germ-line chimera zebrafish using primordial germ cells isolated from cultured blastomeres and cryopreserved embryoids. Int J Dev Biol 54:1491–1499. https://doi.org/10.1387/ijdb.093059yk

Kobayashi T, Takeuchi Y, Takeuchi T, Yoshizaki G (2007) Generation of viable fish from cryopreserved primordial germ cells. Mol Reprod Dev 74:207–213. https://doi.org/10.1002/mrd

Koprunner M, Thisse C, Thisse B, Raz E (2001) A zebrafish nanos-related gene is essential for the development of primordial germ cells. Genes Dev 15:2877–2885. https://doi.org/10.1101/gad.212401

Kurokawa H, Aoki Y, Nakamura S, Ebe Y, Kobayashi D, Tanaka M (2006) Time-lapse analysis reveals different modes of primordial germ cell migration in the medaka Oryzias latipes. Develop Growth Differ 48:209–221. https://doi.org/10.1111/j.1440-169X.2006.00858.x

Lacerda S, Batlouni S, Silva S, Homem C, França L (2006) Germ cells transplantation in fish: the Nile-tilapia model. Anim Reprod 3:146–159
Lacerda SM, Maria S, Costa GMJ, Da Silva Mde A, Almeida Campos-Junior PH, Segatelli TM, Peixoto MTD, Resende RR, De França LR (2013) Phenotypic characterization and in vitro propagation and transplantation of the Nile tilapia (Oreochromis niloticus) spermatogonial stem cells. Gen Comp Endocrinol 192:95–106. https://doi.org/10.1016/j.ygcen.2013.06.013
Leal MC, Cardoso ER, Nóbrega RH, Batlouni SR, Bogerd J, França LR, Schulz RW (2009) Histological and stereological evaluation of zebrafish (Danio rerio) spermatogenesis with an emphasis on spermatogonial generations. Biol Reprod 81:177–187. https://doi.org/10.1095/biolreprod.109.076299
Lee S, Iwasaki Y, Shikina S, Yoshizaki G (2013) Generation of functional eggs and sperm from cryopreserved whole testes. Proc Natl Acad Sci U S A 110:1640–1645. https://doi.org/10.1073/pnas.1218468110
Lee S, Seki S, Katayama N, Yoshizaki G (2015) Production of viable trout offspring derived from frozen whole fish. Sci Rep 5:16045. https://doi.org/10.1038/srep16045
Lin S, Long W, Chen J, Hopkins N (1992) Production of germ-line chimeras in zebrafish by cell transplants from genetically pigmented to albino embryos. Proc Natl Acad Sci U S A 89:4519–4523. https://doi.org/10.1073/pnas.89.10.4519
Linhartová Z, Saito T, Kašpar V, Rodina M, Prášková E, Hagihara S, Pšenička M (2015) Sterilization of sterlet Acipenser ruthenus by using knockdown agent, antisense morpholino oligonucleotide, against dead end gene. Theriogenology 84(7):1246–1255
Maxime V (2008) The physiology of triploid fish: current knowledge and comparisons with diploid fish. Fish Fish 9:67–78. https://doi.org/10.1111/j.1467-2979.2007.00269.x
Mishima Y (2012) Widespread roles of microRNAs during zebrafish development and beyond. Develop Growth Differ 54:55–65. https://doi.org/10.1111/j.1440-169X.2011.01306.x
Mishima Y, Giraldez AJ, Takeda Y, Fujiwara T, Sakamoto H, Schier AF, Inoue K (2006) Differential regulation of germline mRNAs in soma and germ cells by zebrafish miR-430. Curr Biol 16:2135–2142. https://doi.org/10.1016/j.cub.2006.08.086
Nagasawa K, Shikina S, Takeuchi Y, Yoshizaki G (2010) Lymphocyte antigen 75 (Ly75/CD205) is a surface marker on mitotic germ cells in rainbow trout. Biol Reprod 83:597–606. https://doi.org/10.1095/biolreprod.109.082081
Nagasawa K, Fernandes JMO, Yoshizaki G, Miwa M, Babiak I (2013) Identification and migration of primordial germ cells in Atlantic salmon, Salmo salar: characterization of vasa, dead end, and lymphocyte antigen 75 genes. Mol Reprod Dev 80:118–131. https://doi.org/10.1002/mrd.22142
Nóbrega RH, Greebe CD, van de Kant H, Bogerd J, de França LR, Schulz RW (2010) Spermatogonial stem cell niche and spermatogonial stem cell transplantation in zebrafish. PLoS One 5:1–16. https://doi.org/10.1371/journal.pone.0012808
Okutsu T, Suzuki K, Takeuchi Y, Takeuchi T, Yoshizaki G (2006) Testicular germ cells can colonize sexually undifferentiated embryonic gonad and produce functional eggs in fish. Proc Natl Acad Sci U S A 103:2725–2729. https://doi.org/10.1073/pnas.0509218103
Okutsu T, Shikina S, Kanno M, Takeuchi Y, Yoshizaki G (2007) Production of trout offspring from triploid salmon parents. Science 317(5844):1517
Olsen LC, Aasland R, Fjose A (1997) A vasa-like gene in zebrafish identifies putative primordial germ cells. Mech Dev 66:95–105. https://doi.org/10.1016/S0925-4773(97)00099-3
Omoto N, Maebayashi M, Adachi S, Arai K, Yamauchi K (2005) Sex ratios of triploids and gynogenetic diploids induced in the hybrid sturgeon, the bester (Huso huso female x *Acipenser ruthenus* male). Aquaculture 245:39–47. https://doi.org/10.1016/j.aquaculture.2004.12.004
Panda RP, Barman HK, Mohapatra C (2011) Isolation of enriched carp spermatogonial stem cells from Labeo rohita testis for in vitro propagation. Theriogenology 76:241–251. https://doi.org/10.1016/j.theriogenology.2011.01.031
Piferrer F, Beaumont A, Falguière JC, Flajšhans M, Haffray P, Colombo L (2009) Polyploid fish and shellfish: production, biology and applications to aquaculture for performance

improvement and genetic containment. Aquaculture 293:125–156. https://doi.org/10.1016/j.aquaculture.2009.04.036

Pocherniaieva K, Sidova M, Havelka M, Saito T, Psenicka M, Sindelka R, Kaspar V (2018) Comparison of oocyte mRNA localization patterns in sterlet *Acipenser ruthenus* and African clawed frog *Xenopus laevis*. J Exp Zool B Mol Dev Evol 330(3):181–187

Pšenička M, Saito T, Linhartová Z, Gazo I (2015) Isolation and transplantation of sturgeon early-stage germ cells. Theriogenology 83:1085–1092. https://doi.org/10.1016/j.theriogenology.2014.12.010

Pšenička M, Saito T, Rodina M, Dzyuba B (2016) Cryopreservation of early stage Siberian sturgeon *Acipenser baerii* germ cells, comparison of whole tissue and dissociated cells. Cryobiology 72(2):119–122. https://doi.org/10.1016/j.cryobiol.2016.02.005

Saito T, Pšenička M (2015) Novel technique for visualizing primordial germ cells in sturgeons (*Acipenser ruthenus, A. gueldenstaedtii, A. baerii, Huso huso*). Biol Reprod 93(4):96. https://doi.org/10.1095/biolreprod.115.128314

Saito T, Fujimoto T, Maegawa S, Inoue K, Tanaka M, Arai K, Yamaha E (2006) Visualization of primordial germ cells in vivo using GFP-nos1 3′UTR mRNA. Int J Dev Biol 50:691–700. https://doi.org/10.1387/ijdb.062143ts

Saito T, Goto-Kazeto R, Arai K, Yamaha E (2008) Xenogenesis in teleost fish through generation of germ-line chimeras by single primordial germ cell transplantation. Biol Reprod 78:159–166. https://doi.org/10.1095/biolreprod.107.060038

Saito T, Goto-Kazeto R, Fujimoto T, Kawakami Y, Arai K, Yamaha E (2010) Inter-species transplantation and migration of primordial germ cells in cyprinid fish. Int J Dev Biol 54:1479–1484. https://doi.org/10.1387/ijdb.103111ts

Saito T, Goto-Kazeto R, Kawakami Y, Nomura K, Tanaka H, Adachi S, Arai K, Yamaha E (2011) The mechanism for primordial germ-cell migration is conserved between Japanese eel and zebrafish. PLoS One 6:1–8. https://doi.org/10.1371/journal.pone.0024460

Saito T, Pšenička M, Goto R, Adachi S, Inoue K, Arai K, Yamaha E (2014) The origin and migration of primordial germ cells in sturgeons. PLoS One 9(2):e86861. https://doi.org/10.1371/journal.pone.0086861

Seydoux G, Dunn MA (1997) Transcriptionally repressed germ cells lack a subpopulation of phosphorylated RNA polymerase II in early embryos of Caenorhabditis elegans and Drosophila melanogaster. Development 124:2191–2201

Seydoux G, Mello CC, Pettitt J, Wood WB, Priess JR, Fire A (1996) Repression of gene expression in the embryonic germ lineage of C. elegans. Nature 382:713–716. https://doi.org/10.1038/382713a0

Shikina S, Ihara S, Yoshizaki G (2008) Culture conditions for maintaining the survival and mitotic activity of rainbow trout transplantable type a spermatogonia. Mol Reprod Dev 75:529–537

Sutasurja LA, Nieuwkoop PD (1974) The induction of the primordial germ cells in the urodeles. Dev Genes Evol 175:199–220. https://doi.org/10.1007/BF00582092

Takeuchi Y, Yoshizaki G, Takeuchi T (2001) Production of germ-line chimeras in rainbow trout by blastomere transplantation. Mol Reprod Dev 59:380–389. https://doi.org/10.1002/mrd.1044

Takeuchi Y, Yoshizaki G, Takeuchi T (2003) Generation of live fry from intraperitoneally transplanted primordial germ cells in rainbow trout. Biol Reprod 69:1142–1149. https://doi.org/10.1095/biolreprod.103.017624

Takeuchi Y, Yoshizaki G, Takeuchi T (2004) Biotechnology: surrogate broodstock produces salmonids. Nature 430:629–630. https://doi.org/10.1038/430629a

Von Schönfeldt V, Krishnamurthy H, Foppiani L, Schlatt S (1999) Magnetic cell sorting is a fast and effective method of enriching viable spermatogonia from djungarian hamster, mouse, and marmoset monkey testes. Biol Reprod 61:582–589

Weidinger G, Stebler J, Slanchev K, Dumstrei K, Wise C, Lovell-Badge R, Thisse C, Thisse B, Raz E (2003) Dead end, a novel vertebrate germ plasm component, is required for zebrafish primordial germ cell migration and survival. Curr Biol 13:1429–1434. https://doi.org/10.1016/S0960-9822(03)00537-2

Whitington PM, Dixon KE (1975) Quantitative studies of germ plasm and germ cells during early embryogenesis of Xenopus laevis. J Embryol Exp Morphol 33:57–74
Williamson A, Lehmann R (1996) Germ cell development in Drosophila. Annu Rev Cell Dev Biol 12:365–391. https://doi.org/10.1146/annurev.cellbio.12.1.365
Wong T-T, Zohar Y (2015) Production of reproductively sterile fish by a non-transgenic gene silencing technology. Sci Rep 221:1–6. https://doi.org/10.1016/j.ygcen.2014.12.012
Wong T-T, Saito T, Crodian J, Collodi P (2011) Zebrafish germline chimeras produced by transplantation of ovarian germ cells into sterile host larvae. Biol Reprod 84:1190–1197. https://doi.org/10.1095/biolreprod.110.088427
Yamaha E, Kazama-Wakabayashi M, Otani S, Fujimoto T, Arai K (2001) Germ-line chimera by lower-part blastoderm transplantation between diploid goldfish and triploid crucian carp. Genetica 111:227–236. https://doi.org/10.1023/A:1013780423986
Yamaha E, Saito T, Goto-Kazeto R, Arai K (2007) Developmental biotechnology for aquaculture, with special reference to surrogate production in teleost fishes. J Sea Res 58:8–22. https://doi.org/10.1016/j.seares.2007.02.003
Yang X, Yue H, Ye H, Li C, Wei Q (2015) Identification of a germ cell marker gene, the dead end homologue, in Chinese sturgeon *Acipenser sinensis*. Gene 558:118–125. https://doi.org/10.1016/j.gene.2014.12.059
Yasui GS, Fujimoto T, Sakao S, Yamaha E, Arai K (2011) Production of loach (Misgurnus anguillicaudatus) germ-line chimera using transplantation of primordial germ cells isolated from cryopreserved blastomeres. J Anim Sci 89:2380–2388. https://doi.org/10.2527/jas.2010-3633
Yoon C, Kawakami K, Hopkins N (1997) Zebrafish vasa homologue RNA is localized to the cleavage planes of 2- and 4-cell-stage embryos and is expressed in the primordial germ cells. Development 124(16):3157–3165
Yoshizaki G, Ichikawa M, Hayashi M, Iwasaki Y, Miwa M, Shikina S, Okutsu T (2010) Sexual plasticity of ovarian germ cells in rainbow trout. Development 137:1227–1230. https://doi.org/10.1242/dev.051821
Yoshizaki G, Fujinuma K, Iwasaki Y, Okutsu T, Shikina S, Yazawa R, Takeuchi Y (2011) Spermatogonial transplantation in fish: a novel method for the preservation of genetic resources. Comp Biochem Physiol Part D Genomics Proteomics 6:55–63. https://doi.org/10.1016/j.cbd.2010.05.003

Chapter 17
Intraperitoneal Germ Cell Transplantation Technique in Marine Teleosts

Yutaka Takeuchi, Ryosuke Yazawa, and Goro Yoshizaki

Abstract Transplantation of pre-meiotic germ cells, harvested from juvenile or adult fish, into the body cavity of allogeneic or xenogeneic recipients sterilized by triploidization or endogenous germ cell depletion, results in their migration into genital ridges where they are eventually incorporated. These germ cells initiate either spermatogenesis or oogenesis in the recipient gonads, depending on the sex of the recipient. Furthermore, by mating male and female germ-cell-transplanted recipients, viable offspring are derived. Although this technology, called the "surrogate broodstock technology," was established using salmonid fish, we found that it is applicable to a wide variety of fish species including commercially valuable marine teleosts. Thus, we believe that this method will prove to be a novel fish breeding technology for preserving the valuable genetic resources of both aquaculture and endangered fish species. This chapter describes our recent findings regarding the improvements in germ cell manipulation and transplantation in marine teleosts.

Keywords Primordial germ cells · Spermatogonia · Oogonia · Cryopreservation · Fish · Allotransplantation · Xenotransplantation

Y. Takeuchi (✉)
Faculty of Biological Science and Technology, Noto Center for Fisheries Science and Technology, Institute of Science and Engineering, Kanazawa University, Kanazawa, Ishikawa, Japan
e-mail: yutaka@se.kanazawa-u.ac.jp

R. Yazawa · G. Yoshizaki
Department of Marine Bioscience, Tokyo University of Marine Science and Technology, Tokyo, Japan

M. Yoshida, J. F. Asturiano (eds.), *Reproduction in Aquatic Animals*,
https://doi.org/10.1007/978-981-15-2290-1_18

17.1 Introduction

Germ cell transplantation (GCT) is a powerful assisted reproductive technology for the conservation and propagation of elite livestock or endangered wild animals (Ogawa 2001; Dobrinski 2007). It also provides a useful cell transplantation assay system for the functional characterization of germline stem cells (Yoshida 2009). In teleost fish, three different types of GCT techniques have been successfully established to generate donor-derived gametes by transplanting primordial germ cells (PGC), spermatogonia, or oogonia into recipients at various developmental stages, such as (1) embryonic stage (blastula), (2) hatchling stage, and (3) sexually competent adult stage (Lacerda et al. 2013). Development of GCT in fish has created a new avenue of research in the reproductive biology of fish germline stem cells and presents a variety of promising biotechnological applications in the breeding and aquaculture production of commercially valuable fish, preservation of valuable endangered species, and generation of transgenic fish (Okutsu et al. 2006a; Yoshizaki et al. 2012; Yoshizaki and Lee 2018; de Siqueira-Silva et al. 2018).

17.1.1 Germ Cells Can Be Transplanted into the Body Cavity of Newly Hatched Fish Embryos

Intraperitoneal GCT, one of the three GCT techniques mentioned above, was initially established by Takeuchi et al. in salmonids, using allogenic and xenogenic donor/recipient combinations, in 2003 and 2004, respectively (Takeuchi et al. 2003, 2004). It was first applied to the pelagic egg-spawning marine teleost, Nibe croaker (*Nibea mitsukurii*, Perciformes, Sciaenidae) (also known as blue drum) in 2009 (Takeuchi et al. 2009). As shown in Fig. 17.1, the isolated donor PGCs obtained from the rainbow trout (*Oncorhynchus mykiss*) were transplanted into the peritoneal cavities of newly hatched larvae of the allogenic rainbow trout or masu salmon (*Oncorhynchus masou*), using an oil microinjector attached to a micromanipulator (Takeuchi et al. 2004). Intraperitoneally transplanted donor PGCs first attached to the peritoneal wall and then migrated to the recipient genital ridges by extending their pseudopods (Takeuchi et al. 2003). Chemotactic guidance cues secreted from recipient genital ridges and/or extracellular matrix distribution on the migratory route of the endogenous PGCs could support the active migration of donor PGCs towards the recipient genital ridges. In addition, lack of a functional immune system as indicated by the absence of differentiation in both the thymus and T cells at the hatching stage (Manning and Nakanishi 1996) would be key for the survival of exogenous donor-derived germ cells at least until the donor PGCs were incorporated into recipient gonads. These hatchlings upon reaching adulthood produced donor-derived functional gametes and offspring. Moreover, the approach was successfully applied to the trout model wherein spermatogonial and oogonial stem cells from the rainbow trout were transplanted into sterile triploid rainbow trout or masu

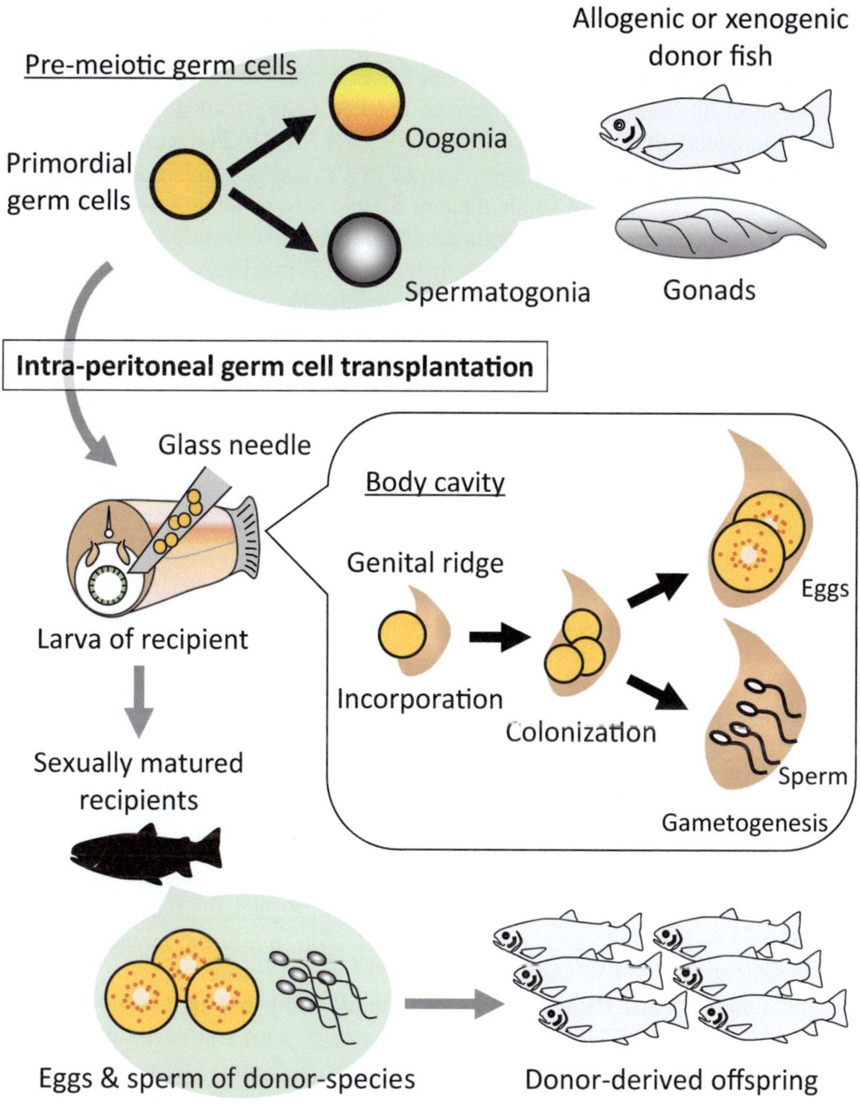

Fig. 17.1 Principle for the production of eggs and sperm of allogenic or xenogenic donor fish in surrogate recipients by intraperitoneal germ cell transplantation

salmon hatchlings, resulting in the establishment of donor-derived gametogenesis in the male and female recipients (Okutsu et al. 2006b, 2007; Yoshizaki et al. 2010a). In addition, the fact that spermatogonia and oogonia transplanted into newly hatched embryos differentiated into either sperm or eggs depending on the sex of the recipient indicated that these gonial germ cells included a cell population that possessed sexual bipotency and could be used in place of the PGCs described above. Because fish embryos possess an extremely small number (approximately 20–100) of PGCs at the hatching stage (Braat et al. 1999; Yoshizaki et al. 2002) and only 10–30 PGCs per donor fish can be isolated through enzymatic treatment (Takeuchi et al. 2003; Kobayashi et al. 2004), large numbers of hatched embryos of donor species are required to isolate PGCs for transplantation (e.g., at least 10 PGCs are required for transplantation into a single recipient). Unlike PGCs, a large number of gonial germ cells can be obtained from a single male or female fish, which allows the isolation of donor germ cells even from commercially valuable marine teleosts, such as the bluefin tuna (*Thunnus orientalis*). Thus, in terms of the practical application of GCT in marine fishes, it was an epoch-making discovery that gonial (pre-meiotic) germ cells that can be found abundantly in the sexually immature gonads of males and females could be adopted as donor germ cells.

As we were unable to directly transplant donor germ cells into the gonads of newly hatched recipients and isolate a large number of PGCs from fish embryos, the development of an intraperitoneal GCT technique using gonial germ cells from the donor was a major breakthrough in our series of studies on salmonids (Yoshizaki et al. 2012). The intraperitoneal GCT technique allows the production of both male and female gametes of the donor species in the gonads of recipient species; this technique was used worldwide by the name of "surrogate broodstock technology" and was applied to several freshwater including endangered species (zebrafish (Wong et al. 2011), tilapias (Farlora et al. 2014), medaka (Seki et al. 2017), sturgeons (Pšenička et al. 2015; Ye et al. 2017), bitterlings (Octavera and Yoshizaki, unpublished)) and commercially valuable marine teleosts (croakers (Takeuchi et al. 2009; Higuchi et al. 2011; Yoshikawa et al. 2017, 2018a), mackerels (Yazawa et al. 2010), yellowtails (Morita et al. 2012, 2015), pufferfish (Hamasaki et al. 2017; Yoshikawa et al. 2018b), tunas (Bar et al. 2015), and flatfish (Pacchiarini et al. 2014)). For this technique to be successful in marine teleosts, the combination and use of several key biological and technical properties of germ cells and the gonadal development process of target species was required. Especially, most marine teleosts spawn small pelagic eggs (diameter, 0.6–1 mm), and these larvae are a few millimeters in total length and have poorly developed sensory systems, skeleton, and musculature. Thus, their larvae are very fragile compared to the newly hatched larvae of salmonids which are large (total length, 15 mm) and have tolerance for handling because of their large egg size (diameter, 5–8 mm) and well-developed external and internal organs. In the present article, we introduce these properties while outlining how we applied the technology to commercially valuable marine teleosts including the small pelagic egg-spawning Perciformes.

17.1.2 Biotechnological Applications of Intraperitoneal GCT in Marine Fish

Marine fish farming for commercial market and stock enhancement has potential for significant growth globally. However, a number of technical impediments including the lack of a consistent supply of juvenile fish, difficulties in broodstock management and spawning induction (Zohar and Mylonas 2001), and the expense of seed production have limited expansion in the past. The potential applications of intraperitoneal GCT techniques in the aquaculture of marine teleosts was first proposed in 2003 (Yoshizaki et al. 2003) and had been partially put into practice for producing marine fish seeds for farming or stocking purposes (Morita et al. 2012, 2015; Hamasaki et al. 2017; Yoshikawa et al. 2018b) (Fig. 17.2).

17.1.2.1 Production of Gametes of Large-Bodied, Commercially Valuable Species (e.g., Tunas, Yellowtails, Groupers) in Closely Related Recipients with Small Body Size and Shorter Generation Time

The bluefin tuna takes 3–5 years to reach sexual maturity, at a body-weight of roughly 50–300 kg. As a result, broodstock management and gamete collection for this species is expensive in terms of time, cost, labor, and space, especially in a land-based tank. However, if bluefin tuna spermatogonia could be transplanted into the Chub mackerel (*Scomber japonicus*), which is a closely related species that reaches maturity in 1–2 years at a body weight of roughly 500 g, bluefin tuna gametes could be produced more easily and rapidly, even in a small fish tank. Indeed, the production of functional donor-derived Japanese yellowtail (*Seriola quinqueradiata*) sperm, using jack mackerel (*Trachurus japonicus*) as "a small-bodied and shorter generation (SS) surrogate" has been demonstrated as the first, successful attempt at the production of xenogenic donor gametes in marine fish (Morita et al. 2015). The production of xenogenic donor derived eggs was first reported in tiger puffer (*Takifugu rubripes*) using the SS-surrogate grass puffer (*Takifugu niphobles*) (Hamasaki et al. 2017). Shortening of the generation period and labor-saving in broodstock management are important requirements to precede the breeding programs of aquaculture-target marine fishes (Fig. 17.2a).

17.1.2.2 Preservation of Elite Breeds of Fish Harboring Desirable Genetic Traits or of Endangered Species, in Combination with Cryo-banking of Donor Germ Cells

Unlike the eggs and embryos of mammals or marine invertebrates, which can be cryopreserved due to their small size and low yolk content, fish eggs are very large (eggs of pelagic egg-spawning marine teleost are about 0.6–1 mm in diameter) and contain large quantities of yolk. Thus, the cryopreservation of fish eggs or embryos

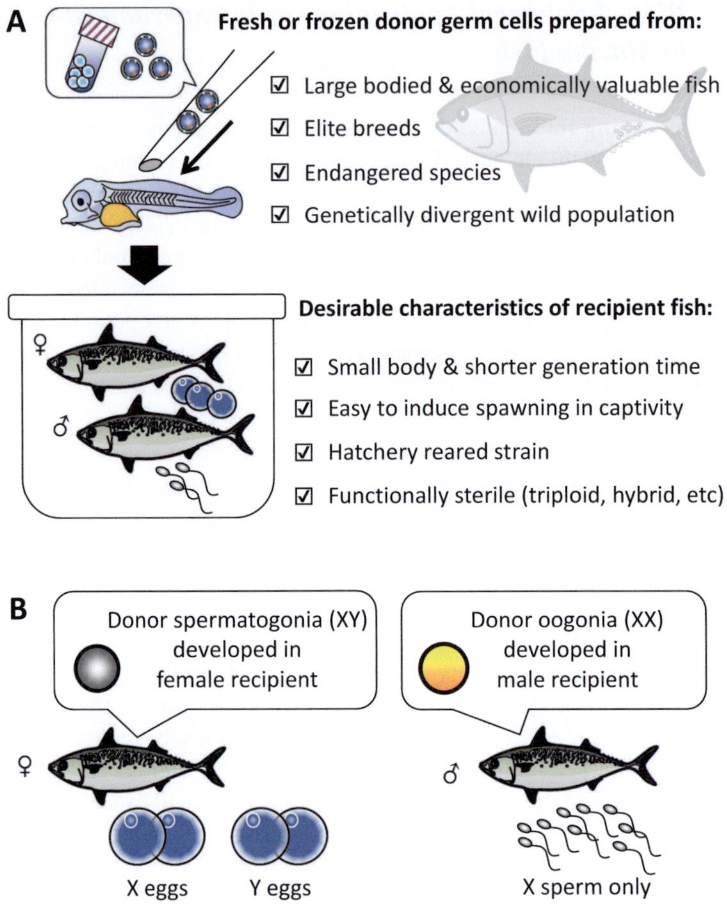

Fig. 17.2 Scheme for novel marine teleost breeding approaches using germ cell transplantation. (**a**) Attributes of potential donor species and desirable characteristics of recipient species. (**b**) Induction of sex-reversal of donor germ cells in the recipient gonads of opposite sex

is still not possible. In salmonids, cryopreservation protocols for PGCs, spermatogonia, and oogonia have been established, and it is possible to revive donor strains or species by transplanting frozen germ cells into recipients to produce the eggs and sperm of the donors, and then to fertilize those eggs and sperm (Okutsu et al. 2006b; Kobayashi et al. 2007; Yoshizaki et al. 2011; Lee et al. 2013, 2016a, b; Yoshizaki and Lee 2018). Recently, production of tiger puffer fish from cryopreserved testicular germ cells using intraperitoneal GCT was published for the first time in marine fish (Yoshikawa et al. 2018b). Notably, transplantable spermatogonia were retrieved from frozen "whole fish" kept for 3 years in a deep freezer at −80 °C (Lee et al. 2015). This implies that in the case of an emergency, such as the sudden death of an economically valuable or endangered fish species at a fish-rearing facility due to disease or accident, the dead fish can be stored in a deep freezer until they can be sent to a laboratory and used as donor for regeneration via GCT (Fig. 17.2a).

17.1.2.3 Production of Mono-sex F1 Offspring by Inducing Sex-Fate Change of Donor Cells in Recipient Gonads

Because of the sexual plasticity of donor germ cells, transdifferentiation of donor spermatogonia into eggs or donor oogonia into sperm can be induced in the recipient gonads following intraperitoneal GCT. In male case of heterogametic fish (XY/XX), spermatogonial cells (XY genotype) developed into functional eggs harboring the Y-chromosome in the recipient ovary and YY-super-males were generated in the donor-derived F1 offspring (Okutsu et al. 2015; Yoshizaki et al. 2010b). Using the sperm of the YY-super male, an all-male population can be obtained. For example, mature testes of the tiger puffer are priced approximately three times higher than their flesh (10,000 Japanese yen per kilogram) as they are a popular ingredient in fish stew or grill. Although the sex of the tiger puffer shows a clear XX/XY system (Kikuchi et al. 2007; Kamiya et al. 2012), the masculinization of tiger puffer using traditional methods such as methyltestosterone treatment (Kakimoto et al. 1994) and aromatase inhibitor treatment (Rashid et al. 2007) is known to be extremely difficult. Therefore, in tiger puffer, the spermatogonial transplantation technique is powerful and novel tool for seed production of all-male populations for aquaculture (Hamasaki et al. 2017). In contrast, oogonial cells (XX genotype) developed to functional sperm only consisted of spermatozoa possessing the X-chromosome in the recipient testis. By using this sperm for fertilization with normal eggs, all-female F1 offspring will be obtained. The production of an all-female population offers an economic advantage for marine fish species due to the higher growth rates of females and as a means of controlling unwanted reproduction (Xu et al. 2018; Budd et al. 2015) (Fig. 17.2b).

17.1.2.4 Production of Fish Seeds with Increased Genetic Diversity for Use in Stock Enhancement Program to Mitigate the Genetic Impact of Released Fish Seeds on Wild Fish Populations

Several studies have suggested that massive releases of hatchery-produced seeds had the potential to alter the genetic structures of local populations (Eldridge and Naish 2007; Kitada et al. 2009). These concerns are based on the fact that hatchery fish populations often exhibit reduced genetic variation caused by the use of relatively small broodstock for seed production and/or inbreeding over multiple generations (Verspoor 1998; Sekino et al. 2002). A mean of producing gametes with a large amount of genetic diversity using only a small number of surrogate recipients has been demonstrated in rainbow trout by using donor germ cells isolated from the testes of different donor individuals and mixed prior to their transplantation into the peritoneal cavity of a single recipient (Sato et al. 2014). The resulting recipient produced gametes derived from several donor individuals, thereby allowing mass production of fish seeds possessing increased genetic diversity (Fig. 17.2a).

17.2 Intraperitoneal GCT in Rotifer-Feeding Stage Marine Fish Larvae

17.2.1 Optimal Developmental Stage of Recipients for Intraperitoneal GCT

Even in marine fish, we predicted that the recipient larvae or juveniles do not have the capacity to reject foreign cells or allografts at least for several weeks because their immune organs are not fully developed until that stage (Zapata et al. 2006). Thus, for the successful migration of intraperitoneally transplanted donor germ cells to the recipient genital ridges, it is important to find the most suitable developmental stage of recipient larvae through detailed histological examination of early gonadal development.

The effect of the recipient body size and age on the incorporation efficiency of donor spermatogonia into the recipient gonads was studied using the Nibe croaker as a marine teleost model (Takeuchi et al. 2009) (Fig. 17.3a). The early gonadal development of total length (TL): 4-mm (12 days post-fertilization [dpf]), rotifer-feeding stage larvae to TL: 12-mm (30 dpf), formula food-feeding juveniles was studied using hematoxylin and eosin (HE)-stained transverse sections of their body cavities (see Fig. 17.2 of Takeuchi et al. 2009). In the 4-mm larvae, endogenous PGCs had settled in the upper part of the peritoneal cavity where the genital ridges form. Genital ridges were first observed in the 6-mm larvae and a single layer of gonadal somatic cells covered the PGCs. Proliferation of PGCs began after they were enclosed in the genital ridges. These data suggested that PGC development in this species passes from the migration phase to the proliferation phase before the larvae reach 6-mm in length. In addition, intraperitoneal GCT in Nibe croaker larvae revealed that no donor cells colonized the genital ridges of 6-mm recipients, whereas donor cells successfully colonized the genital ridges when transplanted into 3- to 5-mm recipients. In conclusion, the migration of endogenous PGCs of Nibe croaker towards their own genital ridges was not completed until the larvae reached around 4-mm in total length, providing an opportunity to transplant donor germ cells near the migratory route of endogenous PGCs. Potential mechanisms explaining the narrow developmental time window for the appropriate recipient are as follows in the peritoneal cavity of 6-mm larvae: (1) chemotactic signals secreted from the gonadal somatic cells during endogenous PGCs migration had ceased, (2) the developed gonadal epithelium blocked the incorporation of donor type-A SG, and (3) niches for germ cells in recipient genital ridges had been already occupied by the proliferating endogenous PGCs. In addition, total body length can be an indicator of the optimal developmental stage of recipient larvae. Although the exact

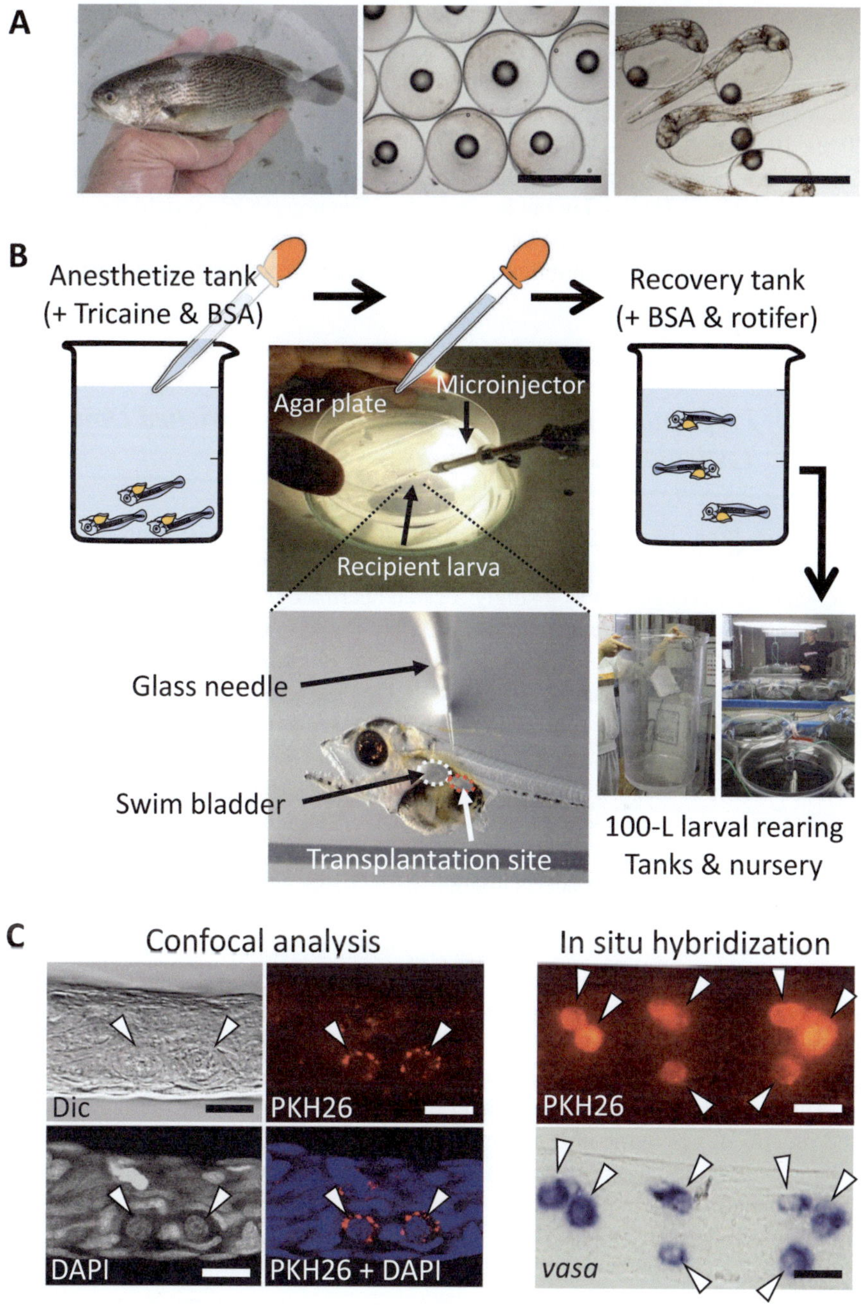

Fig. 17.3 Intraperitoneal germ cell transplantation in pelagic-egg spawning marine teleost. (**a**) Sexually competent adult (left), fertilized eggs (middle), and newly hatched larvae (right) of Nibe croaker *Nibea mitsukurii*. (**b**) Handling, transplantation, and rearing of marine fish larvae. Anesthetized larvae sunk to the bottom of a 1-L beaker are transferred using a pipette onto an agar

timing of genital ridge formation varies in different fish species, the optimal stage of intraperitoneal GCT was determined as the post-swim-bladder-inflated, rotifer-feeding larvae in several marine fish species, e.g., 7 dpf (5.3 mm) in Chub mackerel (Yazawa et al. 2010), 8 days post-hatch (dph) (5.2 mm) in yellowtail (Morita et al. 2012), 10–12 dph (3.9–4.3 mm) in jack mackerel (Morita et al. 2015), and 1–5 dph (2.5–3.3 mm) in grass puffer (Hamasaki et al. 2017). Changes in the morphology of recipient genital ridges and in the number of endogenous PGCs should be carefully monitored to optimize the developmental stage of the recipient for intraperitoneal GCT, because loss of susceptibility to donor germ cell incorporation coincides with these changes.

17.2.2 Preparation of Type-A Spermatogonia Enriched Donor Testicular Cell Suspensions

In rainbow trout, by using GFP-labeled spermatogonia prepared from the p*vasa*-GFP transgenic strain as donor germ cells, only populations of type-A spermatogonia, which are recognized as a singly located spermatogonia with large and round nuclei in the cyst structure of Sertoli cells, were found to contain germ cells having the ability to migrate to and colonize the recipient genital ridges (Yano et al. 2008). Moreover, counting the number of donor-derived gametes produced by a recipient proved that the colonized donor germ cells possessed self-renewal and differentiation capabilities and behaved as germline stem cells in the recipient gonads (Okutsu et al. 2006b; Morita et al. 2012; Yoshikawa et al. 2017). In general, as spermatogenesis progresses, the number of type-B spermatogonia, meiotic germ cells, haploid germ cells, and gonadal somatic cells increases drastically, causing a rapid decline in the percentage of type-A spermatogonia in testicular cell suspensions. In rainbow trout, more than 50% of the cells in suspension need to be type-A spermatogonia for efficient colonization by donor germ cells (Okutsu et al. 2006b). Thus, determination of the most suitable reproductive stage in the testes of seasonal breeding teleosts is important for yielding cell suspensions for successful spermatogonial transplantation. In addition, because the presence of sperm obstructs the enzymatic dissociation of testes, donors with non-spermiogenic testes are preferred.

Autumn siblings of the Nibe croaker males reach maturity in natural conditions at 8 months of age in the next spring. Therefore, pre-spermiogenic testes of 3- and 6-month-old males and post-spawning testis of 16-month-old males were selected and evaluated for their suitability as donors, i.e., through histology for identifying the germ cell types in the testis and enzymatic dissociation of testes for counting the

Fig. 17.3 (continued) plate. A glass needle filled with donor cells was inserted into the body cavity near the posterior end of the swim bladder. (**c**) Confirmation of presence of PKH26-labeled cells in the recipient genital ridges. Germ-cell-specific large and round nuclei were observed in PKH26-labeled cells (arrowheads in confocal images). PKH26-labeled cells expressing the germ cell marker, *vasa* mRNA (arrowheads in in situ images). Bars: 800 μm (**a**), 20 μm (**c**) proportion of type-A spermatogonia in the cell suspension (Takeuchi et al. 2009). Based on the histological studies, spermatogonia that were located close to the basement membrane of the lobule had nuclei larger than 9 μm. Thus, they were counted as type-A spermatogonia. *In situ* hybridization of smear samples revealed that more than 80% of cells with nuclei larger than 9 μm were positive for the germ cell marker, *vasa*. Therefore, the proportions of type-A spermatogonia in the dissociated cells were estimated simply by measuring the nuclear diameters, followed by DAPI staining. Both the proportion and the yield of type-A spermatogonia were significantly high in 3-month-old testis among the tested samples. Testicular cell suspensions of 3-month-old males contained the highest percentage (approximately 40%) of type-A spermatogonia, indicating that 60,000–130,000 spermatogonia could be prepared from 1 mg of 3-month-old testes.

It is often observed that immature gonads of juveniles or young fish of a large-bodied fast-growing fish, such as bluefin tuna, contain large numbers of gonadal somatic cells and a very small number of germ cells (Yazawa et al. 2013). Therefore, in order to enrich type-A spermatogonia, it is necessary to develop a precise molecular marker that can distinguish type-A spermatogonia and a method that can detect and isolate them. Yazawa et al. (2013) reported that the expression of the bluefin tuna homologue of *dead end* (*dnd*) was restricted to PGCs and type-A spermatogonia. Furthermore, the donor cells containing the highest abundance of *dnd*-positive cells were the most effectively colonized in chub mackerel recipient gonads, suggesting that transplantable donor germ cells are enriched in the *dnd*-positive population.

In mammals, the use of antibodies, which can recognize cell surface antigens, could possibly become one of the most powerful tools to identify and enrich specific cell populations (Kanatsu-Shinohara et al. 2004; Altman et al. 2012). In fish, antibodies that specifically recognize cell surface antigens in type-A spermatogonia have been discovered (Nagasawa et al. 2010, 2012). However, flow cytometry (FCM) sorting and magnetic-activated cell sorting (MACS) using these antibodies have not yet become popular as methods for enriching fish spermatogonia. However, the enrichment of transplantable type-A spermatogonia is possible in Nibe croaker (Kise et al. 2012) and in bluefin tuna at several stages of maturation (Ichida et al. 2017) by FCM sorting based on the light-scattering characteristics of type-A spermatogonia. Whole testicular cell suspensions are fractionated based on their forward and side scatter properties, following which ASGs are enriched in a fraction in which the forward scatter signal is relatively high and the side scatter signal is relatively low. The diameter of sorted cells using this fraction was found to be identical to the size of type-A spermatogonia observed in histological analysis, and these cells also expressed the *vasa* gene. In addition, spermatogonial stem cell enrichment using a side-population fraction of Hoechst 33342 staining was established in rainbow trout (Hayashi et al. 2014). Another method for spermatogonial enrichment uses differential plating that exploits the differences in adhesion characteristics among type-A spermatogonia and other gonadal cells (Shikina et al. 2008, 2013; Shikina and Yoshizaki 2010). Since these methods were established based on the characteristics of type-A spermatogonia conserved in a wide range of teleosts, they can potentially be applied to various fish species.

17.2.3 Intraperitoneal Transplantation Procedure and Tracking of Fluorescence-Labeled Donor Cells in Recipient Gonads

The larvae of pelagic egg-spawning marine teleosts are only a few millimeters in length and are highly sensitive to physiological and physical stresses, making it difficult to use them for embryo manipulation. In order to make intraperitoneal transplantation possible in rotifer-feeding stage marine fish larvae, we modified the cell transplantation protocol originally developed for salmonid hatchings (Fig. 17.3b). Transplantation needles were prepared by pulling glass capillaries (GD-1; Narishige, Tokyo, Japan) using an electric puller (PC-10; Narishige). The tips of the needles were sharpened with a grinder (EG-400; Narishige) until the opening attained a diameter of 40–80 μm. Recipient larvae were anesthetized with 0.0075% ethyl 3-aminobenzoate methanesulfonate salt (Tricaine, A5040; Sigma-Aldrich Inc., St. Louis, MO) in seawater. Larvae were transferred to a Petri dish coated with 2–3% agar using a 10 mL glass pipette. Cell transplantation was performed using a micromanipulator (MP-2 or MM-3, Narishige, Japan) and a microinjector (IM-9B, Narishige) attached to a dissecting microscope. After transplantation, recipient larvae were transferred to a recovery tank filled with seawater. Because marine fish larvae possess a large membrane fin, they are easily trapped by the water surface and die upon drying of their body. We observed that the addition of 0.1% bovine serum albumin (BSA) to both the anesthetization and recovery tanks significantly reduced mortality at 2 and 48 h after transplantation. Although positive effects of adding BSA for the reduction of handling stress during larval rearing were already reported (Tagawa et al. 2004), it could be speculated that improved viscosity of the BSA-added seawater prevented the marine fish larvae from water surface-tension-related death. Sagittal sections of the abdomen of larvae showed that genital ridges with enclosed PGCs were found on the dorsal side of the peritoneum near the posterior end of the swim bladder. Thus, the swim bladder was used as a landmark for the transplantation. The white arrow in Fig. 17.3 indicates the position where the glass needle is inserted. Approximately, 15 nL of the donor cell suspension, containing 10,000–20,000 testicular cells was intraperitoneally transplanted into each recipient. Usually, 1–2 hundred recipient larvae were transplanted by a single person in a day. According to published articles, 40–70% and 20–60% of the transplanted marine fish larvae could survive for 24 h and 3 weeks after transplantation, respectively (Takeuchi et al. 2009; Yazawa et al. 2010; Higuchi et al. 2011; Morita et al. 2012, 2015). The transplanted larvae were raised in 100-L seed production tanks (Fig. 17.3b) for further analysis, i.e., tracking of donor cells using fluorescent dyes, molecular markers, and progeny testing.

Germ cell-specific visualization methods including the use of fluorescent-protein genes (Yoshizaki et al. 2000; Takeuchi et al. 2002) or the injection of chimeric RNA designed to be stabilized specifically in germ cells (Yoshizaki et al. 2005; Saito et al. 2008) were used in previous GCT experiments on freshwater teleosts. However, the same methods are not practically applicable to marine teleosts because they require numerous microinjections into the fertilized eggs of donor species and transgenic offspring produced by surrogate broodstock fish will be unsuitable for mariculture. Thus, we tested whether the lipophilic cell membrane dye PKH26 (Sigma-Aldrich Inc.), at five times greater concentration compared to that mentioned in the manufacturer's protocol, could be used to visualize and track marine teleost donor germ cells. Our data showed that more than 95% of dissociated testicular cells were viable and strongly labeled with the PKH26 fluorescent dye. Red-colored fluorescent labeling of donor cells by PKH26 was reduced by dilution due to cell division, but lasted for at least 2–6 weeks post transplantation which was long enough to investigate the incorporation of donor-derived germ cells in recipient gonads (Takeuchi et al. 2009). Thus, PKH26 seems to be a suitable fluorescent dye for GCT in marine teleosts.

Note that the red fluorescence found in recipient gonads is not always from donor-derived germ cells. Non-spermatogonial cells or cell debris co-stained with PKH26 dye and transplanted into the body cavity could be attached to the epithelium of recipient gonads and incorrectly recognized as colonized germ cells. It could also be the result of auto-fluorescence. Thus, identification of the cell types of PKH26-labeled cells found in the recipient gonads following intraperitoneal transplantation is a critical step for confirming the success of donor germ cell colonization. For marine teleost experiments using PKH26 dye, we developed two methods to investigate the incorporation of donor-derived type-A spermatogonia in recipient gonads by combining confocal analysis with molecular marker analysis. As shown in Fig. 17.3c, it was possible to confirm whether PKH26-labeled cells incorporated into the recipient gonads were donor-derived germ cells by merging PKH26-labeling with distinctive nuclear DAPI staining of type-A spermatogonia under a confocal microscope, as well as by in situ hybridization of intact or smear samples of gonads of transplanted recipients for a germ cell marker gene (Takeuchi et al. 2009; Yazawa et al. 2010). Furthermore, it became possible to confirm the proliferation and differentiation of xenogenic donor-derived germ cells in recipient gonads using molecular tools such as species-specific PCR primers or cRNA in situ hybridization probes designed for germ cell marker genes of donor species (e.g., *vasa*, *dnd*) (Yazawa et al. 2010; Bar et al. 2015; Higuchi et al. 2011; Morita et al. 2015; Yoshikawa et al. 2018a).

17.2.4 *Confirmation of Donor-Derived Gametes and F1 Offspring by Progeny Testing*

Once the recipient fish reached maturity, their gametes were collected and crossed with those of their wild-type counterparts. To evaluate whether the germ cells are successfully transplanted and differentiated into functional gametes, genomic DNA PCR is often used to detect the genotype that appears only in donors and not in recipients.

In intraperitoneal GCT experiments with xenogenic donors, species-specific PCR or PCR-restriction fragment length polymorphism (PCR-RFLP) was mostly used to evaluate the genetic identity of donor-derived gametes and offspring (Morita et al. 2015; Hamasaki et al. 2017; Yoshikawa et al. 2018a). In experiments on allogenic donors, microsatellite analysis was used (Morita et al. 2012). For example, when milt samples collected from the recipient were subjected to microsatellite analyses, production of donor-derived spermatozoa would be strongly suggested by the appearance of the same alleles as those in the fin DNA of the donor in addition to the recipient's own alleles in the electrophoretograms. Analysis of microsatellite DNA markers in F1 offspring reveals the average contribution of donor-derived F1 offspring. However, microsatellite PCR is not quantitative and has low sensitivity due to amplification bias. One shortcoming of microsatellite analysis for detecting donor-derived alleles is its relatively low detection limit. When using imitated DNA samples of mixed donor and recipient sperm, we could detect donor-derived alleles successfully when the donor sperm was mixed at more than 10% volume in the recipient sperm (Takeuchi et al., unpublished). Recently, we have adopted digital genotyping of highly polymorphic nucleotide sequences of *major histocompatibility complex* (*MHC*) *alpha/beta* genes using next-generation sequencing (NGS) to detect the appearance of 1% donor-derived sperm in the recipient milt (Kawai et al., unpublished). *MHC* genes function as individual identifiers in the immune system to distinguish self and non-self. Therefore, they are highly polymorphic at the nucleotide sequence level, and it is possible to distinguish over 40 haplotypes using only one pair of primers (Kawai et al., unpublished). In addition, one pair of PCR primers can distinguish many individuals within one species. Thus, NGS is a quite powerful tool to detect small genomic differences such as SNPs, and to quantitatively identify an amount of successfully transplanted genotypes.

17.3 Use of Sterilized Marine Fish as Recipients for Intraperitoneal GCT

A remaining issue in this chapter is that the F1 generation produced by the recipients includes not only donor-derived offspring but also a large number of individuals originating from gametes produced by the recipient itself. Interestingly, a predominance of donor-derived gametes over recipient-derived gametes in normal diploid recipient gonads has been reported with allogenic spermatogonial transplantation in yellowtails (Morita et al. 2012) and Nibe croaker (Yoshikawa et al. 2017); however, a wide variation in the germline transmission rates of donor-derived haplotypes (e.g., 17.0–98.8% and 3.2–90.6% in diploid yellowtail and Nibe croaker recipients) was also noted. Varying the proportion of donor-derived larvae among the F1 offspring of each diploid recipient would involve prohibitive technical difficulties for screening the donor-derived larvae using PCR methods, especially with marine teleost that produce very large numbers of eggs. Thus, to develop practical methods for the surrogate broodstock technology in marine fish aquaculture, it is important to obtain surrogate recipients that only produce donor-derived viable offspring as well as a high frequency of recipients that could produce donor-derived gametes.

As summarized by Piferrer et al. (2009), functionally sterile triploid fish can be mass produced in several species of marine teleosts. Induced triploidy results in infertility in many teleost fishes, and it is believed that this sterility results from a failure in the progress of meiosis (Benfey 1999). Therefore, we aimed to generate triploid surrogate recipients that produce only donor-derived offspring by using functionally sterile triploid Nibe croaker. Both male and female triploid Nibe croakers, obtained by preventing second polar body exclusion using cold-shock treatment of artificially inseminated eggs, showed impaired gametogenesis in two consecutive spawning seasons (Takeuchi et al. 2018). Although the triploids produced small amounts of gametes, no progeny from crosses between triploids and diploids survived later than 24 h subsequent to hatching, indicating that the triploid Nibe croaker was an ideal recipient for surrogate production. In the progeny test, the triploid recipient produced functional eggs and sperm that were only derived from the transplanted allogenic diploid germ cells, but no functional gametes were produced from endogenous triploid germ cells. The gonadal sizes of surrogate recipients were recovered to those of sexually matured diploids. These observations indicated that (1) the gonadal endocrine system of the triploid Nibe croaker could function properly if the recipient gonads harbored diploid germ cells, and (2) the gonadal microenvironment in triploid recipients was capable of producing functional gametes. In addition, the rate at which male and female recipients produced donor-derived gametes became seven- and fourfold higher, respectively, in triploid recipients than in diploid recipients (Yoshikawa et al. 2017). Surrogate recipients of triploid grass puffer produced a large number of recipient-derived sperm and eggs, and surprisingly, they could be fertilized. However, only donor-derived embryos among the F1 embryos were developed and hatched normally (Hamasaki et al. 2013, 2017).

In order to explore another type of sterile surrogates, we focused on hybrid sterility, which is widely known to occur due to an aberrant homologous chromosome pairing during meiosis in hybrid germ cells and to act as a postzygotic reproductive barrier in maintaining speciation (Mallet 2007). Unexpectedly, we found that one of the hybrids among three intergeneric hybrid croakers had small-sized gonads that lacked germ cells (Yoshikawa et al. 2018a). As fertilization in most fish is external, a number of cases of hybrid sterility have been reported (Chevassus 1983; Bartley et al. 2001; Wong et al. 2011; Rahman et al. 2012; Piva et al. 2017). However, this hybrid croaker provided the first evidence that mitotic arrest of hybrid PGCs caused a germ cell-less phenotype in hybrid fish. By using this hybrid croaker larvae as the recipient, we showed that the gonad of the germ cell-less hybrids fully supported the gametogenesis of donor-derived Nibe croaker germ cells and produced only Nibe croaker sperm (Yoshikawa et al. 2018a). Note that, although germ cell-less gonads showed testis-like morphology and gene expression patterns, ovarian differentiation was induced by the colonization of donor-derived spermatogonia in sexually undifferentiated genital ridges of this hybrid fish. Since interspecific hybridization can only be achieved by artificial insemination and the usual rearing of the resulting hybrid embryos, hybrid fish showing the germ cell-less phenotype would be suitable recipients for GCT in marine teleosts.

Germ cell-less fish have also been obtained by knockdown and knockout of genes required for the proper development of germ cells in several freshwater fishes, such as zebrafish (Ciruna et al. 2002; Slanchev et al. 2005; Weidinger et al. 2003; Li et al. 2017), loach (*Misgurnus anguillicaudatus*; Fujimoto et al. 2010), goldfish (*Carassius auratus*; Goto et al. 2012), starlet (*Acipenser ruthenus*; (Linhartová et al. 2015), and salmons (Wargelius et al. 2016; Yoshizaki et al. 2016). Although microinjection experiments require high levels of skill and advanced equipment, genome editing in pelagic egg-spawning marine teleost embryos using CRISPR/Cas9 has become feasible (Kishimoto et al. 2018). Thus, germ cell-less marine fish could be generated by disrupting similar genes already knocked out in freshwater teleosts.

17.4 Evaluation of Successful Intraperitoneal GCT

Previous reports in a wide range of marine teleosts demonstrate that germ cells from xenogeneic species, even from species belonging to a different family, are able to identify the location of, migrate to, eventually be incorporated into, and proliferate in the genital ridges of recipient fish (Yazawa et al. 2010; Higuchi et al. 2011; Bar et al. 2015). In salmonids, production of donor-derived eggs in xenogeneic recipients was only achieved when the donor species and recipient species belonged to the same genus (Takeuchi et al. 2004; Okutsu et al. 2007), whereas production of donor-derived sperm was successful in recipients of different genera. These results suggest that production of donor-derived eggs might be more difficult than production of donor-derived sperm in a xenogeneic recipient. In fact, the inter-generic transplantation of yellowtail spermatogonia into jack mackerel larvae revealed the production of functional donor-derived sperm but not eggs (Morita et al. 2015). Moreover, donor-derived yellowtail *vasa* mRNA was not detected in the 9-month-old jack mackerel recipient ovary. Thus, in marine teleosts as well, we speculated that there might be many factors affecting the oogenesis of donor-derived germ cells (such as vitellogenin and egg envelop proteins) that are obtained from the recipient liver and are indispensable for oogenesis (Lubzens et al. 2010).

The three different levels (Class 1–3) of success reported for intra- and intergeneric and interfamily transplantations in marine teleosts until 2018 are summa-

Table 17.1 Assessment of allogenic and xenogenic donor germ cell development in marine teleost recipients in 2018

Class	1		2		3	
Achievements	Colonization and proliferation of donor germ cells		Sperm production only		Sperm and egg production	
Type of recipient	Non-sterile	Sterile	Non-sterile	Sterile	Non-sterile	Sterile
Allogenic donor	Takeuchi et al. (2009)	–	–	–	Morita et al. (2012)	Yoshikawa et al. (2017)
Xenogenic donor (intra-generic)	–	–	–	–	–	Hamasaki et al. (2017) Yoshikawa et al. (2018b)
Xenogenic donor (inter-generic)	–	–	Morita et al. (2015)	Yoshikawa et al. (2018a)[a]	–	–
Xenogenic donor (inter-family)	Yazawa et al. (2010) Higuchi et al. (2011) Pacchiarini et al. (2014) Bar et al. (2015)	–	–	–	–	–

[a]A germ-cell-less intergeneric hybrid recipient (*Nibea mitsukurii* × *Pennahia argentata*) produced donor-derived *Nibea mistukurii* sperm

rized in Table 17.1. These data suggest that increased phylogenetic distance between donor and recipient species leads to incompatibilities between the transplanted germ cells and the microenvironment of recipient gonads. However, triploid Chinese rosy bitterlings (*Rhodeus ocellatus*), which are small freshwater fish, produced eggs and sperm of the donor species, slender bittering (*Tanakia lanceolate*), belonging to the different genus (Octavera and Yoshizaki, unpublished). Similarly, Atlantic salmon (*Salmo salar*) oocytes and sperm were produced by triploid rainbow trout (*Oncorhynchus mykiss*; Hattori et al. 2019). These suggest that in addition to the genetic distance between these species, ecological and/or physiological similarities also influence the successful gametogenesis of xenogenic donor germ cells in recipient gonads. Prerequisites for the appropriate combination of donor and recipient species will need to be determined in future experiments.

Our ultimate goal for developing intraperitoneal GCT in marine teleosts is to produce bluefin tuna gametes in small-bodied surrogates. Among the scombrid fishes, Eastern little tuna (ELT, *Euthynnus affinis*) could serve as a suitable recipient for bluefin tuna spermatogonia due to its small body size, and the physiological and genetic similarities between the two species (Yazawa et al. 2015a). We also showed that ELT could sexually mature and be induced to spawn in temperature-controlled land-based tanks within a year from hatching (Yazawa et al. 2015b, 2016).

Intraperitoneal GCT using enriched type-A spermatogonia of bluefin tuna into the appropriate developmental stage of functionally sterile triploid or hybrid ELT has been developed and the behavior of donor germ cells in the ELT recipient gonads has been currently studied at the cellular and molecular levels (Kawamura et al. unpublished).

References

Altman E, Yango P, Moustafa R, Smith JF, Klatsky PC, Tran ND (2012) Characterization of human spermatogonial stem cell markers in fetal, pediatric, and adult testicular tissues. Reproduction 148(4):417–427. https://doi.org/10.1530/REP-14-0123

Bar I, Smith A, Bubner E, Yoshizaki G, Takeuchi Y, Yazawa R, Chen BN, Cummins S, Elizur A (2015) Assessment of yellowtail kingfish (*Seriola lalandi*) as a surrogate host for the production of southern bluefin tuna (*Thunnus maccoyii*) seed via spermatogonial germ cell transplantation. Reprod Fertil Dev 28(12):2051–2064. https://doi.org/10.1071/RD15136

Bartley DM, Rana K, Immink AJ (2001) The use of interspecific hybrids in aquaculture and fisheries. Rev Fish Biol Fish 10(3):325–337. https://doi.org/10.1023/A:1016691725361

Benfey TJ (1999) The physiology and behavior of triploid fishes. Rev Fish Sci 7(1):39–67. https://doi.org/10.1080/10641269991319162

Braat AK, Speksnijder JE, Zivkovic D (1999) Germ line development in fishes. Int J Dev Biol 43(7):745–760

Budd AM, Banh QQ, Domingos JA, Jerry DR (2015) Sex control in fish: approaches, challenges and opportunities for aquaculture. J Mar Sci Eng 3(2):329–355. https://doi.org/10.3390/jmse3020329

Chevassus B (1983) Hybridization in fish. Aquaculture 33(1–4):245–262

Ciruna B, Weidinger G, Knaut H, Thisse B, Thisse C, Raz E, Schier AF (2002) Production of maternal-zygotic mutant zebrafish by germ-line replacement. Proc Natl Acad Sci U S A 99(23):14919–14924. https://doi.org/10.1073/pnas.222459999

de Siqueira-Silva DH, Saito T, Dos Santos-Silva AP, da Silva Costa R, Psenicka M, Yasui GS (2018) Biotechnology applied to fish reproduction: tools for conservation. Fish Physiol Biochem 44(6):1469–1485. https://doi.org/10.1007/s10695-018-0506-0

Dobrinski I (2007) Transplantation of germ cells and testis tissue for the study and preservation of fertility. Soc Reprod Fertil Suppl 65:447–458

Eldridge WH, Naish K (2007) Long-term effects of translocation and release numbers on fine-scale population structure among coho salmon (*Oncorhynchus kisutch*). Mol Ecol 16(12):2407–2421. https://doi.org/10.1111/j.1365-294X.2007.03271.x

Farlora R, Hattori-Ihara S, Takeuchi Y, Hayashi M, Octavera A, Alimuddin, Yoshizaki G (2014) Intraperitoneal germ cell transplantation in the Nile tilapia *Oreochromis niloticus*. Mar Biotechnol (NY) 16(3):309–320. https://doi.org/10.1007/s10126-013-9551-y

Fujimoto T, Nishimura T, Goto-Kazeto R, Kawakami Y, Yamaha E, Arai K (2010) Sexual dimorphism of gonadal structure and gene expression in germ cell-deficient loach, a teleost fish. Proc Natl Acad Sci U S A 107(40):17211–17216. https://doi.org/10.1073/pnas.1007032107

Goto R, Saito T, Takeda T, Fujimoto T, Takagi M, Arai K, Yamaha E (2012) Germ cells are not the primary factor for sexual fate determination in goldfish. Dev Biol 370(1):98–109. https://doi.org/10.1016/j.ydbio.2012.07.010

Hamasaki M, Takeuchi Y, Miyaki K, Yoshizaki G (2013) Gonadal development and fertility of triploid grass puffer *Takifugu niphobles* induced by cold shock treatment. Mar Biotechnol (NY) 15(2):133–144

Hamasaki M, Takeuchi Y, Yazawa R, Yoshikawa S, Kadomura K, Yamada T, Miyaki K, Kikuchi K, Yoshizaki G (2017) Production of tiger puffer *Takifugu rubripes* offspring from triploid grass puffer *Takifugu niphobles* Parents. Mar Biotechnol 19(6):579–591. https://doi.org/10.1007/s10126-012-9470-3

Hattori RS, Yoshinaga TT, Katayama N, Hattori-Ihara S, Tsukamoto RY, Takahashi NS, Tabata YA (2019) Surrogate production of *Salmo salar* oocytes and sperm in triploid *Oncorhynchus mykiss* by germ cell transplantation technology. Aquaculture 506:238–245. https://doi.org/10.1016/j.aquaculture.2019.03.037

Hayashi M, Sato M, Nagasaka Y, Sadaie S, Kobayashi S, Yoshizaki G (2014) Enrichment of spermatogonial stem cells using side population in teleost. Biol Reprod 91(1):23. https://doi.org/10.1095/biolreprod.113.114140

Higuchi K, Takeuchi Y, Miwa M, Yamamoto Y, Tsunemoto K, Yoshizaki G (2011) Colonization, proliferation, and survival of intraperitoneally transplanted yellowtail *Seriola quinqueradiata* spermatogonia in nibe croaker *Nibea mitsukurii* recipient. Fish Sci 77:69–77. https://doi.org/10.1007/s12562-010-0314-7

Ichida K, Kise K, Morita T, Yazawa R, Takeuchi Y, Yoshizaki G (2017) Flow-cytometric enrichment of Pacific bluefin tuna type A spermatogonia based on light-scattering properties. Theriogenology 101:91–98. https://doi.org/10.1016/j.theriogenology.2017.06.022

Kakimoto Y, Aida S, Arai K, Suzuki R (1994) Neuters occurred in methyltestosterone treated ocellated puffer *Takifugu rubripes*. Fish Genet Breed Sci 20:63–66

Kamiya T, Kai W, Tasumi S, Oka A, Matsunaga T, Mizuno S, Fujita M, Suetake H, Suzuki S, Hosoya S, Tohari S, Brenner S, Miyadai T, Venkatesh B, Suzuki Y, Kikuchi K (2012) A trans-species missense SNP in *Amhr2* is associated with sex determination in the tiger pufferfish, *Takifugu rubripes* (fugu). PLoS Genet 8(7):e1002798. https://doi.org/10.1371/journal.pgen.1002798

Kanatsu-Shinohara M, Toyokuni S, Shinohara T (2004) CD9 is a surface marker on mouse and rat male germline stem cells. Biol Reprod 70(1):70–75. https://doi.org/10.1095/biolreprod.103.020867

Kikuchi K, Kai W, Hosokawa A, Mizuno N, Suetake H, Asahina K, Suzuki Y (2007) The sex-determining locus in the tiger pufferfish, *Takifugu rubripes*. Genetics 175(4):2039–2042. https://doi.org/10.1534/genetics.106.069278

Kise K, Yoshikawa H, Sato M, Tashiro M, Yazawa R, Nagasaka Y, Takeuchi Y, Yoshizaki G (2012) Flow-cytometric isolation and enrichment of teleost type a spermatogonia based on light-scattering properties. Biol Reprod 86(4):107. https://doi.org/10.1095/biolreprod.111.093161

Kishimoto K, Washio Y, Yoshiura Y, Toyoda A, Ueno T, Fukuyama H, Kato K, Kinoshita M (2018) Production of a breed of red sea bream *Pagrus major* with an increase of skeletal muscle mass and reduced body length by genome editing with CRISPR/Cas9. Aquaculture 495:415–427. https://doi.org/10.1016/j.aquaculture.2018.05.055

Kitada S, Shishidou H, Sugaya T, Kitakado T, Hamasaki K, Kishino H (2009) Genetic effects of long-term stock enhancement programs. Aquaculture 290(1–2):69–79. https://doi.org/10.1016/j.aquaculture.2009.02.011

Kobayashi T, Yoshizaki G, Takeuchi Y, Takeuchi T (2004) Isolation of highly pure and viable primordial germ cells from rainbow trout by GFP-dependent flow cytometry. Mol Reprod Dev 67(1):91–100. https://doi.org/10.1002/mrd.20003

Kobayashi T, Takeuchi Y, Takeuchi T, Yoshizaki G (2007) Generation of viable fish from cryopreserved primordial germ cells. Mol Reprod Dev 74(2):207–213. https://doi.org/10.1002/mrd.20003

Lacerda SMSN, Costa GMJ, Campos-Junior PHA, Segatelli TM, Yazawa R, Takeuchi Y, Morita T, Yoshizaki G, França LR (2013) Germ cell transplantation as a potential biotechnological approach to fish reproduction. Fish Physiol Biochem 39(1):3–11. https://doi.org/10.1007/s10695-012-9606-4

Lee S, Iwasaki Y, Shikina S, Yoshizaki G (2013) Generation of functional eggs and sperm from cryopreserved whole testes. Proc Natl Acad Sci U S A 110(5):1640–1645. https://doi.org/10.1073/pnas.1218468110

Lee S, Seki S, Katayama N, Yoshizaki G (2015) Production of viable trout offspring derived from frozen whole fish. Sci Rep 5:16045. https://doi.org/10.1038/srep16045

Lee S, Iwasaki Y, Yoshizaki G (2016a) Long-term (5 years) cryopreserved spermatogonia have high capacity to generate functional gametes via interspecies transplantation in salmonids. Cryobiology 73(2):286–290. https://doi.org/10.1016/j.cryobiol.2016.08.001

Lee S, Katayama N, Yoshizaki G (2016b) Generation of juvenile rainbow trout derived from cryopreserved whole ovaries by intraperitoneal transplantation of ovarian germ cells. Biochem Biophys Res Commun 478(3):1478–1483. https://doi.org/10.1016/j.bbrc.2016.08.156

Li Q, Fujii W, Naito K, Yoshizaki G (2017) Application of dead end-knockout zebrafish as recipients of germ cell transplantation. Mol Reprod Dev 84(10):1100–1111. https://doi.org/10.1002/mrd.22870

Linhartová Z, Saito T, Kašpar V, Rodina M, Prášková E, Hagihara S, Pšenička M (2015) Sterilization of sterlet *Acipenser ruthenus* by using knockdown agent, antisense morpholino oligonucleotide, against *dead end* gene. Theriogenology 84(7):1246–1255. https://doi.org/10.1016/j.theriogenology.2015.07.003

Lubzens E, Young G, Bobe J, Cerdà J (2010) Oogenesis in teleosts: how eggs are formed. Gen Comp Endocrinol 165(3):367–389. https://doi.org/10.1016/j.ygcen.2009.05.022

Mallet J (2007) Hybrid speciation. Nature 446(7133):279–283. https://doi.org/10.1038/nature05706

Manning MJ, Nakanishi T (1996) The specific immune system: cellular defenses. In: Iwama G, Nakanishi T (eds) The fish immune system. Academic, New York, pp 159–205

Morita T, Kumakura N, Morishima K, Mitsuboshi T, Ishida M, Hara T, Kudo S, Miwa M, Ihara S, Higuchi K, Takeuchi Y, Yoshizaki G (2012) Production of donor-derived offspring by allogeneic transplantation of spermatogonia in the yellowtail (*Seriola quinqueradiata*). Biol Reprod 86(6):176. https://doi.org/10.1095/biolreprod.111.097873

Morita T, Morishima K, Miwa M, Kumakura N, Kudo S, Ichida K, Mitsuboshi T, Takeuchi Y, Yoshizaki G (2015) Functional sperm of the yellowtail (*Seriola quinqueradiata*) were produced in the small-bodied surrogate, jack mackerel (*Trachurus japonicus*). Mar Biotechnol (NY) 17(5):644–654. https://doi.org/10.1007/s10126-015-9657-5

Nagasawa K, Shikina S, Takeuchi Y, Yoshizaki G (2010) Lymphocyte antigen 75 (Ly75/CD205) is a surface marker on mitotic germ cells in rainbow trout. Biol Reprod 83(4):597–606. https://doi.org/10.1095/biolreprod.109.082081

Nagasawa K, Miwa M, Yazawa R, Morita T, Takeuchi Y, Yoshizaki G (2012) Characterization of lymphocyte antigen 75 (Ly75/CD205) as a potential cell-surface marker on spermatogonia in Pacific bluefin tuna *Thunnus orientalis*. Fish Sci 78(4):791–800. https://doi.org/10.1007/s12562-012-0501-9

Ogawa T (2001) Spermatogonial transplantation: the principle and possible applications. J Mol Med (Berl) 79(7):368–374

Okutsu T, Yano A, Nagasawa K, Shikina S, Kobayashi T, Takeuchi Y, Yoshizaki G (2006a) Manipulation of fish germ cell: visualization, cryopreservation and transplantation. J Reprod Dev 52(6):685–693. https://doi.org/10.1262/jrd.18096

Okutsu T, Suzuki K, Takeuchi Y, Takeuchi T, Yoshizaki G (2006b) Testicular germ cells can colonize sexually undifferentiated embryonic gonad and produce functional eggs in fish. Proc Natl Acad Sci U S A 103(8):2725–2729. https://doi.org/10.1073/pnas.0509218103

Okutsu T, Shikina S, Kanno M, Takeuchi Y, Yoshizaki G (2007) Production of trout offspring from triploid salmon parents. Science 317(5844):1517. https://doi.org/10.1126/science.1145626

Okutsu T, Shikina S, Sakamoto T, Mochizuki M, Yoshizaki G (2015) Successful production of functional Y eggs derived from spermatogonia transplanted into female recipients and subsequent production of YY supermales in rainbow trout, *Oncorhynchus mykiss*. Aquaculture 446:298–302. https://doi.org/10.1016/j.aquaculture.2015.05.020

Pacchiarini T, Sarasquete C, Cabrita E (2014) Development of interspecies testicular germ-cell transplantation in flatfish. Reprod Fertil Dev 26(5):690–702. https://doi.org/10.1071/RD13103

Piferrer F, Beaumont A, Falguière JC, Flajšhans M, Haffray P, Colombo L (2009) Polyploid fish and shellfish: production, biology and applications to aquaculture for performance improvement and genetic containment. Aquaculture 293(3–4):125–156. https://doi.org/10.1016/j.aquaculture.2009.04.036

Piva LH, de Siqueira-Silva DH, Goes CAG, Fujimoto T, Saito T, Dragone LV, Senhorini JA, Porto-Foresti F, Ferraz JBS, Yasui GS (2017) Triploid or hybrid tetra: which is the ideal sterile host for surrogate technology? Theriogenology 108:239–244. https://doi.org/10.1016/j.theriogenology.2017.12.013

Pšenička M, Saito T, Linhartová Z, Gazo I (2015) Isolation and transplantation of sturgeon early-stage germ cells. Theriogenology 83(6):1085–1092. https://doi.org/10.1016/j.theriogenology.2014.12.010

Rahman AM, Arshad A, Marimuthu K, Ara R, Amin SMN (2012) Inter-specific hybridization and its potential for aquaculture of fin fishes. Asian J Anim Vet Adv 8(2):139–153. https://doi.org/10.3923/ajava.2013.139.153

Rashid H, Kitano H, Lee KH, Nii S, Shigematsu T, Kadomura K, Yamaguchi A, Matsuyama M (2007) Fugu (*Takifugu rubripes*) sexual differentiation: CYP19 regulation and aromatase inhibitor induced testicular development. Sex Dev 1(5):311–322. https://doi.org/10.1159/000108935

Saito T, Goto-Kazeto R, Arai K, Yamaha E (2008) Xenogenesis in teleost fish through generation of germ-line chimeras by single primordial germ cell transplantation. Biol Reprod 78(1):159–166. https://doi.org/10.1095/biolreprod.107.060038

Sato M, Morita T, Katayama N, Yoshizaki G (2014) Production of genetically diversified fish seeds using spermatogonial transplantation. Aquaculture 422–423:218–224. https://doi.org/10.1016/j.aquaculture.2013.12.016

Seki S, Kusano K, Lee S, Iwasaki Y, Yagisawa M, Ishida M, Hiratsuka T, Sasado T, Naruse K, Yoshizaki G (2017) Production of the medaka derived from vitrified whole testes by germ cell transplantation. Sci Rep 7:43185. https://doi.org/10.1038/srep43185

Sekino M, Hara M, Taniguchi N (2002) Loss of microsatellite and mitochondrial DNA variation in hatchery strains of Japanese flounder *Paralichthys olivaceus*. Aquaculture 213(1–4):101–122. https://doi.org/10.1016/S0044-8486(01)00885-7

Shikina S, Yoshizaki G (2010) Improved in vitro culture conditions to enhance the survival, mitotic activity, and transplantability of rainbow trout type A spermatogonia. Biol Reprod 83(2):268–276. https://doi.org/10.1095/biolreprod.109.082123

Shikina S, Ihara S, Yoshizaki G (2008) Culture conditions for maintaining the survival and mitotic activity of rainbow trout transplantable type A spermatogonia. Mol Reprod Dev 75(3):529–537. https://doi.org/10.1002/mrd.20771

Shikina S, Nagasawa K, Hayashi M, Furuya M, Iwasaki Y, Yoshizaki G (2013) Short-term in vitro culturing improves transplantability of type A spermatogonia in rainbow trout (*Oncorhynchus mykiss*). Mol Reprod Dev 80(9):763–773. https://doi.org/10.1002/mrd.22208

Slanchev K, Stebler J, de la Cueva-Méndez G, Raz E (2005) Development without germ cells: the role of the germ line in zebrafish sex differentiation. Proc Natl Acad Sci U S A 102(11):4074–4079. https://doi.org/10.1073/pnas.0407475102

Tagawa M, Kaji T, Kinoshita M, Tanaka M (2004) Effect of stocking density and addition of proteins on larval survival in Japanese flounder, *Paralichthys olivaceus*. Aquaculture 230(1–4):517–525. https://doi.org/10.1016/S0044-8486(03)00409-5

Takeuchi Y, Yoshizaki G, Kobayashi T, Takeuchi T (2002) Mass isolation of primordial germ cells from transgenic rainbow trout carrying the *green fluorescent protein* gene driven by the *vasa* gene promoter. Biol Reprod 67(4):1087–1092. https://doi.org/10.1095/biolreprod67.4.1087

Takeuchi Y, Yoshizaki G, Takeuchi T (2003) Generation of live fry from intraperitoneally transplanted primordial germ cells in rainbow trout. Biol Reprod 69(4):1142–1149. https://doi.org/10.1095/biolreprod.103.017624

Takeuchi Y, Yoshizaki G, Takeuchi T (2004) Surrogate broodstock produces salmonids. Nature 430(7000):629–630. https://doi.org/10.1038/430629a

Takeuchi Y, Higuchi K, Yatabe T, Miwa M, Yoshizaki G (2009) Development of spermatogonial cell transplantation in Nibe croaker, *Nibea mitsukurii* (Perciformes, Sciaenidae). Biol Reprod 81(6):1055–1063. https://doi.org/10.1095/biolreprod.109.077701

Takeuchi Y, Yatabe T, Yoshikawa H, Ino Y, Kabeya N, Yazawa R, Yoshizaki G (2018) Production of functionally sterile triploid Nibe croaker *Nibea mitsukurii* induced by cold-shock treatment with special emphasis on triploid aptitude as surrogate broodstock. Aquaculture 494:45–56. https://doi.org/10.1016/j.aquaculture.2016.05.030

Verspoor E (1998) Reduced genetic variability in first-generation hatchery populations of Atlantic salmon (*Salmo salar*). Can J Fish Aquat Sci 45(10):1686–1690. https://doi.org/10.1139/f88-199

Wargelius A, Leininger S, Skaftnesmo KO, Kleppe L, Andersson E, Taranger GL, Schulz RW, Edvardsen RB (2016) *Dnd* knockout ablates germ cells and demonstrates germ cell independent sex differentiation in Atlantic salmon. Sci Rep 6:21284. https://doi.org/10.1038/srep21284

Weidinger G, Stebler J, Slanchev K, Dumstrei K, Wise C, Lovell-Badge R, Thisse C, Thisse B, Raz E (2003) *dead end*, a novel vertebrate germ plasm component, is required for zebrafish primordial germ cell migration and survival. Curr Biol 13(16):1429–1434. https://doi.org/10.1016/S0960-9822(03)00537-2

Wong TT, Saito T, Crodian J, Collodi P (2011) Zebrafish germline chimeras produced by transplantation of ovarian germ cells into sterile host larvae. Biol Reprod 84(6):1190–1197. https://doi.org/10.1095/biolreprod.110.088427

Xu D, Yang F, Chen R, Lou B, Zhan W, Hayashida T, Takeuchi Y (2018) Production of neo-males from gynogenetic yellow drum through 17α-methyltestosterone immersion and subsequent application for the establishment of all-female populations. Aquaculture 489:154–161. https://doi.org/10.1016/j.aquaculture.2018.02.015

Yano A, Suzuki K, Yoshizaki G (2008) Flow-cytometric isolation of testicular germ cells from rainbow trout (*Oncorhynchus mykiss*) carrying the green fluorescent protein gene driven by trout vasa regulatory regions. Biol Reprod 78(1):151–158. https://doi.org/10.1095/biolreprod.107.064667

Yazawa R, Takeuchi Y, Higuchi K, Yatabe T, Kabeya N, Yoshizaki G (2010) Chub mackerel gonads support colonization, survival, and proliferation of intraperitoneally transplanted xenogenic germ cells. Biol Reprod 82(5):896–904. https://doi.org/10.1095/biolreprod.109.081281

Yazawa R, Takeuchi Y, Morita T, Ishida M, Yoshizaki G (2013) The Pacific bluefin tuna (*Thunnus orientalis*) *dead end* gene is suitable as a specific molecular marker of type A spermatogonia. Mol Reprod Dev 80(10):871–880. https://doi.org/10.1002/mrd.22224

Yazawa R, Takeuchi Y, Amezawa K, Sato K, Iwata G, Kabeya N, Yoshizaki G (2015a) GnRHa-induced spawning of the Eastern little tuna (*Euthynnus affinis*) in a 70-m^3 land-based tank. Aquaculture 442:58–68. https://doi.org/10.1016/j.aquaculture.2015.01.016

Yazawa R, Takeuchi Y, Satoh K, Machida Y, Amezawa K, Kabeya N, Shimada Y, Yoshizaki G (2015b) Eastern little tuna, *Euthynnus affinis* (Cantor, 1849) mature and reproduce within 1 year of rearing in land-based tanks. Aquac Res 42(11):3800–3810. https://doi.org/10.1111/are.12831

Yazawa R, Takeuchi Y, Satoh K, Machida Y, Amezawa K, Kabeya N, Yoshizaki G (2016) Spawning induction and seed production of Eastern little tuna, *Euthynnus affinis* (Cantor, 1849), in the post- and pre-spawning seasons by hormonal treatment in a semi-closed recirculation system with elevated temperature. Aquac Res 47(12):3472–3481. https://doi.org/10.1111/are.13173

Ye H, Li C-J, Yue H-M, Du H, Yang X-G, Yoshino T, Hayashida T, Takeuchi Y, Wei Q-W (2017) Establishment of intraperitoneal germ cell transplantation for critically endangered Chinese sturgeon *Acipenser sinensis*. Theriogenology 94:37–47. https://doi.org/10.1016/j.theriogenology.2017.02.009

Yoshida S (2009) Spermatogenic stem cell system in the mouse testis. Cold Spring Harb Symp Quant Biol 73:25–32. https://doi.org/10.1101/sqb.2008.73.046

Yoshikawa H, Takeuchi Y, Ino Y, Wang J, Iwata G, Kabeya N, Yazawa R, Yoshizaki G (2017) Efficient production of donor-derived gametes from triploid recipients following intra-

peritoneal germ cell transplantation into a marine teleost, Nibe croaker (*Nibea mitsukurii*). Aquaculture 478:35–47. https://doi.org/10.1016/j.aquaculture.2016.05.011
Yoshikawa H, Xu D, Ino Y, Yoshino T, Hayashida T, Wang J, Yazawa R, Yoshizaki G, Takeuchi Y (2018a) Hybrid sterility in fish caused by mitotic arrest of primordial germ cells. Genetics 209(2):507–521. https://doi.org/10.1534/genetics.118.300777
Yoshikawa H, Ino Y, Shigenaga K, Katayama T, Kuroyanagi M, Yoshiura Y (2018b) Production of tiger puffer *Takifugu rubripes* from cryopreserved testicular germ cells using surrogate broodstock technology. Aquaculture 493:302–313. https://doi.org/10.1016/j.aquaculture.2018.05.016
Yoshizaki G, Lee S (2018) Production of live fish derived from frozen germ cells via germ cell transplantation. Stem Cell Res 29:103–110. https://doi.org/10.1016/j.scr.2018.03.015
Yoshizaki G, Takeuchi Y, Sakatani S, Takeuchi T (2000) Germ cell-specific expression of green fluorescent protein in transgenic rainbow trout under control of the rainbow trout *vasa*-like gene promoter. Int J Dev Biol 44(3):323–326. https://doi.org/10.1387/ijdb.10853829
Yoshizaki G, Takeuchi Y, Kobayashi T, Ihara S, Takeuchi T (2002) Primordial germ cells: the blueprint for a piscine life. Fish Physiol Biochem 26(1):3–12. https://doi.org/10.1023/A:1023388317621
Yoshizaki G, Takeuchi Y, Kobayashi T, Takeuchi T (2003) Primordial germ cell: a novel tool for fish bioengineering. Fish Physiol Biochem 28(1–4):453–457. https://doi.org/10.1023/B:FISH.0000030628.91607.2d
Yoshizaki G, Tago Y, Takeuchi Y, Sawatari E, Kobayashi T, Takeuchi T (2005) Green fluorescent protein labeling of primordial germ cells using a nontransgenic method and its application for germ cell transplantation in salmonidae. Biol Reprod 73(1):88–93. https://doi.org/10.1095/biolreprod.104.034249
Yoshizaki G, Ichikawa M, Hayashi M, Iwasaki Y, Miwa M, Shikina S, Okutsu T (2010a) Sexual plasticity of ovarian germ cells in rainbow trout. Development 137(8):1227–1230. https://doi.org/10.1242/dev.044982
Yoshizaki G, Okutsu T, Ichikawa M, Hayashi M, Takeuchi Y (2010b) Sexual plasticity of rainbow trout germ cells. Anim Reprod 7(3):187–196
Yoshizaki G, Fujinuma K, Iwasaki Y, Okutsu T, Shikina S, Yazawa R, Takeuchi Y (2011) Spermatogonial transplantation in fish: a novel method for the preservation of genetic resources. Comp Biochem Physiol Part D Genomics Proteomics 6(1):55–61. https://doi.org/10.1016/j.cbd.2010.05.003
Yoshizaki G, Okutsu T, Morita T, Terasawa M, Yazawa R, Takeuchi Y (2012) Biological characteristics of fish germ cells and their application to developmental biotechnology. Reprod Domest Anim 47(Suppl 4):187–192. https://doi.org/10.1111/j.1439-0531.2012.02074.x
Yoshizaki G, Takashiba K, Shimamori S, Fujinuma K, Shikina S, Okutsu T, Kume S, Hayashi M (2016) Production of germ cell-deficient salmonids by *dead end* gene knockdown, and their use as recipients for germ cell transplantation. Mol Reprod Dev 83(4):298–311. https://doi.org/10.1002/mrd.22625
Zapata A, Diez B, Cejalvo T, Gutiérrez-De Frías C, Cortés A (2006) Ontogeny of the immune system of fish. Fish Shellfish Immunol 20(2):126–136. https://doi.org/10.1016/j.fsi.2004.09.005
Zohar Y, Mylonas CC (2001) Endocrine manipulations of spawning in cultured fish: from hormones to genes. Aquaculture 197(1–4):99–136. https://doi.org/10.1016/S0044-8486(01)00584-1